Engineering Management

FRAIDOON MAZDA

ADDISON-WESLEY

Harlow, England ■ Reading, Massachusetts ■ Menlo Park, California
New York ■ Don Mills, Ontario ■ Amsterdam ■ Bonn ■ Sydney ■ Singapore
Tokyo ■ Madrid ■ San Juan, Milan ■ Mexico City ■ Seoul ■ Taipei

Addison Wesley Longman Limited
Edinburgh Gate
Harlow
Essex CM20 2JE
England

and Associated Companies throughout the World.

Published in the United States of America by Addison Wesley Longman Inc., New York.

Cover designed by odB Design & Communication, Reading
Cover photograph by Joel Simon © Tony Stone Images
Text design by Claire Brodmann
Typeset in 10/12pt Times by 32
Produced by Longman Singapore Publishers (Pte) Ltd
Printed in Singapore

Publisher's acknowledgements
The Primo Levi quote on page 18 is reprinted with the permission of Simon & Schuster from *The Drowned and the Saved* by Primo Levi. © 1986 Guillo Einaudi editoire s.p.a., Torino. Translation copyright © 1988 Simon & Schuster, Inc.
Figure 3.17 is reprinted with the permission of the European Foundation for Quality Management.

First printed 1997

ISBN 0-201-17798-6

British Library Cataloguing in Publication Data
A catalogue record for this book is available from the British Library

Library of Congress Cataloguing-in-Publication Data
Mazda, F. F.
Engineering management/Fraidoon Mazda.
p. cm.
Includes bibliographical references and index.
ISBN 0-201-17798-6 (alk. paper)
1. Engineering–Management. I. Title.
TA190.M39 1997
659–dc21 97-22139
CIP

Contents

Foreword

The engineering profession can be justifiably proud of its achievements. Engineering skills alone, however, are not sufficient to guarantee success; industry requires competent managers to ensure that products are developed and brought to the market-place to meet the ever increasing needs of sophisticated consumers.

It has been said that successful companies are run by accountants. This is a fallacy; in the modern technological era, where information can span the globe in a fraction of a second, travelling as pulses of light or as radio waves, engineering companies need to be led by managers who understand the technology they are controlling. Engineers are therefore finding that the paths to executive positions are now open to them and it is essential that they take advantage of these opportunities.

This does not mean that the requirement for good engineers has diminished. On the contrary; the pace of development has meant that, in parallel with the demand for good engineering managers, the need for good engineers has also increased. No longer should engineers feel that the only way to promotion is by moving into management. Most organizations now operate a dual career ladder so that engineers can stay and progress within their discipline, and only those who have a genuine desire to move into management can do so.

Management is not easy, however, and too many engineers make the move without first considering what is involved and without any preparation for the new role. There are many differences between engineering and management; the two that the new engineering manager is most likely to encounter are the need to manage people, and to interface with a wide variety of different functions. People management extends to those who report directly to the new manager, as well as his or her peers within the company, and also to its customers and suppliers. The engineering manager must acquire a range of new people-related skills such as team building, communications and motivation. The new manager must also learn that time is now the most difficult commodity to manage as greater and greater demands are made on this scarce resource.

In addition to managing people, engineering managers find themselves interfacing to a greater extent with the many different functions within an organization – such as finance, manufacturing and marketing – and in order to communicate effectively they must understand the basic principles of these functions. Budgets and project plans need to be prepared and managed, and managers who do not understand these concepts will not be successful in their positions.

I am very pleased to note that *Engineering Management* meets all the basic needs of the new engineer moving into management. Theory has been combined with good practical case studies, to which I am certain engineers within industry will be able to relate. It covers the basic people skills which engineering managers must learn and also provides a very comprehensive introduction to topics such as business strategy, decision making, financial management, project management, manufacturing operations, marketing and sales. I am confident that this breadth of coverage will ensure that *Engineering Management* becomes a reference manual for students and for new and practising engineering managers within industry.

Peter Schuddeboom
Assistant Vice President
Product Development
International Broadband Networks
Nortel Ltd
August 1997

Preface

Congratulations on buying this book; a wise choice! I wonder, however, what your objectives are in obtaining the book? If you believe that by reading it you will automatically become a 'manager', then you may be disappointed. Perhaps it would be better to donate the book to your local charity shop and see if you can pick up a second-hand copy of Gray's *Anatomy*. You probably have a greater chance of becoming a surgeon by reading that book, and I am reliably informed the pay is better!

I have written *Engineering Management* with two types of reader in mind: (i) the student who is about to enter the field of management and wants to acquire a basic understanding of its theory and practice; and (ii) the practising manager who drifted into management, without any formal training, and now wishes to acquire this essential knowledge. If you fall into either of these categories, then read on. This book is for you!

There are many aspects of management which need to be studied and learned, and *Engineering Management* provides information on all of these. However, to become a 'good' manager also requires hands-on practical experience in management. Not only is theory important but managers must also learn how to apply this theory in practice and – of equal importance – how to fit the theory to their own characteristics and style.

Engineering Management relates theory to practice by presenting a large number of case studies throughout the book. These case studies are based on a single company, the FloRoll Manufacturing Company. This is a fictitious company, but one that represents a typical medium-sized industrial organization. The situations presented are those which arise in real companies and are based on the author's own experience of working in four different organizations over a 25 year period. Case study exercises are also included at the end of each chapter for use in classroom exercises and discussions.

Management covers a wide range of disciplines and skills and this book presents an insight into all the functions within an organization which the Engineering Manager is likely to come into contact with. It is divided into seven parts. Each part is largely self-contained and may be studied, or taught, in any order, to suit individual needs or course requirements.

Part 1 introduces the FloRoll Manufacturing Company and the basic principles of management: the levels of management found within a company; the types of tasks undertaken by managers; and the different management styles adopted in different situations.

Part 2 describes the business environment within an organization. It discusses the dynamics of organizations and looks at the following areas: organization structures; formal and informal organizations; global organizations; the importance of total quality management within all disciplines; and the management of change, which must occur continually if the organization is to survive and grow.

Part 3 covers the concepts of corporate strategy and the processes used in arriving at decisions. It describes the elements of a strategic plan, including strategic alliances and acquisitions, and how it can be developed and implemented. Because strategic plans require the collection of relevant data to aid decision making, the techniques for data collection and analysis are also covered in detail.

Part 4 deals with financial management within an organization. It describes the basic financial environment, including the following areas: the concepts of accounting; budgets and controls; profitability; obtaining finance; valuating a company; and the importance of and techniques used for costing and cost control. This section concludes with a chapter on the important topic of investment decisions; that is, how to spread limited financial resources so as to obtain maximum benefit.

Part 5 describes the techniques used for planning and control of large projects, and the aims and techniques of Operations Management, including a description of the use of information technology in different parts of an organization.

Part 6 covers the critically important subjects of marketing and sales management. It describes the role of marketing within an organization, introducing concepts such as: the marketing mix; marketing intelligence; market segmentation; consumer and industrial markets; product introduction and management; the management of customers; and the sales and distribution functions.

Part 7 provides an insight into the skills needed by a competent manager in the following areas: leadership and motivation of staff; the importance of and techniques for building effective teams; and different methods of effective communication that can be used in various circumstances, including meetings and presentations. The book concludes with a short chapter on the importance of that very scarce commodity, time, and how it can be used effectively.

I have been asked why I called this book *Engineering Management*, and whether or not the book is applicable to management within any other discipline, such as in a service or non-engineering environment. I have worked within the engineering profession for over 25 years – in many different junior, middle and senior management positions – managing teams from 10 to over 100 engineers, responsible on some occasions for multi-million pound projects. I have been fortunate to do this in several different disciplines, as an Engineering Manager, a Sales Support Manager, a Product Marketing Manager, and as a Project Manager. However, in spite of this wide range of management experience, it has all been for engineering organizations, and the case studies within the book are closely related to an engineering company. Consequently, I do not feel qualified to claim that the book meets the requirements of industries other than engineering, although I do believe that it does.

Fraidoon Mazda
Bishop's Stortford
August 1997

Part

Introduction

1 The FloRoll Manufacturing Company

Introduction

This chapter introduces the FloRoll Manufacturing Company, part of a multinational organization operating in most of the major markets of the world. This company is used for case studies throughout the book.

1.1 The past

The floroll started its life in similar fashion to many high-technology products; in a garage belonging to David Swinton. Both David and his brother Edward were responsible for its early development, but it was not until some years later, when the product was shown to have military applications, that the floroll began to take its present form.

David and Edward founded the FloRoll Manufacturing Company and obtained a contract from the Royal Air Force to carry out the early experimental work. They moved from their garage workshop to secure premises and soon built the team up to 25 skilled engineers and technicians. It was some years later that the product came off the classified list and the company was free to look for non-military applications. The company then went public, although David and Edward still maintained a significant amount of the equity.

David died suddenly while the company was still in its growth stage, and soon after that Edward was forced to retire due to ill health. Control of the FloRoll Manufacturing Company passed to Edward's son, Percy Swinton, who was of mature age but was neither an engineer nor a businessman.

Years of mismanagement followed until Percy Swinton left the board, selling all his shares in the company. The FloRoll Manufacturing Company lost its lead within the market-place and was strongly challenged by several local and overseas competitors, the largest being Winger Inc.

1.1 The present

Figure 1.1 shows the board of the company as it exists today. The FloRoll Manufacturing Company is now a fully owned subsidiary of RGU International, with headquarters in Chicago, Illinois. Jake Topper II, Chief Executive, reports to Al Big, back at headquarters, but is given considerable autonomy to run the company. Jake has a small staff function reporting to him, consisting of Mary Bennett, Technical Director, Harry Dean, Special Project Director, and Fred Steel, Quality Director. Fred joined the company from General Motors and has done much to improve the quality of the processes used within the company.

Jane Wilmore is the Human Resources Director and is also responsible for site services, training, safety and medical services. Part of her organization is shown in Figure 1.2. Vic Smith, Training Manager, deserves special mention, since he has been instrumental in introducing many training schemes into the company. Employees were given very little incentive to train before Vic joined; the only training method being used then was the 'sit by Sally' scheme (learning by watching others perform the job).

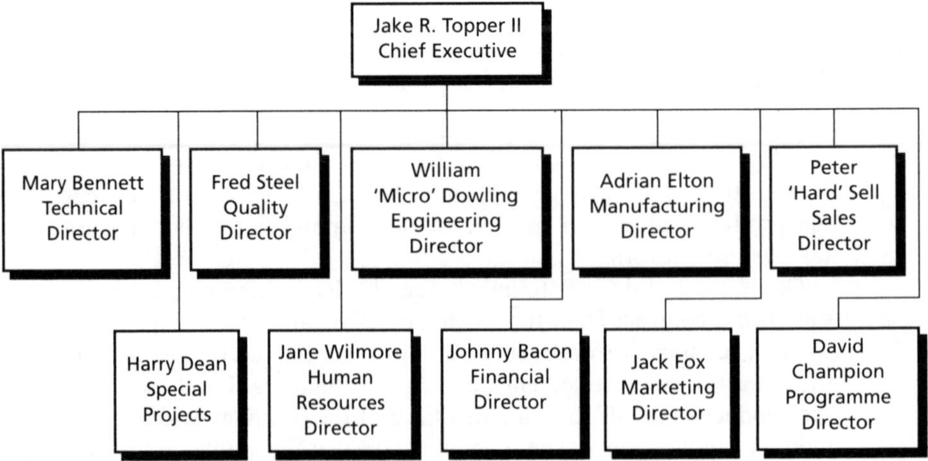

Figure 1.1 The Board of the FloRoll Manufacturing Company.

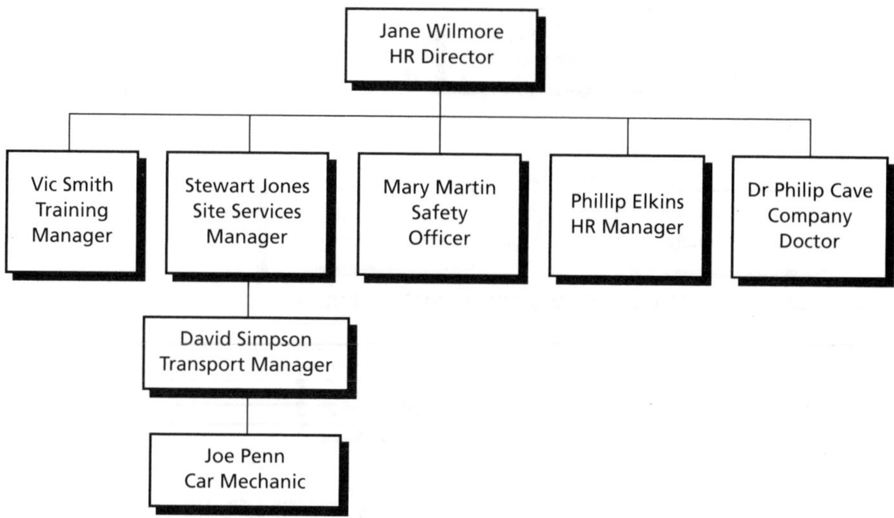

Figure 1.2 Part of the Human Resources (HR) organization.

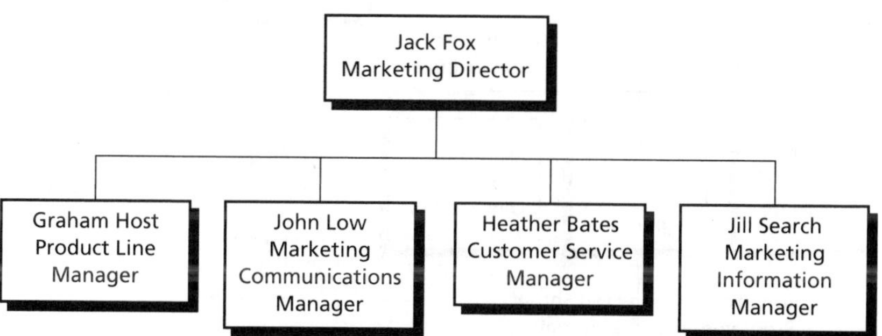

Figure 1.3 Part of the Marketing organization.

Jack Fox, Marketing Director, is a relatively new member of the Board. He took over from Peter Fool, who left under a cloud after a major marketing flop. Figure 1.3 shows part of Jack's first line organization. Graham Host, Product Line Manager, and Jill Search, Marketing Information Manager, were both recruited by Jack and have proved to be invaluable to the company.

The Programme Office, under David Champion, Programme Director, is primarily responsible for managing large customer projects. Part of this organization is shown in Figure 1.4; it consists of several senior project managers, each with a small team of assistants.

Manufacturing Operations has come a long way since Percy Swinton's time. It is now recognized as one of the company's strengths and its source of competitive advantage. Adrian Elton, Manufacturing Director, who took over from John

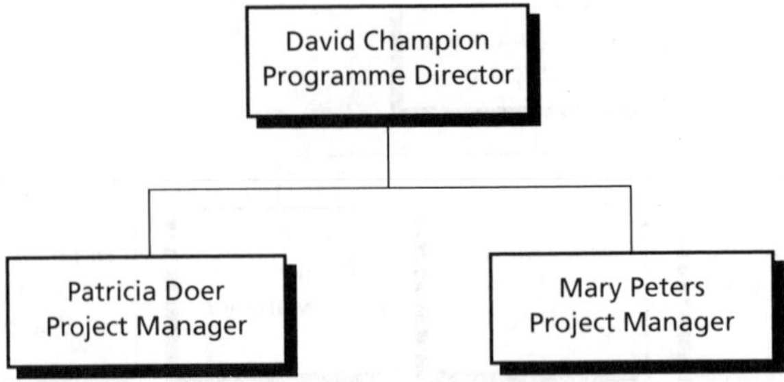

Figure 1.4 Part of the Programme organization.

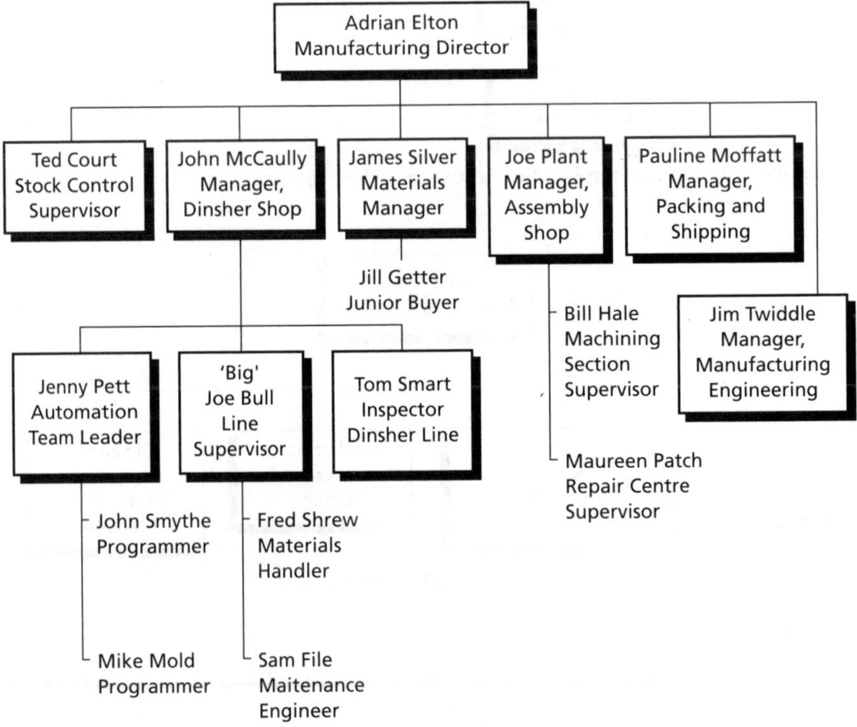

Figure 1.5 Part of the Manufacturing organization.

Strongbow, has largely built up the management team and has been the drive behind the automation programme which has made the company one of the most efficient manufacturers of florolls in the country. Figure 1.5 shows part of Adrian's organization.

Florolls are assembled in the Assembly Shop, under Joe Plant, while dinshers, which are a key component of florolls, are made in the Dinsher Shop, under John McCaully. Most of the other components that go into a floroll (some 600 items)

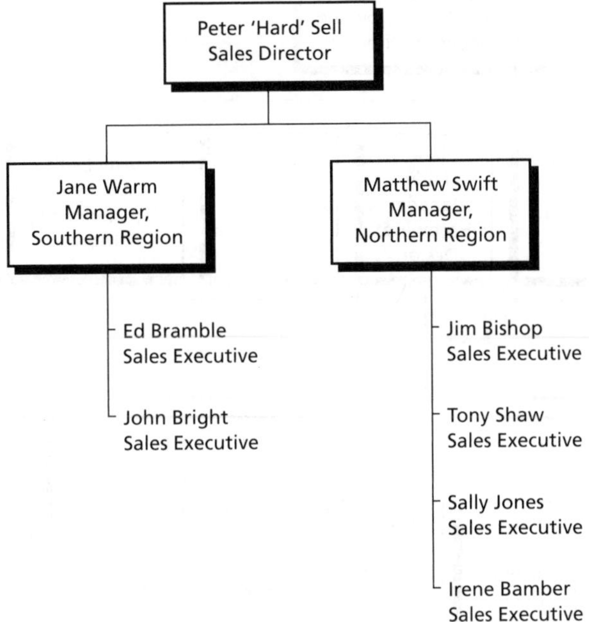

Figure 1.6 Part of the Sales organization.

are bought in from several suppliers. The company at one time also owned a subsidiary, called Stand-U Ltd, making stands for florolls. However, this proved to be uneconomic and was sold off.

The sales force, under Peter 'Hard' Sell, Sales Director, is organized into two regions, as shown in Figure 1.6, with regional offices. The Southern Region is run by Jane Warm and the Northern Region by Matthew Swift. Sales executives are responsible for large customer accounts, often cutting across geographical boundaries.

Ever since the pioneering days, the FloRoll Manufacturing Company has prided itself on being in the forefront of technology. It lost this momentum for several years, under Percy Swinton, but the arrival of William 'Micro' Dowling, as Engineering Director, soon ensured that this reputation was re-established. 'Micro' Dowling's organization is shown in Figure 1.7.

The Engineering Department is organized along traditional lines. The high level of hardware and software within a floroll has resulted in separate hardware and software groups. Both groups are supported by a Planning function, under Peter Hardcastle, a Systems group under Jill Tait, and a Tools group under David Laith. It is a fairly young organization, and the group managers and their staff are relatively new. The one exception is David Laith, who has been with the company for some time and seen the changes it has gone through, resenting much of this change.

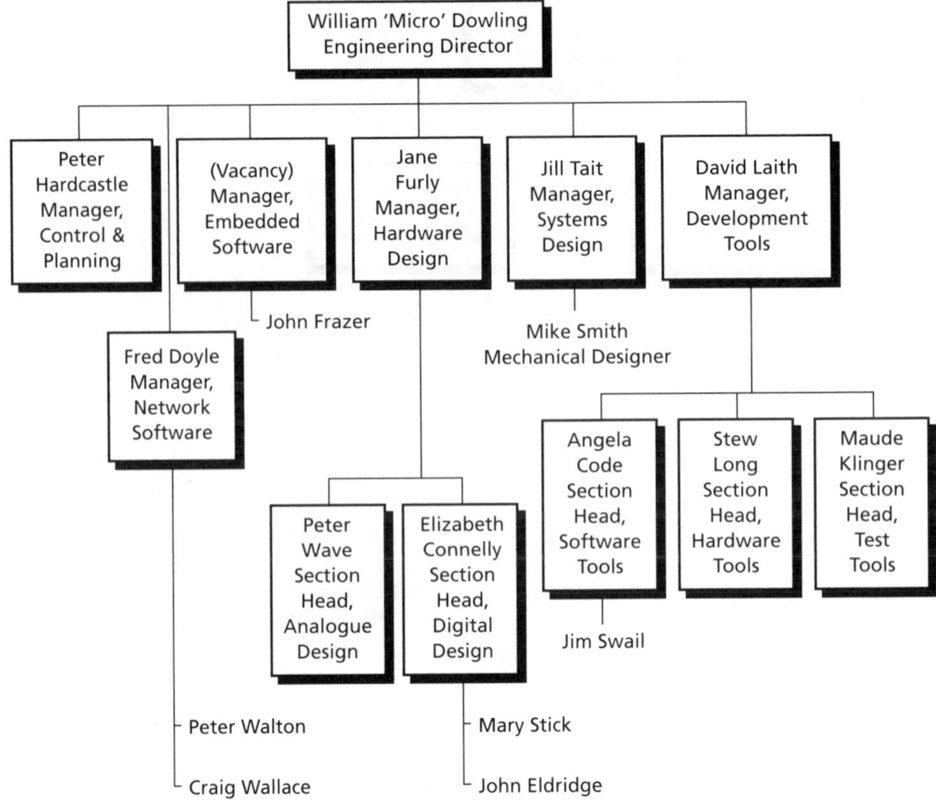

Figure 1.7 Part of the Engineering organization.

The organization described in this section will be used in subsequent chapters of the book, to illustrate case studies.

Summary

The FloRoll Manufacturing Company was a sole-owner establishment which, like several other technology-based companies, started life in a garage. After the death of one of its founders, and retirement due to ill health of the other founder, control of the company passed to the son, who had very little knowledge of running a business, or indeed any interest in doing so.

The FloRoll Manufacturing Company is now part of RGU International, a multinational corporation with headquarters in Chicago, Illinois. The company is structured along conventional lines, similar to many other manufacturing organizations. It has a board of directors, with a Chief Executive appointed by RGU International. All the functions of the company are represented on the board, such as Engineering, Manufacturing, Marketing, Sales, Human Resources and Finance. Organization charts for many of these functions, where they are referred to in subsequent case studies in the book, are given in the chapter.

2 Definitions of management

Introduction

This chapter seeks to answer the question 'What is management?' by looking at the management styles and responsibilities and the tasks which managers frequently carry out. The special problems faced by engineers moving into management are also considered.

2.1 What is management?

There are several answers to the question, 'What is management?' and for a very good reason: there is no single answer (Harris, 1994). Is management an art or a science? Neither, or both! Management is the use of techniques, based on measures, artfully applied.

The saying 'there are those who do things (workers) and those who talk about things (managers)' is not true. A more accurate statement would be 'there are those who do things and those who get things done'. That would certainly be the case in modern industry. Scientific ideas usually come from individuals, but it needs a large multi-disciplined team to develop it into a product and to take it to market. The activities of this team need to be coordinated and 'managed'. Much greater leverage can be obtained from a well-organized team than by its members working as individuals.

To young engineers or scientists, management is something 'they' do, a world full of time-wasting effort spent mainly on covering up one's mistakes and stabbing peers in the back in a bid to reach the top of the corporate ladder. In their minds managers behave as shown in the flow chart of Figure 2.1 when something goes wrong. The alternatives on the chart give the impression that mistakes are not tolerated in a modern organization, and that the opportunity to 'light a fire' under a peer must not be missed.

Needless to say, this impression of management is incorrect, but then what is management and how does one become a 'good' manager?

The one golden rule about management is that there are no golden rules. There are no 'rights' or 'wrongs' in management. It is not an exact science, like engineering. Management has often been compartmentalized as the management of 'people' and of 'tasks', the implication being that tasks are predictable and people are not. This is incorrect; everything in management eventually leads to people, so most things are unpredictable, and will vary with time and from situation to situation. In management one cannot even predict the result of the sum of two numbers (see the case study 'Two plus two makes?')

As there are no fixed answers in management, all that one can hope to do is to learn the basic techniques – the 'language' of management – and then to be guided by experience, bearing in mind that management requires a constant reappraisal of all that has been learnt, since it is almost certain that changes will occur over time.

After all, to learn to play football it is not sufficient to read a book on the subject. This is an essential first step: learning the fundamentals, the 'rules' of the game. The only way to become a proficient player, however, is to go out on the field and actually play the game, learning from the bruising one receives!

If students of management bear this in mind they will not be too disappointed when they find that the reality of management does not fully match the text that they read in this and other books. The most that a management book can hope to achieve is to explain the basic principles: the tools of management; the aspiring manager must use common sense in applying these to real-life situations.

Remember, managers are not the 'bosses'; they are simply employees of the organization, like everyone else, and are doing a job for which they have acquired certain skills. The engineer who goes into management expecting to be a leader, to issue commands and have subordinates follow without question, will be disappointed.

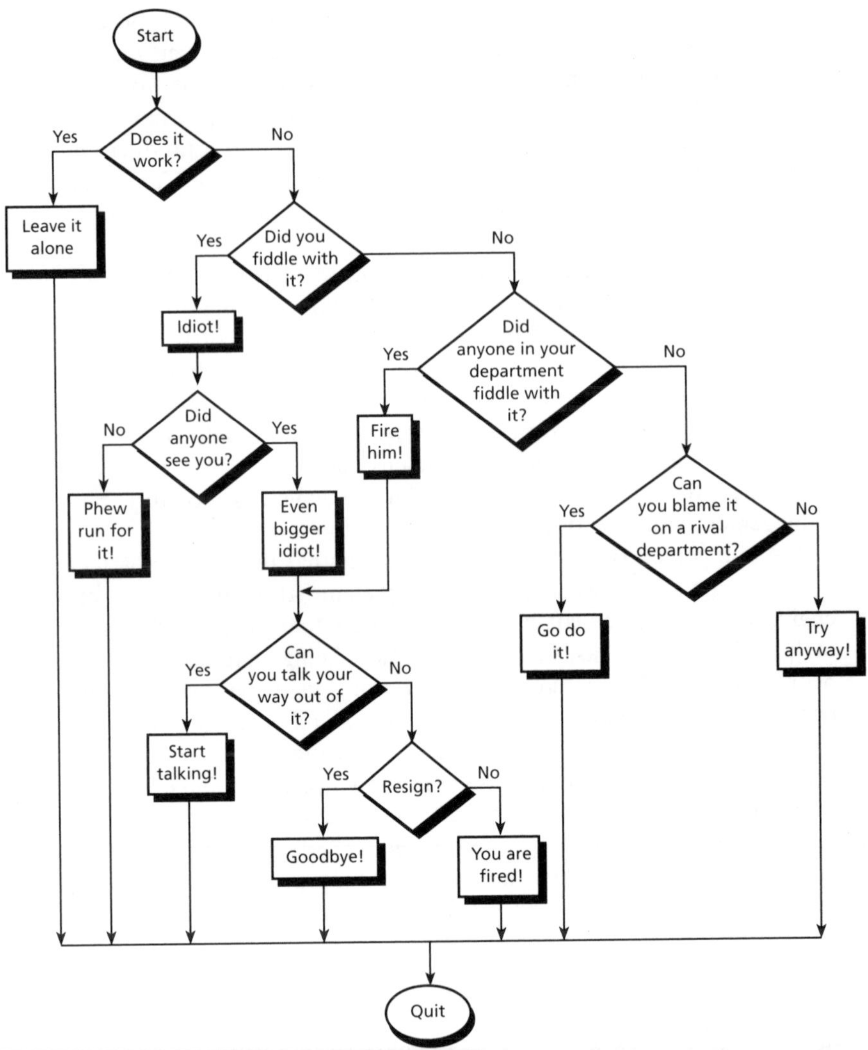

Figure 2.1 Impressions of management behaviour.

Case study

Two plus two makes?

When Jake R. Topper II took over the reins at the FloRoll Manufacturing Company, as Chief Executive, he decided to have a one-to-one interview with his directors, to get to know them better. The first in was Jane Wilmore, the Human Resources Director.

'Jane,' said Jake, towards the end of their interview, 'one last question. How much is two plus two?'

Jane was taken aback by the question and suspected a trap. 'Well,' she said, 'the labour market is currently pretty slack, partly due to the McDowell Enterprise closure of their R&D facility close by. I think that we should be able to staff our new project with the right people without too much trouble. I should not make two plus two any more than three and a half.'

'Excellent!' cried Jake, rising to show Jane out.

Second in was William 'Micro' Dowling, the Engineering Director. The same question was put to him. Micro paused for a long time, his mini-computer of a brain buzzing away.

'That's a difficult question, as you know, Jake,' he replied at long last. 'We thought we knew the speed of light once, but as techniques improve we find that it can be measured with greater accuracy. I think that two plus two probably equals 4.002 456 77, but I will take a few more measurements and give you a more accurate answer next week.'

Last to be interviewed was the Financial Director, Johnny Bacon.

'Johnny,' said Jake, 'it is important that we two can work together. Tell me, what does two plus two make?'

Johnny's jaw dropped to the floor and the colour drained from his face. He jumped up from his chair, ran to the door, opened it, looked outside, shut the door carefully, and walked quickly up to Jake's desk.

'What would you like it to be?' he whispered.

2.2 The history of management

The practices of engineering and of management may be considered to have grown hand in hand. The pyramids of Egypt represent a remarkable engineering feat, but they are equally notable for the management skills that went into their construction. Many thousands of people were involved: the customers (the Pharaohs), the subcontractors (the artisans) and the employees (the slaves). All these had to be coordinated, controlled and monitored; quite a remarkable feat.

However, as a profession, management is relatively new, if one applies the definition of a professional as one having a minimum standard of educational or work achievement, administered by an independent body to which members belong. In the UK the IEE set up a professional group on Engineering Management in 1970, which was almost one hundred years after the IEE itself was founded.

This is understandable, since the earlier industrial organizations were primarily family businesses in which the family head was also the head of the business and his sons took over after him. There was very little promotion of outsiders, and even the most competent employee could only start at the bottom and stay there.

The growth of large corporations in the 1950s, particularly multinationals, provided a boost to the professional manager. These organizations had good working conditions compared with the public sector (although lower security of employment), and gave their staff better training and experience. Promotion was based on merit, and with leavers moving to other companies, cross-fertilization of ideas followed, with the expansion of management techniques and practices.

Management theory was first popularized in the USA and spread slowly to Europe. The first business school was founded at the University of Philadelphia in 1881 and a hundred years later there were over 500 business schools in the USA. By contrast the first two British business schools were formed in 1965, in London and Manchester.

The American Management Association (AMA) was founded in 1923 and in the UK the Institute of Management (formerly the British Institute of Management) was formed in 1948.

Management techniques have also changed over the years, emphasis on different techniques being largely dictated by the industrial scene at the time. The post-war period saw a steady expansion in the developed countries, and the emphasis was on corporate planning for growth.

The techniques used were simplistic and assumed that the future would continue to grow in a similar pattern to the past. This period also saw the application of operational research (OR) techniques to many things, whether the subject was suitable or not (for example human behaviour).

The oil crisis of 1973 changed this, by changing the external environment. Although the emphasis had previously been on growth, with many companies borrowing heavily to finance this growth, survival now became the key consideration. Companies with large debts went bankrupt as interest rates jumped, and many others were forced to reappraise their activities and pull out of non-profitable sectors, in effect shrinking their operations. Short-term payback became the key consideration, not long-term growth.

Recent years have shown characteristics that include those of both the above periods, following the industrial cycles of growth and consolidation.

2.3 Types of manager

Managers come in every shape and size, with very different personalities. This is not surprising, since managers are, after all, human. It should also be remembered that all employees within an organization, whether they have the job title of manager or not, are in reality managers, being responsible for their own tasks if nothing else.

In looking at the types of manager we shall first consider management levels and then the styles that are used by managers.

2.3.1 Management levels

Management levels can be considered as those that are shown on organization charts; line and staff management; and corporate and divisional management.

The interaction between the various levels of management within an organization is illustrated by Figure 2.2. Assuming a middle management level, each individual has above him or her a manager, followed by the manager's manager, and so on up the chain. Communications can occur to any number of levels up or down the organization, but the most frequent communication is usually limited to two levels. Communication upwards also occurs with other managers who are at the same level as the person's own manager.

There will be numerous peers, both within the same function and within different functions. Similarly, down the chain there are likely to be subordinates and subordinates' subordinates.

Communications radiate back and forth like spokes in the organizational wheel. All these interactions need to be managed, not only those that occur between a manager and a subordinate.

Usually, as managers progress up the corporate ladder, they will zigzag through the organization; very rarely does one move vertically upwards, taking over one's manager's job.

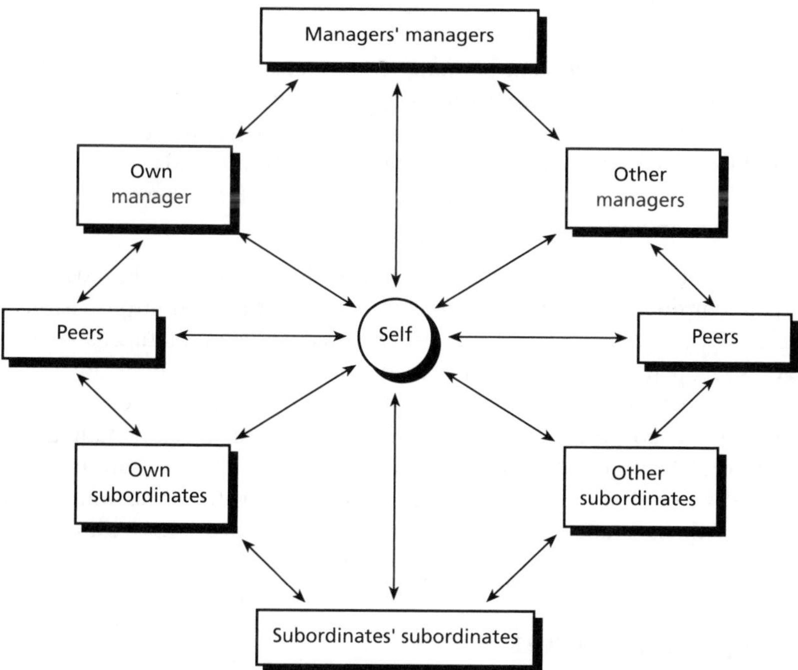

Figure 2.2 The management wheel of communications, showing the interaction between various levels of management.

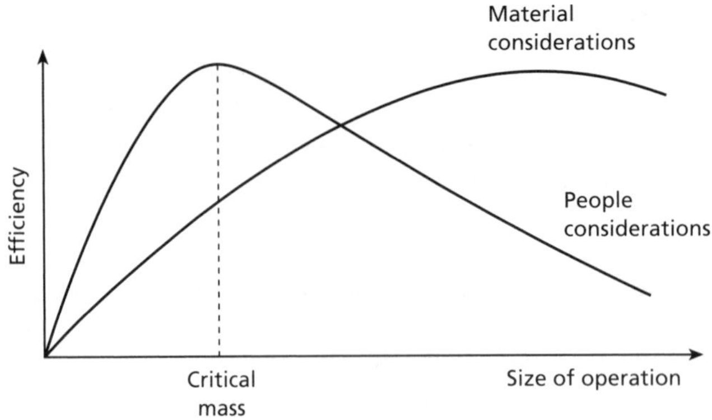

Figure 2.3 The effect of the size of an operation on its efficiency.

The sizes of organizations vary, and there is debate as to the optimum size for maximum efficiency. Figure 2.3 illustrates the most accepted curves for variation of efficiency with size. Material considerations favour large sizes, which can result in economies of scale within manufacture. Even then efficiency will begin to decline if the organization is too large and so difficult to manage.

People considerations favour much smaller units, critical mass being reached much sooner. Thereafter efficiency falls off as size increases. This is usually because:

- In large teams there can be many communication problems and time wasted in ensuring that communication occurs smoothly.
- Staff working on large projects are less able to see the contribution that their individual efforts make towards the organization's goals.
- The larger the team the greater the amount of overheads it needs, such as supervision and coordination.

To overcome the people problems while still maintaining the advantages from material considerations, many organizations are structured into divisions or profit centres. One such division could be devoted to manufacturing only, giving a larger size operation from the materials viewpoint. Each division has a level of autonomy for its profit and loss, and the divisional general manager is answerable to corporate headquarters. A small team of staff managers are now employed at headquarters, with the tasks of: formulating corporate strategy and setting divisional goals; monitoring divisional performance; and carrying out centrally those functions that benefit from size, for example material procurement.

2.3.2 Management styles

There are many, many different management styles. This is not surprising, since the style used is often determined by the personality of the manager concerned, and this can vary through every shade between black and white. In this section seven styles will be described.

A manager often displays many of the attributes of these styles. Also, most experienced managers have learned to vary their style to suit the situation (McClelland and Burnham, 1995); for example, the organization level at which the communication is taking place and the task that is being addressed. It is, however, difficult to adopt a style that is diametrically opposite to one's own personality, since it will then be unconvincing and ineffective.

Administrators

Administrators look to company rules and regulations for solving all problems. They live 'by the book' and are usually very good employees. They show total loyalty to the organization and have probably been with the company for many years.

Administrators are very formal in their approach and work with strict lines of demarcation between departments and functions. They are usually not very good communicators, using the official company channels for all communications, which are often limited to one level upwards and downwards. They protect their department and status and look after their staff. They are not good at resolving conflict, looking to company rules for resolving these.

Administrators expect everything to be black and white, and for practical situations to match theory; they are at a loss when this does not happen. They are very logical and practical and have good planning skills. In spite of their rather mechanistic approach they are generally respected by their staff, and by peers, for their organizational loyalty and knowledge.

Time Servers

These are generally older managers who have lost interest in their job and environment, and are marking time until retirement or moving to another job. They take all necessary action to avoid stress, and maintain a low profile within the company.

Their low personal motivation is reflected in the people who report to them. Conflict at all levels is avoided at any cost.

Although these managers are not generally lazy, their low motivation means that they do the minimum amount of work needed to hold down a job. Decisions are avoided since they could lead to mistakes. Personal status is very important to them.

Time Servers usually have good management experience, and if motivated can become a very valuable asset to the organization. They often consider themselves to be 'father or mother figures'. They understand people and can build an effective team if they try. They recognize achievements in others and are ready to acknowledge them.

Climbers

These managers are driven by extreme personal ambition and will sacrifice everything, including self and family, to get to the top of the corporate ladder. They want to achieve and to be seen to have achieved, especially by those in a superior position.

Climbers will pursue personal advancement by fair means or foul. However, they become demotivated if this does not show quick results, and this can eventually lead to stress. Personal knowledge is very important to them, as a means for advancement, and they will learn from their staff, pushing themselves at their staff's expense, if it suits them. However, Climbers look after those reporting to them, knowing that they are measured on the output from their department.

Self interests come before those of the organization, and peers will be fought in order to gain an advantage and to build an empire. Status is important, but only as a sign of seniority.

Generals

This is usually a younger person who exhibits lots of energy. The General likes to rule and manipulate power, but is achievement oriented: power is used to get tasks done. Generals work extremely hard, driving themselves and those under them.

Generals are sociable and mix well at all levels. They usually get their way with peers by overwhelming them, although peers can resent this if it is done too often.

Status is important to Generals, but for the luxury associated with it, not as a symbol of seniority. They are strong-willed individuals, often with the same characteristics as a self-made entrepreneur. Usually they are optimistic about the future, sometimes wrongly.

Supporters

Supporters maintain a balanced view about the world, the organization, subordinates and themselves. They are usually experienced managers who are knowledgeable in management techniques and apply them where they can.

Supporters work through people in achieving their aims. They are good at delegation and develop their subordinates by giving them responsibility. The people working under them are highly motivated.

Supporters' personal technical knowledge is usually lacking, but this is compensated for by the support they themselves receive from the specialists within their department.

Supporters are good facilitators and very good at managing change. They recognize achievement and reward it. They are deep thinkers and have excellent imagination. Often this can lead to a clash between the goals of the organization and what they believe to be right. However, Supporters are good compromisers and exhibit effective intuition. They are flexible but very persistent in carrying out tasks which they believe need to be done. They can handle stress.

Supporters tend to be loners and do not mix well with peers. This means that they can often miss out on information from the grapevine, so that they are not always well briefed on organizational matters.

Nice guys

These managers are usually weak-willed and are more interested in being liked, by peers and subordinates, than in achieving targets. They do not criticize their subordinates, even when they are poor performers, and may in fact support them too much, so unconsciously retarding their development.

The productivity of the group under the Nice guy is low and conflict often simmers under the surface, waiting to burst out. When it does the manager does not know how to handle it. Very few decisions are made and usually they are very poor, since the manager is ready to yield to pressure from almost any source.

Bosses

These managers are bullies! They like to have their own way and bully their staff (especially their secretaries) in order to enhance their own sense of power. They are a living example of the effect of power on people, as stated by Primo Levi in his book *The Drowned and the Saved*: 'Power is like a drug; the need for either is unknown to anyone who has not tried them, but after the initiation . . . the dependency and the need for ever larger doses is born; also born is the denial of reality and the return to childish dreams of omnipotence.'

Bosses occur at every level, often quite low within the organization. They operate in Administrative mode, playing things by the book where it suits them. They use the power of their position, real and imaginary. They drive the people under them but not themselves. They expect recognition from peers, but often do not get it.

Bosses are extremely inflexible and are often mistaken for strong-minded people. Usually, however, they are only strong talkers, and hide behind abusive language. They try to terrorize subordinates and peers, creating conflict to emphasize their own power.

Managers in the Boss category are often brought into a company to act as 'hatchet men'. In the short-term they can show results, but long-term they are very destructive, causing more harm than good. They are insecure in themselves and get security by humiliating others in public. They advance by pointing out the mistakes of others, and not by their own achievements.

2.3.2 Tasks and people

Although many different and complex factors have been considered in describing the various management styles, it is usual to use two main factors when comparing them: the strengths of the various styles in the management of people and tasks. This is very simplistic, and can only provide a first level guide.

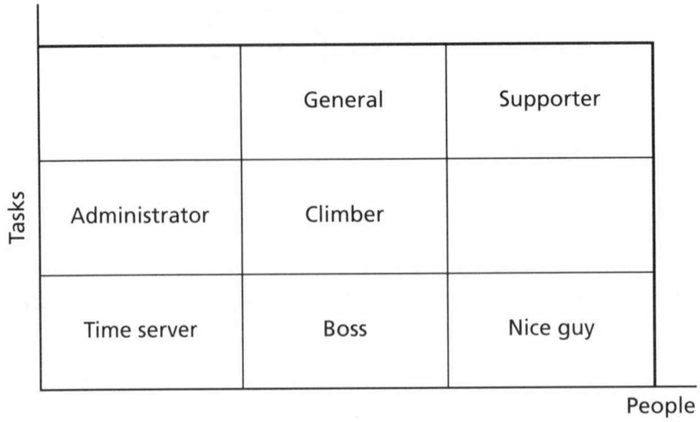

Tasks			
		General	Supporter
	Administrator	Climber	
	Time server	Boss	Nice guy

People

Figure 2.4 Compartmentalization of management styles.

It is also wrong to 'compartmentalize' management behaviour, but in Figure 2.4 we do just that! Each of the styles described in earlier sections is placed neatly into its own little box, with no overlap between boxes. This shows how wrong it is to think of one style as better than another since, on this scale, the most balanced style is that of the Climber. Instead each style must be considered to have its strengths and weaknesses, and each style has its place in various circumstances. Obviously some styles are met more often than others and can be used more frequently.

2.4 Management responsibilities

The functions that a manager performs are complex, which is partly why it is so difficult to define and record them accurately (Mintzberg, 1990; Uyterhoeven, 1989). This is not surprising, since managers are people whose main task is to deal with other people, and human behaviour is complex and difficult to predict.

Generally, managers operate in an environment in which they voluntarily accept certain responsibilities as part of their jobs, while being constrained by other factors. These are shown in Figure 2.5.

Managers have a responsibility to the shareholders of the organization in which they work. These shareholders have invested their money in the corporation and they expect to receive a return for this, measured in terms of dividends and growth. Generally the shareholders are 'faceless', and the average manager is unlikely to meet them, except perhaps at an AGM (Vliet, 1995), but nevertheless their interests must not be forgotten.

Employees of the company are also the manager's responsibility, even though managers are themselves employees of the same organization. Employees expect rewards for their labour, good working conditions, and a job that meets their career aspirations. In return, the employee is expected to show loyalty to the company and look after its interests.

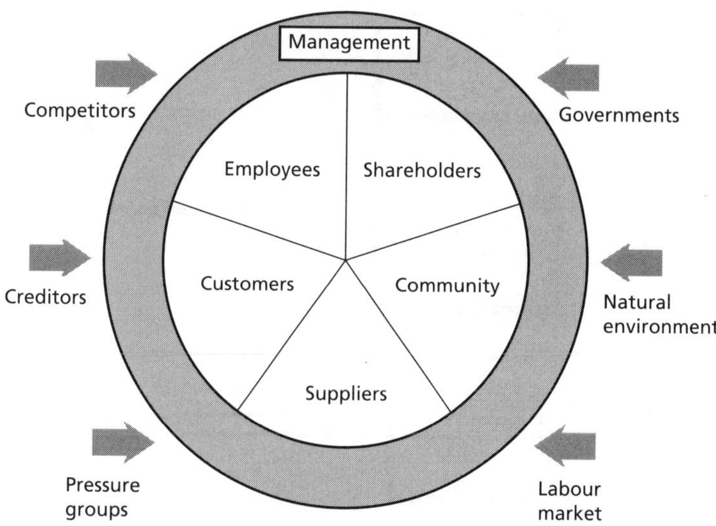

Figure 2.5 Management responsibilities and constraints.

It is customers who make it possible for the organization to exist. Customers expect to receive the goods they need at the right price and at the right time. They should feel that the company abides by its motto: 'The Customer is King!'.

Managers sometimes forget that there are two very distinct and important types of customer: external and internal. External customers are those who buy the company's products, and are the ones usually thought of when the term 'customer' is used. Internal customers are those who depend on the manager and his or her department to provide a product or service to which they can add value before it reaches the external customer. There are many examples of internal customers within an organization, a few of these being shown in Figure 2.6.

In the example shown, Design Engineering is a customer of the Procurement Department, since it is dependent on this department to purchase all the tools and equipment it needs, carrying out negotiations with suppliers in order to obtain the most favourable terms. However, Procurement is also an internal customer of the Design Engineering Department, since it depends on it to produce the technical procurement specifications to which the supplier will deliver the goods. Procurement also looks to Design to verify that the goods produced meet the specifications. In the same way, Manufacturing is an internal customer of Procurement and Procurement depends on Manufacturing.

Marketing, which includes product management, contains the greatest number of internal dependencies, as expected. The Design Engineering Department is dependent on Marketing for the product or customer specification to which it is to design the product. In turn, Marketing is dependent on Design Engineering to provide it with information on product development time-scales and design costs, so that it can develop its business plans, and for technical support in answering

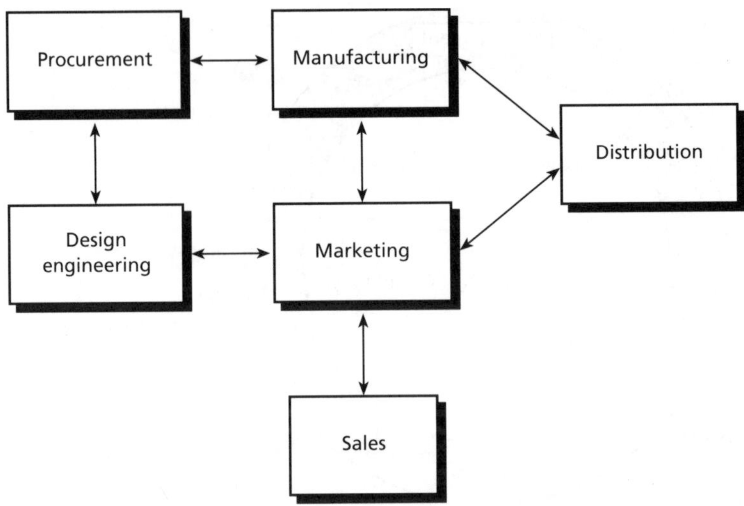

Figure 2.6 Examples of internal customer relationships within an organization.

customer queries. Manufacturing is also dependent on Marketing for product related information, and Marketing requires product manufacturing costs from the Manufacturing Department. Marketing depends on the Sales Department to make the sales to external customers and for market intelligence data, and it depends on Distribution to get this to the customers. In turn, Sales depends on Marketing for product-related information, such as price, and for support in making the sale, such as presentation material and product handbooks.

The relationship between the organization and its suppliers is also vital. No organization can survive without a source for all the components and raw materials it needs, at the right price and on time. In this the organization is itself a customer of its suppliers. The relationship between a company and its suppliers should be very close and is summed up by the phrase: 'getting into bed with one's suppliers'. Managers have the responsibility of ensuring that suppliers are kept fully informed of the company's future plans and requirements, so that they can plan their workload. In addition, suppliers expect prompt payment for their goods or services; they also have to generate profits and growth for their own shareholders.

Managers have a responsibility to the community in which their company operates. This is especially the case if it is a large organization and, as such, the major employer in the region. The organization must ensure that the environment is protected and the employment balance maintained. In addition, managers should look at other ways in which the community can be supported, such as staff visits to local colleges and community work.

Several factors impede managers in their tasks. Competitor activity can result in loss of market share and revenues, although competition can sometimes have the effect of forcing the organization to become more efficient to meet this threat.

Competitors can also help to grow a market, so that although market share falls the actual revenue may increase.

Governments act as a constraint in two ways, by direct and indirect action. Direct action covers regulations imposed on the company, such as by health and safety acts; by the taxes on profits; and by constraints to prevent monopoly actions and so open up the market to competition. Indirect government action is the effect that the government has on currency fluctuations and on interest rates, both of which can critically affect a company working in international markets.

The manager has very little control over the natural environment, the 'acts of God', the fire that closes down production for a week, or the snow that disrupts transport and prevents workers from reaching their workplace. However, the natural environment also presents opportunities, such as the unexpected cold period that increases the sale of electric heaters.

The labour market can be a severe constraint, especially as there is so often a shortage of skilled staff just when a manager is trying to get a new project off the ground. It is, however, the responsibility of every large organization to ensure that the labour market is encouraged to grow, by providing training for its staff and sponsoring courses in schools, colleges and universities.

There are many pressure groups, all of whom strive to influence the manager so that the aims of their members are met. These include environmental groups, which act out of community spirit, and trade and user organizations, which are concerned with obtaining the best products, services and prices for their members.

Finally, most organizations fund part of their growth or operations by borrowing on the market and are restricted by their creditors, who wish to obtain short-term returns on their loans. In addition, a company has indirect creditors, such as the payment it owes to its supplier for goods already received and invoiced.

The importance of the constraints described above changes with time. For example, the labour market ceases to be a constraint at times of recession and high unemployment, but a poor economic climate may also mean that competitors are active, prices being cut and margins eroded.

2.5 Management tasks

The tasks that a manager carries out are many and varied, so that classification becomes difficult. However, for convenience, they will be considered here to fall into four groups, mainly linked to the activities of a project. These are: planning, organizing, integrating and monitoring. These tasks are interrelated, as shown in Figure 2.7.

The main aim during the planning phase is to define the long-term goals, objectives and policies of the corporation and the implementation steps and performance standards. The organizing phase splits the task into workable packages and sets up the organization to carry out the work. Integration covers

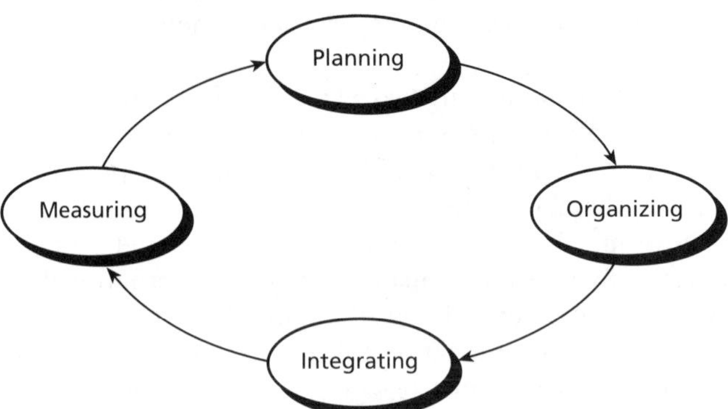

Figure 2.7 The interaction of management tasks.

the tasks required to complete the work, and measuring is a process for determining what was accomplished and how successful this was in meeting the original plan.

2.5.1 Planning

The planning phase starts from the definition of corporate goals, as shown in Figure 2.8. In a large organization these are normally defined by corporate planning staff with input from the various divisions. These corporate goals are further broken down into goals for the divisions and then to departmental goals and goals for each individual. These individual goals often take the form of job descriptions and yearly targets, particularly in those companies that use management by objectives (MBO) techniques.

From the goals a strategic plan is prepared, which defines the strategy that will be used to reach the goals. It sets the long-term direction for the company and must include a what-if type of activity to ensure that, where possible, unexpected factors have been considered, enabling the company to take prompt action should it be needed.

Strategic plans often tread the fine line between survival (short-term goals) and growth, which is the long-term aim. In times of depression the company may have to take short-term action that damages its long-term growth aims, such as pulling out of markets that do not have near-term sales potential and developing products that show an immediate payback but are not part of the company's long-term product portfolio. It is during the strategic planning phase that the company will carry out an appraisal of its strengths and weakness, analyzing the opportunities that exist for it and the threats that constrain it. This is commonly known as a SWOT (Strengths, Weakness, Opportunities, Threats) analysis.

The strategic plan is translated into a plan of action, which comprises the tasks managers need to carry out to meet the objectives set for them. The plan of action must therefore be related back to the set of goals, as in Figure 2.8.

Figure 2.8 The formulation of a plan.

The planning phase establishes processes and sets performance standards, which are used later in the measuring phase. The plan of action needs to include budgets, covering expenses, capital and human resources. Training plans and costs have to be included, especially if the work is new and staff are to be recruited. These plans have to be communicated to all involved on the project.

Two of the key items to be covered during the planning phase are the need and process for change (Kotter, 1995). A business must continually adapt to a changing environment, such as changes in the political scene, in legislation and even to its own goals. Some changes are easy to implement, such as a change in the company's policies. However, some changes take a long time to implement and may even meet with resistance from within the workforce. This is especially so when a company is planning to change its whole culture. There is considerable inertia to change, and in some cases even resistance to change when employees feel threatened, such as with a loss of jobs or status. This should not deter the manager from planning and implementing change. Furthermore, it is essential that the manager is proactive, foreseeing change and taking early action, rather than reacting to change; that is, changing to meet a circumstance after it has occurred, such as the loss of an account due to competitor activity.

2.5.2 Organizing

The prime activity during the organization phase is splitting of the work into manageable tasks and allocating these to groups or individuals. By acting as the person who sanctions all the work within the department, the manager ensures that tasks are coordinated and duplication is avoided.

Jobs and responsibilities need to be clearly defined, which is not always easy, since in most cases there are no fixed boundaries between them, and tasks often overlap. Overlap will also occur between departments, causing managers to fight for the interests of their departments and staff. However, this must never be done at the expense of the overall goals of the organization.

The organization needed to carry out the plan of action is set up; people are recruited, teams established and team leaders appointed. The skill mix within teams becomes an important issue, some of this being acquired by training.

Managers need to ensure that all their staff are focused on common goals. Managers must understand the needs of their customers and ensure that these are communicated to the teams. It is important that this is not a one-off activity; customer requirements are continually changing and must be accommodated.

Effective delegation is important in this phase as it is in every other phase of a manager's activity. There is much that managers cannot delegate, so everything that can be delegated lightens the management load and helps towards developing subordinates.

2.5.3 Integrating

The integration phase normally occupies the longest time. During this phase the plan of action is carried out to achieve the company's goals. To the outsider this phase often appears to be chaotic, with the group plunging from crisis to crisis. The manager has very little time to think before taking decisions and needs to shift rapidly between resolving today's problems, evaluating and modifying long-term plans, and resolving personal conflicts both within and outside the department.

It is important that, in spite of the pressures on their time, managers are able to 'see the whole picture', differentiating the wood from the trees, so that correct decisions can be made. The problem is obtaining accurate information on which to base a decision, especially as the situation is often complex and is continually changing. Information is collected from a variety of sources, much of it informal. (See the case study 'A lucky meal'.)

Clearly, the longer a decision can be put off the better the information on which it is based, but too long a delay will mean that opportunities are missed and a problem can get worse. Risks must be taken, not avoided. However, risk management is important, with risks being minimized if possible and 'what-if' plans put in place.

Managers act as the acknowledged leaders of their groups, especially in times of crisis. They must set the direction, but be flexible enough to change if their decision proves to be wrong, or if circumstances change. Involving the group in decision making is important, yet there are times when managers need to make decisions on their own. (See the case study 'Micro's dilemma'.)

Good communications are important, and the manager must ensure that facts flow down as well as up through the group, and that team members are in communication where their work affects each other. In some circumstances the manager must, however, act as a filter, so as to protect the team from information that will lead to misunderstanding or divert the team from their immediate task.

It is also important that the manager act as impartial arbitrator in disputes within the team, while at the same time showing warmth and understanding when personal problems are involved. The manager must follow organizational rules, but apply them intelligently, knowing when they can be bent to suit extreme circumstances. It is the responsibility of the managers to look after the interests of the staff under their control.

Throughout the implementation phase the manager needs to ensure that the eventual goals are kept continually in front of the team and they are motivated to achieve them. Their teams are the most valuable resource that managers have and they must ensure their maximum efficiency. It is said that the morale within the team working on the Apollo XI mission's Lunar Module programme was very high for the simple reason that the eventual goal was very visible to them all every cloudless night! There must be synergy within the team.

Case study

A lucky meal

When the Adlaigh Aircraft Company went out to tender for a transport mechanism to fit into its next range of cargo planes, Jack Fox, Marketing Director at the FloRoll Manufacturing Company, set up an urgent meeting with Peter Sell, Sales Director, to discuss tactics.

'As you know, the tender is for supply over five years', said Jack Fox, 'and the quantities involved are large. It is crucial that we get this if we are to continue to invest in our next range of florolls.'

'I agree', said Peter Sell. 'It is not only the money, although that will come in very useful now that we are no longer in the running for the Saudi Arabian contract. Much more important is that we will be getting a reference site in the civil aviation field, which we desperately need.'

'What do you think our chances are?' asked Jack.

'Very good', replied Peter. 'I have put our best sales executive, Sally Jones, on the account; she tells me that there is only one other company in the running and we know that they do not have a product that even remotely meets the specification.'

'How about Winger Inc.?' asked Jack.

'We have not seen anything of them', said Peter. 'Sally Jones thinks that they have internal problems and are keeping a low profile until these have been sorted out.'

'That sounds promising', said Jack. 'In that case I think our strategy should be to lead on quality and specification and to bump the price up as high as possible.'

'I agree', said Peter.

Over the next few weeks there was intense activity within the FloRoll Manufacturing Company, as they prepared their response to the tender for the

Adlaigh Aircraft Company. Many long hours were spent by management arguing over technical detail and delivery dates. The customer was forgotten; after all, there was very little competition to worry about.

It was three days before the tender submission date. Matthew Swift, Sales Manager of the Northern Region, in which the Adlaigh Aircraft Company had its headquarters, visited a local restaurant one evening with his wife, to celebrate their wedding anniversary. They were sitting in the corner of the restaurant, giving Matthew a good view of everyone who went in and out. Half-way through the meal he was surprised to see the Marketing Director of Adlaigh enter with the Sales Director from Winger Inc., whom Matthew had once met at a function.

Throughout the meal Matthew had a good view of them both, but they were so wrapped up in their discussion that they did not notice him. From the material that was being handed around it was clear to Matthew that business was being discussed.

That same evening Matthew contacted Peter Sell on his mobile telephone. An urgent meeting was called for the next morning at which the strategy for the Adlaigh bid was revised. Price was now an important consideration. Peter Sell and Jack Fox were no longer confident of winning the contract, but they discovered, once again, that important management information comes from a variety of sources, much of it accidental.

Case study

Micro's dilemma

William 'Micro' Dowling, Engineering Director at the FloRoll Manufacturing Company, ran his fingers through his thick beard and looked out of his office window. The sun was shining and the sky was a deep blue, changing to a lighter shade where it met the tops of the distant houses. A few clouds 'with sails of silver' drifted by.

But Micro Dowling did not notice any of this. He was concentrating on a much greater problem, which he knew he could not put off any longer. Jane Wilmore had insisted that a decision was needed today.

Once again he looked over the organization chart of his department (partly shown in Figure 2.9). The problem lay in the Embedded Software section. The previous manager of that section, Ed Merrick, had retired two months earlier and Micro had not as yet appointed a successor. True, he had received plenty of notice that Ed was going to retire, but somehow things seemed to slip by. He had thought of making an appointment while Ed was still with the company, so that the new person could 'learn the ropes' before Ed left. He had thought better of that; Ed was a bit slack with discipline in his section and it was desirable that the new manager did not pick up any bad habits.

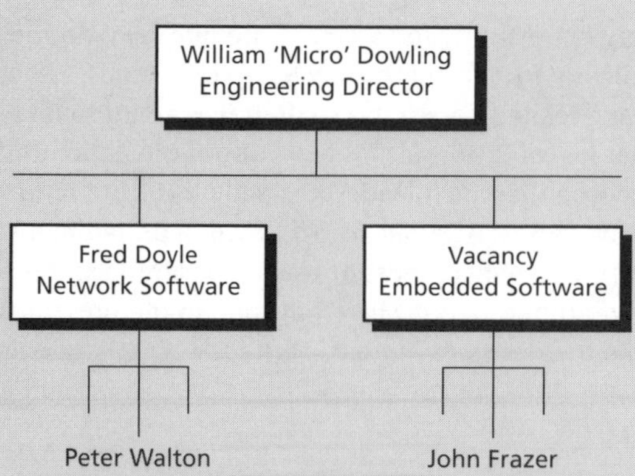

Figure 2.9 Part of the organization tree of the Engineering department.

The other reason for waiting until Ed left was to see how the section settled down without Ed's influence. Over the past two months Ed's second in command, John Frazer, had been acting head of the section and things seemed to have gone quite well; he had heard no complaints from his staff or from other department managers.

As far as Micro could see there were two candidates for the job. There was John, of course, but also Peter Walton, who was deputy to the manager of his second software group. Whereas John had been with the company for over thirty years, and would be retiring in eight years' time, Peter was only 34 years old, and had been recruited into the company six months ago. He showed considerable promise as an engineer, although some of the other managers had complained that he tended to be a little abrasive, and could upset people by his outspoken views.

If John was appointed then the software team would carry on much the same as before, or so Micro thought. They were a good team, but perhaps they needed shaking up a bit to see if they could perform even better. Peter would do just that. However, would John accept the appointment of such a young man over his head?

John would not leave the company and go anywhere else after so many years, but Micro quite liked John, had known him for many years, and would not want to see him upset.

There were two other possibilities: bring in another manager from outside or combine the two software groups under Fred Doyle. The first alternative was tempting, but there were no suitable candidates within other departments in the company. This meant that he would have to go outside, and his managers could accuse him of not promoting from within the company. The Personnel Director, Jane Wilmore, was equally anxious that more promotions were made from within the company.

The two software groups could be combined, but then it would be large and would put Fred Doyle in a very powerful position. Micro Dowling did not like to admit the fact that he was a little worried about Fred. He was a very good engineering manager and had ambitions to become a board director. Micro did not like the way that Fred took every opportunity to communicate directly with the MD, Jake Topper, sometimes conveniently forgetting to copy Micro on memos.

Micro Dowling shut his eyes and leaned back in his chair. It was going to be a difficult decision. He wished that he could consult his team about the appointment. It would be good to let them choose their next leader; certainly they could not complain then if things did not work out. After all, Micro Dowling prided himself on running a very democratic department, where most of the important decisions were made as a team. It was a tempting thought, but Micro felt that, in the present instance, he would have to make the decisions himself, perhaps with a little help from Jane Wilmore.

2.5.4 Measuring

Measuring primarily involves taking stock of achievements and putting in corrective actions, including process changes if necessary. Measures must be taken continually throughout the implementation phase, not only towards the end, when faults can be expensive to correct.

Several items need to be measured and controlled during the life of the project, such as:

- The costs, which include salaries, expenses and capital. These need to be compared with the budget values and variances understood and accounted for.
- The use of skills and specialized equipment. To the manager these are often more important than the monetary values of labour and capital, since budgets can be increased, if justified, but skills and specialized equipment usually cannot be acquired quickly. If these items are not available the project may be delayed, with resulting loss of revenue and increased development costs.
- Progress on the project, measured on a time-scale. This will indicate whether there are delays in parts of the project that will affect its ultimate timely completion. Progress must be measured against deliverables. It is often the case that the content of the project will change, either because the market requirements change, or because it would take too long to achieve the original goal, and the requirements are reduced to bring the product to market earlier. Concessions are now needed against the original plans, which must be clearly recorded and communicated to all who are affected.
- The quality of the product. Quality is a key consideration during all phases of the project, not only towards the end in manufacturing. Quality must be built into a product during the early design stage, and this quality needs to be measured. It is for this reason that all departments and every project usually has its own quality plan, which defines its stated standards and set of measures.

The methods for taking measures vary from the formal weekly measures chart, showing progress against a set of agreed deliverables, to the informal measures a manager is continually making throughout the day. There is no substitute for MWA, Management by Walking About, to enable the manager to gauge the way a project is progressing (see the case study 'Mr. Walkabouts'). MWA has the added benefit of improving communications and raising morale within the team, since the manager is being 'seen' by all.

Individuals must be set measurable targets that are in line with company goals. They should be achievable, but should also aim to stretch them. The targets must be used for growing staff and rewarding them, and not for finding defects and determining punishment. Measures must also be flexible so that they can be changed to meet changing circumstances. The prime purpose of a measure is to record and provide feedback, resulting in corrective action and process improvement.

Three mistakes are commonly made during the measuring phase:

1. Measures are put in place because they are required by the company, for example as part of the annual appraisal. Managers apply them grudgingly, believing that they take up too much time, and therefore the measurements lose credibility with their staff.
2. Managers feel that a poorly performing team reflects badly on them (as a backward class reflects on the quality of the teaching) and therefore they set easily attainable targets or help their staff unduly with them, so that they do not grow.
3. Managers do not fully trust their staff and apply very tight measures and controls, in the form of day-to-day supervision. The amount of supervision needed generally varies with the quality of the staff concerned. The higher the skill level the less the supervision required. This is illustrated by the supervision curves of Figure 2.10. Productivity is poor at very low levels of supervision (A), since the tasks are not then clearly defined and some

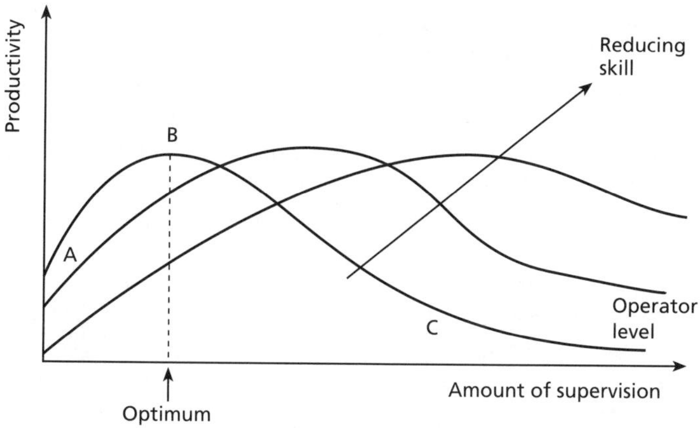

Figure 2.10 The supervision curves. Productivity peaks at an optimum level of supervision, this level varying with the skill of the people being supervized.

minimum guidance is needed. As supervision increases it also results in greater productivity until an optimum level is reached (B). After this productivity actually falls, since staff lose initiative and motivation. Eventually the productivity settles at some low level (C), called the operator level, where staff basically wait for detailed instructions before carrying them out. The lower the skill the higher the optimum supervision level and the greater the productivity at which the operator level settles.

Case study

Mr Walkabouts

Graham Host, Product Line Manager with the FloRoll Manufacturing Company, was better know as Mr Walkabouts because of his reputation for always being out of his office. Graham loved to walk around the development and production areas, chatting to the engineers.

'In my job I have to know what is happening to my product line', Graham used to say, 'and what better method than to go and see for myself.'

Quite early one morning Graham wandered into the design laboratories and struck up a discussion with Jane Furley, Hardware Design Manager.

'How is the new design for the Adlaigh drive mechanism going?' asked Graham. It was now six months since the company had won the contract to supply florolls to the Adlaigh Aircraft Company, and Graham liked to keep in touch with how the new design was progressing.

'Very well', said Jane, beaming. She liked Graham and was pleased that a Product Manager was taking such interest in the design work.

'Finished the rudder section?' asked Graham.

'Yes', said Jane, 'yesterday. We have now moved on to the tail spinner. The carbon hinges will cause us some problems, I suspect.'

'Carbon hinges?' asked Graham, surprised.

'Yes', said Jane. 'If you remember that was one of the items we were most concerned about during the tender stage, because we had no experience in this form of fixing.'

'But the revised specification from Adlaigh reverted back to metal hinges, as we had recommended', said Graham.

'Really? Well the design documents prepared by our systems engineers still refer to carbon hinges.'

Graham Host went to see Micro Dowling later that morning and Micro asked for the design documents to be reviewed. It was discovered that the systems group was working from an earlier specification issued by the customer. Quite a few changes had been made in the new specification, on which the company's contract was based. Graham Host's Management by Walking About had saved the company a lot of money and embarrassment by discovering the error! It had also revealed that the design was running later than that indicated by the formal measures.

2.6 The engineering manager

How often has one heard the saying: 'Management is management; it does not matter whether it is an engineering department or a grocery shop'. To some extent this is correct – many of the management techniques used are similar – but to manage in a technologically oriented company requires technical skills, which is why so many engineers and scientists are now on the boards of engineering companies, where previously they were dominated by accountants and marketers.

Unfortunately, many engineers drift into management, rather than having started out planning for it as a career. They may attend a business course prior to this move, but even that does not always happen. Generally the feeling is that management requires common sense and nothing much else, unlike technical work, for which a college degree is essential.

Usually the move to management is seen as a promotion and an engineer becomes a manager as a reward for good engineering work. Unfortunately, a good engineer does not always make a good manager! To make matters worse, engineers are usually at the peak of their engineering creativity when they are promoted into management, and they leave a career in which they are competent for one in which they have no experience. Furthermore, engineering managers still consider themselves to be engineers, often rolling up their sleeves at the first opportunity for getting down to the bench. This is, of course, a mistake, since managers must learn to work through other people; otherwise they will not succeed. It is a pity that so few organizations operate a dual career ladder, so that good engineers can continue to do what they are best at and can be assured that they are progressing as rapidly up the technical ladder as they would if they made a switch to management.

2.6.1 Technical considerations

Engineers and scientists do carry out many tasks that are done by managers. For example, they spend some 80 per cent of their time on management-related activities, such as communications in meetings, influencing people, supervising technicians and peer interaction. Good engineers must not only be masters of their technology; they also need an awareness of other issues, such as the political and social impact of their work, the customer and market requirements, and the costs of the product they are developing.

The difference between engineers and engineering managers is the emphasis that is placed on these activities. The higher up the organization tree one progresses the more one is concerned with financial and people issues and less with detailed technical matters. The coverage is much wider in scope and shallower in depth. However, management is as creative, if not more creative, than pure science, since it is dealing with rules that are not fixed, and one is operating in an environment that is uncertain and continually changing.

Engineers working on technical projects want to be managed by someone who can understand their technical problems and can act as a sounding board for new ideas. Engineering managers must be able to understand the tasks within their group and to explain and defend these to other parts of the organization. When passing information up the management chain it must be phrased in terms that can be readily understood, so that its importance is appreciated. It is the responsibility of technical managers to get resources approved for their projects, and for this they have to understand the resources they need, such as development tools.

Engineering managers do not give up their technical skills when moving into management and resort to 'paper pushing'. Instead they take on a much broader view of the activities they are involved in, such as the effects on company profits and customers.

The engineering manager is often a middle manager, who must receive goals from above and translate these into a strategy and plan of action. Because of their technical knowledge and closeness to the coal face they are in the best position to do so. However, these plans must take into account many non-technical issues, such as economic, marketing, competition and political considerations. Unfortunately, the middle manager frequently has more responsibility than authority, being dependent on other departments for the successful completion of projects, and this can lead to frustration. If projects run over a long period of time it is also inevitable that market conditions will change, and so will the original goals, so the engineering manager must continually monitor these and must accept the changes that sometimes seem to give the feeling of 'two steps forward and one step back'.

2.6.2 The time factor

Time management is one of the factors that the engineering manager, especially the newly appointed manager, finds difficult to cope with (Adair, 1987; Covey, 1994; Fraser, 1992; Lowson, 1995; Murdock, 1993).

Engineering tasks require thought and time. An engineer is used to spending weeks and sometimes months of continuous effort on a new design. When this is completed and successfully tried out there is considerable job satisfaction from knowing that a task has been completed.

The theory states that engineering management also requires time for reflection and thought and for strategy and plan formulation, but in practice this is rarely available. Managers are called upon to maintain an open door policy, so that they in effect invite interruptions. Small matters, such as personnel problems, require immediate attention, often over more weighty tasks such as planning. The need to interact with others is also important and means that managers cannot shut themselves away for any length of time.

Engineering management requires the ability to shift rapidly from one task to the next, from one person to another, often in the space of a few minutes. Very rarely is a job worked on continuously for more than half an hour and many jobs must be abandoned when they are partly completed, being replaced by another of higher priority. Therefore the new engineering manager may suffer from a sense of never achieving anything. Time management is considered further in Chapter 20.

2.6.3 The people factor

Engineering managers usually find people problems the most difficult to tackle, partly because technical staff are highly educated and independent individuals who do not like to be told what to do. As an engineer, one is primarily concerned with tasks. As an engineering manager, one is concerned with people, the tasks being achieved through people, whether they are within one's direct control or not.

The new manager expects to command and lead, but will soon find that managers can only achieve by negotiation at all levels, and this requires interpersonal skills. Many engineers find this wasteful of time and would rather do things themselves, something which the engineering manager must never do.

Managers must delegate technical tasks, even those that they feel they are most capable of doing, owing to their past experience. It is only by delegation that they can develop their staff, so that they are not perpetually the 'specialist' for the department.

Reviews and discussions with staff should not be limited to tasks, but should cover personnel issues as well, such as development. Many engineering managers feel uncomfortable in such a discussion, especially if they are counselling a poor performer or giving a bad appraisal. Show firmness but be constructive. Stress strengths while striving to improve weaknesses. Take pride in growing people, just as an engineer takes pride in seeing a design in production.

New managers need to be sensitive of their peers. As engineers they could wander into another department without attracting too much attention. As engineering managers they cannot go into other departments and talk to the staff there, especially asking them to do some work, without first informing the department manager concerned. It also goes without saying that a manager cannot poach staff from another department, although transfer of staff between departments should be encouraged where these represent career development moves.

Engineering development is a precarious business, and many tasks are started which never reach the market-place. Figure 2.11 illustrates the probability of failure as the product goes through its various phases, and also the costs associated with each phase. Engineering managers must plan for product failure and be able to reorganize their team when this occurs, while still maintaining high motivation, especially if engineers have had to be reassigned to new projects.

Usually the probability of failure reduces after the initial concept stage. It remains relatively constant as development progresses. Expenditure rises sharply once a product is committed to manufacture, so the decision to go to that phase is usually a difficult one.

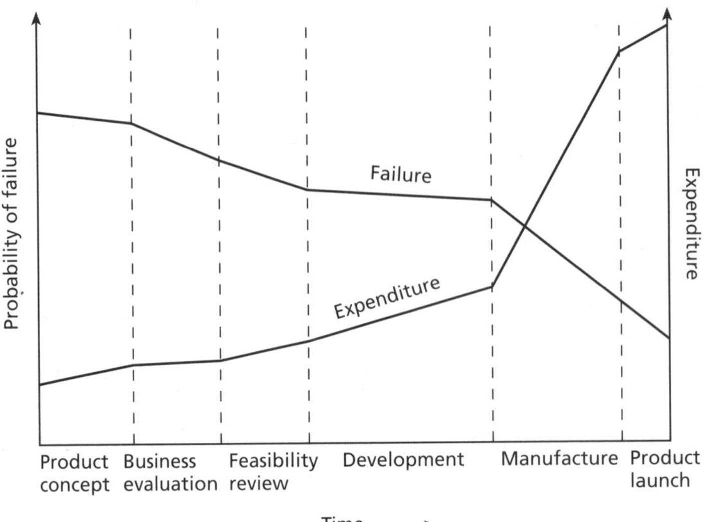

Figure 2.11 Project failure and expenditure.

Once in manufacture the probability of failure declines steeply. This curve indicates the importance of shortening the time between the product concept and launch phases so as to get quickly to the market-place, ahead of the competition, and so reduce the risk of product failure.

Summary

Management is not an exact science, such as engineering. There are usually no right and wrong answers to management questions and the best the student can do is to study the basic techniques (the 'jargon' of management) and then learn by experience to apply them to real-life situations.

There are many different levels of management within an organization and managers need to communicate at these levels, which are summarized by the management wheel of communications shown in Figure 2.2. Management styles vary depending on the personality of the manager and the situation in which the manager is operating. Some of the styles are the Administrator, who works strictly to company rules and has a very formal approach to work; the Time Server, who is generally marking time to retirement, but has good management experience which can be applied to certain circumstances; the Climber, who is highly motivated and driven by personal ambitions; the General, who is full of energy and very achievement oriented; the Supporter, who works through people to achieve tasks; the Nice Guy, who is more interested in being liked than in meeting targets; and the Boss, a bully who likes to have his or her own way.

A manager is responsible to a wide range of groups, such as employees, customers, shareholders, suppliers and the community, as illustrated in Figure 2.5. In addition, managers need to work under several constraints, such as competitors, creditors, pressure groups, governments, the labour market and nature.

Management tasks are varied but can be grouped into four main areas: planning the tasks; organizing these tasks into manageable units and allocating them to groups or individuals for action; integrating the activities of the groups so that their output meets the company's goals; and measuring the activities and putting in place corrective actions, if necessary. Engineers carry out many of the tasks done by managers. The prime differences found by engineers when they move into management are the need to work through people, rather than doing the detailed technical tasks themselves, and the requirement to interface with a wide range of people, such as those who report to them, their peers and customers.

Case study exercises

2.1 Johnny Bacon, Financial Director, presented his recommendations for the new accounting system to the Board of the FloRoll Manufacturing Company. The system had been devised very largely by Sally Smith, one of Johnny's staff, but Johnny omitted to say this during his presentation. Jake Topper complimented Johnny on the new system, in the belief that most of the work was done by Johnny himself. Johnny did not correct him, but resolved to give Sally a bonus for her work. Which management styles is Johnny exhibiting?

2.2 The FloRoll Manufacturing Company wants to start a new development for a low-cost version of florolls. Market information is currently sketchy, so the exact size of the market is uncertain. However, if the company does not enter this market soon it is possible that low-cost imports may come in, which could eventually cause them to lose market share. Should the company start the development work or wait for more market information? Is there an alternative for them to follow?

2.3 Adrian Elton, Manufacturing Director at the FloRoll Manufacturing Company, is trying to decide whether to make the tail mechanism for the new floroll in his manufacturing plant, or to buy this section in from another company. He would like to take the advice of his production supervisors, especially since they will need to work around any decision he makes. However, he fears that the supervisors will want to make the product, irrespective of any other considerations, in order to keep work for their staff. Should Adrian make the decision himself or consult his supervisors?

2.4 Ed Bramble, Sales Executive with the FloRoll Manufacturing Company, has been told by his boss to take an order from a new customer for a batch of florolls. Ed knows that the florolls will only partially meet the customer's requirements and has told his manager so, but she has insisted that the order is too valuable to lose. Ed is torn between obeying his manager and acting in the best interests of his customer. What should he do?

References

Adair, J. (1987) *How to Manage Your Time*, McGraw-Hill Book Company, UK.

Covey, S.R. (1994) *The Seven Habits of Highly Effective People*, Simon & Schuster Ltd, UK.

Fraser, H. (1992) Take control of your time, *EDN*, 7 May, pp. 264–8.

Harris, L. (1994) What is a manager? *Professional Manager*, January, pp. 13–14.

Kotter, J.P. (1995) Leading change: why transformations fail, *Harvard Business Review*, March–April, pp. 59–67.

Lowson, R. (1995) Time: panacea or predicament, *Professional Manager*, July, pp. 38–39.

McClelland, D.C. and Burnham, D.H. (1995) Power is the great motivator, *Harvard Business Review*, January–February, pp. 126–39.

Mintzberg, H. (1990) The Manager's Job; Folklore and Fact, *Harvard Business Review*, March–April.

Murdock, A. (1993) *Personal Effectiveness*, Butterworth-Heinemann, UK.

Uyterhoeven, H. (1989) General Managers in the Middle, *Harvard Business Review*, September–October.

Vliet, A. van de (1995) No meeting point, *Management Today*, June, pp. 74–6.

Part II

The business environment

3 The organization

Introduction

This chapter looks at the key aspects of organizations: how they are changing; the dynamics within modern organizations; the organization structures that can exist, both formal and informal; the importance of Total Quality Management at all levels within the organization; and the need and process for organization change and improvement.

3.1 Defining the organization

3.1.1 Changing organizations

The term 'organization' is probably as difficult to define as 'management', since both are commonly used to mean many different things. An organization may be considered to be a group of people with defined relationships to each other, for example a family, a club or the boy scout movement. By an organization one may also be referring to a collection of human and material resources, which are gathered together for a stated aim, for example a factory, a firm of solicitors, etc. At a more general level an organization may also refer to a structure defining the division of work and interaction between individuals, groups and resources (Dunderdale, 1994).

The common view of an organization is a system that contains one or more of the following elements:

- A collection of people in formal and informal groupings.
- Individuals who have defined tasks and responsibilities, some of which may consist of specialization.
- The manner in which these tasks interact and relate to each other is defined.
- The tasks all lead to achievement of a common aim.

The traditional view of an organization is shown in Figure 3.1. In the hierarchical structure commands pass down the chain from superior to subordinate while information on task progress flows upwards. Authority is delegated to lower levels from the one above, until it reaches the bottom. No commands go upwards and no information flows down. The organization may be flattened, as in Figure 3.1(b), but the basic principle remains unchanged. However, in these instances, information may also flow horizontally, across the various levels, because of the wide spans of control involved.

The problem with this view of the organization is that it is impersonal and based on ideal assumptions regarding people. It assumes that employees are units of work, doing what has been assigned to them and therefore furthering the aim of the organization. Employees are expected to react to situations in terms of work to be done and not personal likes or dislikes.

Changing technologies and competitive pressures are forcing a change in the way organizations are perceived (Lawler, 1994). Hierarchies are disappearing; horizontal communication is replacing the traditional flow of information and command up and down the organization; alliances are being formed with other companies, and with suppliers and customers, and they frequently operate as part of the formal organization (Gross, 1993; Gretton, 1993). These changes have been driven by:

- The emphasis in organizations on the attainment of results rather on the process used in achieving them.
- The emphasis on horizontal activities within the organization in order to gain influence and information, rather then vertical activity.
- Many more opportunities for action and exerting influence within an organization.
- The realization that external contacts are becoming an important factor in being able to wield internal influence and power.
- Rapidly disappearing formal control mechanisms between managers and subordinates (Simons, 1995).
- No clear career progression paths within the organization, but many more opportunities for advancement.

Many managers are finding it difficult to work in this changing organizational environment. (See the case study 'Am I the boss?')

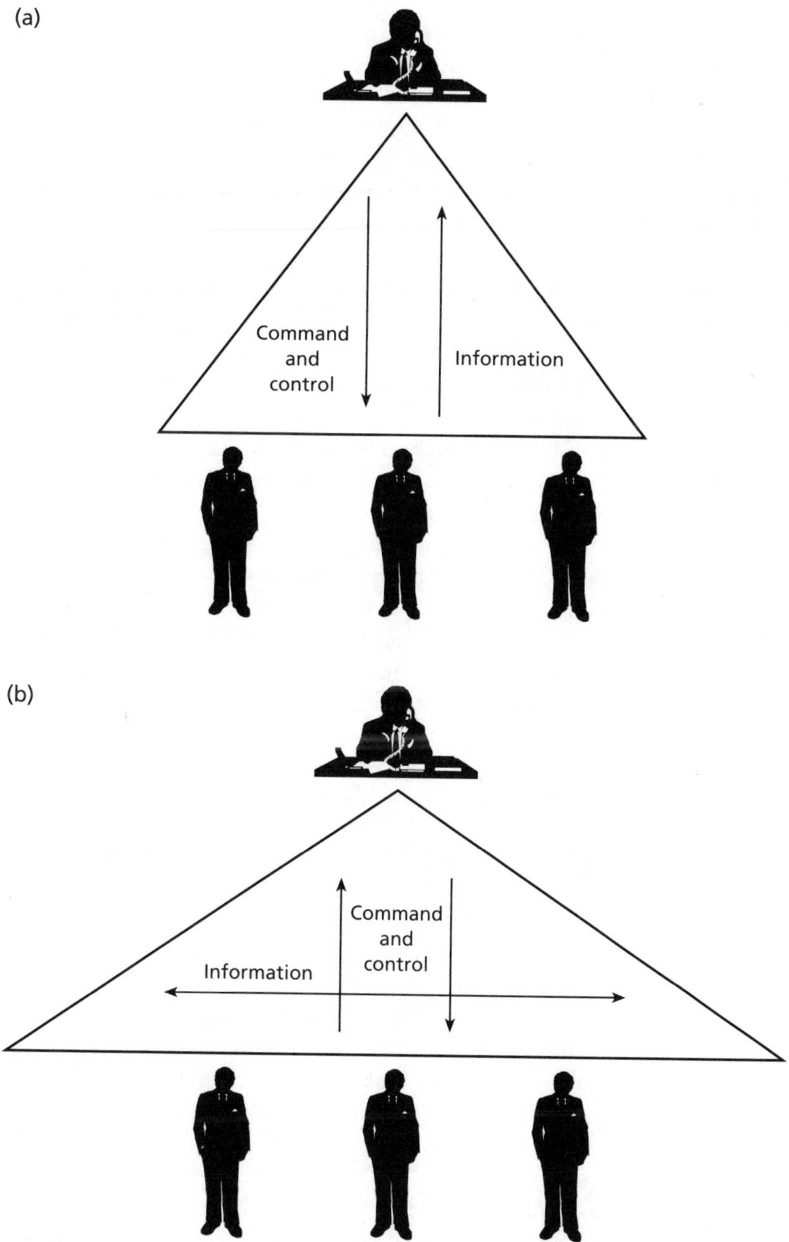

Figure 3.1 Traditional view of an organization: (a) hierarchical; (b) flattened.

Case study

Am I the boss?

For The FloRoll Manufacturing Company to continue to operate in a relatively dominant position within the UK, a more reliable but cheaper material had to be found for the main drive mechanism.

'This is a major challenge for us all', Jake Topper, Chief Executive, said at his Executive Staff Meeting. 'Not only do we have to come up with a brand new concept but we have to make sure that we introduce it to the market-place on a much shorter time-scale.'

Following long discussions it was agreed that a project team would be set up to develop the new material. Engineering, Marketing and Manufacturing would each contribute staff to this new team. Because of the importance of the task, David Champion would himself act as Programme Manager for the project.

Jane Furley, from Micro Dowling's Engineering Department, was asked by Micro to allocate two of her key staff to the project.

'But will I still be responsible for them?' asked Jane.

'Yes', replied Micro. 'This is very much a matrix structure. You have to make sure that you provide technical guidance to your staff, where this is needed, and that you look after all personal matters relating to them, such as appraisals and counselling.'

As the project progressed it became evident that several of the managers, including Jane Furley, were having problems coping with the new organizational structure. After four months Jane sent a confidential memo to Micro Dowling, expressing her doubts and those of other managers in her position, in the new organizational structure. Her main concerns were:

- The members of the task force were becoming very powerful within the organization and it was difficult for the managers to exert any meaningful control over them.
- Because of their work the team members were dealing directly with senior people within the organization, often many levels above themselves, and the managers were not party to these discussions.
- Team members were attending senior level meetings to which their managers were not being invited.
- The line managers were not being briefed on what their staff on the task force were doing, although they felt that they should be so briefed. It was therefore going to be extremely difficult to conduct any form of meaningful appraisal of their work at a later date. Under these circumstances the managers also felt that they could not accept responsibility for the quality of their subordinates' work.
- Because of their position, the subordinates were in direct contact with important customers and suppliers. They often knew more than their managers about what was happening or going to happen in the organization.

Micro Dowling voiced his concerns regarding the task force at Jake Topper's next Executive Staff Meeting.

'But the task force is a tremendous success', said David Champion. 'Strong bonds are being formed within the group and for once we are getting away from the "them" and "us" organization, where marketing throws its requirements over the wall at Engineering and Engineering throws the drawings over the wall at Manufacturing. We are close to a breakthrough. A few more months and we should have the makings of a new product.'

'I agree', said Jake Topper. 'This task force is a very good idea; pity we did not do it sooner. I don't see what all the fuss is about.'

The problems of the matrix organization were however recognized by David Champion. He ensured that the task force produced regular reports, which had a wide distribution within the organization, and that regular reviews were held to which line managers were invited.

3.1.2 Organizational dynamics

Organizations are dynamic feedback systems that need to be creative if they are to survive (Stacey, 1993). Managers cannot easily predict cause and effect, leading to difficulties in controlling the long-term direction of their organizations.

Management's main task is in creating conditions for the organization to learn and develop (McGill and Slocum, 1994). Staff in the organization will take actions based on their past experience and memory, repeating actions that were known to have worked in previous similar situations. However, situations change, and staff must be willing to experiment, learning from new experiences. For effective learning the manager must create an open, questioning environment, very different from the structures of the past, where staff obeyed orders either out of fear or blind loyalty.

An organization is a complex web of feedback structures, where one activity leads to another according to some predetermined relationship. Most organizations tend to oscillate between two states: integration and disintegration.

In the integration state all activities are closely controlled, usually from a strong centre. The aim is to minimize waste by re-use of technology and processes. Overlaps in markets are eliminated, and the whole organization drives forwards towards common goals using a common culture. Tight rules are implemented and rule breakers are punished. Such an organization will eventually lead to stagnation, where all initiative is lost and the organization dies on its feet.

In the disintegration state tasks are divided into smaller and smaller units, as are the organizations and markets which they serve. Individuals are empowered and motivated to act in the best interests of themselves and the organization as a whole. Communications, procedures and processes are entirely informal, replacing any formal ones that existed. Cultures are fragmented and there is no central control. This organization will proceed along merrily towards chaos, until it eventually splinters apart.

It is clearly important to find a happy medium between the integrated and disintegrated states. Chaos is, however, very important since it can lead to complex new organizations which are much more effective than the ones they replace (Prigogine and Stengers, 1984; Peters, 1992). The trick is to catch the disintegrating organization before it splits apart, and this is usually best done by those who are actually taking part in the organizational change. Once a new state has been reached it is also important to service it continually to make sure that it survives. Eventually further chaos will be needed to reach an even higher and more complex state.

Many companies introduce chaos artificially into their organizations. This can be done in several ways, for example by ensuring that the company culture is continually being challenged. Canon and Honda did this by recruiting a significant number of new managers at the same time from external organizations, so as to create a culture pocket different from the current company culture. Another technique is to seek out challenging and risky situations for the organization to face and to grow out of. One such example is taking on a more powerful competitor head-on (Porter, 1991).

Self-organizing groups are a very effective mechanism for organization development and learning. These usually consist of groups which form spontaneously with their own objectives and goals. If a manager sets out to form such a group then it is important that the group is left to set its own objectives, or else it becomes nothing more than a task force. The group must not be subjected to tight controls. Normal hierarchy will no longer be operative within such a group, although the group may grow its own control structure. If senior members of staff are part of such a group then they must be prepared to act as any of the other members.

There have been many attempts to drive towards a 'boundaryless organization' (Hirschhorn and Gilmore, 1992) which has no strict functional demarcations. However, it is important to appreciate that removing boundaries on organization charts will still not remove them from the organization – the boundaries will exist in people's minds.

In traditional organizations the organizational boundaries were clearly defined: reporting structures were indicated and boundaries drawn around functions and people. Individual behaviour was conditioned by these boundaries. Removal of these boundaries, caused by the need for greater flexibility in present-day organizations, has replaced them by behavioural barriers. There are still differences in authority, knowledge, skill and so on, which cause individuals to exploit and set up boundaries, often without realizing it. (See the case study 'I know my place'.)

Two of the major causes of boundary creation are authority, formal and informal, and knowledge. A manager must always be prepared to use the formal authority given to him; he cannot abdicate it. However the authority should be used to create an open organization, with staff participation, empowerment and so on, and not for tight control or giving orders and direction.

Case study

I know my place

Sam File, Maintenance Engineer in the Dinsher Shop, had heard that the company was encouraging employees at all levels to take more interest in the products being developed and to make suggestions on how they could be improved. When he spotted John Eldridge, from the Engineering Department, on the shop floor, taking measurements for a new design relating to the floroll lift mechanism, he approached him. John was busy adjusting some fixings under the flap, and as his drawings were spread out on the table next to him Sam studied these while he waited for John to finish.

'Hello, Mr Eldridge', said Sam cheerily when John came out. 'This drawing looks a bit funny. Can you tell me how you came to determine the exact strength of the flap?'

Sam's question was quite innocent; he genuinely wanted to learn. However, John misunderstood his intentions. What he saw was a young man in a boiler suit from the shop floor questioning his design skills.

'I don't think I have the time', replied John, irritated. 'Anyway, I don't think you will understand if I did.'

Sam was taken aback. Instead of John Eldridge he saw an older man in a sports jacket, who had clearly been to university and considered himself to be too senior to talk to the workers on the shop floor. Sam turned and walked back to his machine. He resolved never again to believe the company's propaganda about breaking down barriers between different groups and encouraging everyone to participate in improving the quality of products.

3.2 Organization structures

3.2.1 Defining a structure

An organization structure is the way the organization allocates its resources towards meeting its strategic aims. It is commonly defined by organization charts.

There are many structures that are in use, and these cover a variety of functions, as illustrated in Table 3.1. This table also gives the groupings that are used in the sections which follow to describe the organizations. It is important to realize that there is no single structure that is best for every company and for

Table 3.1 Groupings for common organizational structures

Organization structures	Grouping
Line	Hierarchical
Staff	
Project-based	
Functional	
Flexible	
Matrix	Matrix
Informal	Informal

every period in its operation. Structures will change and develop to meet the aims of the organization at any given time.

Structures can also be pluralist, where overlaps occur between groups, each representing different activities or spheres of interest. A person can belong to several such groups.

All the organizations given in Table 3.1 may be defined as either operational or representative. An operational organization is set up by a company to produce a unit of output, while a representative organization generally represents a separate sphere of interest, although this interest does not necessarily have to clash with that of the company. The best known representative organization is a trade union, which may have a variety of structures of its own, such as line, formal or informal. It is important that the representative organization is treated as separate from the operational organization if is to succeed. (See the case study 'The un-representative representatives'.)

Case study

The un-representative representatives

When the FloRoll Manufacturing Company was still a relatively small organization it determined that it would try to keep its factories free from trade union influence. The Union of Trade Workers (UTW) was actively campaigning for members, and was obtaining agreement from many organizations in the neighbourhood for the exclusive right to represent their shop floor workers.

John Strongbow, the Manufacturing Director at the time, sought the assistance of Dave Morgan, the then HR Director.

'The only way to keep the unions out, if that's what you want', said Dave, 'is to convince your workforce that they are better off without union representation.'

'That may be difficult', replied John. 'In the past several of my supervisors have asked that they have some say on working conditions on the shop floor, and I have refused.'

'Why?' asked Dave, surprised.

'I don't trust them,' said John. 'Give them an inch and before you know it they will be asking for all sorts of privileges. How am I expected to run an efficient manufacturing plant if I let the workers, who know nothing about management, poke their oar in?'

Dave shrugged his shoulders. 'Well, it's up to you, John. I suggest you set up a joint management and worker representative committee, or you will find that they turn to the unions for help.'

John pondered over this suggestion for a few days. He then called his supervisors together for a briefing.

'In the past you have all asked that you have more say in the running of the shop floor', he said. 'I agree. I think that it is important that this is done and you will see that this gives you much more say than if you were represented by some screwball union officials who are based in London and know nothing about your local conditions.'

The supervisors were suspicious and asked for clarification.

'I propose to set up a management–worker consultative committee, which will determine the working conditions on the shop floor. The committee will consist of six members, three from management and three from the shop floor. I propose that the committee has its first meeting in two weeks time.'

The supervisors were pleased and went away to inform their staff. They set about discussing how to select the three shop floor representatives to the committee.

However, John Strongbow had other ideas. He called three other supervisors from the shop floor, who were know to him personally as being sympathetic to management views, and informed them that they were to be the three shop floor representatives to the consultative committee. The other supervisors were furious when they learnt of this selection, but John persuaded them to give the committee a chance.

They reluctantly agreed. After three months it was clear to all that the consultative committee was nothing more than a rubber stamping operation for management decisions. The workers turned to the unions and soon all the members of the shop floor had joined the UTW. A lengthy battle followed, with several crippling strikes, before the UTW was finally recognized by the company. John Strongbow's failure to recognize the importance of a strong representative organization, alongside the operational one, cost the company dearly.

3.2.2 Hierarchical structures

Hierarchical organizational structures are probably the most common and are often considered to be essential for control of large organizations (Jaques, 1990). Hierarchical organizations, also know as bureaucratic structures, are characterized by the following:

- A hierarchy of authority. The amount of authority, and related rewards, is clearly defined for each level within the hierarchy. Authority is related to the position within the organization and not to individuals.
- Specialization of tasks. Once again this specialization applies to the position within the structure and not to individuals. Therefore, when selecting people for tasks it is important to select the person to fit the job and not tailor the organization to fit the person.
- A system of rules which is tightly enforced throughout the organization, to ensure conformity. Allocation of privileges, authority and so on is defined by these rules.
- Impersonality in the application of rules to ensure that the organization functions as an efficient, impersonal operation. The system must be fair and must be seen to be fair.

Figure 3.2 shows a hierarchical organization which operates on four levels. As one moves up the hierarchical tree the tasks increase in complexity, become more varied, and have longer completion times. It is important that the layers are separated by wide differences to avoid the feeling of bureaucracy, where each layer adds very little value to the layer below it. Too many levels within an organization usually result in the manager and subordinate being too close in experience and skills.

Long time-scales for tasks sometimes have disadvantages. For example, the bottom rung of the management ladder is usually occupied by the supervisor (Wentworth, 1993). The position has direct access to productive resources and a very short time frame for actions. This means that the supervisor can spot immediately when things start to go wrong and can take corrective action.

Although the upper layer usually manages the layer below it, it is important that the managers in lower layers recognize the importance of managing their bosses (Gabarro and Kotter, 1993). Subordinates must work with their manager to ensure that the objectives of the company are met. (See the case study

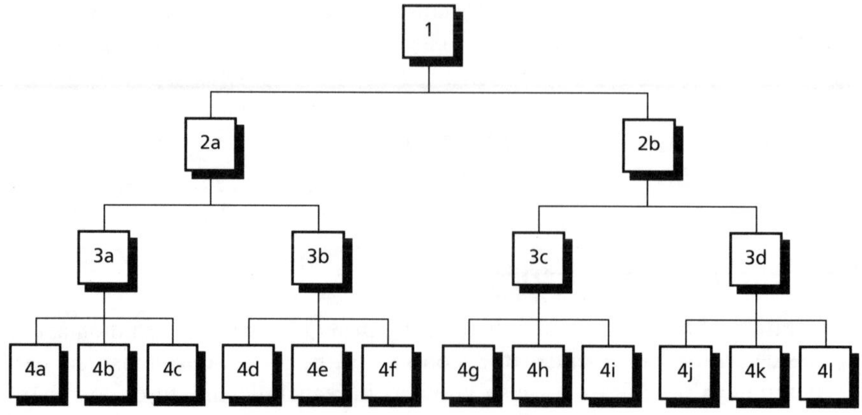

Figure 3.2 A hierarchical organization.

'Managing the boss'.) Managers are also instrumental in: helping subordinates obtain the resources they need; providing a link to upper layers; and checking that priorities match those of the organization. Subordinates must recognize that their managers are busy doing their own jobs and so cannot be expected to know all the problems and achievements at lower levels unless they are told.

Higher levels within the organization have greater authority than those below them. Authority can be of three types:

- Legal or organizational authority, which is the formal authority granted to a position within the organization. It is the authority granted to enable it to achieve the goals set for that position. It is related to the position and not to the person involved. It covers the area of responsibility and those within it. People obey the position and not the person, and follow the rules defined by the position. Therefore in Figure 3.2 level 2a has authority over levels 3a, 3b and those below them, and not over 3c or 3d.
- Traditional authority, which is again related to the position rather than the individual. This is authority that is not specified by the organization but has grown up out of custom or tradition. For example, the manager's secretary may be responsible for determining who uses the copier machine located on the secretary's floor.
- Personal authority, which is related to the individual rather than the position held within an organization. This may occur at any level within the organization and it cuts across official organization charts. The authority and span of control are not defined, being determined by the charisma of the individual.

In a hierarchical organization the span of control at each level, that is, the number reporting to the position, is usually limited to six or seven. This is on the basis that it is difficult for any one person to supervise a larger number of subordinates. However, this number should be considered to be a very rough guide. Much depends on the capabilities of the managers concerned and of their subordinates, and on the type of work being done. In some instances spans of fifteen to twenty are quite workable.

Closely related to span of control is the size of any department. Status is often linked to size, and empire building is common in large organizations. Often the number of levels of hierarchy under a manager is taken as denoting seniority. However, larger departments are often counterproductive. People who are underemployed can become disruptive and waste others' time. Specialists who manage other people have less time to practise their specialities. Larger departments also have bigger communication problems and are often involved in rivalry between departments.

Advocates of a hierarchical structure claim that it is an efficient communications medium. It was certainly used very effectively in the Roman legions, which were based on the principle that each level had ten reports. Therefore a commander giving an order could be certain that it would very

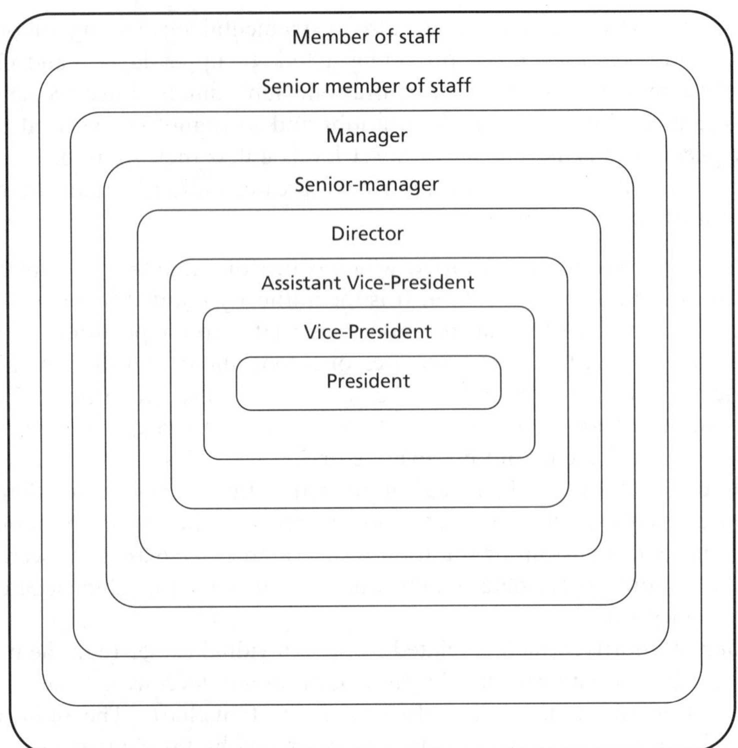

Figure 3.3 A concentric or 'hive' hierarchy.

quickly reach 1000 men by only having to go down three levels. However, in the modern technological age, a command can be sent electronically, almost instantaneously, to many thousands of people. Hierarchical structures are therefore not necessarily good for effective communications, especially when information needs to flow up as well as down an organization.

Figure 3.3 shows an alternative representative of the hierarchical organization structure, where power flows from the centre out. This figure shows the levels that actually exists in several large multinational organizations. It illustrates the problems of effective communications and the 'cocoon' effect on the chief decision maker, the President.

One of the aims of an organization is to break down the total task and responsibility into smaller units, which can then be allocated to groups and individuals to allow specialization. The hierarchical organizational structure is claimed to be the only one that allows large numbers of people to be managed and still preserves accountability (Jaques, 1990). It ensures that accountability exists at each level and that it is allocated to individuals and not groups.

However many hierarchies have overlaps, as in Figure 3.4, with people belonging to several groups, superior, subordinate and peer. Allocation of strict

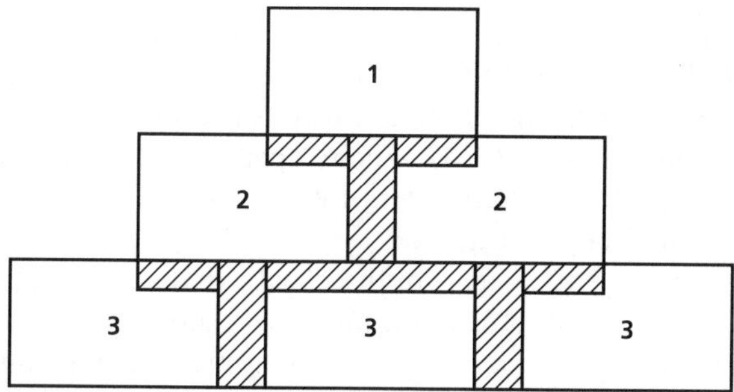

Figure 3.4 An overlapping hierarchical organization.

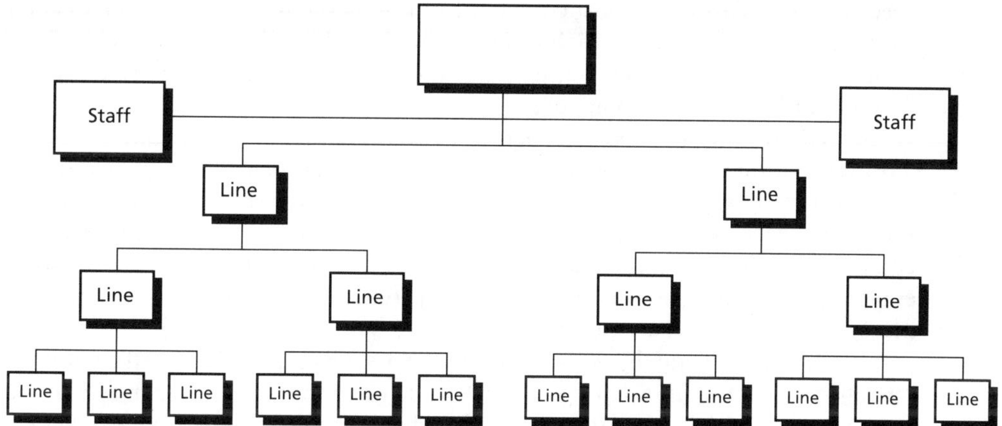

Figure 3.5 A line and staff organization.

personal responsibility and accountability is now difficult. It is, however, important that everyone sees their jobs as challenging, important and meaningful. Responsibilities must be seen to contribute to the organization's goals, and the jobholders must feel personal importance and worth.

Figure 3.2 represents a line organization, with each line manager being responsible for producing a unit of output. In most organizations staff functions also exist, as in Figure 3.5, who have an advisory role, supporting the line organization and ensuring that uniform policies are applied across an organization. These staff functions are usually small, for example having responsibility for strategy formulation. They may consist of a department, organized on a hierarchical basis, or an individual. Often there is conflict between the line and staff organizations, since the line managers are responsible for achieving goals and delivering products and they may resent interference from staff functions. This is especially the case if the staff organization is given a controlling function, such as approving capital spending.

Organizations sometimes have a large headquarters staff function and then smaller staff functions within each of their divisions. The trend, however, is to dismantle the headquarters staff organization, and to push responsibility down to many more smaller divisions (Foster, 1994). These operate autonomously at a local level and are controlled from headquarters by a set of objectives, such as meeting a level of profit or market penetration.

Sometimes the rôle of service groups gets confused with that of staff functions. Service groups are part of the line organization and are responsible for carrying out some supplementary activity, such as preparing customer documentation on looking after facilities.

Hierarchies can be formed in several ways. For example Figure 3.6 illustrates a simplified functional hierarchy, where each group is responsible for a specialization. This is the most common form of hierarchy, since it enables people to specialize and learn from each other within their discipline. It also builds closer bonds between professionals, although the danger is that loyalty is given to the function rather than the company.

Hierarchies can also be product- or project-based, as in Figure 3.7. The advantage of this is that it provides a very clear project focus and is probably best applied to very large projects. Its disadvantage is that, for smaller projects, it does not allow each specialization to have the critical mass needed to invest in equipment and skills, and creates inefficiencies by having to duplicate expensive equipment. Usually projects may have some functions under direct control of a project manager but share others.

A geographical hierarchy is shown in Figure 3.8. Once again this is efficient if the operation in each region is large enough to allow duplication of engineering and manufacturing. Global organizations often have several sites carrying out these functions, but ensure close coordination to obtain leverage from R&D and processes.

Hierarchies can also be developed on other bases, such as markets and customers; for example, telecommunications, information technology and consumer.

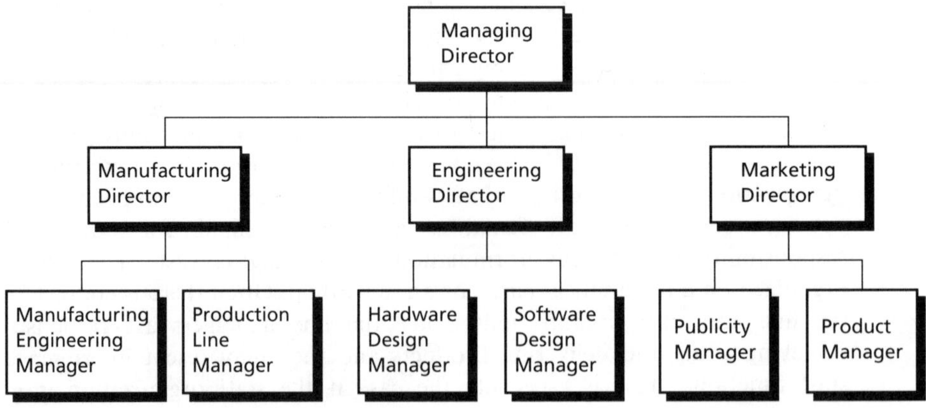

Figure 3.6 A functional hierarchy.

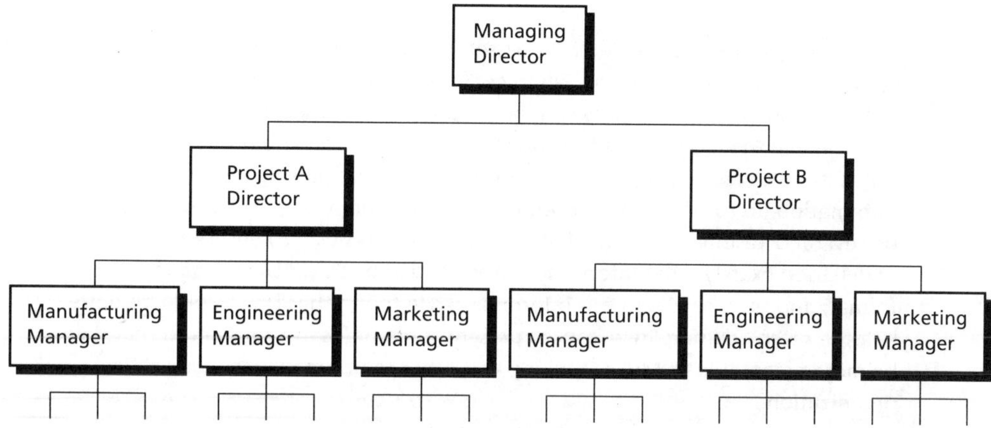

Figure 3.7 A project hierarchy.

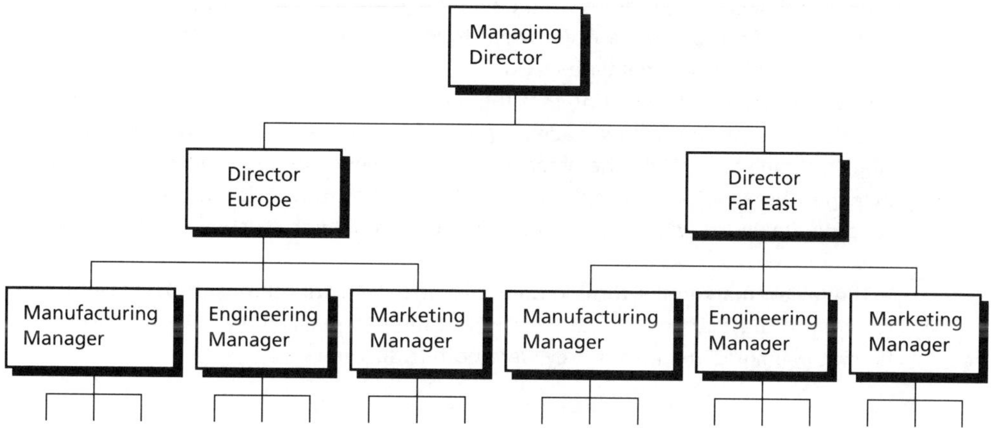

Figure 3.8 A geographical hierarchy.

Case study

Managing the boss

When David Laith was appointed Manager of Development Tools, his subordinates found a significant change from their previous manager, Peter Carp. Peter had an easy-going, casual style of management, asking to be briefed in exceptional circumstances only and leaving his reports to get on with their defined tasks.

David Laith needed much more detail and insisted on having a written report every week, which clearly indicated the progress made against all tasks with problem areas highlighted.

'He doesn't trust us, like Peter Carp used to', grumbled Angela Code to her peer, Stew Long. 'What does he want all this information for, anyway?'

'He obviously wants to make sure he is fully briefed', soothed Stew. 'After all, he is fairly new in the job and must be wanting to learn what is going on.'

As time progressed Angela Code found that she was passing less and less information up to David Laith, even though David pointedly asked for further data on several occasions. By contrast Stew went out of his way to ensure that he understood exactly what information was needed by David Laith, and in what format, and then gave him this information. He found that David could be quite helpful, often advising Stew how to present some of his information so that it would be more acceptable to Micro Dowling and other senior managers within the organization.

The crunch came at an important programme review where both Angela Code and Stew Long had to present to the board. David Laith asked for a pre-review with both his staff. Stew Long went through all his material with David and made several changes following suggestions from David. Angela Code, on the other hand, made excuses about not being ready and went to the presentation without having first gone through the material with David.

It was obvious at the meeting that there was a vast difference between Stew's and Angela's presentations. Stew made his points forcefully and was supported by David. Angela completely missed the objectives of the review and upset several of the senior managers by some of her proposals. David did his best to provide support to his subordinate, but it was obvious that he too disagreed with much of what Angela said.

The review marked the turning point in Angela's relationship with David Laith. She recognized that her boss was autocratic, but she also realized that she had to work hard at managing her boss if they were both to succeed.

3.2.3 The matrix organization

The matrix organization aims to capture the best of both the functional and project organizations. It is hierarchical in structure, except that two parallel hierarchies exist.

Figure 3.9 shows a simple matrix management structure consisting of functional and project lines. Either the functional or project line could have been drawn in the vertical; they both have equal strength in a full matrix structure. As seen from this, individuals at lower levels tend to have two reporting lines, one to their functional manager and the other to the project manager.

Figure 3.10 shows the responsibilities of these two areas. The project manager is responsible for directing day-to-day work, setting priorities, meeting tight programme time-scales and delivering products to the customer.

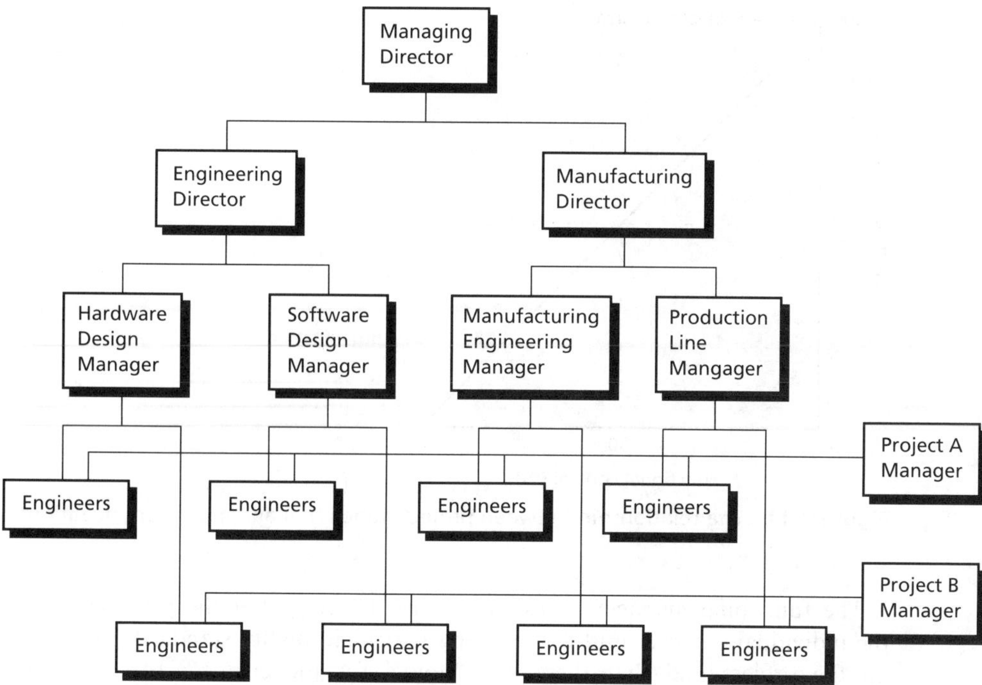

Figure 3.9 A simple matrix management scheme.

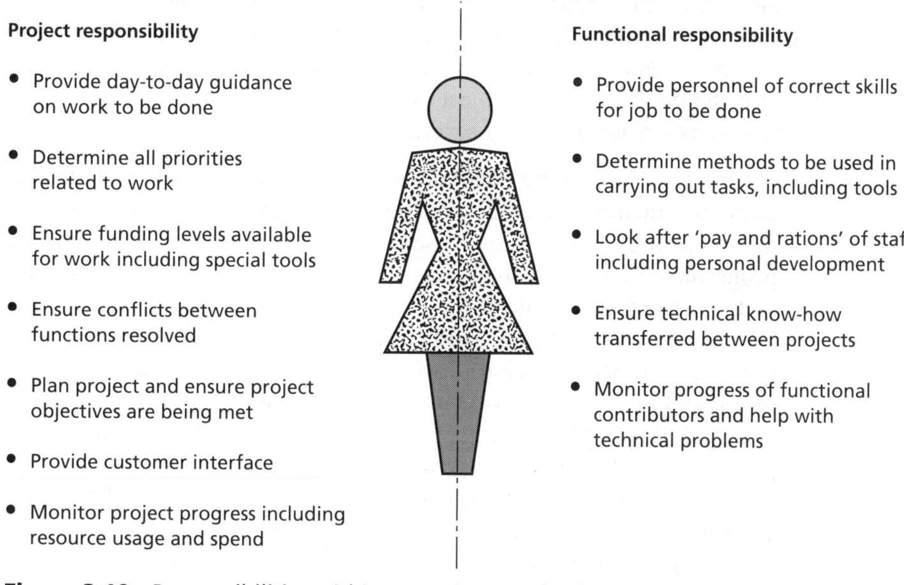

Project responsibility
• Provide day-to-day guidance on work to be done
• Determine all priorities related to work
• Ensure funding levels available for work including special tools
• Ensure conflicts between functions resolved
• Plan project and ensure project objectives are being met
• Provide customer interface
• Monitor project progress including resource usage and spend
Functional responsibility
• Provide personnel of correct skills for job to be done
• Determine methods to be used in carrying out tasks, including tools
• Look after 'pay and rations' of staff, including personal development
• Ensure technical know-how transferred between projects
• Monitor progress of functional contributors and help with technical problems

Figure 3.10 Responsibilities within a matrix organization.

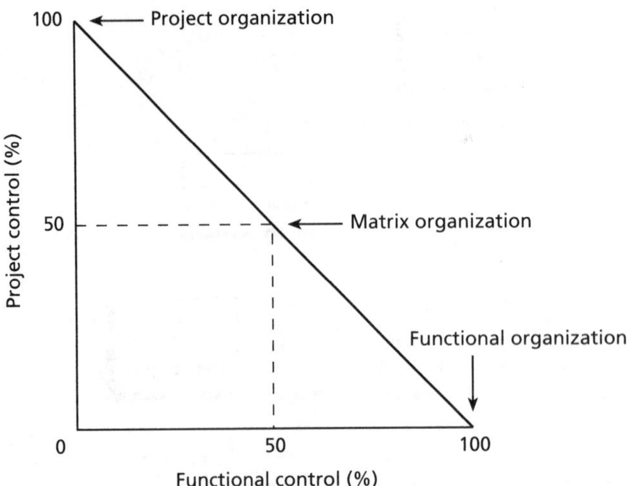

Figure 3.11 The relationship between project, functional and matrix organizations.

The functional manager is responsible for looking after the personal needs of the individual engineer and for ensuring that high quality standards are followed on the project and that there is a 'home' for the engineer to go to at the termination of the project. This may only be a temporary home from which the engineer goes on to another project. Functional organizations must look across project boundaries to unify staff of the same discipline, developing them in their technical skills and ensuring that they learn from each other.

Figure 3.11 shows the level of control that exists in functional, project and matrix management structures. The matrix organization provides a balance between the functional and project organizations.

The matrix organization has several aims:

- To allow projects to be formulated, grow and deliver products to the customer, including post-sales support, with minimum time and expense. This means having a pool of expertise that can be drawn on at the start of a programme and to which engineers can return at the end, so minimizing time and expense to the project.
- To provide staff levelling, as illustrated in Figure 3.12. Each project needs a different number of engineers with different skills, depending on the stage that they have reached. By drawing these from a pool, the overall total level of any skill needed can be kept relatively constant. This requires long-range planning to determine the overall staffing level. Problems can also arise when programmes start to run late and are therefore reluctant to free their resources for subsequent programmes. This results in conflict between functional managers and the various project managers, which needs resolving.

- To provide a project focus on all the work being done. Functional groups usually only focus on a portion of the total work, while the project organization has a view of the total job. However, the functional organization satisfies the need that professionals have for belonging and working with others of their own profession.
- To provide a focus on the customer. Large customers often ask for copies of the project organization when placing contracts, and the project manager becomes their prime interface with the work being undertaken.
- To provide a view across all the various functional areas in order to determine the impact of developments or changes in one area on another. To some extent a matrix organization also provides a mechanism for highlighting and resolving differences which occur between functions.
- To provide a platform for faster decision making, especially in relation to the impact on the customer.
- To allow the project to concentrate on delivering the product to the customer, by moving some of the day-to-day issues to the functional organization.

Matrix management is suitable for use in situations needing multiple simultaneous management capabilities (Bartlett and Ghoshal, 1990). Parallel reporting within this structure provides a method for resolving the conflicting needs of functions, projects, products and so on. Multiple communication channels ensure multiple capture of information. Overlapping responsibilities provide flexibility and faster response to change. It has been suggested that the actual structure used is not important; the attitudes and skills of staff are more critical. The matrix must exist in the minds of mangers: they must be receptive to the pull which exists in several directions in order to have flexibility.

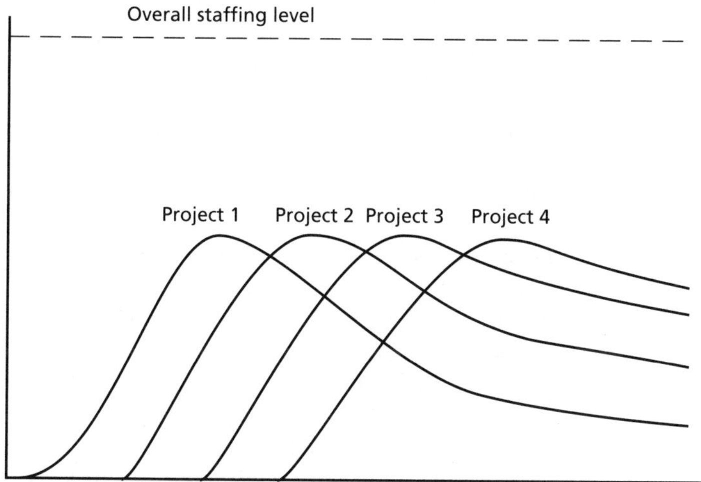

Figure 3.12 Staff levelling.

Matrix organizations have often been criticized since they require that staff have more than one manager. It is interesting to note that in these circumstances it is usual that conflict arises between managers rather than between the staff concerned. All managers wish to have total control over their staff.

Everyone also wants the best people for their project and will demand this from the functional managers. The functional managers also want to keep their best staff to work on some pet developments of their own. Conflict will arise and it will need resolution. It is important that the responsibilities of the various groups involved in the matrix are very clearly defined to minimize conflict.

People usually accept the concept of two managers quite easily. In a normal family children readily accept the equal authority of both parents. There are also several advantages to be gained for the engineer working in a martix organization:

- Experience is gained of working in a function and in a project.
- The engineer can learn from the various managers.
- Appraisals and promotions are based on the input from two managers, so they are more likely to be related to merit rather than the whim of individual managers. However, anomalies can arise when engineers are promoted within a project and then return to their old jobs within the functions, when the project ends.
- It is easier for the engineer to move between jobs.

It has been suggested that project managers should be allowed to go outside their organizations to get resources if they find that the internal functions are too expensive or are not ready to meet their requirements. For example, consultants can be used (Chaulkin, 1997; Wake, 1992) or telecommuters from different geographical areas (Skarratt, 1991). Similarly, the functional manager should be able to take on work from outside the organization, for example get funds to develop a new technology. Generally this is not favoured except for a small percentage of the overall task. However, it is usual for the functional organizations to be granted a level of discretionary funds in order to do work that may not be immediately needed by projects, but which might be required in future.

Matrix management structures are often confused with task teams, also known as task forces. (See the case study 'The reorganization'.) The aims of both are similar, although the task team lasts for a much shorter time. Task forces are set up for a specific task. Usually the problem to be solved is critical to the company. It needs to be tackled by a multidisciplinary team. It is a unique occurrence and not likely to occur again, so the task force is dissolved after its aims have been met. Matrix management structures, however, are designed to last for a long time and will tackle several projects in their lifetime.

Case study

The reorganization

When Jake Topper became Chief Executive of the FloRoll Manufacturing Company, the organization was in the middle of a crisis. Florolls were experiencing a very high level of failure at customer sites, primarily due to quality problems. Several major customers had threatened to cancel existing contracts and go to other suppliers.

'How can we build a quality product?' grumbled Adrian Screw, Manufacturing Director. 'The designs are so complex it would take us all week to build a single floroll if we followed all the processes.'

'It is a complex product', retorted Micro Dowling, Engineering Director, 'it has to be to meet the tight military specifications. If you employed proper manufacturing engineers, however, they would have no problems tailoring your processes to make the product.'

It was clear to Jake Topper that urgent action was needed.

'We are going to reorganize', he told his reports at a special meeting. 'It is a matrix structure and here is the organization chart.'

Nobody liked the new organization, least of all the functional managers who had to give up a large part of their authority to the new project organization, which was set up to resolve the quality problem. Mary Peters, the project manager, was given considerable authority, with a direct line to Jake Topper if she needed any help. Mary set up a strong team of specialists, drawing the best talents from the functional units. The group was highly motivated to work on the project, which had such high visibility within the company.

Within a few months the new organization started to show results. The design was modified slightly to help in manufacturing and quality levels rose steadily. New manufacturing engineering techniques and the introduction of a new tool also increased the output by a factor of 20 per cent. Jake Topper was very pleased and pointed to the success as a shining example of the effective use of matrix management.

Within six months all the major problems had been resolved. The members of the project found that they no longer had a real job to do. Morale fell and one by one they left the project team, either to rejoin their functions or to go to jobs outside the company. Other developments, which had been deprived of their best resources, started to run late. It became clear that the new matrix organization was not such a big success as had been supposed.

Jake realized that what he had needed to solve the quality problem was a task force and not a new matrix organization. A second reorganization followed, which restored the previous functional structure.

3.2.4 The informal organization

Organizations which exist within a company are essentially pluralist in structure; that is, they consist of an overlap of the structures defined earlier: functional, project, geographical, staff, matrix and so on. In addition, there is an overlap between formal organizations (those set up by the company and published as organization charts) and informal organizations.

Informal organizations can occur at all levels and usually grow up within a company for several reasons:

- The failure of the official organization to provide leadership. For example, the official organization in a company may be as in Figure 3.13(a), where all the managers report to a director. It is possible that the director concerned may not be very effective, whereas one of the managers reporting to the director (say Manager B) is a better leader. For example, this manager may be more knowledgeable and helpful, even advising the director on many occasions. Under these circumstances the unofficial organization functions as in Figure 3.13(b), where the other managers, with the agreement of the director, take their lead from Manager B.
- The official organization fails to provide the employees concerned with a feeling of self-respect in their work and a sense of achievement.
- The official organization does not make the employee feel 'accepted' and part of the wider group.
- The employees do not feel that they are receiving sufficient recognition and development in their work.

(a)

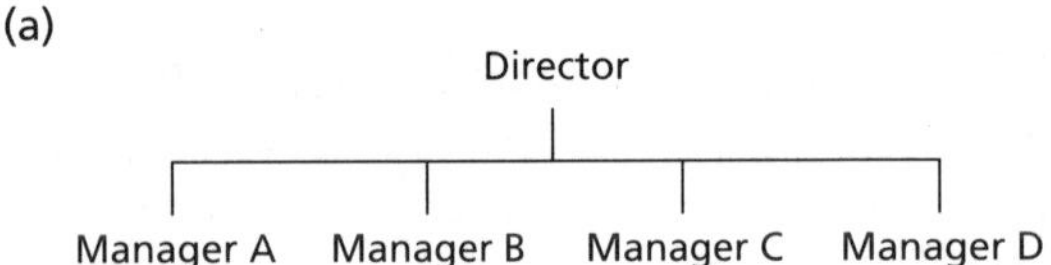

(b)

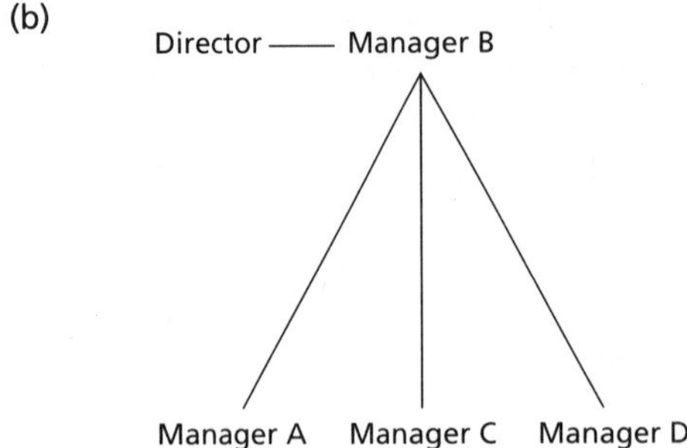

Figure 3.13 (a) Formal and (b) informal organizational structures.

- The official communication channels are ineffective and the employees gain more from the unofficial channels – the grapevine. Informal organizations can communicate suspicion and opinions, which cannot pass easily through official channels. The informal (often horizontal) links that exist between formal line structures are also important and beneficial in getting work done smoothly, for example by unofficial 'deals' between groups.

Under these circumstances employees may turn to unofficial organizations to gain the sense of satisfaction, recognition, and belonging which they do not get from the official organization. These are really organizations within organizations: gossip groups, alliance groups and so on. The informal organization maintains its own disciplines, sets its own standards and provides its own rewards, which is usually no more than the privilege of being a member of the group.

A large organization is usually composed of many smaller groups consisting of between eight and ten people. These are called primary groups, and the smaller size enables easy face-to-face communication between its members. Close interaction occurs, leading to interpersonal relationships, both favourable and unfavourable, between members of the primary group, each person having defined attitudes towards other members. The larger secondary group is more formal and impersonal. The feelings of the members is conditioned by how closely the goals of the secondary group coincide with those of the primary group.

The primary group is the main area where the individual acquires opinions, attitudes and goals and is subject to discipline and control. The primary group is self-sustaining; its main goal is its own existence. A secondary group, on the other hand, has more specific aims, such as the production of goods and services, and such a group can disband after its main aims have been achieved.

Informal organizations may modify the formal organization to fit a given circumstance. They can be very powerful and beneficial if they are made to work in the interests of the whole formal organization, rather than against its aims. Managers often look on informal organizations with suspicion, and try to break them up in order to reduce the influence of the unofficial leader. However, if possible, ways should be sought to make use of the informal organization.

3.2.5 Global organizations

There have been many changes in the global environment, which have forced multinational organizations to rethink the way in which they organize to tackle world-wide markets (Brandel, 1997; Lettice, 1996; Maruca, 1994). Traditional multinational organizations were structured on a strict pyramid, as in Figure 3.14, where a sense of order prevailed. World headquarters was usually based in the country of origin, and it also housed the largest manufacturing plant, R&D centre and all the senior officials. Commands were issued from this headquarters to all subsidiaries around the world, who were controlled closely by the senior, centrally based, executives.

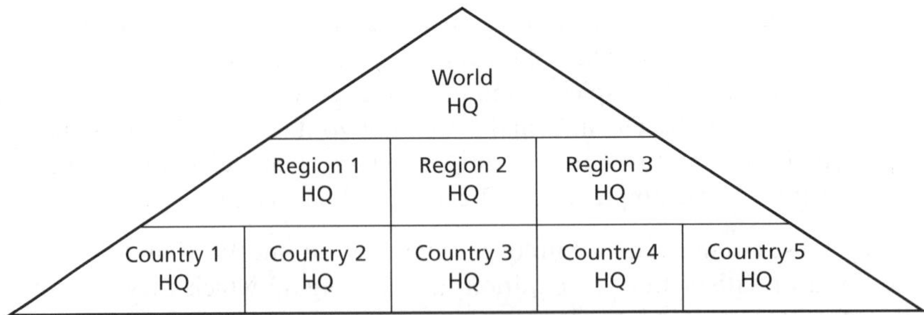

Figure 3.14 The pyramid global organization.

Rapid changes in market conditions soon made it difficult for companies to succeed with this form of organization. They found that, for success, they needed to pay attention to the local environment (Ohmae, 1989). This environment, such as employee culture, consumer behaviour, political considerations and competitive strategy, could not be exported from the home country.

It also became impossible to gauge local conditions by operating at long distance, and so make effective decisions. The temptation was to carry out activities for 'world markets'. This drive towards meeting the needs of an average of what was perceived to be a global market often resulted in the requirements of no specific market being met, with resulting failure.

New corporate headquarters are now much smaller and may be located anywhere in the world, not necessarily in the country of origin. Some companies have a World HQ which may consist of nothing better than a suite of rooms in a hotel, close to an international airport. These headquarters can also be mobile, moving between locations at frequent intervals. Some organizations prefer to have several headquarters, on a regional basis, rather than one central one.

Associated with this diffusion of headquarters is the spread in major activities, such as manufacturing and R&D. These will be done on several sites in many countries, with the largest or most important ones located in any country, not necessarily in the country of origin of the parent company (Blacklock, 1995).

Global managers have allegiance to their company, not to their country of origin or country in which they work. They are driven by economic considerations, looking for the greatest competitive advantage. Decisions such as where to locate the major plants, or where to carry out certain activities, are based on economic, business and political considerations; what is best for their organizations is of paramount importance (Reich, 1991). Customer needs have to be met world-wide, such as the requirement for local R&D and support, or local sourcing of components and materials. Skills are used wherever they are most available and cheapest. In their private lives, however, these same managers may be highly patriotic towards their own countries.

The original pyramid organization has been replaced by a 'treacle' organization, where a multitude of operations and headquarters are bound

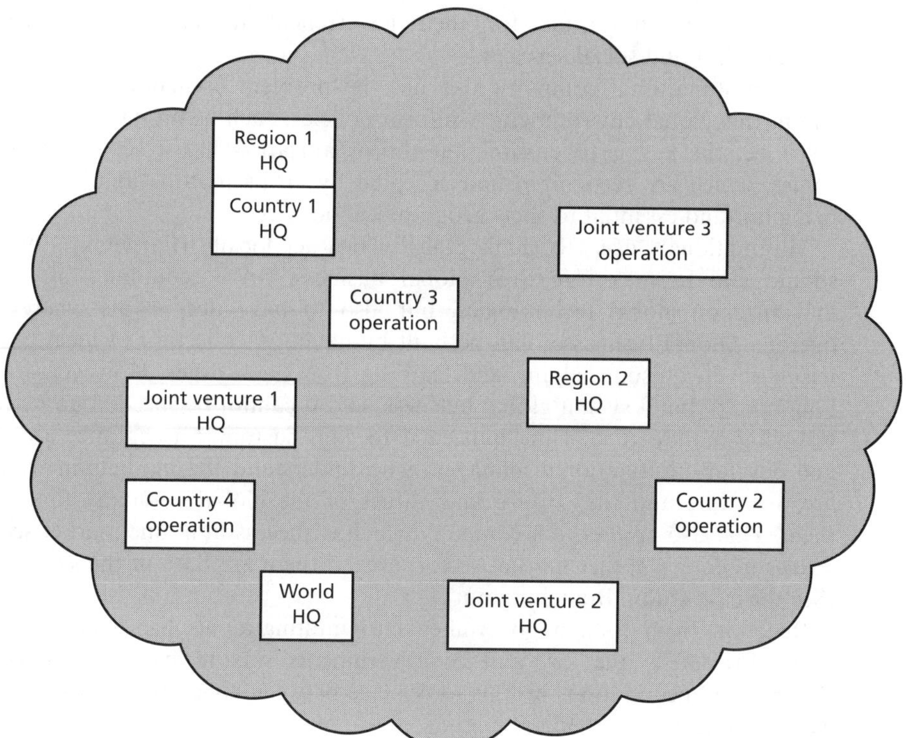

Figure 3.15 The 'treacle' global organization.

together into an amorphous mass, as in Figure 3.15. In the past, companies did everything by themselves, but the modern organization often relies on global strategic alliances and joint ventures, and these can be encompassed much more easily by the treacle organization than could be done by the strictly hierarchical global organization of the pyramid.

Previously it was common practice for the key managers in foreign operations of a company to be appointed from the home country. The trend now is to have a very cosmopolitan management structure, where, for example, a US company operating in Italy may have a French Chief Executive. Also, the company's headquarters staff may have a mixture of several nationalities in its top positions. It is essential that an organization operates this cosmopolitan management structure if it hopes to attract and keep the best of management talent from many countries.

Employee goals within a global organization do not always coincide with those of the multinational managers. Employees want to attract investment into their own countries, since they are aware that this will have other added value for themselves, such as increase in general pay and living standards in the country, and generating more tax for their government to spend on infrastructure improvement and social services.

The multinational company also has the problem of maintaining its identity within this global environment. Uniform policies, such as on pay and conditions, help, but the key is to ensure that the organization has a set of defined core values, which are recognized and accepted by all its multinational employees as being fair and essential to their geographical activity.

Multinationals need to think globally but act locally (Barnevik, 1992). They should aim to maximize their global business, drive economies of scale and maximize on global technologies, but also to have deep roots wherever they operate. Global businesses can be built by having a collection of local businesses which act in unison, helping and learning from each other. However, it takes a long time to build such a global business, and it cannot be speeded up by force or throwing money at it. Time is needed to gain customer acceptance and to find and develop multinational managers who understand the markets in which they are operating and the culture and values of the global company that employs them. This also applies to a company which is successful in one market sector but wishes to enter another market sector, even if these are both in the same country. (See the case study 'Branching out'.)

Reaction from foreign governments to multinationals has changed over the years. Previously the key aim of governments was to increase the country's exports, and negotiations on this took place between governments. Multinationals were considered as outsiders.

It has now been recognized that the extent of foreign investment within a country is as important as its exports, and that negotiations for these take place between governments and the global managers of the multinationals. Trade may now take the form of intangibles, such as services that include engineering design, finance and marketing. The multinationals may export ideas into their foreign branch for the manufacture of goods, which then generate revenue and may result in exports out of the host country.

Since multinationals recruit local managers and do essential value-added work around the globe, they receive equal treatment with local companies in most countries. Governments usually employ the carrot and stick policy in dealing with multinationals. The carrot takes the form of tax concessions and grants, and many countries have ended up bidding against each other to attract work into their countries in order to increase local employment. Managers have not been slow to recognize this and to play off one country or state against another, in order to obtain the best deals for their company. Since most companies have plants in several countries the threat to close a plant and move work to another area is a very real one. This same threat has also been used very effectively against the trade unions.

Multinational organizations also face threats from governments, to induce them to set up operations in their countries. For example, the government may exclude all imports of a certain type unless a certain proportion of the goods is manufactured locally. Generally this threat is very effective, especially in the larger developed markets of Europe, Japan and the USA. Companies are also being attracted by side benefits, such as access to a large market, a well-trained labour force and good infrastructure, rather than by direct incentives such as grants.

Case study

Branching out

The first application for florolls was very clearly in the military field. The original concept came from the UK Ministry of Defence and the early florolls were developed in conjunction with the Royal Air Force. The RAF knew precisely what it wanted and was willing to wait over five years while the product was developed and built to meet its specification.

It was this early success which helped the FloRoll Manufacturing Company to become established. Eventually the company began to do its own designs and, with the reduction of tensions between East and West, the floroll came off the restricted military list.

It was at this stage that the FloRoll Manufacturing Company decided to tackle overseas markets. Rather than try to sell the floroll to other governments the company decided to develop a cheaper version for commercial use and to sell this in Japan, which clearly had the largest market for this product.

The company did its homework well, to avoid the earlier problems they had when trying to enter the Chinese market. Its market research showed that considerable demand existed for the product and there was very little competition. It put in place a high-profile engineering and manufacturing programme to develop a product for this market. There were delays, of course, but after two years the product was ready. A suitable site in Japan was found for marketing, the plan being to expand to local assembly and support as demand grew. Japanese managers were also recruited locally and even the Chief Executive of this branch was Japanese.

Unfortunately, in spite of considerable investment, the venture was not a success. In the two years it had taken to adapt the product, consumer requirements had changed. The locally appointed Chief Executive made over fifty visits to the UK Headquarters during this period to report on progress, but his input could not be acted on by headquarters staff, who lacked first-hand knowledge of the market. Competitive pressure began to build up, and after a further year the Chief Executive resigned, since he found that he could no longer align himself with the aims and values of the parent company. The FloRoll Manufacturing Company had given itself three years to enter the Japanese market and since it had failed to do so it concluded that the market did not exist. The Japanese branch was closed and the company decided to try again in the future 'when conditions improved'.

3.3 The quality organization

Total Quality Management (TQM) is a key phrase within every organization (Burrows, 1991; Fenwick, 1991; Heller, 1993; Hiam, 1993; Jordan, 1993; Sanderson, 1992; Shores, 1992). The three words associated with this title are worthy of further consideration.

'Total' has several implications. 'Total' refers to the total organization, embracing everyone including suppliers and customers. 'Total' also covers all activities within the company, such as engineering, manufacturing, sales and customer support. This concept differs considerably from the old notion that quality is only associated with a manufactured product. 'Total' also refers to the total commitment from the organization towards quality, which is driven from the Chief Executive all the way down the organization.

The word 'quality' refers to delivering value. Quality is not a skill but a value (Cormack, 1992). Some basic principles can be taught, but to make it effective the employees within the organization need commitment, motivation, loyalty and other such important values.

Finally, the 'management' within TQM refers to the important fact that the key is management of the quality process, which will lead to actual quality improvement.

TQM requires continual change (Logothetis, 1992) in order to continually improve. The management of this change is covered in the next section. Senior management must recognize this and be committed to change, leading to improving quality. Everyone within the organization must be involved in this change process.

Quality can be measured in many ways, for example:

- The quality of produced goods, measured in terms of: long-term failure rate; or cost of the product; or whether it was produced on time; or its performance parameters.
- The innovations which are generated within the organization at all levels and in every activity.
- The profit or turnover of the company.
- Market penetration or new markets entered.

Measurement of TQM is carried out by many companies around the world, using a variety of schemes. These have the advantage of enabling comparisons to be made between companies, as well as providing an absolute indicator of the progress that a company is making towards continuous improvement (Vliet, 1996).

In 1983 President Reagan of the USA was keynote speaker at a White House Conference on Productivity, whose aim was to explore how the slowdown in productivity of the USA could be reversed. This conference was followed by a long and detailed debate, involving government and industry, which resulted in the recognition of the need for an award to encourage higher productivity within industry. This award, which was launched in 1988, became know as the Malcolm Baldrige National Quality Award, after the Commerce Secretary who took part in the original 1983 conference.

The Baldrige Award is now presented each year by the President of the USA to recognize those companies who operate on best business practice, as measured by a model shown in Figure 3.16.

The key goal is Customer Focus and Satisfaction, measured in several ways, such as gain in market share, the results of customer satisfaction surveys and customer retention. Measures of progress are applied, indicated by the Quality and Operational Results, which include items such as productivity improvements, waste reduction and supply quality improvement.

The system that makes these improvements possible consist of four factors: Management of Process Quality; Human Resource Development and Management; Strategic Quality Planning; and Information and Analysis. The driver for making all these happen is the Leadership provided by the Executives within the company, who create the climate for productivity improvement as well as providing the resources and focus to enable this to occur.

Several other systems have been established for quality improvements, which recognize the need for TQM within organizations. The European Foundation for Quality Management (EFQM) was established in 1988 with the mission of improving the quality of companies within Western Europe and so enhancing their competitiveness. They identified an achievement award, similar to the Deming Prize and the Malcolm Baldrige National Quality Award. This was launched as the European Quality Award in 1992. Figure 3.17 shows the EFQM model, which is the basis for determining the allocation of the awards (EFQM, 1996).

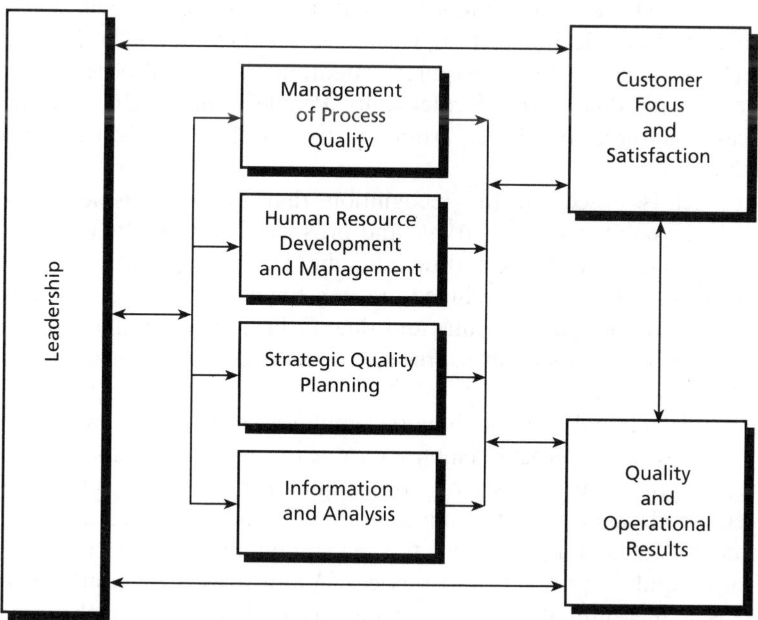

Figure 3.16 The model used for measurement in the Malcolm Baldrige Award scheme.

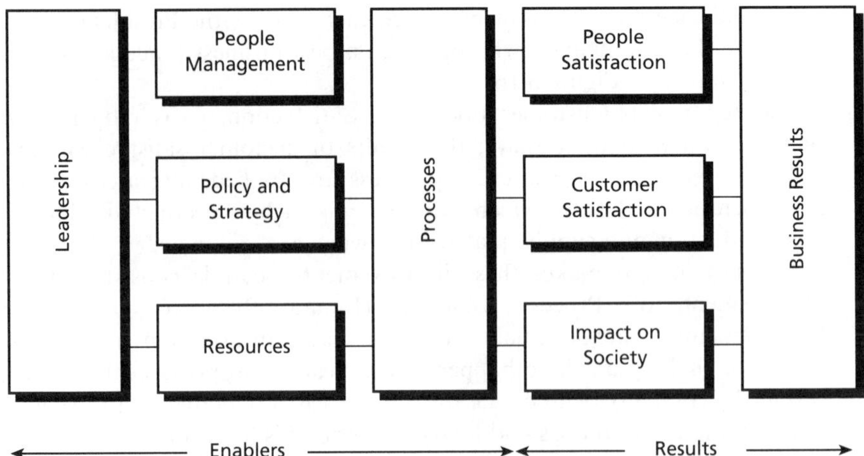

Figure 3.17 The model used for the European Foundation for Quality Management (EFQM) scheme.

The model consists of two parts, the Results section and the Enablers section, as indicated in Figure 3.17. The Results seek to measure what the organization has achieved, both in terms of its actual performance, compared with the industry, and in terms of performance against its own targets. The Enablers are, to a large extent, those that determine how the organization sets about achieving the Results.

The basic premise of the model is that Business Results will follow if the requirements of People Satisfaction, Customer Satisfaction and Impact on Society are met. To achieve these requires attention to the Processes within the organization, its Policy and Strategy, its People management and its use of Resources. All these are driven from the top of the organization by effective Leadership.

The model is based on the recognition that quality Business Results are essential if the company is to survive and prosper. These results include financial measures such as profit, cash flow, shareholder returns, working capital and return on investments. Also included are non-financial measures, such as market share, waste, product delivery time and time to bring new products or services to market. These non-financial measures are as important as financial measures in a competitive environment.

People Satisfaction determines how the employees view the organization in which they work. It seeks to measure employee satisfaction from items such as results of employee surveys, awareness of job requirements, employment conditions, absenteeism and sickness, staff turnover, grievances and ease of recruitment.

Customer Satisfaction considers the perception of external customers of the organization and its products and services. Again it uses the results of customer surveys as a measure of this satisfaction, as well as items such as complaint handling procedures, customer returns, warranty payments and repeat business.

Impact on Society considers the perception of the society at large to the company, especially of those within the community in which the organization is based. It looks at the organization's approach to environmental issues as well as to quality of life issues. Measures include how active the organization is within the community, such as in charity work, education and training, sports and leisure, and the steps the organization is taking to preserve global resources. The actions the organization takes to reduce and prevent nuisance and harm to the community are also important, such as reduction of noise and pollution.

The Processes criteria measure how successful the organization is in the management of its key processes. They seek to determine whether these are identified and reviewed, and then revised, if necessary, so as to result in continuous improvement of the organization's business. They look at several factors, such as:

- How the processes critical to the success of the organization are identified.
- How these processes are systematically managed.
- How the processes are measured to determine their effectiveness. Whether targets are set for improvement and whether feedback is used to review and improve the processes.
- How the organization encourages employees to apply innovation and creativity in process improvement.
- How process changes are implemented and the benefits of these changes are measured.

People Management measures how successful the organization is in managing its employees, and in releasing their full potential to continually improve the business in which they work. It considers several factors, such as:

- How the process of continuous improvement is achieved within the organization.
- How the capabilities and skills of employees are developed and preserved.
- How employees and teams within the organization agree improvement targets and how these are reviewed for success.
- How all staff are encouraged to take part in continuous improvement activities and how they are empowered to take action in support of these activities.
- How communication, both upwards and downwards, is achieved within the organization, in order to make continuous improvement effective.

The Policy and Strategy criteria look at the organization's strategic direction, its mission, and its values and vision. They seek to measure how these reflect the concept of TQM, and how TQM principles are used in their review, deployment and improvement. Several factors are considered, such as:

- How business improvement concepts are incorporated into the organization's policy and strategy.
- How policy and strategy are used as the basis for the organization's business plans.

- How the organization's policies and strategies are communicated to those within and outside the organization.
- How the organization regularly reviews and improves its policies and strategies.

The Resource criteria are used to measure how the organization manages, preserves and uses its resources, and in particular whether these resources are effectively deployed in support of its policies and strategies. Areas covered include how effectively the organization manages its financial resources, its information, its fixed assets and material resources, its technology base and its vital data.

The Leadership criteria aim to determine how the Executive team, and all the managers within the organization, inspire and drive TQM throughout the organization, treating it as the fundamental process for continuous improvement. Areas considered include:

- How the managers are visibly involved in leading business improvement within their organization.
- How managers ensure that a consistent business improvement culture exists within the organization.
- How managers provide resources and support to those taking part in business improvement activities.
- How managers recognize the efforts and successes of those involved in business improvement.
- How managers involve customers and suppliers in the business improvement.
- How managers promote their company, to outsiders, as being a quality organization.

The EFQM model is weighted, as shown in Table 3.2. Customer satisfaction carries a maximum of 200 points out of a total of 1000 points (20 per cent). Next comes Business results with 150 points or 15 per cent, and so on. Guidelines are laid down within the EFQM documentation on the weights to be applied to each sub-criterion in arriving at the final scores.

Table 3.2 Weighting used in the EFQM model

Criteria	Number of points	Percentage of total points
Leadership	100	10
People management	90	9
Policy and strategy	80	8
Resources	90	9
Processes	140	14
People satisfaction	90	9
Customer satisfaction	200	20
Impact on society	60	6
Business results	150	15

The process that a company would use in carrying out an evaluation against the EFQM model is illustrated in Figure 3.18. It is a self-assessment process in which the assessors are employees of the company. External assessors are involved only when an external award is being applied for.

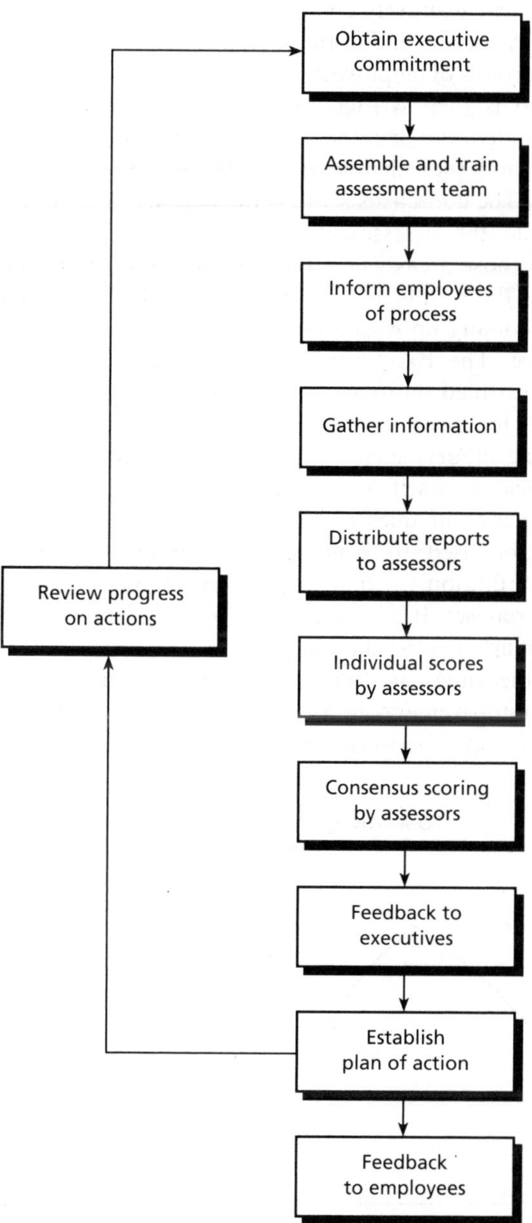

Figure 3.18 Process used for completing a self-assessment using the EFQM model.

The requirement for the assessment is driven by the Executive team, and they need to commit to it. Once this has been agreed the assessors are selected, usually from several different functions and levels within the organization, and they are trained in the use of the EFQM model. All employees are then informed that the assessment is taking place and the assessment team sets about gathering information, usually by analyzing data from company surveys and of the company's financial results, and by conducting interviews with groups of employees.

All the data so gathered is then written up by the assessment team and is shared within the team. Individual members of the team mark the company against the EFQM model and they then meet to compare their marks and to arrive at a consensus score. The consensus score is the one that is used in the final analysis. At the same time the assessment team looks for areas where the company has strengths and those areas where there is room for improvement.

Following the evaluation the results are presented to the Executive team, who have the responsibility for coming up with a plan of actions which should lead to better business performance. The Executive team owns these actions, although the EFQM assessors may be called on to assist and monitor progress. Finally, the results of the assessment are fed down to the employees.

As stated before, by using models such as those produced by the EFQM, not only does the company measure itself against a set of criteria, but it can also benchmark itself against other companies within the industry in which it operates. Different companies may be used to benchmark in different areas, such as manufacturing, sales or distribution, so that the best in any given area is always used in the benchmark exercise. It is important to look beyond one's own industry and also to benchmark against the competition.

Figure 3.19 illustrates the time lag that exists between quality or process management and actual improvement in quality (Tsuda and Tribus, 1991). Therefore some time t_1 will elapse between the start of quality management and

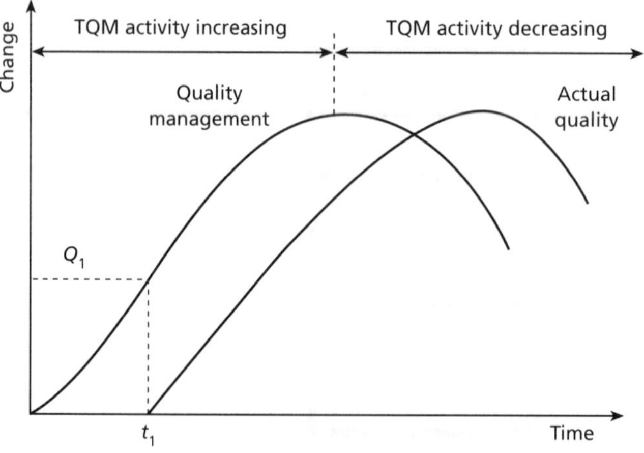

Figure 3.19 The quality time-lag.

its effects becoming evident. More important, if quality management activities are decreased at a later date there may at first be no outward sign, and quality may go on increasing, until later, when quality begins to deteriorate rapidly.

3.4 Organizational change

Engineers and technologists create change. Their profession requires them to explore new methods and techniques, driving the frontiers of technology and science. This results in change in every facet of our private and working lives. Very few engineers appreciate that, in the same way, they need to drive change when they become managers.

An organization must continually change to survive; one can move forwards or backwards, but no company can hope to mark time and remain where it is for very long (Chapman, 1993; Warren, 1992). Change should therefore be seen as an opportunity for growth and development, although it is often looked on as a threat both to the corporate organization and to individuals within it.

3.4.1 The need for organizational change

There are many pressures on organizations to change, as illustrated in Figure 3.20. Some of these are internal and some are generated by external factors. Internal pressures include:

- New investments in plant or products, requiring a new organization or different skills.
- Staff developing and being promoted, creating a need for an organization change to accommodate them or to fill their previous positions.
- Career moves within the organization, which create opportunities for reappraisal of job functions.
- Leavers and new joiners to the organization, which again require organization adjustments. Every leaver presents the organization with an opportunity to assess its needs and to see if it can become more efficient by reassigning tasks. Similarly, new recruits should not be chosen simply to fill a vacancy left by a leaver; they should be chosen for the value that they can add in a changed, more effective, organization.

External pressures on organization change include:

- Changing markets, such as new markets that the organization wishes to enter, or existing markets that it plans to withdraw from.
- Changing customer requirements, including changes in the customer base.
- Changing legislation and regulatory conditions in the countries in which the organization operates.
- Economic considerations, such as the availability of capital or the level of taxes, which are often driven by government activity.

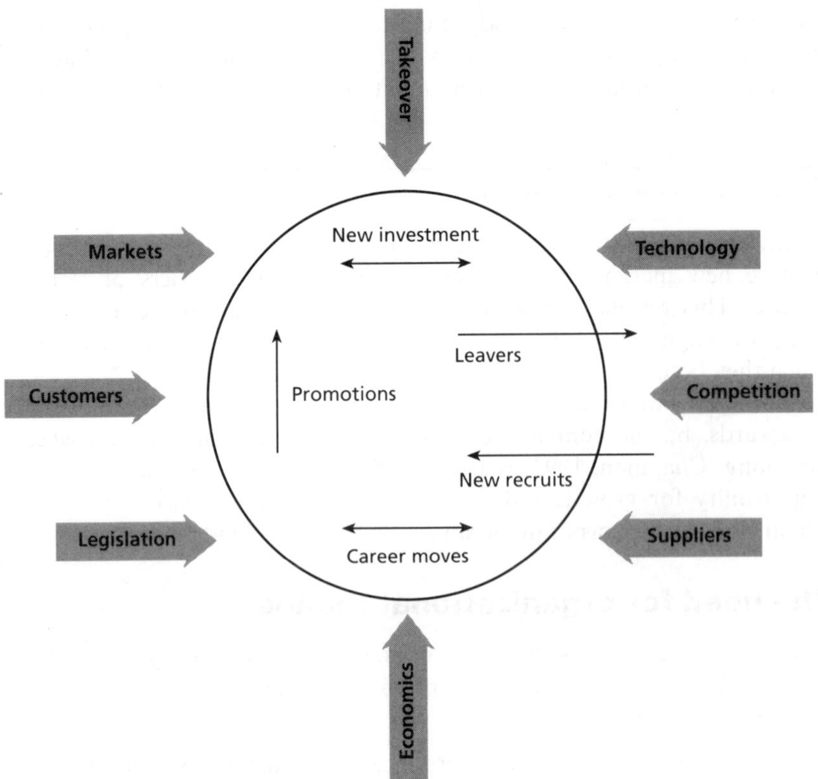

Figure 3.20 The pressures for organizational change.

- Supplier changes, such as price fluctuations, new suppliers being added or existing suppliers being removed.
- Competitive pressures, causing a shift in the organization to meet the threat. This could include cost cutting within the company, changing the product base or changing markets.
- Technology changes, which ripple through into many other areas, such as changes in products, competitors, customers and suppliers.
- Takeovers and mergers between organizations. This probably has the most direct impact on change within the organization, creating the most disruption and posing the greatest threat to employees. Usually rationalization takes place between the two organizations, resulting in widespread and rapid changes.

It is important that change is accepted as being a continuous process. This change should result in significant alterations, often referred to as re-engineering the organization. Incremental small changes can be implemented by a TQM programme, but re-engineering is needed to get big results quickly. It also requires a radical change in the way the organization thinks. (See the case study 'A case for re-engineering'.)

Case study

A case for re-engineering

Concern was being expressed within the FloRoll Manufacturing Company that their response to customer service requests was very slow.

'All our customers are complaining about the length of time they spend on the phone, chasing us up to come and repair one of their machines', moaned Peter Sell, Sales Director, at Jake Topper's Executive Staff Meeting. 'Unless we do something about cutting down our response time I am afraid we will start to lose customers.'

It was agreed that Heather Bates, the Customer Service Manager, would be tasked with benchmarking the company against a selected number of other suppliers.

Following the benchmarking exercise, the company was alarmed to find that it took them twice as long as the average company within their industry to respond to customer service request calls. An intensive TQM exercise was put in place and resulted in the service time being cut by 40 per cent.

'It was hard work,' said Heather Bates, as she presented the final results of the work to the Executive Board, but well worth it. Not only have we cut response time by 40 per cent but we hope to reduce it by another 40 per cent next year and all this is being achieved at no extra cost.'

Peter Sell was less impressed. 'That may be', he said, 'but feedback from my sales engineers is that the customers are still not satisfied and are threatening to take their business elsewhere.'

The company decided to set up a small team of executives to discover what the real problem was. They went and spoke to the customers in greater detail and found that customers did not want to have to make service calls in the first place; they wanted a reliable product that did not need servicing.

The organization of the company was changed (re-engineered) to strengthen the engineering function, and to ensure that field service had direct control over the specification of the product being developed, so that it was much more reliable and also allowed customers to carry out minor servicing themselves. The new product, when it was deployed, eliminated service calls by 50 per cent, resulting in greater customer satisfaction.

3.4.2 Barriers to change

Many of the barriers which a manager faces when change is being implemented within an organization are related to personal issues (Lockhart and Young, 1990). These are:

- The inertia within the organization, caused by the norms that have been operating over several years. Norms are shared values within the organization ('that's the way we always do things around here') and they prevent its employees from accepting that a different set of values, resulting from the change, may be better. It is important that all organizations have some norms, for example time-keeping or conduct at meetings. However, these need to be flexible, and employees must be aware that there may be better ways of behaving (Firth, 1994).

- Employees feel threatened by the change. Managers may also feel that the change will result in loss of control over some of their staff and a reduction in their status.
- Employees may feel that they are no longer in control of their own career paths. This will especially be the case if the change has been enforced without any prior consultation. All employees need to be prepared for a change and it should not come as a surprise.
- There could be uncertainty about future roles. Employees may feel that they would no longer have a meaningful job in the new organization.
- Employees who are put into a new role by the change may feel that they would not be able to cope. This may be true of those members of staff and management who have reached a plateau of growth and learning. Generally, however, this barrier can be overcome by suitable training and counselling.
- Past experience within the organization often determines future behaviour. The change would therefore be judged against this experience and may not be viewed in a favourable light. (See the case study 'It did not work before'.)

Case study

It did not work before

Jack Fox, Marketing Director, thought the time was right for another attempt at entering the Asian market.

'The market for florolls, especially for our new low-cost design, is very big in several of these countries, such as India and Pakistan', he said. 'I know that it will be difficult to service them at that distance, but the new design allows parts to be replaced very easily by the customers themselves, so the amount of support we need to provide is much reduced.'

'Not India again!', moaned Peter Sell, Sales Director. 'You must have relatives there, Jack! We have already made three attempts at entering India and failed. Sure, there may be a big market out there, but it is not for us.'

'I think I agree with Peter', said Jake Topper, the Chief Executive. 'We lost quite a lot of money in our previous three attempts and it has probably also affected our reputation.'

'But things have changed since then,' said Jack. 'The last time we tried to sell there we had no experience of overseas operations, but we have gained quite a lot of experience now. We also had an inferior product. It cost too much and every time some little thing went wrong we had to fly out an engineer to fix it. Now that the product cost has been significantly reduced, and the customers can carry out minor repairs themselves, we only need to tackle the major faults, which rarely occur.'

'It is too competitive a market to sell into',complained Peter Sell. 'We could never hope to gain anything but a very small market share, even if we were lower in price.'

'I disagree,' said Jack. 'When we last tried to enter the market there were three competitors who between them had 90 per cent of the market in the region. However, since then two of them have merged, so that customers are actively looking for a third source. In addition, one of the suppliers has moved its manufacturing plant from Japan to Taiwan and is having considerable problems getting it up and running. Not only is quality very poor, but they are also failing on their delivery schedules. I believe that, if we went in today, we could pick up several large orders quite easily.'

Unfortunately, Jack Fox failed to convince the Board that circumstances had changed and that they should re-enter the Asian market. Past experience of the three failed attempts was too great to be forgotten easily. They decided not to make a fourth attempt. A year later they learned that another competitor had entered the Asian market and had taken 25 per cent share from the two major players.

3.5 Managing change

Change cannot be implemented by the 'wild man' technique: swinging the club to make indiscriminate changes right across the board. For example, faced with the need to cut costs, many executives issue a directive for a proportionate cut across the organization. This results in ineffective operations, which should have been shut down, continuing but becoming even less effective, while effective operations suffer and lose some of their effectiveness.

There are several rules that should be followed in implementing change within an organization (Jeans and Bishop, 1993), as described in the following sections.

3.5.1 Setting the vision

The first requirement for change is to set the vision of what the organization is to change to (Kotter, 1995). This must be clearly defined and communicated to all levels within the organization. In setting the vision the Chief Executive must ensure that the team is looking forward, not backwards. What worked for the company in the past may not do so in the future.

Leadership must be provided from the top. The Chief Executive must be seen to be squarely behind the proposed change. A transitional management team may be set up, consisting of senior executives from across functions, to make sure that the change is a success and fits together across the organization, but the Chief

Executive eventually must accept full responsibility for it. When the change has stabilized the transitional management team should be disbanded. Generally employees accept that change will occur. They have been accustomed to it by activities such as downsizing and layoffs, and realize that there are no guarantees regarding jobs or careers. However, employees do expect consistency and predictability in such things as the process or rules applied within an organization, and the goals and vision of the company. If they can predict and influence these then they are more likely to accept the consequences of any change.

Smaller structures can usually implement change much faster than larger ones. Large organizations often have a considerable amount of momentum behind them, in the form of layers of hierarchy, numbers of processes, and longer lines for communications and decision making. If change is to be a regular process the organization needs to be smaller and swifter on its feet, a state which can be arrived at by the 'virtual organization' concept of small autonomous organizations within a larger one.

Technology has added a further dimension to the virtual organization, where an organization does not physically exist. The concept of virtual reality has gained popularity, particularly within games arcades, where the use of real-time computer simulation creates the illusion of moving around within a building, in space and so on.

Computers, with their considerably enhanced processing power, combined with the transmission capability of the information superhighway, have also made the concept of remote working a reality. Most companies now have a proportion of their staff who work from home: so-called telecommuters.

Technology has also added a new dimension to mobility, where people can be contacted anywhere at any time on mobile terminals and can access remote databases of information. It is now commonplace to see people working on trains, in hotels and in other remote or mobile areas, as effectively as if they were sitting at their own desks within the office.

The concept of the virtual organization is not new; salespeople have been working within it for many years. The sales force usually works from home, keeping in touch by mail, phone, computer terminals and videoconferencing, and coming into the office only occasionally. When in the office they do not possess their own desks but use one of several desks which are available for general use, a process known as hot-desking.

The advantages of the virtual organization are many, such as the saving on building and facility costs. However people within it miss the social environment created by work, and loyalty to the organization may be reduced. It also becomes difficult for managers to manage in an environment where they rarely see their staff and cannot control them closely. This calls for a high degree of trust (Handy, 1995) and it becomes even more important that the vision of the organization is widely communicated and accepted by all those within this virtual organization.

3.5.2 Communicating the change

The aims and reasons for the change must be clearly communicated throughout the organization (Martin, 1993). Communications must take the form of 'saturation cover'. It is important to explain the change and then to explain it again and again so that everyone understands it. People need to hear the message over and over again, so that they believe in it and begin to accept it.

Change can also only be implemented successfully if it changes each individual, usually one by one (Duck, 1993). This will take a long time and require considerable resources.

Change will start to occur from the minute that it is expected, even if it has not been formally announced. Those who expect to be affected by the change must be kept continually informed, even if full details are not available. It is worse when they have to guess the impact of the change, for example when a task force is set up to implement the change and works behind locked doors in closed sessions. People must buy into the change.

3.5.3 Project managing the change

Changes usually result in a hive of activity which needs to be coordinated to ensure that all energy is channelled into meeting common goals. An organization can be likened to an interconnected web, where change in one area affects another area, even if that area was not intended to be directly affected by the change. Project management is required to ensure that all these tasks fit together to make up the total picture.

It is also important that sufficient resources are allocated to make the change happen and that these are managed. Full time resources are often required to implement changes.

Resource savings should be sought, such as by closing down old activities which no longer have a place in the changed environment, and moving the resources so freed onto the new activities. Progress must be measured against the original goals and vision, and corrective action taken if required.

Change should be timed to coincide with some other discontinuity in the organization, when the people who need to change feel that they have a problem that needs resolving, even though this problem may be different from the one the change is addressing (Kotter, 1995).

Customers and suppliers must be managed carefully during the change so that they also see the benefit the change will bring to them. Managing change, however, needs a different model from managing other operational issues within a business (Duck, 1993).

All change must be results oriented, not activity oriented (Schaffer and Thomson, 1992). One can have a considerable amount of activity, such as TQM or continuous improvement, but these will not result in effective change unless the results that are expected from the change are clearly defined and aimed for.

Many organizations spend a considerable amount of time and money in training all their employees in quality procedures, but find that they receive very little benefit because they have omitted to define what the end result is to be.

3.5.4 Flexible policies

It is important to encourage feedback and comment from all employees, customers and suppliers on the change. It is equally important that the policies are flexible enough so that they can be modified or changed if good suggestions are received. Staff empowerment should be implemented and be seen to work. Employees should be given the information to enable them to make sensible decisions and recommendations. Good suggestions and effective participation in the change process should be rewarded.

For effective change to occur employees need to accept the company vision, so changing their attitude. Most people in organizations have been through many changes and can be cynical if a change is being 'sold' to them as being the best thing since sliced bread. It is wrong to try to get them excited about the change and then expect their attitudes to change.

It is better to start with the attitude change; excitement will then follow. Employees will believe in change when they see the benefits for themselves, not when they are told of the benefits.

3.5.5 Putting people first

In any change, as discussed in the last section, people problems represent the biggest obstacle. All employees must be treated as individuals and not as groups. It is important to be sensitive to the effect of change on an individual's security and status. Changes may be made in what people have been doing and the way in which they have being doing it.

Uncertainty and insecurity are the biggest barriers to change, especially where values need to be altered. Extensive counselling and training may be needed to help individuals over the change. The display of emotions, usually suppressed within the work environment, should be allowed for and encouraged, since these normally help the change process.

Summary

Organizations are changing in several ways. For example, the emphasis is shifting from a process-oriented approach to a results-oriented approach. In addition there is now a much greater need for interaction at horizontal levels within an organization, instead of the more traditional up and down interaction. Formal control mechanisms are disappearing between managers and subordinates, and the need for contacts outside the organization, at all levels, is growing. Employees are finding many more opportunities for career advancement, while at the same time not being able to identify any clearly defined career path up the organization ladder.

An organization is a complex web of interrelated structures. To survive the organization must continually change and renew itself, oscillating between a state of integration and one bordering on disintegration. Formal boundaries between hierarchies and functions are disappearing, leading to the concept of a 'boundaryless organization'.

Several organizational structures exist, depending on the type of business or project and also the stage which the activity has reached. These organizations can be operational or representative. Operational organizations are those set up formally by the company for a clearly defined aim, such as the production of a unit of output. Representative organizations are also formal structures, but they serve a different sphere of interest, such as a trade union representing its members. A manager must recognize the need for a strong representative organization alongside the operational one, and must ensure that the two work together for mutual benefit.

The most common formal organization is that based on a hierarchy structure. This can include line and staff positions and may be subdivided by the different projects within the organization, the different functions, different geographical locations and so on. The key, however, is that a clearly defined chain of command exists and information flows up and down the organization. Authority is related to the position occupied within the hierarchy.

The matrix organization usually consists of a combination of two or more hierarchy structures. The most common are functional and project-related. Although the matrix organization has been criticized on the basis that it is 'difficult to serve two masters', in reality most engineers find it easy to have dual reporting; usually it is the managers who find it difficult to manage staff unless they fully control their work. Matrix organizations have many advantages, such as their project focus and ensuring a closer fit to customer needs, while at the same time providing a 'home' for engineers within their disciplines or functions. The matrix organization also helps to bring teams together, since it draws on people from different, and often rival, functions.

Informal organizations supplement the formal organization. They are usually smaller in size than formal organization and exist within a formal structure. They exert stronger ties on members than the formal organization. Usually the informal leader has qualities which supplement those of the formal department manager, so it is important that the manager uses the informal organization to enhance the output from the group, rather than treating it as a threat.

Modern multinational companies are often comparable in size to some of the smaller countries in which they operate, so it is inevitable that they are drawn into economical and political affairs. Global organizations tended to be pyramidal in structure, as shown in Figure 3.14, with all the power residing in the home country. This has changed to a more fluid organization, as shown in Figure 3.15, where plants and offices are located where they can best serve the customers, and senior executives are recruited locally to run these concerns.

Quality considerations are vitally important at all levels and in all functions within an organization. A company must continuously improve and this is done by setting improvement targets, measuring its activities, benchmarking against these internal targets and against other companies, and then putting in plans for improvement. Several quality award schemes exist which include models for measurement and benchmarking. Two such systems are the Baldrige Quality Award (Figure 3.16) and the European Foundation for Quality Management (Figure 3.17).

Improvements require change and all companies must change to survive. Internal factors naturally result in change, such as new investments needing new skills or organizational structures, and movement of staff to different areas of the organization for career moves. In addition staff will leave the company altogether and new starters will join, which naturally result in an element of change. External factors also require change, such as competitive pressures, changes in market requirements or the customer base, legislation changes, and technology changes.

There are many barriers to change and these must be understood and overcome if the change is to be successful. Company inertia constitutes one of these barriers ('we have always done it that way'), as well as the norms and past experiences of people, and the processes that exist within most large organizations. People also feel threatened by change, which introduces uncertainty, and some may even feel unable to cope with the new changed situation.

All change must be carefully managed by ensuring that a clear vision of what is to be achieved is articulated by the executive team and then effectively communicated to all those effected by the change. The change process must be carefully managed, including the definition and measurement of success criteria. The process used must be flexible so that it can be rapidly modified, based on feedback received. Above all, it is important that people considerations are placed foremost, since their behaviour and attitudes will be the most difficult element to change, but at the same time it is crucial that this is done or the change will not be successful.

Case study exercises

3.1 The FloRoll Manufacturing Company is already fully stretched developing a new product for a major customer order. Peter Sell, Sales Director, has also secured a smaller, but highly profitable contract, which needs a low level of development and a slight change to the existing manufacturing process. Should the company put in place a different organization to deliver this contract?

3.2 It is important that close cooperation exists between Engineering, Marketing and Manufacturing if the FloRoll Manufacturing Company is to succeed in meeting the delivery time required on a new product. What organization should it put in place, remembering that the product calls for a high level of specialization within functions?

3.3 Pauline Moffatt, Manager of the Packing and Shipping department of the FloRoll Manufacturing Company, realized that the group under her were working very effectively, primarily due to the informal leadership provided by a charge hand. The problem was that she was being excluded from this informal organization. The work was getting done and all the targets within the group were being met. Should Pauline take any action to take control of the group?

3.4 James Silver, Materials Manager in the FloRoll Manufacturing Company, suspected that the quality of the work being done within his department was steadily deteriorating. He wanted to introduce a programme of change to improve quality. What steps should he take and should he consider reorganization to obtain results?

References

Barnevik, P. (1992) Think global act local, *Compass*, Autumn, pp. 6–9.

Bartlett, C.A. and Ghoshal, S. (1990) Matrix management: not a structure, a frame of mind, *Harvard Business Review*, July–August, pp. 138–45.

Blacklock, David (1995) What hope for the Euro company? *Professional Manager*, September, pp. 12–14.

Brandel, M. (1997) Think global, act local, *Computing Global Innovations*, 10 April, pp. 8–9.

Burrows, P. (1991) Corporate culture is the next quality frontier, *Electronic Business*, 7 October, pp. 64–6.

Caulkin, S. (1997) The great consultancy cop-out, *Management Today*, March, pp. 33–6.

Chapman, R. (1993) Change and challenge, *Professional Manager*, January, pp. 12–14.

Cormack, D. (1992) Vision to regain lost values, *Professional Manager*, November, pp. 16–18.

Duck, D.J. (1993) Managing change: the art of balancing, *Harvard Business Review*, November–December, pp. 109–18.

Dunderdale, P. (1994) Analysing effective organizations, *Professional Manager*, September, pp. 23–5.

EFQM (1996) *Self-Assessment Guidelines for Companies*, European Foundation for Quality Management, Brussels.

Fenwick, A.C. (1991) Five easy lessons: a primer for starting a total quality management programme, *Quality Progress*, December, pp. 63–6.

Firth, D. (1994) Leading people in a time of change, *Professional Manager*, July, pp. 15–16.

Foster, G. (1994) The central question, *Management Today*, April, pp. 56–61.

Gabarro, J.J. and Kotter, J.P. (1993) Managing your boss, *Harvard Business Review*, May–June, pp. 150–7.

Goss, T. *et al.* (1993) The reinvention roller coaster: risking the present for a powerful future, *Harvard Business Review*, November–December, pp. 97–108.

Gretton, I. (1993) Striving to succeed in a changing environment, *Professional Manager*, July, pp. 15–17.
Handy, C. (1995) Trust and the virtual organization, *Harvard Business Review*, May–June, pp. 40–50.
Heller, R. (1993) TQM: Not a panacea but a pilgrimage, *Management Today*, January, pp. 37–40.
Hiam, A. (1993) Visualizing quality, *IEEE Engineering Management Review*, Spring, pp. 51–9.
Hirschhorn, L. and Gilmore, T. (1992) The new boundaries of the 'boundaryless company', *Harvard Business Review*, May–June, pp. 104–15.
Jaques, E. (1990) In praise of hierarchy, *Harvard Business Review*, January–February, pp. 127–33.
Jeans, M. and Bishop, D. (1993) Wringing in the changes, *Professional Manager*, pp. 8–9.
Jordan, P. (1993) Lasting quality, *IEE Review*, November, pp. 269–71.
Kotter, J. (1995) Leading change: why transformation efforts fail, *Harvard Business Review*, March–April, pp. 59–67.
Lawler, E.E. (1994) From job-based to competency-based organizations, *IEEE Engineering Management Review*, Fall, pp. 83–90.
Lettice, J. (1996) The not-so-great dictators, *The VAR*, April, pp. 16–19.
Lockhart, H. and Young, J. (1990) Management of change – Theory and practice in the network administration implementation programme, *British Telecommunications Engineering*, October, pp. 165–71.
Logothetis, N. (1992) *Managing for Total Quality*, Prentice Hall International, Englewood Cliffs NJ.
Martin, R. (1993) Changing the mind of the corporation, *Harvard Business Review*, November–December, pp. 81–94.
Maruca, R.F. (1994) The right way to go global, *Harvard Business Review*, March–April, pp. 135–45.
McGill, M.E. and Slocum, J.W. (1994) Unlearning the organization, *IEEE Engineering Management Review*, Summer, pp. 36–43.
Ohmae, K. (1989) Planting a global harvest, *Harvard Business Review*, July–August, pp. 136–45.
Peters, T. (1992) *Liberation Management*, Macmillan, London.
Porter, M. (1991) *The Competitive Advantage of Nations*, The Free Press, New York.
Prigogine, I. and Stengers, I. (1984) *Order Out of Chaos; Man's Dialogue with Nature*, Bantam Books, New York.
Reich, R.B. (1991) Who is them? *Harvard Business Review*, March–April, pp. 77–88.
Sanderson, M. (1992) BS7850: 1992 – Where does it fit? *BSI News*, November, pp. 8–9.
Schaffer, R.H. and Thomson, H.A. (1992) Successful change programs begin with results, *Harvard Business Review*, January–February, pp. 80–9.
Shores, A.R. (1992) Improving the quality of management systems, *Quality Process*, June, pp. 53–7.
Simons, Robert (1995) Control in an age of empowerment, *Harvard Business Review*, March–April, pp. 80–8.
Skarratt, J. (1991) Telecommuting – reality or hype? *Lines of Communication*, May, pp. 16–17.
Stacey, R. (1993) Strategy as order emerging from chaos, *IEEE Engineering Management Review*, Fall, pp. 106–12.
Tsuda, Y. and Tribus, M. (1991) Planning the quality visit, *Quality Process*, April, pp. 30–4.
Vliet, A. van de (1996) To beat the best, *Management Today*, January, pp. 56–60.
Wake, A. (1992) Consultants: a user's guide, *Mobile and Cellular*, February, pp. 33–4.
Warren, L. (1992) Dynamics of change, *Integration*, December, pp. 13–16.
Wentworth, F. (1993) It's time we took supervisors seriously, *Professional Manager*, January, pp. 15–17.

Legal and ethical considerations

Introduction

This chapter looks at some of the responsibilities that managers voluntarily accept, due to professional, ethical or social considerations, and those that are enforced on managers, usually by legal statutes.

4.1 Management obligations

Government industrial legislation has increased steadily since the start of the industrial revolution. This has to some extent been influenced by trade union pressure, which has been cyclical in its strength. Management within an organization clearly needs to be aware of this legislation, but managers must also understand the aims and reasons behind Government regulations, and be able to work with government agencies to satisfy these requirements. In addition, modern management needs to maintain a greater awareness of the environmental and social obligations placed on organizations by the communities in which they

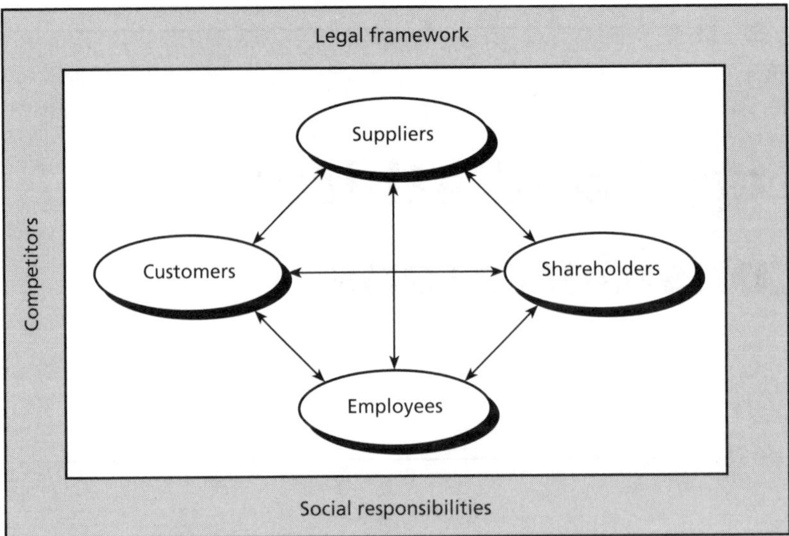

Figure 4.1 Management responsibilities and constraints.

operate (Smith, 1994). The requirements and norms of present and future society must be understood and the needs of present and future generations fulfilled.

This is not easy, especially since, as shown in Figure 4.1, the manager at the same time needs to resolve the conflict which can exist between the needs of society, employees, shareholders, customers and suppliers (Walley and Whitehead, 1994). All these are stakeholders in the organization and their interests must be kept in mind, while the manager works within the confines of the legal framework, social responsibilities and competitive pressures.

The obligation to employees of the company is often forgotten. It is important that employees are adequately trained and developed. The climate should be such that they enjoy their work and have the opportunity to express themselves, while contributing to the success of their organization.

Some of the obligations placed on management are easy to define and relatively easy to follow, such as the need to avoid prejudice on the grounds of race, religion or sex (Echiejile, 1995). Others are more difficult to quantify and legislate against, such as those relating to ethical matters. (See the case study 'Intelligence gathering'.)

Case study

Intelligence gathering

All new starters within the FloRoll Manufacturing Company go on a two-day quality course, during which the company's policies on ethical conduct are fully explained.

This relates to all aspects, such as the social obligation of employees and conduct towards competitors.

Unfortunately, no other reference is made to this ethical code, and there is no published documentation or guidelines. However, it is clearly understood that the company will not enter into any unethical practices in order to win business from competitors.

Irene Bamber, Sales Executive within the Northern Region, recently visited the company's headquarters and had a short meeting with Jill Search, Marketing Information Manager.

'We know that Winger Inc. are developing a new floroll especially for the European market', said Jill, 'and we even know that this is being done in the UK. We also know that it is code named 'Pingo'; why, God only knows! What we don't know is what its special characteristics are and the markets to which it is being targeted.'

'I realize that their research laboratories are situated very close to our Northern offices', said Irene Bamber, 'but we know very little about the work that is going on there.'

'Nothing in the local papers?' asked Jill Search. 'They must be recruiting new staff; very often quite a lot of information can be obtained from job advertisements.'

'Nothing that I have noticed', said Irene, 'but I will keep my eyes open.'

It was a few days after this conversation that Irene Bamber noticed a job advertisement in the local paper from Winger Inc., looking for sales staff to lead their drive into Europe on 'a revolutionary new product'. With her recent conversation with Jill Search still fresh in her mind, she decided to apply for the job.

A week later Irene received an invitation to attend the nearby Winger Inc. laboratories for an interview. It was obvious that the company had invited her primarily to get as much information as they could about the activities of their competitor, the FloRoll Manufacturing Company. Irene was taken around the offices and met several senior executives who questioned her about the work she was doing and her sales accounts.

Irene was careful only to give information which she was certain was already known to the competitors. While in one of the offices she also noticed a computer printout titled 'Pingo Sales Forecasts'. It was partly hidden under a pile of other papers, carelessly piled up on the desk. It was an easy matter for Irene to slip the paper into the portfolio which she had in her hands and to take it away with her when she left.

The next day Irene went to see Jill Search and proudly gave her the sales projections for Pingo. Jill was horrified when she learned how Irene had come by the information. She had no option but to report it to her immediate manager, Jack Fox, who passed in on to the Chief Executive, Jake Topper.

The executive debated the case of Irene Bamber. It was not an easy decision, but eventually they felt that Irene had breached the company's ethical code of practice. The paper containing the sales projections for Pingo was returned to Winger Inc. with an apology, and Irene was severely reprimanded.

4.2 Social and professional responsibilities

Three factors differentiate a professional group from a society or club:

- Whether members who are admitted to it are required to reach a specific academic level.
- Whether standards of performance have been specified for members, both technical and ethical.
- Whether a body exists to set and enforce these standards, to provide ongoing guidance to members and to administer membership.

Most professional bodies have rules of conduct and can expel members if these rules are broken. However, in practice very few bodies exercise this right to expel members; only the medical profession enforces it effectively.

Management is a new profession and has professional associations which are relatively new. Management also cuts across several other professions, so that members who belong to management professional bodies also often belong to other professional associations, such as in engineering, physics, accountancy, marketing and personnel.

None the less, the management professional bodies are active and effective in several countries, laying down clear guidelines for their members to follow in both the academic and professional areas. We shall consider four areas, those of environmental protection; responsibility to employees; ethics; and consumer protection, which also relates to the duty to customers.

4.2.1 Environmental protection

It is the responsibility of all managers to be sensitive to the human, natural and physical environment and the impact which their actions have on it (Institute of Management, 1994; Clarke *et al.*, 1994; Donohue, 1995; Brown, 1995; Porter and Linde, 1995; Crosbie, 1997; Castle, 1997). Governments in many countries have introduced legislation to protect the environment. Managers must be aware of these and ensure that not only the letter of the law but also the spirit of the law is followed (Brown, 1997). In addition, standards bodies have published guidelines, such as that from the British Standards Institute, BS7750: 1992 on the Specification for Environmental Management Systems.

There are three main environmental considerations:

- Pollution, usually created as a by-product in the day-to-day operation of the company, must be eliminated. It is important that processes used during manufacture do not create waste by-products which harm the environment, and that the products themselves do not create an environmental hazard when being used. Environmental considerations must be paramount at every stage in a product's life cycle, from design through to distribution to the customer and eventual scrap. Ideally the product itself should be made from disposable material, so as not to cause an environmental hazard when it reaches the end of its life.

- The use of natural resources in the manufacture of the product, particularly those that are non renewable, should be minimized (White, 1996). This applies to all forms of natural resources whether they are used in the product itself or in the processes used to make it.
- Health and safety issues must be considered, relating to the work environment and to the product. This is an area where there is strict government legislation, but often, especially in the case of a new product or process, this cannot be clearly defined. It is a management responsibility to ensure that all possible precautions are taken in these circumstances and one errs on the side of safety.

Figure 4.2 illustrates the three major steps in the introduction of an environmental policy into an organization. First the awareness of all employees within the company to environmental matters must be increased. This can only be done if senior management is willing to provide leadership from the top and to lead by example. All employees must be involved and consulted, with training and awareness sessions run throughout the organization. This is clearly an expensive commitment and it is not surprising that only a few organizations are willing to make the investment. However, once made it can pay back ample dividends in terms of employee satisfaction, enhanced reputation within the community and greater customer satisfaction.

The second step involves developing the strategies the company is going to follow in its environmental programme. In doing this the company will be influenced by any government legislation that exists, by codes of practice issued by professional organizations and by competitor activity, especially by the competitive advantage it can provide.

Figure 4.2 Steps in introducing an environmental policy within an organization.

The strategic aims of the organization must be clearly defined and communicated to employees, bearing in mind the short- and long-term implications of these and the effect they will have on the organization and the community. In addition, the methods and processes which are to be used have to be developed and buy-in obtained from employees who are to operate these new systems. Tasks must be prioritized, since only a few of these can be done initially; choosing those that can be completed quickly may be desirable to ensure early success.

Once the environmental policy is in operation the company must conduct an environmental audit. To do this it needs to set measures and to conduct audits against these, feeding back results to all concerned. The aim must be to ensure that the targets are being revised so that the process is one of continuous improvement.

4.2.2 Responsibilities to employees

In the 'good old days' it used to be said that if employees performed satisfactorily then they would do well within the organization and would have a job. Unfortunately, in the present fiercely competitive environment many companies and employees have discovered that there is no job which is for life (Caulkin, 1995a). Even organizations such as IBM, who once prided themselves on not having to lay off employees, have been forced into massive redundancy programmes and drastic downsizing in order to survive.

Employees are finding that they are moving from long-term employment relationships with their companies to short-term 'contractor' relationships, where employers and employees stay together as long as the relationship is for mutual good (Egan, 1994; Caulkin, 1995b). Employees are stakeholders in the organization and have a call on the company, in common with other stakeholders, such as shareholders, customers and suppliers. Employees and employers must form alliances for mutual good. 'Them' and 'us' do not exist in an organization that is working in alliance to create wealth for all stakeholders. There is no longer any 'loyalty', only a contract to work together for mutual good. The relationship is no different from that between competitors who form an alliance for mutual benefit, such as the Rover and Honda alliance, and that between IBM and Apple Computer.

Employees are expected to grow with the organization, to learn and develop so that they can continue to contribute to the success of the organization. Employees must change even as the business is forced to change, but companies have an obligation to provide the opportunity for employees to learn and develop.

Flatter organizations mean that there is less chance of promotion. However, opportunities must still be provided to employees to move laterally within the organization, to gain experience and to develop, at the same time adding value to the business. Achievement must be rewarded, not seniority.

It is acknowledged that an organization can no longer provide lifetime careers, and a company may sometimes need to lose employees. Employees may also move of their own will for better opportunities. The company must not hold employees back and must ensure that they go onto the job market with the latest skills, even if these skills may later be used in the service of a competitor.

In estimating the worth of an employee to an organization there are no easy answers; it usually depends on circumstances at the time concerned. (See the case study 'Which one to go?'.)

Case study

Which one to go?

John and Mary joined the FloRoll Manufacturing Company at the same time, straight from university, on a graduate training scheme. They spent a year moving between departments, gaining experience, and then settled into design roles within Engineering.

John spent the next ten years within the design group. He started as a software development engineer and worked his way up to section leader of one of the software sections. Mary, on the other hand, left the Engineering department after three years and moved into Manufacturing. This gave her valuable experience in production techniques and reinforced what she had learned in design, especially relating to design for test.

After two years in Manufacturing Mary felt that she would like to gain commercial experience and moved to Marketing. She spent three years in product line management and reached the position of section leader. After this Mary returned to Engineering, where her wide experience was rewarded by promotion to section leader of the other software section.

Two years later the organization found itself in financial difficulties and had to reduce its engineering budget. It meant that redundancies were necessary, and one obvious method was to combine the two software sections. Mary was the obvious candidate to lead the new group. Her wide experience was a clear advantage in dealing with Marketing and Manufacturing and gave her a much better understanding of Engineering problems. However, her knowledge and experience in software systems was significantly less than John's, and the company was in the middle of a new product development which required John's software skills.

It was not an easy choice, but after much soul searching Micro Dowling, the Engineering Director, decided to appoint John as the leader of the software group, and to let Mary go. Short-term 'survival' requirements had outweighed longer term considerations on what would be best for the company, once it began to prosper again.

4.3 Ethical considerations

Many companies believe that an organization is not accountable for the ethical behaviour of its employees. However, management cannot wash its hands of responsibility if employees behave unethically. Unethical business practices by employees reflect the values, attitudes and beliefs of an organization's operating culture (Paine, 1994).

Sometimes employees are guilty of positively following unethical practices (see the case study 'Cheaper by the dozen') but usually this is unintended, although still very real (see the case study 'Selling to survive').

Many organizations have the 'if it is legal it is ethical' attitude. However, this can lead to unethical behaviour, especially in the multinational environment in which companies work. Organizations should set their own internal values, which lead to laws of conduct, and these should go beyond minimum legal requirements (see the case study 'Legal sweat shops').

Case study

Cheaper by the dozen

In the early days, when the FloRoll Manufacturing Company was struggling to establish itself, product cost was a critical factor. A high proportion of the cost of a floroll was determined by one material, Zal 22, which was being bought from a single supplier. The product from this supplier was 40 per cent cheaper than from the next cheapest supplier.

John Strongbow, who was the Manufacturing Director at that time, suspected that the product might have been inferior to that being provided by other suppliers, but it seemed to do the job and there were no customer complaints, so he was prepared to turn a blind eye.

True, the Engineering department were not happy with the use of Zal 22. In their test they had found that Zal 22 failed in specific circumstances, causing florolls to dip suddenly and unexpectedly in mid-flight. However, the probability of this occurring under normal usage was less than 1000:1.

'With odds like that we can sleep easy', boasted John Strongbow. 'And it's no good bleating like a nanny goat', he jeered at the Engineering Director who complained. 'There is nothing unethical about it. The material may be a bit below specification, but the end product, our floroll, still does the job it is supposed to. If I did not use my initiative and have the guts to take a few risks we would price ourselves right out of the market and this company would not be selling any products.'

Unfortunately for John Strongbow, a year later that 1000:1 jackpot turned up. A floroll took a dive during a training flight and almost killed an operator. Investigations by the Ministry of Defence engineers traced the fault to Zal 22. Fortunately, no criminal charges were pressed, although John Strongbow left the company soon after that, under a cloud.

Case study

Selling to survive

It was felt by the Board at the FloRoll Manufacturing Company that the repair centre was costing too much and was unprofitable.

'It all depends on how you measure profit in the case of repair', said Adrian Elton, Manufacturing Director. 'We need the repair centre to meet our warranty obligations, and when we do a repair we only charge for materials used, not for labour. Under the terms of our warranty we also have to provide a two week turnaround, which means having a minimum number of operators available. However, since our products are fairly reliable, it means that there is waiting time in-between jobs. I think some of the costs of the repair centre should be offset as the sales cost of the product.'

This argument was understood by the Board, but it was still felt that the repair centre should pay its own way. It was decided to turn it into a profit centre. If the repair centre did not break even within one year, measured by sales made in piece parts, then it would be shut and the repair function put out to subcontractors.

Adrian Elton communicated this message to the shop floor, through Joe Plant, Manager of the Assembly Shop. The operators were despondent but decided to do their best to make a profit for their centre. Without realizing it they fell into the habit of fixing the immediate problem on the florolls which were returned for repair and then carrying out preventive maintenance. This usually meant changing out any parts which looked as though they could fail in the near future.

Piece part sales increased dramatically and so also did the profits for the repair centre. After a year the centre was showing a healthy profit and the operators' jobs looked secure. However, soon after this the Ministry also carried out an audit of its expenditure and realized that the cost of repairs on florolls had trebled over the past year. An investigation was ordered and the change in practice within the repair centre was identified. It was recognized that the operators had not knowingly acted unethically and were therefore not to blame. Their judgement had been affected by the undue pressure placed on them by the company ultimatum, and they were unable to determine accurately where preventive maintenance stopped and unnecessary service began. Also, no management guidance was available for them on this.

Jake Topper, Chief Executive, accepted responsibility for causing the problem and agreed to repay a large amount of the repair costs incurred by the Ministry during the year. He also changed the system to that initially proposed by Adrian Elton, so that a proportion of the cost of the repair centre was covered as a marketing expense in the warranty costs. The repair centre showed a profit on this basis.

Case study

Legal sweat shops

When James Silver, Materials Manager at the FloRoll Manufacturing Company, wanted to find a cheap source for Ovilite he decided to look to the Far East. The material was used in large quantities in the manufacture of florolls, but it did not need to be of a high quality, although low cost was important. James soon found a supplier operating from India and asked for a sample delivery.

The material was received on time and testing showed that it was of acceptable quality. A contract was arranged and deliveries were made.

This arrangement continued successfully for a year. It was after this time that James Silver learnt that Fred Steel, the Quality Director, was going on a business visit to the region and he asked Fred to drop in on his supplier of Ovilite.

'They seem to be very good', said James Silver. 'So far all their deliveries have been on time and we have had very few rejects. The price is also well below anything we can get from our local suppliers.'

'What seems to be the problem then?' asked Fred Steel.

'No problem, certainly not worth a special visit. However, since you are going to be there I would be grateful if you could carry out a brief audit for us, just for the records.'

Fred Steel agreed. What he found was a sweat shop in every sense of the word.

'But is it legal?' asked James, when Fred Steel reported back on his return.

'Oh, absolutely!' said Fred. 'Sweat shops are quite legal out there, although they don't call it that. And it is also ethical by their standards; after all, these people have jobs which they would not otherwise have.'

James Silver was not happy with what he had learned and consulted Adrian Elton, the Manufacturing Director. Adrian brought the topic up at Jake Topper's Executive Staff Meeting. It was agreed that although the supplier met the practices in their country of operation it did not meet the ethical requirements of the FloRoll Manufacturing Company. The Board's recommendation was to see if the source could be maintained by improving the conditions in the plant.

Fred Steel revisited the company in India, this time accompanied by a manufacturing engineer. Advice was provided to the supplier and a small investment was made by the FloRoll Manufacturing Company in the supplier's factory, to install air conditioning and to modernize some of the machines. This not only improved the working conditions of staff, but also greatly improved their morale and productivity. Within a year the company was producing Ovilites of much higher quality, greatly reducing inspection costs and so reducing the overall cost of the product, which was passed on as a lower price to the FloRoll Manufacturing Company.

By maintaining its ethical standards the FloRoll Manufacturing Company was rewarded by a loyal supplier who was able to supply at even lower cost than before.

4.3.1 Consumer protection

Governments have produced legislation to protect consumers, but this can only go so far in ensuring that the public is treated fairly by organizations, and there remain many social issues in areas such as marketing. It is the responsibility of managers to ensure that in their products and advertising their customers' needs are kept in sharp focus.

Customers should be given adequate information during the marketing stage, to ensure that they are capable of making a correct decision when choosing a product which satisfies their needs. This means that sales literature should be more factual and less like an invitation to enter paradise with the purchase of a cheap toothbrush. Information on less obvious subjects, such as warranty and consumer reports, should be readily available. Product safety is a legal requirement, but even here there are many social and ethical issues. For example, the safety of a product often depends on how it is used, and no amount of government action can legislate against improper use. It is important that clear and easy to use information is provided with the product in order to ensure that it is used correctly.

Products also must not be designed with a strategy of planned obsolescence. In its simplest form this means designing a product that fails after a specific time, which is longer than the warranty period. However, planned obsolescence can also be implemented by marketing a product that has deliberately been designed with limited features and then bringing out an update of the product with additional features, so that owners of the older product are tempted to change to the new one.

Ethical conduct in advertising is often controlled by legislation and by voluntary standards boards. Unethical advertising seeks to create a phantom need in its target audience, such as for example in a cigarette advertisement which appeals to the young smoker by implying that it is socially desirable to smoke and that not smoking may result in rejection by friends. Children can also be targeted by unethical advertisements, such as a breakfast cereal which is not very nutritious but which uses advertising to appeal to the very young and so reach parents.

4.4 Government regulations

Governments influence the day-to-day running of organizations by passing laws which, although they vary in detail depending on the country in which the company operates, all have similar objectives and content (Ryley, 1995). Generally these can be classified as those laws which seek to effect the organization's conduct towards its employees; those which protect consumers; environmental protection laws; and laws on the company's trading relationships.

4.4.1 Employee regulations

Employee-related regulations cover a wide spectrum, from rules governing the employment of people to safety and conditions at work and conditions of employment (Institute of Management, 1994). A few of these are given in Table 4.1.

The Factories Act of 1833 was an early law, passed to define the conditions of employees. It prohibited the employment of children under nine years of age and restricted the hours worked for children between nine and thirteen years to 9 hours per day, and those for children between thirteen and eighteen years to 12 hours per day. If these regulations seem harsh by present day standards, it should be set against the conditions which existed within factories at that time.

The Factories Act of 1961 laid down the minimum conditions which were to exist wherever people were employed in manual labour, the equivalent law for office workers being the Shops, Offices and Railway Premises Act of 1963. Other countries have similar legislation, such as the Occupational Safety and Health Act

Table 4.1 Some UK employee related legislation

Year	Act
1833, 1961	Factories Acts
1944, 1958	Disabled Persons Employment Acts
1963	Shops, Offices and Railway Premises Act
1963	Contracts of Employment Act
1965	Redundancy Payments Act
1970	Equal Pay Act
1971	Fire Precautions Act
1974	Health and Safety at Work Act
1974	Rehabilitation of Offenders Act
1975, 1986	Sex Discrimination Acts
1976	Race Relations Act
1978	Employment Protection (Consolidation) Act
1980, 1982	Employment Acts
1983	Equal Pay Regulations
1984	Data Protection Act
1985	Control of Industrial Major Accident Hazard (CIMAH) Regulations
1989	Electricity at Work Regulations
1989	Noise at Work Regulations
1992	Health and Safety (Display Screen Equipment) Regulations
1992	Provision and Use of Work Equipment Regulations
1992	Personal Protective Equipment at Work Regulations
1992	Manual Handling Operations Regulations
1992	Management of Health and Safety at Work Regulations
1994	Control of Substances Hazardous to Health Regulations
1995	Reporting of Injuries, Diseases and Dangerous Occurrences Regulations
1996	Health and Safety (Safety Signs and Signals) Regulations
1996	The Construction (Health, Safety and Welfare) Regulations

(OSHA) of 1970 in the USA, which defined safe working conditions for employees.

The Factories Act can be considered to affect four areas:

- **Health of workers.** This defines factors such as cleanliness of premises; overcrowding; temperature and ventilation; and sanitary accommodation.
- **Safety of employees.** This defines factors such as the procedure for recording and reporting accidents; use of protective guards around machinery; supervision and training of operators; and inspection and maintenance of machinery.
- **Welfare of workers.** This defines factors such as provision of drinking water; availability of washing facilities; provision of lockers for outdoor clothing; and availability of first aid kits.
- **Employment of women and young persons.** This defines factors such as hours worked.

In order to comply with these regulations, management must follow several principles in its dealing with employees:

- Company policies must be applied consistently and fairly to all concerned. For example, if the company policy is to pay a certain grade of staff for working overtime, then the manager cannot discriminate and pay some of the staff within this grade for overtime worked and not pay others who qualify. (See also the case study 'Fair treatment'.)
- Managers and supervisors represent the company in their dealing with subordinates. Therefore information which they pass down represents company policy as far as their staff are concerned, and will be acted upon. This puts the onus on managers to ensure that they are briefed on all aspects of company policy and adequately trained to deal with their subordinates.
- All actions which the manager takes relating to employees' personal performance must be documented. If this information is to go onto employees' records a copy must be given to them.
- All employee grievances must be investigated promptly, fairly and thoroughly.

There are several government regulations protecting the interests of minority groups. These laws are strongest in the USA, where it is even illegal to discriminate against people on the grounds of age.

The Disabled Persons Acts of 1944 and 1958 place an obligation on employers, above a certain size, to employ a quota of disabled people, and to incorporate a policy for training and promotion (Echiejile, 1995). The Rehabilitation of Offenders Act of 1974 is aimed at preventing people being discriminated against, or dismissed, on the basis of past convictions. This is to ensure that offenders, once they have served their term, are treated as any other employee.

The Race Relations Act of 1976 makes it illegal to discriminate against people on grounds of race. This applies to potential recruits and to employees. An industrial tribunal was set up to make awards to people who have been discriminated against.

Several laws also exist to prevent discrimination on grounds of sex. These include the Sex Discrimination Acts of 1975 and 1986, the Equal Pay Act of 1970 and the Equal Pay Regulations (Amended) of 1983. They outlaw sex discrimination relating to employment conditions, promotion opportunities, recruitment and so on.

The Data Protection Act of 1984 regulates the information about living people stored on computers and similar data processing equipment. The law imposes certain duties on users and computer bureaux and gives rights to individuals affected by the stored information.

The Contracts of Employment Act of 1963 was intended to ensure that employees have a clear understanding of their rights and obligations under their contract of employment. This was enhanced by the Employment Protection (Consolidation) Act of 1978, amended by the Employment Acts of 1980 and 1982. These also defined items such as sick leave, notice periods and unfair dismissal.

There are two instances when employers may have cause to dismiss employees: to enforce discipline and for redundancy. Both of these are controlled by government regulations, specified in employment acts. It is essential that the manager ensures that employees within the company know of the discipline and grievance procedures which exist and that they understand them.

Sometimes employees need to be disciplined, either because they are consistently breaking company regulations or because, due to poor motivation, they are failing to meet minimum performance standards. Figure 4.3 illustrates a possible series of steps which should be followed in these instances, which would generally be taken once counselling had failed.

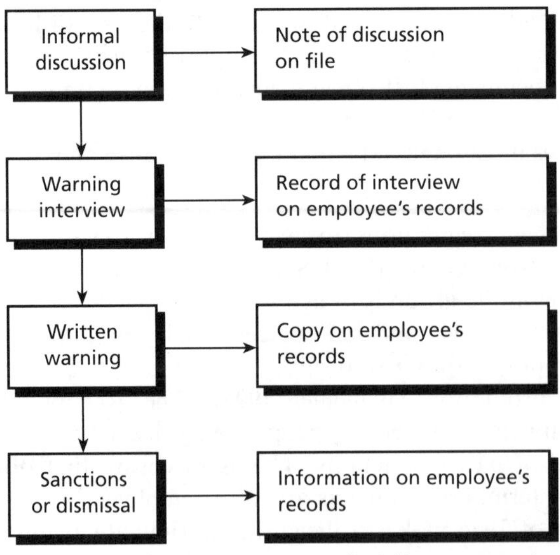

Figure 4.3 Possible steps in a disciplinary procedure.

In the first step the manager has an informal discussion with the subordinate and explains the problem areas and the actions which would need to be taken if the employee did not improve. Notes of this interview should be put on the company's files in case they need to be referred to later, and the Human Resources (HR) function should be notified.

If this fails the next step is to conduct a warning interview, with a member of the HR staff present. A formal record of this interview should be kept, taken by the HR representative, and this should be lodged in the employee's official company records. A copy should also be given to the employee.

The next step in the disciplinary procedure is a formal written warning to the employee. The sanctions the company would enforce should performance not improve must be given. A copy of this must also be put in the employee's records. Finally the sanctions are applied. This could take the form of financial sanctions, such as loss of bonus (if applicable) or no annual salary increase, although the final sanction is dismissal from the company.

Dismissal notices must always come from the HR department. The reasons for dismissal must be clearly stated and recorded, and must be such as can stand up under scrutiny in a court of law, if the employee wishes to challenge it. Throughout the process the employees should also be apprised of their rights and may be represented, should they so wish.

Redundancy is different from dismissal, since it is not caused by poor employee performance, but by poor company performance, forcing it to reduce the number of its staff in order to cut costs. The end result is however the same: employees are forced to leave the company. Government legislation also defines the procedure for redundancy, such as in the Redundancy Payments Act of 1965, which specifies the basis for calculating redundancy payment.

Figure 4.4 shows some of the steps that a company may go through when carrying out a redundancy programme. The number of people to be made redundant and the criteria for selection need first to be determined. The severance terms and overall costs to the company should also be determined at this stage. The individuals who are to receive redundancy notices can then be selected. It is important that this selection is done fairly and that it is recorded.

A general announcement can then be made to the workforce about the company's intentions. This can be done as a general announcement, with a statement that the people affected will be notified within a given time, or it can be done as a general announcement with the people affected being informed by their managers at the same time in face-to-face interviews.

This should then be followed by a meeting with HR, resulting in an 'at risk' letter giving a week's formal notice that the company intends to make the employee redundant unless circumstances change. A week later the formal redundancy letter can be given. This must clearly state the benefits that the employee will receive on leaving the company and it must meet or exceed the minimum statutory value.

Organizations often like to ask employees who are being made redundant to

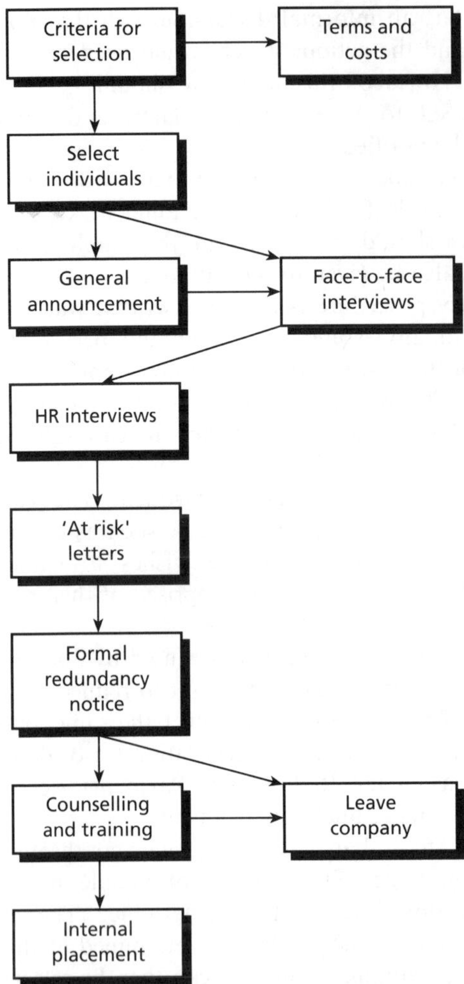

Figure 4.4 Possible steps in a redundancy programme.

leave as soon as possible. This is on the basis that these employees are likely to have a disruptive influence on other colleagues. Usually the reverse is true, since the company is judged on how it treats the people who are being made redundant. Every attempt must be made to counsel the people concerned, and to train and help them to find jobs on the open market. The company's telephone and secretarial services should be made available to them during their notice period. Also, the HR department should actively try to find them suitable employment in other parts of the company, which may not have been affected by the redundancy, or in outside companies. If a successful internal placement is found the redundancy notice can be withdrawn.

Governments generally pass laws covering relationships between trade unions and employees. The aim of this legislation is to ensure that employees have the freedom to join unions if they wish, without any reprisals from employers. In addition these laws cover topics such as trade disputes, picketing and ballots, to protect the rights of individuals within a union and the rights of employers.

There may be a single union covering all the employees within an organization (forming a union 'closed shop') or several unions, each negotiating terms and conditions for their members. Usually the separate unions would cover classes of employees, such as factory workers, office staff and managers. A manager who is dealing with a subordinate who is a union member must become familiar with the topics negotiated between the company and the union, in order to deal effectively with any requests or grievances.

Unions usually negotiate with the company on terms and conditions of employment for their members. An arbitration service is normally available in most countries to help resolve disputes; in the UK this is the Advisory Conciliation and Arbitrations Service.

There are several acts which cover the health and safety of employees at work. The Health and Safety at Work Act of 1974 provides a basis for all the subsequent additional legislation. It also set up the Health and Safety Commission and the Health and Safety Executives to enforce the Act. Under this Act the employer has responsibility for the health, safety and welfare at work of its employees. This includes consultation with employees on health and safety matters and for setting up safety committees. Under this law managers have direct responsibility for all personnel and installations under their direct control.

The Noise at Work Regulations of 1989 require the protection of employees from excessive noise, and refer to three action levels of noise for daily personal noise exposure: 85dB(A), 90dB(A) and 200 Pascals. The daily personal noise exposure is measured over the whole working day and needs to be assessed by a competent person making the relevant measurements and calculations. At the first action level of noise the employer must have the risk assessed by a competent person and must make hearing protection freely available. At the second action level and above, noise managers are required to do all they can to reduce the exposure of employees to noise, apart from providing hearing protection. They are also duty bound to ensure that employees are wearing hearing protection.

The Management of Health and Safety at Work Regulations of 1992 introduced most of the European Directive No. 91/391/EEC and the Council Directive No. 91/393/EEC. These form additions to the requirements under the 1974 Health and Safety at Work Act. For example, the employer is required to make an assessment of risks to employees, taking into account specific categories such as young persons and expectant mothers. They also need to appoint competent persons to assist with the implementation of the legislation.

The Display Screen Equipment Regulation of 1992 requires employers to ensure that sufficient analysis is carried out to assess health and safety risks associated with workstations that are used by their employees. Activities must also be planned for users so that the work being done is stopped periodically throughout the day, to enable rest periods.

Under the Provision and Use of Work Equipment Regulation of 1992 the employer is responsible for ensuring that equipment provided to employees is suitable for the purpose for which it is intended; that is, it is properly maintained and a log is kept of its maintenance. Health and safety information must also be provided to users and supervisors and they should be trained in its use.

The Personal Protective Equipment at Work Regulations of 1992 covers the provision of suitable personal protective equipment to employees where it has been found, by assessment, that it is needed. Employers must ensure that employees know of the risks that the equipment will avoid or limit, and that they are trained in their use. They must also provide appropriate accommodation for storing the personal protective equipment when it is not being used. Managers are required to monitor this equipment and to ensure its proper use. Loss or damage to protective equipment must be rectified immediately.

The Manual Handling Operations Regulations of 1992 define manual handling operations as those which require human effort to manipulate a load. The regulations seek to prevent injury to all parts of the body. Therefore the external physical properties of loads, such as sharp edges, should also be taken into account. Transporting or supporting a load in a static posture with any part of the body is also defined as manual handling. Applying some mechanical assistance, such as a power hoist, may reduce but will not eliminate manual handling. The Regulations require the employer to avoid hazardous manual handling operations and to reduce risk of injury as far as is reasonably possible. Managers are also required to continually monitor manual handling operations to ensure assessments are kept up to date.

The objective of the Control of Substances Hazardous to Health Regulations of 1994 is to prevent disease or injury occurring from hazardous substances, which can include dust in sizeable quantities. Employers are required to carry out regular assessments of the risks created and put in place measures to control it. They are also required to publish the results of assessments to employees and to monitor the health of employees, where appropriate. The employer must ensure that risk to employees is minimized and must measure the effectiveness of these measures. Managers need to be aware that they are directly responsible for all personnel and installations under their sphere of control.

The Reporting of Injuries, Diseases and Dangerous Occurrences Regulations of 1995 require injuries, diseases and other occurrences in specific categories, which occur at work or can be attributed to it, to be reported to the relevant authority. Failure to do so is a criminal offence. Examples of injuries include loss of sight in an eye, injury requiring medical treatment due to electric shock, acute illness or loss of consciousness, injury causing hospitalization for more than 24 hours, and three days or more absence. Disease includes poisonings, some skin diseases and occupational cancer. Dangerous occurrences include an explosion of fire resulting in suspension of normal work for more than 24 hours and electric short circuit or overload causing a fire or explosion. Employers are responsible for reporting any injury, disease or occurrences and for keeping a record for a minimum of three years.

The Safety Signs and Signals Regulations of 1996 provide the legal system for implementation of the British Standards applicable to signs, so as to ensure a uniform approach to signs for employees. The Construction (Health, Safety and Welfare) Regulations of 1996 consolidate, modernize and simplify previous construction regulations.

Case study

Fair treatment

Mick Kelby had been an operator within the FloRoll Assembly Shop for ten years, which was a relatively short time compared with the number of years most of the other operators had been with the company. However, he was very quick to learn and was soon one of the best operators in the Assembly Shop, often helping the others with their problems.

Mick was a special favourite of Bill Hale, the Machining Section Supervisor, and the two of them often took their lunch together, discussing the day's schedule or a knotty production problem.

It was during the end of year production schedule that a problem arose with one of the machining planes. This caused a week's loss of production, which needed to be made up with overtime working. All the operators worked hard, but none harder than Mick Kelby. For several weeks he was the first man in and the last one out, often taking on jobs which were not his direct responsibility in order to meet the production schedule. Eventually the rush was over and customer orders were met.

'That was excellent work', said Bill Hale. 'The Assembly Shop met its quota and the gaffer, Joe Plant, is very pleased.'

'I expect he will get a good bonus', joked Mike Kelby. 'Still, I guess my wife will also be pleased, now that it is over and she sees a bit more of me. I had planned to take her away for a break last month but had to cancel it.'

'That was very good of you,' said Bill. 'Why didn't you tell me?'

'The work had to be done,' replied Mick, shrugging his shoulders.

'Well, the work would not have been completed if you had not been here. Thank you!' said Bill. Then he had a thought. 'Look, Mick, you worked yourself into the ground this past month, doing all the dirty jobs for everyone. The company owes you quite a lot more than the overtime money you received. Things are going to be slack around here for the rest of the week. Why don't you take the next two days off, on full pay, as a 'thank you' from the company, and treat your wife to a long weekend break?'

Mick was persuaded by Bill to take the time off, and he and his wife had an enjoyable break. However, the rest of the operators got to hear of what had happened and objected. They selected a representative to put their case to Joe Plant.

'We worked as hard as Mick and he got paid overtime the same as we did', said the representative. 'If the company is going to give two days' paid holiday to one of us it must do the same for everyone else.'

Joe Plant consulted Adrian Elton, Manufacturing Director, and the HR representative. It was agreed that Bill Hale had committed an error of judgement in giving preferential treatment to one of the operators, although the company was sympathetic with his reason for doing so, and he was not reprimanded. The company agreed with the operators that they would each receive two days' pay in lieu of holidays, to make up for the two days' holiday given to Mick.

4.4.2 Consumer protection

Some of the laws relating to consumer protection are given in Table 4.2. The Sale of Goods Act of 1893 was a landmark in the field of consumer protection. It legislated that goods supplied must be as described, that it must be of 'merchantable quality' and should be fit for its intended purpose. This was enhanced by a series of further acts, such as the Trade Descriptions Acts of 1968 and 1972 (which specified that goods supplied must be as described, whether this description was on the package, on point of sale material or given verbally by the seller) and the Consumer Protection Act of 1982 (Sumner, 1991). The Office of Fair Trading was established by the Fair Trading Act of 1973 in order to look after consumer interests and to administer these acts. Other countries introduced similar consumer protection legislation, such as The Federal Fair Packaging and Labeling Act (1966) in the USA.

In the UK, product liability and consumer safety are part of the Consumer Protection Act. The aim of safety legislation is to keep unsafe goods off the market. Failure to obey this part of the Act is a criminal offence, punishable by imprisonment or a fine or both.

To meet product liability requirements the goods sold must be fit for purpose, whether this purpose is explicitly stated, implied or can be reasonably expected by the buyer. The EEC Product Liability Directive, which came into force on 31 July 1988, also removed barriers to compensation claims for defective products, so companies can now be sued anywhere within the EEC under similar laws. This applies to producers of the goods, own-branders and importers of the product

Table 4.2 Some UK Consumer protection legislation

Year	Act
1893	Sale of Goods Act
1968, 1972	Trade Descriptions Acts
1982	Consumer Protection Act
1988	EEC Product Liability Directive

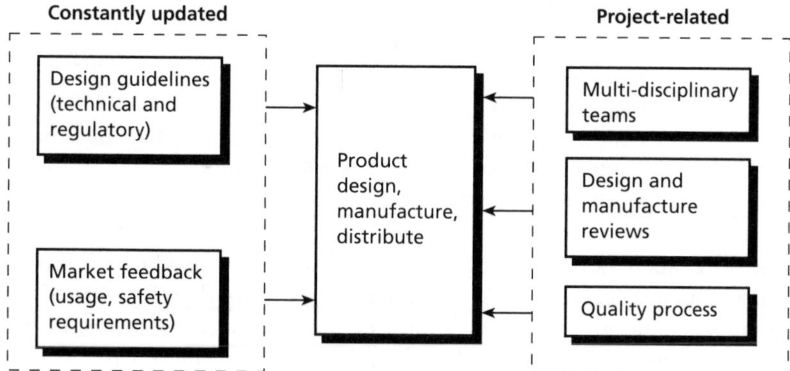

Figure 4.5 A process for minimizing product liability.

into the European Community. The onus is also now on the defendant to prove that the defect in the product did not exist at the relevant time.

Figure 4.5 illustrates a process which can be used for minimizing product liability. Designers should be aware of relevant regulations and standards which could affect the safety and performance of the product. These should be incorporated into the design guidelines and fed forward into product manufacture and distribution. Feedback should also be received from the market in the form of customer usage practice and comments, which can be incorporated into design guidelines. These guidelines should constantly be updated to take into account changes in technology and legislation.

Multidisciplinary teams need to be set up on a project basis to work on performance and safety aspects and they should take part in design and safety reviews. Quality systems and controls should also be kept current to ensure that the product requirements are met throughout the operation. This could be based on ISO 9001, 'Quality system – model for quality assurance in design, development, production, installation and servicing'. This has been issued in the UK as BS5750: Part 1: 1987.

Documentation forms an important part of the process, and it should show that reasonable care has been taken throughout the product life cycle. These documents must be stored for a relatively long time, to meet product liability criteria. For example they should cover the maximum lifetime for an item's circulation (say about 10 years), plus the maximum time after an injury has occurred that an action can be brought (3 years), plus the maximum time for the case to be settled (2 years), plus the maximum time for an appeal to be settled (2 years), making 17 years in all!

4.4.3 Environmental protection

Environmental laws are relatively new, but they now exist in most countries. Their aim is to ensure that organizations do not harm the environment in which

Table 4.3 Some UK environmental protection legislation

Year	Act
1968	Clean Air Act
1974	Control of Pollution Act
1988	Control of Substances Hazardous to Health (COSHH) Regulations
1989	Control of Industrial Air Pollution Regulations
1990	Environmental Protection Act

they operate, such as by discharging harmful wastes or producing excessive noise, and also conserve scarce natural resources. These laws therefore benefit the whole community, including customers, suppliers and employees (Hutton, 1995).

Table 4.3 shows some of the laws in the UK designed to protect the environment. Other countries have similar regulation, such as The National Environmental Policy Act (1969), The Noise Control Act (1972) and The Resource Conservation and Recovery Act (1976) in the USA.

4.4.4 Trading regulations

There are several laws governing contracts, which need to be followed by organizations and which are usually administered by the company's legal advisers. Table 4.4 illustrates some of these UK laws. Much of this legislation is complex and needs careful study.

An example is the law relating to transfer of ownership of goods (Peterson, 1993). Usually this transfer occurs from the seller to the buyer in exchange for money (the price). However, the Sale of Goods Act says that once equipment has been sold or contracted to be sold then the title of the equipment passes to the buyer, even if payment has not been made. Therefore if the buyer goes into liquidation the supplier of the equipment ranks as an unsecured creditor and is likely to get only a small part of what is owing.

This situation can be avoided by inserting a retention of title clause into the agreement for sale, so that the seller maintains ownership until full payment has been made. However, even then complications can arise, such as if the buyer resells the product to a third party, without a retention clause, before paying for it and then goes into liquidation.

Company directors can also be personally affected by trading legislation (Treadwell, 1993). For example, Schedule 24 of the Companies Act of 1985 lists over 200 offences which a director can commit that can lead to a fine or disqualification from holding office for a number of years. Similar coverage is given in the Insolvency Act of 1986 and the Company Directors Disqualification Act of 1986. Offences include fraudulent trading and wrongful trading.

Fraudulent trading occurs if the company does business with fraudulent intent or to defraud its creditors. Directors are then personally liable for the company's debts and can also be criminally liable. Wrongful trading arises if the company goes into insolvent liquidation and the directors know or should have known that

Table 4.4 Some UK legislation covering trading

Year	Act
1985	Companies Act
1986	Insolvency Act
1986	Company Directors Disqualification Act

the company could not pay its debts. This law also applies to those who were past directors of the company. The directors can be disqualified for up to 15 years.

4.5 Protecting ideas

Industry is moving from the age of manufacturing to the age of ideas (Whiting, 1992) and this is especially true in high-technology industries. This is placing greater emphasis on the need for companies to protect their ideas. In the USA, for example, the number of patents issued between 1980 and 1990 grew by 46 per cent, while the number of patent-related lawsuits grew by over 50 per cent.

A company's intellectual property rights (IPR) are similar to rights available with conventional property. The legal owner of an IPR is allowed to use and exploit it for gain, until all or part of the IPR is transferred to another owner (Adeloye, 1994). In modern high-technology industry, however, an IPR is often more precious than property, since it can give an organization considerable competitive advantage.

4.5.1 Trade and service marks

Trade marks and service marks are used to differentiate a product or service originating from one organization from those which come from another company (Treadwell, 1992). They are very important since consumers often buy on the basis of the trade mark; for example, computers from IBM.

A trade or service mark must be distinctive to be registered at the Trade Marks Registry. The registration certificate gives the registered owner the exclusive right to use the mark. This right does not have a time limit, so long as renewal fees are paid.

4.5.2 Patents

Patents are connected with inventions that have practical applications (McIntyre and Miller, 1997). In order to get a patent on a product idea it must meet several criteria:

- It must be new and must never have been disclosed to the public in any part of the world.
- It must be inventive; that is, it must not be obvious to a person with state of the art knowledge in the field concerned.

- It must be capable of practical applications. This application could be in any kind of industry, including agriculture.

Countries have developed well-defined procedures for having patents granted, those for the UK, USA and Japan being compared in Table 4.5.

A specification of the patent needs to be filed with the Patent Office, which will usually take the form of drawings, a written description and a claim of the inventive steps covered by the patent application. The Patent Office would then do a search of existing patents, which could take several years to complete. If this does not reveal any prior patents then a patent will be granted and made public.

A national monopoly is then granted to the originator, typically lasting twenty years, as long as renewal fees are paid. Most countries observe the international Patent Co-operation Treaty, which recognizes the initial filing date in one country as binding in other countries.

A large portfolio of patents form an important commodity for an organization to own. They serve several purposes:

- They provide a strong bargaining position when the company is negotiating with another company to carry out a joint operation, or to come to cross-licensing agreements on technology.
- They provide a competitive edge over other companies, often by keeping them out of a given field.
- They protect the company from allegations that it is infringing another company's patents.

Every organization should have an internal process for filing patents. Employees should be educated on the importance of protecting the firm's IPR, and should be encouraged to file patents.

A company may license some of its patents to others, often because they are in a better position to exploit them. This could be limited to a given application or geographical area. Commercial arrangements would need to be agreed, such as the payment of a fee, or a royalty on every product sold that was based on the IPR.

Often companies need to license their IPR to competitors. This arises when customers need two sources of supply before they are prepared to buy a product. Also, when a technology is new the company may wish it to be adopted as an international standard. It would then need to licence others to gain their support. In these circumstances the originating company still has a competitive edge, since its competitor's costs would be increased by the licence fees involved.

Table 4.5 Comparison of patent procedures in the UK, USA and Japan

Item	UK	USA	Japan
Time to granting	2–4 years	1.5–2 years	4–6 years
Approximate cost	£10 000	£7 000	£10 000
Protection granted	20 years	17 years	20 years
Criteria for award	First to file	First to invent	First to file
Publication of patent	1.5 years after filing	When awarded	1.5 years after filing

Summary

All managers have the responsibility to act ethically in everything they do, even though it may not always be easy to differentiate between ethical and unethical actions. In addition, managers must ensure that staff under them behave ethically and must not place them in a position where they are forced to react unethically.

Professional and social responsibilities are also placed on the manager. One of these concerns environmental issues. Although there are several pieces of legislation which relate to the environment, such as the Environmental Protection Act, these cannot always cover every aspect of this complex subject. The manager must be aware of this and ensure that the spirit as well as the letter of the law is met. Three aspects of environmental protection must be considered: the effect of pollution generated by the company on the environment, on its employees and on the community at large; the protection of natural non-renewable resources, by ensuring that energy saving measures are implemented; and the health and safety of employees and the community in which the company operates.

There are many government regulations relating to health and safety at work, the prime ones being the Factories Act and the Health and Safety at Work Act. Along with other smaller regulations, such as the Manual Handling Regulations, the Display Screen Equipment Regulations and the Noise at Work Regulations, these are primarily aimed at protecting employees at work. The manager also has other responsibilities to employees, which cannot be legislated for, such as ensuring that they develop while at work so that they meet their career aims. This must be done in spite of the 'contract' status which all employees currently occupy within organizations, in which employees and employers come together for mutual benefit and they will part when this no longer occurs.

There are several consumer protection laws, such as the Consumer Protection Act and the EEC Product Liability Directive, but once again these cannot always protect the consumer in complex situations. The manager must always act ethically and ensure that the consumer's interests are protected, such as by not building obsolescence into products and by using ethical standards when advertising their products.

There is government legislation to cover trading regulations, such as the Companies Act and the Company Directors Disqualification Act. These form the basis for company law as well as guidelines for company directors. Failure to meet these can result in the directors being fined and disqualified from holding office.

An important aspect of the legal framework is the protection of intellectual property, either in the way of trade and service marks or patents. It is important that managers encourage their staff to file patents, since these serve three aims: providing a strong bargaining position in cases where companies are planning to work together, such as in joint ventures or for cross-licensing agreements; providing the organization with a competitive edge over other companies; and protecting the company against claims that it has infringed another company's patents.

Case study exercises

4.1 The Board of the FloRoll Manufacturing Company was faced with a dilemma. The company had been approached by a large and well-known Middle Eastern state with a proposition that they set up a floroll assembly plant in that country. There was no danger of the company losing control of the product since all the key components would still be made in the UK and shipped out for final assembly. The financial terms were also very attractive. What troubled the Board was a moral issue. The country concerned was run by a dictator and had an abysmal record on human rights, openly suppressing its own citizens. The floroll manufacturing plant would clearly boost the country's economy and the political position of its dictator. Should the company accept the proposal?

4.2 Another moral issue faced the Board of the FloRoll Manufacturing Company. The sales team, led by Peter Sell, the Sales Director, had recently closed a multi-million dollar deal with an African state for the supply of florolls. However, the first group of the country's elite engineers had just completed the company's two month training course on floroll maintenance and operations and Vic Smith, the Training Manager, had reported to the board that the level of skill and knowledge available in the country was so low that no amount of training would ever enable them to properly operate and maintain equipment as complex as a floroll. It was Vic's opinion that the country would be much better off buying a less featured but simpler product, such as sold by a competitor. Should the Board disclose this information to the customer?

4.3 Research papers, published within the industry, seemed to indicate that a more advanced version of a floroll could be built using hydraulic drives instead of the more conventional electrical mechanisms. Unfortunately, the FloRoll Manufacturing Company did not have any in-house expertise to carry out research into hydraulic drives, so it decided to recruit a specialist in this field. Three months after the new engineer joined the company, however, papers began to appear showing that hydraulic drives were inferior to electrical mechanisms after all. The company continued to carry out its own research in this area for a further six months, but their research only confirmed the published material. The FloRoll Manufacturing Company no longer needed a hydraulic expert and, because of the high level of specialization of the new recruit, she could not be retrained to carry out any other function within the company. Micro Dowling, the Engineering Director, was therefore faced with the fact that he might need to make her redundant. What should Micro do?

4.4 Increasing automation within the manufacturing plant, coupled with static orders, meant that the FloRoll Manufacturing Company needed to introduce a programme of redundancy. The Board held a meeting to determine the basis to be used for selecting candidates for redundancy. Some of the options were:
(a) voluntary redundancy, which the Board felt could be attractive to several bright young manufacturing engineers, which the company did not wish to lose;
(b) first in, first out, which would again result in the loss of a high proportion of the younger, more versatile, operators; (c) early retirement for those within ten

years of normal retirement age, which would cause the loss of some engineers and operators with valuable product knowledge; (d) on the basis of past performance, which unfortunately could not be operated fairly since it was only recently that the company had introduced a uniform performance appraisal system. Information on past performance would therefore be subjective and influenced by prejudice. What method do you think the company should use to select candidates for redundancy?

References

Adeloye, A. (1994) The Uruguay round: discussion of the trade-related aspects of intellectual property rights, *IEEE Engineering Management Review*, Summer, pp. 44–7.

Brown, M. (1995) Greening the bottom line, *Management Today*, July, pp. 73–5.

Brown, M. (1997) A green piece of the action, *Management Today*, May, pp. 84–8.

Castle, P. (1997) Standards needed on environmental reporting, *Professional Manager*, January, pp. 12–13.

Caulkin, S. (1995a) Take your partners, *Management Today*, February, pp. 26–30.

Caulkin, S (1995b) The New Avengers, *Management Today*, November, pp. 48–52.

Clarke, R.A. *et al.* (1994) The challenge of going green, *Harvard Business Review*, July–August, pp. 37–50.

Crosbie, L. (1997) Looking beyond the boundaries to create sustainable business, *Professional Manager*, January, pp. 8–9.

Donohue, W.A. (1995) Seeds of Eco-Management are still struggling to take root, *Professional Manager*, May, pp. 10–12.

Echiejile, I. (1995) The business case for diversity, *Professional Manager*, July, pp. 8–11.

Egan, G. (1994) Hard times contracts, *Management Today*, January, pp. 48–50.

Institute of Management (1994) *Environmental Guidelines for the Professional Manager*, Institute of Management, London.

Hutton, W. (1995) Red alert on green concerns, *Management Today*, January, pp. 27–30.

McIntyre, I. and Miller, W. (1997) How to protect your patents, *Laser Focus World*, January, pp. 117–121.

Paine, L.S. (1994) Managing for organizational integrity, *Harvard Business Review*, March–April, pp. 106–17.

Peterson, R. (1993) A question of title, *Mobile News*, April, pp. 16–17.

Porter, M.E. and Linde, C. (1995) Green and competitive, *Harvard Business Review*, September–October, pp. 120–34.

Ryley, M. (1995) Rising tide of European law, *Professional Manager*, November, pp. 20–1.

Smith, C. (1994) The new corporate philanthropy, *Harvard Business Review*, May–June, pp. 105–16.

Sumner, L. (1991) Standards, product liability and the consumer, *Engineering Management Journal*, February, pp. 19–25.

Treadwell, R. (1992) Intellectual property to have and to hold, *Professional Manager*, November, pp. 24.

Walley, N. and Whitehead, B. (1994) It's not easy being green, *Harvard Business Review*, May–June, pp. 46–52.

White, P. (1996) Setting standards for managing energy, *Professional Manager*, March, pp. 24–5.

Whiting, R. (1992) Protecting the power of the idea, *Electronic Business*, February, pp. 24–32.

Treadwell, R. (1993) What price a directorship? *Professional Manager*, March, pp. 22–3.

Part III

Strategy and decision making

5 Strategy formulation

Introduction

This chapter looks at the elements of corporate strategy, its importance to an organization and the process and techniques that are used for its development and operation. An important consideration in formulating strategy is the need to consider alliances and acquisitions, and these are also described.

5.1 The elements of corporate strategy

The requirements for strategy are well known and accepted by every engineer and manager. One needs to decide, often some time in advance, what is to be achieved, what action is to be taken, when it is to be completed, how it is to be done, and by whom.

A corporate strategic plan is no different, except that it is at a much higher level and involves the whole organization (Hamel and Prahalad, 1989). It comprises the actions that the corporation needs to take to get from its present

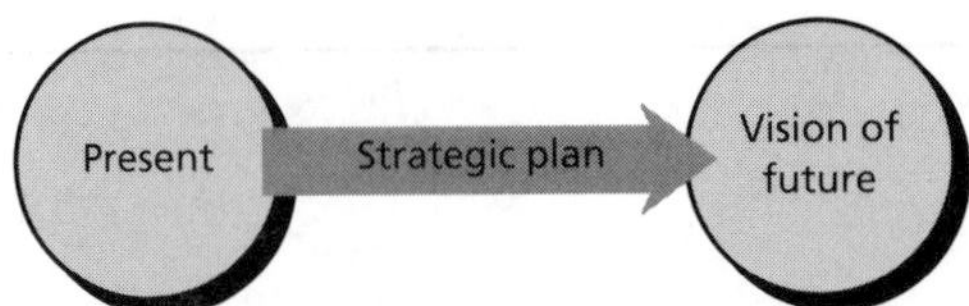

Figure 5.1 Linking the present and the future with a strategic plan.

position to where its vision (Collins and Lazier, 1993; Morley, 1997) of the future says it wants to be (Figure 5.1). The company is deciding on what it wants in the future and takes appropriate action today to get there (Hamel and Prahalad, 1994; MacVicar, 1996).

A strategic plan is often long range, for several reasons:

- It usually involves high levels of investment, in terms of capital and people, which need time to develop.
- Businesses are much more global than they used to be, and entering new markets or planning to take on a foreign competitor in a home market needs long-term planning.
- Technological changes need to be predicted and intercepted. Higher levels of technology also require greater investment in product development. However, reduced product life cycles, caused by technology change, and reduced margins, due to competitor activity, mean that payback periods are longer.

The difference between a strategic plan and a tactical plan needs to be understood. Strategic plans are longer term and they determine the actions which the company intends to follow in order to meet its long-term mission and objectives. This strategic plan would often contain shorter term actions which lead it towards achievement of the overall strategic plan. These actions form part of a tactical plan.

Strategic plans must be under constant review (Donaldson, 1995). The environment in which the company operates may change and so may its vision of where it wants to be. Often changes in technology may mean that product market windows are likely to be missed, so the company will need to revise its strategy. It is this change which nullifies the policy, often used in the past, of waiting to see what the leaders in the industry do and then copying them.

Formulating a strategic plan is relatively easy when changes in the business environment are evolutionary and predictable. One can then extrapolate what the future is likely to be and set one's strategic vision for the expected best and worst cases. This is the trap most managers fall into, where the strategic plan is set to extrapolate the present into some time in the future, with a factor added to account for 'hoped for' growth.

Unfortunately changes are normally unpredictable and occur as step functions.

5.1.1 Organizational variances

Every company needs to develop its own strategic plan, which will be different from that of another company, even if they are both operating in the same business and market sector. The prime reason for this is that the strategic plan needs to meet the cultural background of the company concerned (Egan, 1994) and the image it wishes to project to its customers and employees.

Strategic plans will also vary depending on the stage in the company's development. There are generally four distinct stages, not all of them relevant to all companies (Moore, 1993):

- **The entrepreneur or birth stage**. This usually applies to start-up companies, who are 'lean and mean'. It may also apply to an established company launching a new product or entering a new market. The organization needs the ability to take decisions quickly and move rapidly to capitalize on market opportunities. The company will often work closely with its customers and suppliers to define the product or service. These companies usually plan for growth, operating within niche markets.
- **The expansion and consolidation stages**. This is a period of rapid growth. Following it the company needs to take stock of its position, to pause to establish its policies and processes, and to ensure that it has effective controls in place for the next stage of growth. The company will continue to work with its customers and suppliers for maximum coverage in its chosen markets. Usually the company will have a central organization for strategy formulation.
- **The diversification stage**. The strategy adopted by the company will be such as to allow it to capitalize on its main product, but also to grow in several other related areas. Usually this may be the only way that the company can get bigger, since its primary market may be saturated. Others may wish to diversify in order to reduce risk; that is, reduce dependency on a single product, customer or market segment. Diversification also involves risk: the risk of entering new markets and of neglecting one's core business. The strategic plan may be one of minimizing risk, although usually it is better to aim to maximize opportunities at acceptable risk. The issues of good corporate parenting in large diverse organizations must also be understood and kept in mind (Campbell *et al.*, 1995).
- **The decline and renewal stages**. No organization actually plans to reach a stage of decline, but it has happened to many companies. In these circumstances the organization needs to develop a strategy that will enable it to survive and to grow again; that is, renew itself. It is important that opportunities are explored in these circumstances, and that the strategy is not over-biased towards avoiding risk in the belief that this is the best way to survive.

Long-term strategy needs to take into account the stage that the organization is in, as well as other external factors, such as environmental, political and technological changes. Another consideration is the business the organization wishes to undertake. Examples are shown in Table 5.1.

Table 5.1 Types of organization and business sectors

	Type of Company									
Product stage	**Marketing and sales**	**Marketing only**	**Sales only**	**Design**	**Manufac-turing**	**Partially integrated**				**Fully vertically integrated**
	(1)	**(2)**	**(3)**	**(4)**	**(5)**	**(6a)**	**(6b)**	**(6c)**	**(6d)**	**(7)**
Specification	In house	In house	Subcontract	Subcontract	Subcontract	In house	In house	In house	In house	In house
Design	Subcontract	Subcontract	Subcontract	In house	Subcontract	In house	In house	In house	Subcontract	In house
Manufacture	Subcontract	Subcontract	Subcontract	Subcontract	In house	Subcontract	In house	In house	In house	In house
Marketing	In house	In house	Subcontract	Subcontract	Subcontract	In house	Subcontract	In house	In house	In house
Sales	In house	Subcontract	In house	Subcontract	Subcontract	In house	In house	Subcontract	In house	In house

The marketing and sales company would probably specify the product to meet a given market need. It would go to outside suppliers to have it designed and built, and would then market and sell the product.

The marketing company would only specify and market the product, depending on an outside sales team, usually working on a commission basis, to sell it.

A design house is usually a consultancy which designs the product to a given specification and delivers it ready to be manufactured. Contract manufacturing companies are also available, who will build to a specification.

Partially integrated companies carry out most of the activities in house, going to external suppliers for one or more of the tasks, whereas a fully vertically integrated company does all the key activities in house.

It should be noted that a company can operate simultaneously in several of the columns of Table 5.1. For example, it may make one product line completely in house; go outside for design of a second product line, for which it does not have its own expertise; and have a third product line manufactured outside to supplement its own capacity. Even within a single product line the organization can operate in several of the columns; for example, it may go outside to have critical parts designed or built.

Buying in products or services generally reduces the profit margin due to the mark-up by other suppliers. However, if the supplier can deliver the goods much more efficiently than can be done in house, then this disadvantage can be overcome. The other problem of going outside is that the supplier may be a competitor in other markets and one may become dependent on a competitor to provide a critical commodity.

5.1.2 Strategic ownership

The owner of the strategic plan is the Chief Executive, although every level within the organization must contribute to it. Unfortunately, this task is usually carried out by a few corporate staff, since most managers and directors within the operations divisions would claim to be too busy with day-to-day issues to develop a vision or strategy for the company.

Corporate strategy has many elements to it and covers all the functions within a company. The many layers which go to make up the overall strategy are shown in Figure 5.2. Within each area there are also many considerations, such as technology, processes, environmental factors (including competitors and customer needs) and political considerations (which include economic and legislative factors). The strategy within the various areas affects others, such as product development having to consider the needs of manufacturing, marketing and finance (profit and loss).

A key aim of corporate strategy is to improve the overall capabilities of the company (Stalk *et al.*, 1992). The business processes are the prime

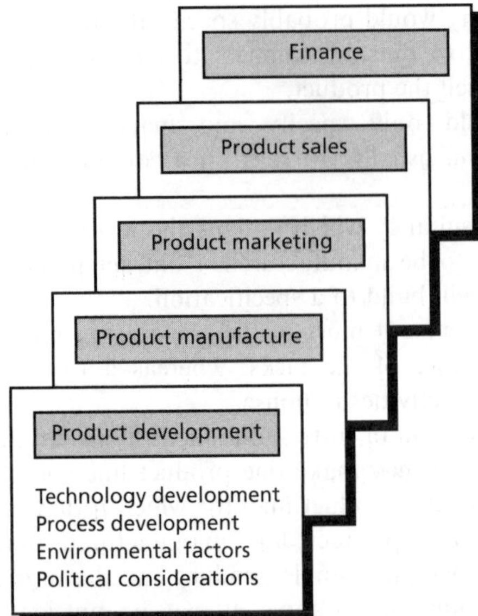

Figure 5.2 Layers of corporate strategy.

elements of a strategic plan and not the products being made. A company therefore needs to invest in its processes and develop a support infrastructure that can deliver value to the customer. The key requirements of an effective process are:

- The process is geared to meeting the customer's needs.
- The process must allow fast response to changes in the environment, such as competitor activity or political changes.
- The process must be transferable, for example from one company to another when mergers or acquisitions occur.
- The process must be expandable, to cope with changes in the company's operations.

Strategic processes cut across departmental boundaries and must therefore be viewed on an organizational basis if they are to be successful (see the case study 'The Helper'). It is because of this that these processes are owned by the Chief Executive.

Case study

The Helper

Shulster Corporation was a large user of florolls. Unfortunately they had bought these almost exclusively from the FloRoll Manufacturing Company's rival, Winger Inc.

Peter Sell, Sales Director of the FloRoll Manufacturing Company, believed he had made a major breakthrough when he persuaded Shulster Corporation to install some of his company's florolls on a trial basis.

After a few weeks some florolls started to malfunction and the procurement manager from Shulster Corporation threatened to discontinue the trial.

'We have got to support this trial', Peter said at Jake Topper's Executive Staff Meeting. 'Shulster Corporation are a major user of florolls and I am convinced that, if we perform, we can get most of their business. I think we should put a top notch engineer on the customer's site so that we can deal with problems as they arise.'

'I can't spare any of my engineers', protested Micro Dowling. 'They are all too busy with the new design modification that you wanted for the Saudi market. Besides, I don't know what all the fuss is about; I am sure that they are having just as many problems with the florolls supplied by Winger Inc.'

After much discussion Micro was eventually persuaded to put one of his best engineers on the customer's premises. He decided on John Kent, who was a very good engineer and had experience in dealing with customers.

John spent nine months in Shulster Corporation. He developed a reputation for being very competent and always being willing to help, so much so that he was soon nicknamed 'Helper' John. For the first three months John spent most of his time ensuring that the florolls that were on trial were in top condition. He slowly began to get a better understanding of what the customer's requirements were and he fed these back to Product Marketing, so that they could define the next generation of florolls. In time John also learned about the weaknesses in the competitor's products and was soon being called in to fix minor problems which occurred in their product. This information was also fed back to the Sales, Engineering and Marketing groups.

John's presence on the customer's site benefited all the departments in the FloRoll Manufacturing Company and within a year the company had signed up an exclusive contract with the Shulster Corporation for the supply of a modified range of florolls which met their requirements exactly.

5.2 The strategy formulation process

5.2.1 The vision

The strategy formulation process, often known as a strategic cycle, usually commences with an assessment of the future, as in Figure 5.3. The single most important aspect of this is the formulation of the organization's vision of the future, taking into account its mission, business aims, the environment in which it is working and its current position.

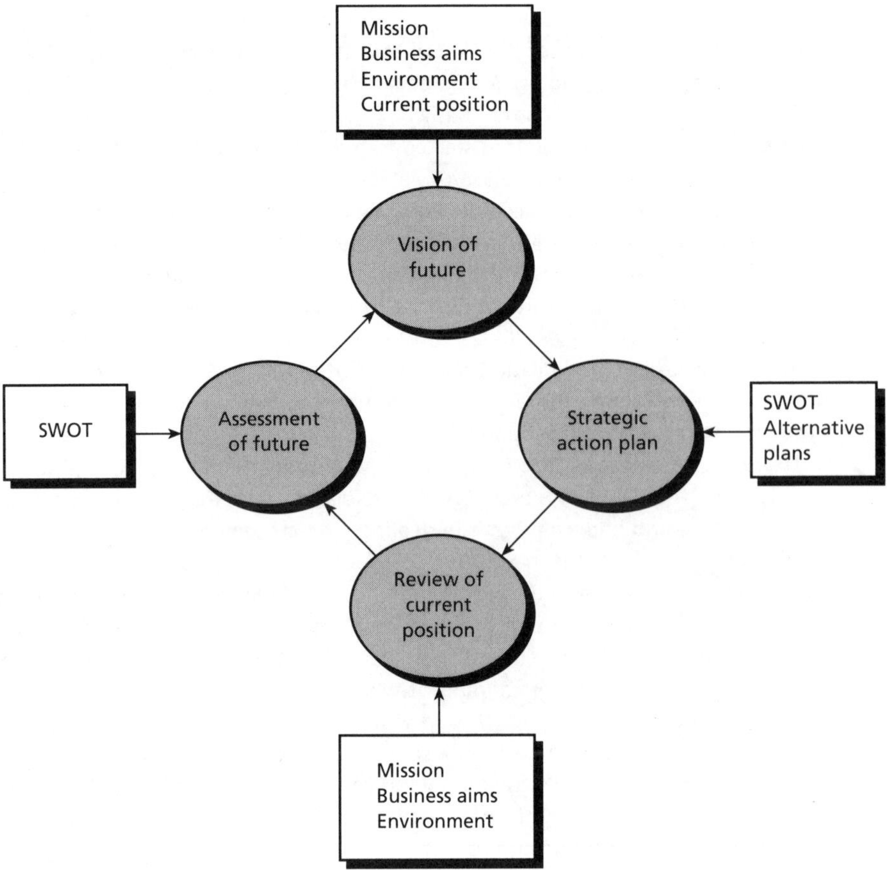

Figure 5.3 The strategic cycle.

In its determination of this vision the organization must consider what it believes to be important for its future. For example, it may wish to:

- Establish itself as a technology market leader, being the first into the field with innovative products, before the competition, so that it can establish itself in emerging markets.
- Be second into the market-place with a new product, gaining market share on the basis of a better engineered product at lower price. This way the costs of early failure and of educating the market can be avoided.
- Be a narrow supplier of specialist products and services. Alternatively, the organization may wish to broaden its product and service base, so as to tackle a much larger market sector.
- Concentrate on its home market, gaining greater share. Alternatively, it may wish to penetrate several overseas markets.

- Ensure that it is vertically integrated, producing all its critical components in house. Alternatively, this may be considered to be too costly and the organization may conclude that it needs to form strategic alliances with its suppliers.

The vision provides the guidelines for decision making throughout the organization. It must take account of the company's business aims and how these affect customers, suppliers, employees and shareholders.

5.2.2 Action plan

The vision must be translated into the strategic action plan, which defines the steps needed to achieve the organization's vision. Whereas the vision is usually formulated by a small number of the organization's top executives, the strategic action plan needs to have wider participation. It is usually generated by a mix of the following methods:

- As a directive from the top; from the Chief Executive or a very small group of executives.
- On a top-down principle, the plans being generated at the executive level but then being elaborated on and adopted by the different levels within the organization.
- On the basis of consensus among a committee of members, representative of the various functions and levels within the organization and set up for this purpose.
- On a bottom-up principle, where the ideas go up the organization, being modified and added to until they are finally adopted by the organization.

In preparing a strategic action plan it is important to stretch the organization (Hamel and Prahalad, 1993); its aspirations should be high, often exceeding the available resources. This will ensure that the organization is always aiming for the highest levels of excellence.

Resources should always be levered (obtaining more from existing resources) and not downsized (doing the same amount using less resources). Resources can be levered by several methods, such as:

- Prioritizing objectives, so as always to focus on key tasks first. Strategic goals should be aimed at and resources not squandered by following up side issues and projects.
- Using other people's resources wherever possible, such as customers and suppliers. It is important, however, that each contributor creates value (Normann and Ramirez, 1993; Heijden *et al.*, 1993), and that the organization helps this to occur.

- Extensive re-use of technology, experiences and ideas. The company must look both inside and outside its organization for items to re-use. It is vital not to reinvent the wheel each time a new task is to be performed. Examples of re-use are advertising the company image or logo, rather than individual products made by different divisions of the company, and flexible manufacturing, so that the production line can be switched quickly from making one type of product to another.
- Breaking down specialities within the company, both in people and in machinery. This should reduce the down-time associated with specialized machines or people waiting for their speciality to be required. It should also improve the motivation and productivity of staff. Resources should always complement each other. A company which is good in one area alone, such as R&D or manufacturing, cannot succeed unless it acquires the other elements, for instance marketing and sales, by other means, for example by forming strategic alliances.
- Forming alliances with organizations which complement each other (Brandenburger and Nalebuff, 1995). Even competitors can be drawn into alliances against a common foe, or for some other mutual benefit.
- Taking every opportunity to conserve one's scarce resources, similar to that done by a general during war. For example, it may not be prudent to attack stronger competitors head on; it would be much more prudent to attack their weak areas first and then to spread from this. For example, Canon did not tackle Xerox in the high-volume copying area when it first entered the field of copying. It started with low-volume desktop copiers, and then moved into high-volume copiers, which was Xerox's main focus.
- Looking for quick paybacks, which will allow the resources to be redeployed. Smaller organizations cannot normally afford to wait for long periods between making an investment and it becoming profitable. This requires shorter development cycles, perhaps with limited functionality. This also has the advantage of ensuring that subsequent releases of the product are able to track market changes more effectively.

A strategic action plan usually has three main considerations:

- The need to maintain current levels of, for example, customers and markets; revenues while a new product is being developed; and existing processes until new and improved processes are available to replace them.
- Improving existing items, such as the performance of an existing product or the revenue from an existing customer.
- Introducing change, such as a new product, or entering a new market or dealing with a new customer. Change is the key to the success of any organization, and all strategic plans must have a high change content.

A strategic action plan can cover many areas, although it usually includes the following:

- Financial considerations, such as the need to increase profitability; or increasing finance by borrowing, issuing equity or by internal investment.
- Market development, both home and overseas, with considerations such as the extent of vertical integration; strategic alliances, acquisitions and joint ventures; and the deployment of an internal sales force, the use of distributors and the appointment of agents.
- Technology, product and service development, such as development of new products or extending existing services.
- Process improvements, to increase productivity, which include the process that exists between the organization and its customers and suppliers.
- Developing the culture of the organization through the attitudes of its managers and staff. Employees should be encouraged to think for themselves and to act in the interest of the company. A strategically managed company is a confederation of entrepreneurs (Hinterhuber and Popp, 1992).
- Responsibility to the public and to the society in which the organization operates.

Most of these areas are interconnected. For example, increasing profitability must not be done at the expense of the employees and of the public, such as by increasing pollution. Also, items such as customer care and quality are implied within these considerations, since profitability and market penetration cannot otherwise be sustained.

Following completion of the strategic action plan the activities are carried out and results monitored. What is measured is important (Kaplan and Norton, 1992), since the measurement system affects the behaviour of the organization. Financial measures (return on investments (ROI), earnings per share and so on) and operational measures (defect rate, time to market, customer satisfaction and so on) will often be included, although they are linked together. A few key measures are required, rather than a large number of mediocre ones, to ensure that measures do not become confusing and that attention is focused on key issues.

At periodic intervals a review is conducted of the new position reached by the company, as shown in Figure 5.3. As before, the mission, business aims and environment must be taken into account. Environmental factors include regulations and competitors. Competitor activity is not always a disadvantage, since it can often help to grow and educate a market. For example, the UK Telepoint communications market started out with four consortia being granted licences. Very soon all but one pulled out. The remaining applicant found it difficult to educate the market, with the result that subscriber numbers were too small to be economic, causing the company to withdraw also, even though it had no competition. Competition in lateral markets must also be considered, such as the development of videoconferencing reducing the need for extensive business travel.

Following a review of the new position an assessment is again made of the future and a new or modified vision of the future formulated, completing the strategic cycle.

5.2.3 SWOT analysis

An analysis of the strengths, weakness, opportunities and threats (SWOT) facing the company is conducted at all stages of the strategic cycle. These items must always be considered in the context of the future position of the company. For example, the strength of the company today may be that it has a large and strong internal sales force. However, if it is proposing to move into an area which requires an extensive distributor network, then the internal sales force may be a weakness rather than a strength.

Often a company will recognize its present weakness and put plans in place to eliminate it as it moves into the future, so that the weakness becomes a strength. However, undue optimism must be avoided in making these plans. (See the case study 'Elusive productivity'.)

Opportunities can be found in many ways, one method being identified by the opportunities triangle of Figure 5.4 (Jones, 1991). Market and customer characteristics vary depending on the business sector in which the company seeks to operate. Even related business sectors can have different characteristics, for example telecommunication switching is primarily associated with PTTs, which have similar characteristics geographically; telecommunications private networks are associated with individual organizations, which have distinct characteristics according to geography and the service they provide; and telecommunications terminals, such as telephones, are sold to the public and represent a consumer market, with very different characteristics. Each of these, however, represents a market opportunity for the company.

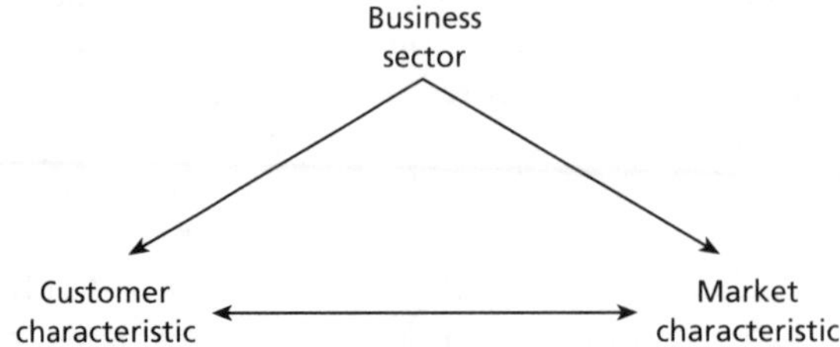

Figure 5.4 The opportunities triangle.

Customer characteristics also change with time. Organizations employ producers of goods and services and they have consumers who use these products. However, all organizations and their employees are usually consumers of these same products and services, so that the interests of the three groups often converge, as in Figure 5.5. This illustrates the continual shift which occurs in customer characteristics.

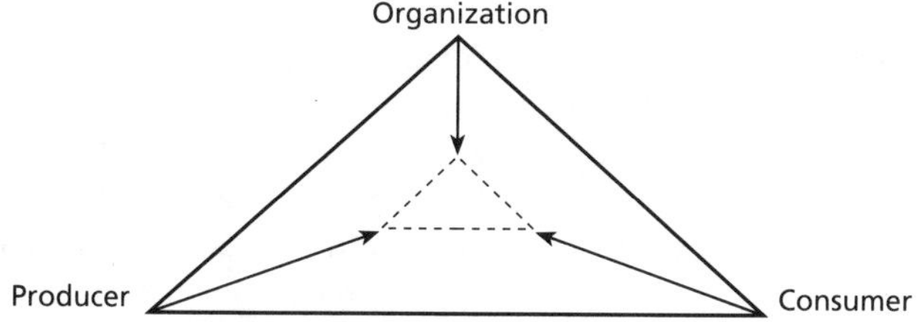

Figure 5.5 The imploding triangle.

Case study

Elusive productivity

As part of the strategic plan, Peter Trenchard (who was then Materials Manager in the FloRoll Manufacturing Company) produced a forecast for productivity improvements in the materials procurement department. This was to be achieved by the introduction of computer-based material management systems. These would initially result in a lowering of productivity, due to the introduction and learning problems involved, but then productivity was expected to rise sharply. Peter illustrated this by the curve AXY in Figure 5.6, productivity falling at the end of the first year, but then increasing to the target value at the end of the third year.

At the end of the first year productivity was found to have declined to B, primarily since the new material management systems had not yet been introduced and workload had increased. For the new strategic plan Peter forecast a curve BPQ to reach the target level of productivity again by the end of the third year.

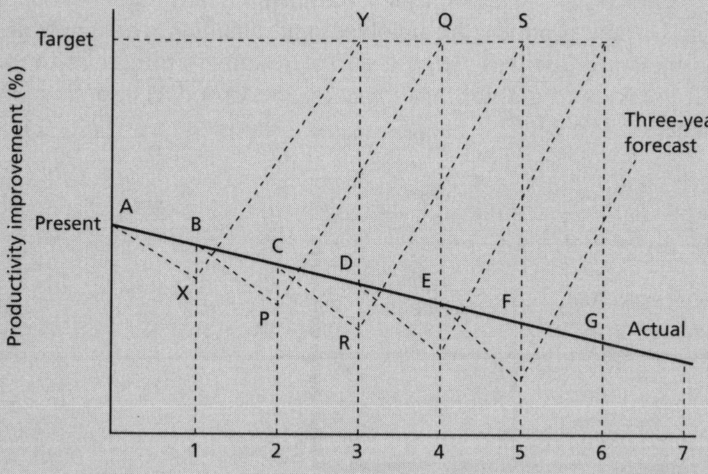

Figure 5.6 The optimistic forecast.

At the end of year two productivity had fallen to C, again because the material management systems had not as yet been introduced. Peter's new forecast was CRS. Clearly the curves in subsequent years were likely to follow the pattern shown in Figure 5.6, productivity gain forecasts each year being greater than in previous forecasts in spite of increasing workload, a situation which was clearly very optimistic and could not support the strategic action plan of the organization.

Peter retired soon after this and was replaced by James Silver, who succeeded in introducing the new materials management system, achieving a lower target productivity increase.

5.3 Alliances and acquisitions

5.3.1 Alliances or acquisitions?

Alliance (which includes a merger) and acquisition considerations are important elements of any modern corporate strategy (Almassy and Baatz, 1992; Cave, 1996; GuideLines, 1996; Hamel *et al.*, 1989). They both have an equal chance of success, which is about 50 per cent (Bleeke and Ernst, 1991) although the chances of success are increased if a few simple rules are followed. For example, generally alliances are more successful if they are used to enter markets in new geographical areas, or to develop products that are not core to either party's business strategy (Bleeke and Ernst, 1995). Acquisitions, on the other hand, should be in geographical areas where the company already has a strong presence, or where the product is core to its operation.

Figure 5.7 illustrates this principle. Acquisitions are used where existing geography and prime products are involved, and alliances where new geography and new business are involved. High levels of resource are needed to enter new geographical areas or to develop new business (and products), a situation which favours alliances.

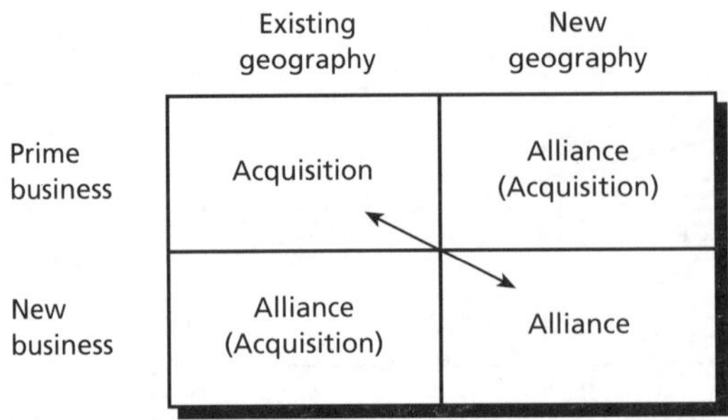

Figure 5.7 The alliance–acquisition matrix.

In the other two squares alliances should generally be considered, although the strength of either one aspect could swing the balance in favour of an acquisition.

5.3.2 Reasons for alliances

Two thirds of cross-border alliances run into serious problems in the first two years of their existence (Bleeke and Ernst, 1991) and one third result in failure. However, there are many reasons why companies seek these strategic alliances, and there have been some spectacular successes. For example, Rank Xerox and Fuji Films set up a joint venture, Fuji Xerox, to develop copier sales in Japan. The company was soon to become one of the world's largest manufacturers of photocopier machines.

Strategic alliances can involve several groups of companies (Gomes-Casseres, 1994) and may be set up to develop new products (Bidault and Cummings, 1994); to enter new markets; to share in the cost of new research; for one of the partners to obtain political credibility in gaining a foreign licence; and so on. The alliance may be short-term, each partner making the knowledge part of their own organization, or it can be long-term, such as a joint venture, where the alliance has an identity of its own and provides revenue to the partners.

Sometimes a large firm will form an alliance with a much smaller company, and tap into the smaller partner's innovative and entrepreneural skills, while the smaller company gains access to world markets, volume manufacturing and much greater resources (Peridis, 1993).

Reasons for forming alliances depend largely on the activities of the partners (Lorange *et al.*, 1993). The main considerations are whether the company is a market leader or market follower in the activity being considered for the alliance, and whether it forms the prime business or secondary business of the potential partners. This is illustrated in Figure 5.8 and includes four considerations:

- If the subject of the alliance is a primary business for any of the partners and they are already a leader in the field, then the alliance would mainly be formed as a means for that company to defend its lead position.

	Market leader	Market follower
Prime business	Defend lead position	Drive for lead position
Secondary business	Maintain status quo	Change position

Figure 5.8 Reasons for forming alliances.

- If the subject of the alliance is a primary business and the company is a market follower, then the alliance may be formed to enable the organization to move to a market leader position.
- If the subject of the alliance is a secondary business, but the company is a market leader in it, then the reason for the alliance may be to maintain the status quo. The company does not wish to devote a high level of management resource in this area, which is non-core business, but sees advantages in maintaining its presence in it. It will look to the strategic alliance to help it to manage this business.
- If the subject of the alliance is a secondary business and the company is a market follower, then it would look to the strategic alliance as a means of changing this situation. It may well be that the company is considering the alliance as a way of allowing it to exit this business area gracefully, by transferring expertise to one of the partners before selling the business to it.

5.3.3 Strong alliances

Alliances formed between two partners of equal strength have the greatest chance of success. Clearly it is better if the two are strong companies, rather than both being in a weak state. Unfortunately, many alliances are formed for the wrong reasons. Often a weak company will chose a stronger partner in the hope that this will help it to get out of its problems. The stronger company will accept the weaker partner, since it believes that it can then dominate the partnership and get its own way.

The management of a company which is in a weak state is often too preoccupied with the day-to-day aspects of surviving to be able to give the alliance the attention that it deserves, with the result that the weaker company drags down the stronger one.

An alliance needs to use the skills of both companies in equal amounts and both must be willing to share in its investment and not lumber any one company with an unfair share of the burden (Kanter, 1994). This may result in that company not performing its part of the bargain, to the detriment of both parties in the alliance. (See the case study 'The weak link'.)

It is important always to bear in mind that alliances can only succeed if they are considered to be a win–win situation for all participants, and are not treated as win–lose games. If one of the partners fails to achieve its objectives then all parties to the alliance must also fail.

Case study

The weak link

The FloRoll Manufacturing Company had a good working relationship with the Zing Ting Company, one of the major manufacturers of zing, a product used in the manufacture of florolls.

Although Zing Ting had been very successful in the past, recently it had been running at a loss, and Peter Smythe, Managing Director of Zing Ting, realized that something had to be done if the company was not to go into liquidation.

Peter Smythe approached Jake Topper, Chief Executive of the FloRoll Manufacturing Company, with a view to discussing an alliance. The logic was simple: FloRoll Manufacturing Company was Zing Ting's biggest customer, and with an alliance it had a low-cost supply of one of the key components which went into making florolls.

'And that's not all', said Peter, concluding his presentation to the FloRoll Board. 'Zings are used in the manufacture of many products apart from florolls, some in the defence field. With my company's technical know-how and your organization's marketing and sale muscle, we could form a joint company which really exploited the potential of zing, world-wide.'

Jake Topper was not so sure. The FloRoll Manufacturing Company was about ten times bigger than Zing Ting, and with Zing Ting's recent low stock price Jake was sure he could successfully mount a takeover bid if he wished. However, he agreed to a working party to look into the prospects of a joint venture, and when the party recommended that the joint venture should go ahead he reluctantly agreed.

'But I want to make sure that Zing Ting carries its fair share of the venture', he insisted. 'All costs must be split 50-50.'

It was quite easy for costs to be apportioned as the venture went along. Because of its small size Zing Ting had very few financial personnel, and they were soon overwhelmed by the army of financial experts from the FloRoll Manufacturing Company. Jake Topper's wishes were carried to extremes, and extra costs were loaded onto the Zing Ting Company.

The joint venture agreement required that Zing Ting contribute engineering and manufacturing to the venture, while FloRoll Manufacturing Company supplied the marketing, sales and finance groups. Unfortunately, the Zing Ting Company plunged further into the red under the financial burden of the joint venture start-up costs. Many of its staff became disillusioned and left. Zing Ting's management was too preoccupied with keeping its own company afloat to think too much about the problems of the joint venture, with the result that development and manufacturing times started to slip.

Although the salesmen from the FloRoll Manufacturing Company got several orders for zings, they found that the product was not being produced in sufficient volume to meet customer orders.

After two years of mounting losses Jake Topper decided to pull out of the joint venture. Zing Ting was close to bankruptcy, but with its technical and manufacturing base severely weakened it was no longer an attractive target for an acquisition. It could not even provide all the zing required by the FloRoll Manufacturing Company, which had to go out into the open market and buy the extra material it needed at much higher prices.

In all honesty Jake Topper had to admit to himself that the joint venture had been a mistake and both companies were much worse off for it.

5.3.4 Forming a joint venture

Joint ventures are the ultimate form of alliance; a separate company is set up to which each partner contributes a given amount of finance, personnel and skills.

It is vitally important that the joint venture has full autonomy; its management team must be given authority for solving the joint venture's problems without interference from its parents. The management team must act in the best interests of the joint venture, and not that of the parent companies.

The joint venture is a separate entity and not part of either of its parents. It must have its own operational responsibility, although often the parents retain equity financing and overall structural control. Financing and financial control must be shared equally between the parents, and this must be kept separate from operational control, which rests with the board of the joint venture. This way, each parent will be equally concerned with the success of the venture.

The objectives for the joint venture must be as broad and flexible as possible, so that it can follow up opportunities as they arise and grow beyond its original mandate. It is important to remember when setting up a joint venture that the future is difficult to predict and even the objectives of the parent companies will change with time.

Employees of the parent companies may sometimes feel threatened when a joint venture is being formed. Management must be aware of this and ensure that the threat of job losses is minimized. Also, the employees of the joint venture must have loyalty to the joint venture and not to any of the parent companies. This is difficult, since often a joint venture is formed by secondment of senior staff from the parent companies.

It is important that all parties to a joint venture make internal plans to cater for the time when the joint venture may be terminated or fail. Even if it is successful a joint venture may end once it has met all its objectives. Often, however, ventures are ended due to failure, and this can often leave one or both parties in a dangerous position. This is especially true if the two partners are competitors in some other area, since the suspicion is always there that one of the partners will go it alone after it has acquired sufficient know-how from the joint venture. (See the case study 'Partners in competition'.)

Usually, three-quarters of failed joint ventures are acquired by one of the partners, and in these circumstances problems can arise regarding ownership of items such as intellectual property rights.

Case study

Partners in competition

The FloRoll Manufacturing Company did not always manufacture its own dinshers in house. There was a time when it used to import large quantities of dinshers from the USA, since there were very few local suppliers.

Efforts were made to start internal manufacturing, but it ran into many process-related problems. Fortunately, at about this time, a new start-up in the USA, Dinsher Manufacturing Inc., was looking to expand into the UK and a joint venture was formed with the FloRoll Manufacturing Company. The USA company provided the know-how, while the UK partner provided premises and labour.

The alliance worked well; the excess material was even sold to other floroll manufacturers who were not in direct competition with the FloRoll Manufacturing Company.

After three years things began to change. Dinsher Manufacturing Inc. acquired a small company in the USA which made florolls, and it grew the business and started to attack the UK market. The joint venture began to feel the effects of the competition which now grew up between its parents. Directors from the two companies, which largely made up the board of the joint venture, grew suspicious of each other. The FloRoll Manufacturing Company was in a strong position; it owned the premises and most of the staff. It had also now acquired all the technology to manufacture dinshers.

The joint venture was eventually dissolved and the FloRoll Manufacturing Company started to make dinshers in house. Unfortunately, Dinsher Manufacturing Inc. lost all its manufacturing and marketing facilities in the UK with the demise of the joint venture, a situation for which it had not planned adequately. It was another three years before it could re-enter the UK market.

5.4 Strategy formulation tools and techniques

5.4.1 The need for techniques and tools

Corporate strategy formulation can be considered to have started in the mid-1950s. Prior to this managers used planning, programming and budgeting techniques that had been introduced earlier in the 1940s. Strategy formulation was found to require enhanced techniques to carry out detailed analysis and diagnostics of situations and strategies; some of these are described in this section.

There are several advantages in the use of these formal tools and techniques, such as:

- They ensure that a systematic thought process is followed in the analysis of the strategy. All the pros and cons are considered logically and a disciplined and rigorous approach is applied to problem solving and looking for opportunities.
- They ensure that the emphasis is placed on facts rather than on hunches. It is important to follow up hunches, but these must eventually be supported by facts.
- They ensure a more detailed analysis of the problem and so result in better decision making. The effect of personal bias is minimized. Application of these

techniques requires that the situation, cause and effect, is clearly understood, resulting in better insight into markets, competitors and customers.
- They provide fall-back strategies, which may be necessary if circumstances make the selected strategy unworkable.

5.4.2 Planning tools and techniques

Many tools and techniques have been described by writers (for example, Webster *et al.*, 1989), only a few of these being considered in this section. These tools are required at all stages in a strategy formulation process, although some techniques are better applied at some stages rather than at others.

Value analysis

In this technique the driving forces in the business are itemized and analyzed. Activities which add value, such as manufacturing, marketing, sales and R&D, need to be itemized and clearly identified.

The key in this area is shareholder value analysis (Day and Fahey, 1990). However, it is important to remember that there are other stakeholders in the company apart from shareholders, for example creditors and employees. What is good for one party may not be good for another, certainly not in the short-term (see the case study 'Value tradeoff').

Customer satisfaction is high on the list of value analysis considerations, and many organizations regularly carry out customer satisfaction surveys, although these are best conducted on behalf of the company by independent organizations.

It is important to identify the critical success factors (Boynton and Zmud, 1984). These should be limited to a few key areas, which can be analyzed and acted on. They should be the areas where high performance by the company would show greatest benefit and payback.

Market analysis

This covers the analysis of many different market-related factors which affect the supply and demand for products or services provided by the organization. The aim is to identify market niches and new opportunities.

A key element is competitor analysis. This should be broken into several areas, and each of these analysed in detail. Examples of the different areas are:

- Finance and control
- Processes used by the competitors
- Product portfolio
- Marketing aspects
- Sales and distribution
- Manufacturing
- Strategic alliances
- Supplier relationships
- Customer relationships
- Company personnel

It is important to identify the competitive advantage that exists in each area, compared with the competition. SWOT analysis can be used to compare the competitor against the company. The key success factors for the period covered by the strategic plan need to be identified.

Identifying company strengths and competitive advantage is not easy, and the company can be lulled into a false sense of security. For example, the company may have a very good and reliable product, but if it is operating in a mature market where all the competitors have equally good products, then this is no longer a competitive advantage. Rôle playing is important in order to avoid underplaying the competitor's strengths. Good market intelligence is also essential, which requires considerable time and resources to obtain.

Also key to market analysis is benchmarking (Mennon and Landers, 1987). The organization needs to benchmark its various operations not only against the competition but also against others who are operating in similar industries to itself. Some of these organizations may be customers or suppliers. Benchmarking will highlight strengths and weaknesses and will point the way towards improvements.

Another aspect of market analysis is product life cycle analysis (Day, 1986). This is covered in more detail in Chapter 16, as it is a key element of product management.

Technological and market forecasting is important when making strategic plans. This is covered in Chapter 9. Environmental changes which will affect operations need to be identified (Nair and Sarin, 1979). Computer-based tools are available for forecasting (Klein and Newman, 1980; Fahey *et al.*, 1981).

All plans must include sensitivity analysis: an indication of what the results would be like if items such as costs, time-scales and market share were different from the values assumed. How much do these need to change before the strategic plan no longer becomes attractive, and what is the probability that these unattractive changes will occur?

Gap analysis

Gap analysis is the technique for identifying the difference between the aims of the corporation and the expected results if no action is taken (Ansoff, 1972). This is illustrated in Figure 5.9. The aims of the corporation are directly related to its future vision, while the expected results are the expected outcome based on current plans. Note that gap analysis can be done in different areas; for example, the 'achievement' of Figure 5.9 could cover the profits that the company aims to achieve or the level of customer satisfaction that it plans to get.

Gap analysis deals with the following:

- Corporate aims. These must be realistic and attainable. (How many corporate aim curves slope downwards; that is, the organization aims to achieve less than it does at present? This may be required at times.) Aims must, however, also be ambitious, the corporation striving continually to better itself.

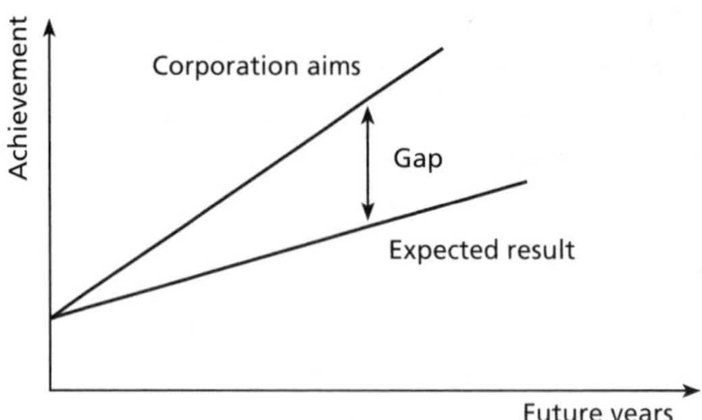

Figure 5.9 Gap analysis.

- Expected results. These are the results that are expected to occur on present performance standards, including any improvements planned for in the corporate strategy.
- Size of the gap. This is obtained by the difference in the corporate aims and the expected results. The gap is a key element, since it focuses attention on the size of the difference that has to be made up. If the gap is very large the corporation may decide that it does not wish to pay the price of taking action to reduce the gap. In these circumstances the corporate aims need to be downsized to reduce the gap.
- Method of filling the gap. This will depend on the item being covered by the gap analysis, although usually it will include growth alternatives, such as internal funding, acquisitions and alliances; investment considerations, such as in technologies and markets; and financial and operational considerations. Lateral thinking is essential when looking at ways of filling the gap (see the case study 'The sell off').

Financial analysis

This is a well-established technique and several tools exist (Day, 1986) to enable the company's financial position to be determined and understood. Tools are also available for cost, funding and control mechanisms, and to plan operating budgets (Varadarajan, 1984).

Financial considerations are covered in greater detail in Chapters 10, 11 and 12.

Group decision making

Decision making by groups is often more effective than by individuals (Delbeeq *et al.*, 1975; Johnston and Mendleson, 1982) and this is described in Chapter 6. The key is to ensure that different points of view are developed, debated, analyzed, refined and finally accepted by the group.

This technique can take up a considerable amount of management resource at all levels, and it requires participants with good analytical skills. However, it is good at eliminating bias and can also get past the problem where people withhold information because it would work against them, or because it supports a view which they do not agree with.

Case study

Value tradeoff

In its early trading days the FloRoll Manufacturing Company sold stands for florolls through a wholly owned subsidiary, Stand-U Ltd. It had a small share of the UK market for floroll stands and made a small but useful addition to the overall operating profit of the company.

Following the yearly corporate strategy review it was decided that Stand-U would aggressively attack the UK market in order to significantly increase its market share. Since the product was already priced competitively the decision was taken to gain competitive advantage by improving customer care. Stand-U had a modern factory which was currently used to less than 40 per cent capacity, so there was plenty of scope for increasing manufacturing capacity without needing further investment.

The number of salespeople for the product was sharply increased and a new department, called Customer Care, within the Customer Service Division, was set up with the aim of following up each customer complaint within twelve hours, with a personal visit.

It was recognized that these actions would add significantly to costs and would reduce the profit margin still further, but the increased volume of sales would compensate for this, driving down costs and increasing profits.

The policy worked for a short time. Market share went from 15 per cent to 17 per cent within six months. However, during that time other competitors were also preparing to increase market share and several had decided to cut prices. Although the customers were pleased with the improved service they received from Stand-U, they did not find this attractive enough to compensate for higher prices. Stand-U found that its market share slipped back to 15 per cent and, because it had raised expectations with customers regarding quality of service, it could not return to the original mode of operation.

Clearly the customers, one of the key elements of the strategic plan, were a beneficiary, but the shareholders, another element of the plan, lost out, since profits were now lower.

Case study

The sell-off

In order to improve the situation at Stand-U (see the case study 'Value tradeoff') the management of the FloRoll Manufacturing Co. decided to carry out a gap analysis to determine how its corporate aims regarding profit levels could be met. The gap was very large, and after several intensive sessions the team decided that the gap was unbridgeable. Since corporate guidelines regarding the level of profit which should be obtained from any operation could not be changed, it was concluded that the only alternative was to find a buyer for Stand-U.

The management team also did not have much time to devote to the problem, as they were busy thinking up a method for increasing the manufacturing capacity for florolls to meet a large five-year contract they had recently received.

The management were surprised when interest in the purchase of Stand-U was expressed by another manufacturer of florolls. It was obvious that the management of this competitor did not understand the intense competition which existed in the market for floroll stands. However, the sale was completed at a fair price, considering the relative low profit level currently being made by Stand-U.

The new owners of Stand-U decided to retain the policy of good customer care, which was obviously popular with customers. They even improved it by investing in an automatic logging and paging system, so that customer calls could be routed automatically to a duty engineer's home number. This meant that the company help desk did not need to be staffed on a 24 hour, 365 day a year basis, saving costs.

The biggest problem was reducing the manufacturing cost of the floroll stands. Stand-U had a modern factory which was considerably underutilized, and since the process used to make florolls was very similar to that for florolls stands, the company transferred some of its production of florolls to this factory.

This immediately reduced the overheads involved in making floroll stands, and therefore reduced their cost. Prices could now be dropped to match those of competitors without eroding profit margins.

Within one year Stand-U, under its new owners, had increased its market share from 15 per cent to 25 per cent and was turning in a very healthy profit.

Clearly the board of the FloRoll Manufacturing Company had missed an opportunity when they carried out their gap analysis of the Stand-U company; they could also have transferred the manufacture of some florolls to the Stand-U factory, which would also have solved their problem of how to meet the increased volumes for florolls, required on the new contract.

5.5 Plan implementation

The final steps in the strategy formulation process are to define the actions, allocate these within the organization and measure the results. The various goals which were set by the strategic plan can be considered to be long-term, medium-term and short-term in nature, and they will fall into the goals implementation pyramid of Figure 5.10. Long-term goals and actions will be few and of a relatively general nature, for example the goal to increase market share by 5 per cent.

Medium-term goals are more numerous and of a shorter term, for example the goal to develop two new products in order to increase market share by 5 per cent.

Short-term goals are more specific and can cover several actions, all supporting the long-term goal. For example: define products to meet market requirements; agree budgets; acquire software development skills for the new products; design new drive mechanism; prepare manufacturing plans; and define which technologies are to be bought in.

The strategic plan needs to be communicated to all levels within the organization, in a form which is meaningful to each level, showing their relevance and contribution towards it. An effective technique for this is Management by Objectives (MBO), as described in the next section.

It is also important to realize that if the plans call for change (and all good strategic plans should have a high element of change associated with them), greater effort is needed to ensure that it is accepted by all levels within the organization. There may be resistance to change, especially from those who feel threatened. If staff are involved early on in the decision-making process, this resistance is reduced since no surprises are involved. Relevance of any change must be clearly demonstrated and staff helped with accepting it.

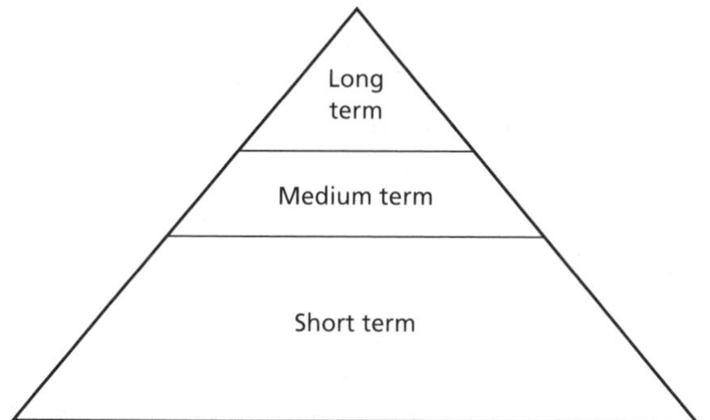

Figure 5.10 The goals implementation pyramid.

5.5.1 Management by objectives

This is an established technique (see also Chapter 18) for implementation of company strategic plans (Odiorne, 1965) and it ensures that management at all levels within the organization is committed to the same goals. The steps involved are:

1. Organizational goals are set by the executive, with involvement by others within the organization.
2. Objectives are agreed for all levels within the organization, which map into and complement the organizational goals. For example the organization's goal may be to enter the Middle East market. The Engineering Director may be set an objective to develop a product which is suitable for the Middle East market. The Design Engineer may then receive an objective to develop a material which can stand up to the high temperatures expected in the Middle East.
3. Progress is monitored against each objective and objectives are changed if the corporate aims change, or if the environment changes such that a different course of action is needed to meet the original aims.
4. Individuals are appraised against their objectives and feedback provided. Reward is usually associated with attainment of objectives.

In completing their objectives all personnel must work within company processes and procedures. Most procedures ensure equal treatment for all staff, such as personnel policies or buying goods from selected suppliers. Usually these will also help in attaining objectives, since new procedures will not be required for such activities as hiring staff or ordering materials. However, sometimes the processes and procedures may hinder, and if this occurs they need to be evaluated, since it is possible that the process or procedure is out of date and must be changed to keep pace with changes in the organization, strategy and structure.

5.5.2 Monitoring and measuring

Progress in the implementation of the strategic plan needs to be measured. In addition, the processes used should also be measured to ensure their continual effectiveness. It is important to have a mix of measures, both operational and financial, to get a balanced view of performance and to focus attention on key areas of the business (Kaplan and Norton, 1992).

The measures must be able to provide quick feedback to management on progress being made. In addition, they should be supported by more detailed information so that problem areas can be explored in greater depth.

Three key performance measures are considered here: customer-related, process-related and financial. It is important to appreciate that no measure can be considered in isolation since they affect each other. Also, by considering them all together management can see whether gains in one measure are real gains or whether they are being made at the expense of losses in some other measure.

Customer-related measures

The key company strategy could be to get closer to the customer so as to gain preferred supplier status. This can be translated into several actions which can be measured. Examples are:

- Reducing the time from receiving an order to delivering the product to the customer. This means that company internal processes, such as the order processing system, manufacturing, packing and delivery to the customer, must all be improved. Manufacturing will cover many items, such as allocating machines to the batch, holding stock and ordering raw materials.
- Improving the quality of products delivered to the customer in terms of reduced defects in goods inwards inspection. Other aspects of quality are delivering the product at the time promised; delivering the correct items; and delivering the correct amount.
- Reducing the cost to the customer. This involves whole life cost: the initial product cost, installation cost and maintenance cost.
- Improving customer care, such as responding to customer complaints in a short time and following up other customer queries, often with an on-site call.

Process-related measures

Generally these relate to internal company processes that ensure that customer needs are met. The company strategy may call for its processes to be one of the best in the industry sector in which the organization operates. Aspects of process improvement are:

- Reducing the time to market for new products. This involves all company processes, such as product development, manufacture and marketing. Clearly products can be brought quicker to market if they consist of incremental product releases only, rather than totally new products. However, linking this in with customer-related measures will show that the customer is not likely to be satisfied with incremental product releases and genuine process improvements will eventually be required.
- Quality improvements throughout the organization. Once again, when this measure is taken with others, it can be seen whether genuine quality improvements are being made or not. For example, manufacturing quality may increase if volumes fall. This will, however, reflect on the financial measures. True manufacturing quality improvements call for enhancements while maintaining full capacity.
- Lower costs, again in all stages of the development and manufacturing process, including overhead costs associated with indirects such as marketing and management. Again, measures taken together are important. For example, if the volume of manufacturing goes up the cost may fall, since the

overheads are now amortized over a greater volume. However, this should be disregarded to get at the true improvement in costs caused by process changes.
- Ensuring that the product meets customer requrirements, which is primarily a measure of the marketing department. It can be measured by considering the percentage of sales that comes from new products, discounting other factors such as sales lost due to late product development or delivery.
- Development of internal core competencies. These are vital to enable the company to survive and grow in a competitive market-place. Examples are technical knowledge of fibre optic technology and developing the capability to manufacture subminiature electronic assemblies.

It is important that the company strategy puts plans into place to cater for the effects of process improvement. For example, if the manufacturing process is improved this may result in excess capacity, which may require a programme of layoffs among staff. Clearly this is undesirable, since it penalizes employees for becoming more efficient. It would be better to put in place a sales and marketing plan to fill the excess capacity with new products.

Financial measures

The key goal of the financial strategy is to meet shareholder expectations. This can be translated into several measures, such as:

- Profit, measured in terms of return on investment
- Sales growth in terms of turnover
- Growth in market share
- Adequate cash flow to fund ongoing developments

Financial measures must be forward looking, not only based on historical results. They should look at activities which drive the financial results.

It has been argued that financial measures are not important since they would follow if the other two, customer measures and process measures, are successful. However, this is not always the case. For example, a process improvement can significantly improve the quality of the product, but if the customer is satisfied with the product as it is then this will not show through as improved financial measures, such as sales growth. The strategy then needs to be reassessed and the measures adjusted.

Summary

The corporate strategic plan aims to move the organization from its present position to where it wishes to be to meet its vision of the future. It is inevitably long range and may include shorter term tactical plans.

Strategic plans vary depending on the organization's aims and where it is in its development cycle. Examples of these are the entrepreneur stage, the expansion and consolidation stage, the diversification stage, and the decline and renewal stage.

Three key steps are involved in setting up a strategic plan: setting the vision, putting an action plan into place, and implementing the plan. When setting the vision it is necessary to take into account the mission of the company, its business plans, the environment in which it is operating, and its current position.

It is essential to ensure that the plan is realistic and attainable, while at the same time being ambitious enough to stretch the company. Resources to accomplish the plan will usually be the biggest problem, and every attempt should be used to leverage resources as much as possible, for example by prioritizing activities, reusing technology wherever possible and forming alliances. The strategic plan should include several aims, such as improving financial position, market development, technology and service development, process improvement, developing the company culture, and responsibility to the public. SWOT analysis should be undertaken to help in developing the plan.

Alliance and acquisition considerations should form part of the plan. Both need careful selection and management if they are to succeed. Alliances are best if the joint company is to operate in a new geographical area and is dealing in a new business. Acquisitions are preferred if the joint company is to operate in the same geographical area as its parents and the product under consideration forms a prime business of one of the partners.

There are several tools and techniques that may be used for strategy formulation, such as value analysis, market analysis, gap analysis, financial analysis and group discussions. In value analysis the driving forces in the business are identified, itemized and analyzed to arrive at the critical success factors, which are then acted on. For market analysis the key market-related factors are identified to determine new market niches or opportunities. Gap analysis seeks to determine the gap that exists between the company's vision of where it wishes to be and the actual position the company thinks it will occupy if no action is taken. Analysis of this gap is then undertaken to determine the actions that need to be put into place to close the gap. In financial analysis the company's financial performance is analyzed to determine strengths and weaknesses, and these are used in formulating actions. Group decision making can finally be used to arrive at a consensus view of the actions needed.

The final stage of strategic planning involves implementation of the plan of action. Short-, medium- and long-term goals should be identified and implemented. These need to be communicated to all levels within the organization and should form part of the objectives at these levels (for example Management by Objectives). The plan needs to be monitored to determine progress towards meeting these actions, based on a set of measures. These measures should be such that they can determine the effectiveness of the actions and are not masked by other changes. The key measures are customer-related, process-related and financial. Customer-related measures include reducing order processing time, improving quality of equipment delivered to the customer, reducing the whole life costs to the customer and improving customer care. Examples of process-related measures are reduced time to market for a new product, quality improvement throughout the organization (TQM), lower internal costs, ensuring that the product meets customer requirements and the development of internal core competencies. Financial measures include increased profit (return on investment), sales or turnover growth, increased market share and increased cash flow.

Case study exercises

5.1 The FloRoll Manufacturing Company has traditionally concentrated on the high-quality, high-price end of the market. Jake Topper, the Chief Executive, has been told by his market intelligence department that smaller competitors are poised to enter the lower cost end of the market. Jake Topper knows that there is very little profit to be made in this low end and is tempted to ignore it, especially as the FloRoll Manufacturing Company's cost base is such that it is not competitive at this end of the market. What should he do?

5.2 The FloRoll Manufacturing Company makes most of its sales to a few large customers. Jake Topper, the Chief Executive, wishes to increase the company's customer base, and realizes that this will require a change in the culture of the company. He needs to set policies for the company to follow. Should Jake set these policies on his own or involve other members of his staff, and if so whom? What steps should he follow in setting the policies and what areas should these cover?

5.3 The FloRoll Manufacturing Company wishes to enter a new market, which is dominated by a single large competitor, having 80 per cent market share. There are several other smaller competitors, who are primarily active in niche areas. What strategy should the FloRoll Manufacturing Company follow? Should it attack the main supplier, in the hope of getting a small share of its market, or should it concentrate on a niche area?

5.4 The FloRoll Manufacturing Company wishes to develop a new floroll range to replace its existing line. The cost of developing the technology for this product will be high and the company is considering its strategic options; that is, forming an alliance with a smaller company that has developed the technology it wants, or taking it over. It intends to continue to sell into its current markets with the new product. What should it do?

References

Almassy, S.E. and Baatz, E.B. (1992) 455 electronics execs say rugged individualism is fading, *Electronic Business*, 30 March, pp. 38–44.

Ansoff, H.I. (1972) *Corporate Strategy*, McGraw-Hill Book Co., USA.

Bidault, F. and Cummings, T. (1994) Innovating through alliances: expectations and limitations, *IEEE Engineering Management Review*, Fall, pp. 116–24.

Bleeke, J. and Ernst, D. (1991) The way to win in cross-border alliances, *Harvard Business Review*, November–December, pp. 127–35.

Bleeke, J. and Ernst, D. (1995) Is your strategic alliance really a sale? *Harvard Business Review*, January–February, pp. 97–105.

Boynton, A.C. and Zmud, R.W. (1984) An assessment of critical success factors, *Sloan Management Review*, No. 25, Summer.

Brandenburger, A.M. and Nalebuff, B.J. (1995) The right game: use game theory to shape strategy, *Harvard Business Review*, July–August, pp. 57–71.

Campbell, A. *et al.* (1995) Corporate strategy: the quest for parenting advantage, *Harvard Business Review*, March–April, pp. 120–32.

Cave, P. (1996) The lure of acquisitions, *Telecommunications*, May, pp. 34–6.

Collins, J.C. and Lazier, W.C. (1993) Vision, *IEEE Engineering Management Review*, Spring, pp. 61–75.

Day, G.S. (1986) *Analysis for Strategic Market Decisions*, West Publishing Co., USA.

Day, G.S. and Fahey, L. (1990) Putting strategy into shareholder value analysis, *Harvard Business Review*, March–April, pp. 156–62.

Delbeeq, A.L. *et al.* (1975) *Group Techniques for Program Planning: A Guide to Nominal Group and Delphi Process*, Scott, Foresman and Co., New York.

Donaldson, G. (1995) The strategic audit, *Harvard Business Review*, July–August, pp. 99–107.

Egan, G. (1994) Cultivate your culture, *Management Today*, April, pp. 39–42.

Fahey, L. *et al.* (1981) Environmental scanning and forecasting in strategic planning – the state of the art, *Long Range Planning*, February.

Gomes-Casseres, B. (1994) Group versus group: how alliance networks compete, *Harvard Business Review*, July–August, pp. 62–74.

GuideLines (1996) Making joint ventures work, GuideLines, Spring.

Hamel, G. and Prahalad, C.K. (1989) Strategic Intent, *Harvard Business Review*, May–June, pp. 63–76.

Hamel, G. and Prahalad, C.K. (1993) Strategy as stretch and leverage, *Harvard Business Review*, March–April, pp. 75–84.

Hamel, G. and Prahalad, C.K. (1994) Competing for the future, *Harvard Business Review*, July–August, pp. 122–8.

Hamel, G. *et al.* (1989) Collaborate with your competitors – and win, *Harvard Business Review*, January–February, pp. 133–9.

Heijden, Kees van der, *et al.* (1993) Strategy and the art of reinventing value, *Harvard Business Review*, September–October, pp. 39–51.

Hinterhuber, H.H. and Popp, W. (1992) Are you a strategist or just a manager? *Harvard Business Review*, January–February, pp. 105–13.

Johnston, D.L. and Mendleson, A.H. (1982) *Using Focus Groups in Marketing Planning*, West Publishing Co., USA.

Jones, W.G.T. (1991) Company strategy – a role for the electrical engineer? *Engineering Management Journal*, February, pp. 12–18.

Kanter, R.M. (1994) Collaborative advantage: the art of alliances, *Harvard Business Review*, July–August, pp. 96–108.

Kaplan, R.S. and Norton, D.P. (1992) The balanced scorecard – measures that drive performance, *Harvard Business Review*, January–February, pp. 71–9.

Klein, H. and Newman, W. (1980) How to use SPIRE: A Systematic Procedure for Identifying Relevant Environments for strategic planning, *Journal of Business Strategy*, Summer.

Lorange, P. *et al.* (1993) Building successful strategic alliances, *IEEE Engineering Management Review*, Summer, pp. 4–10.

MacVicar, D. (1996) Why your company needs a strategic plan, *Laser Focus World*, October, pp. 105–10.

Mennon, L.J. and Landers, D.W. (1987) *Advanced Techniques for Strategic Analysis*, The Dryden Press, USA.

Moore, J.F. (1993) Predators and prey: a new ecology of competition, *Harvard Business Review*, May–June, pp. 75–86.

Morley, M. (1997) What business are you in? *Professional Manager*, March, pp.12–13.

Nair, K. and Sarin, R. (1979) Generating future scenarios – their use in strategic planning, *Long Range Planning*, June.

Normann, R. and Ramirez, R. (1993) Designing interactive strategy, *Harvard Business Review* July–August, pp. 65–77.

Odiorne, G.S. (1965) *Management by Objectives: A System of Management Leadership*, Pitman Publishing Co., London.

Peridis, T. (1993) Strategic alliances for smaller firms, *IEEE Engineering Management Review*, Fall, pp. 66–71.

Stalk, G., Evans, P. and Shulman, L.E. (1992) Competing on capabilities: the new rules of corporate strategy, *Harvard Business Review*, March–April, pp. 57–69.

Varadarajan, P. (1984) The sustainable growth model: a tool for evaluating the financial feasibility of market share strategies, *Strategic Management Journal*, October–December.

Webster, J.L., Reif, W.E. and Bracker, J.S. (1989) The manager's guide to strategic planning tools and techniques, *Planning Review*, November–December, pp. 108–118.

6 Decision making

Introduction

This chapter introduces the key elements involved in management decision making, and describes how this differs from decision making within engineering. The process used, along with the techniques that aid in decision making, are described.

6.1 The nature of management decisions

Decision making is usually required to solve a problem. To an engineer or scientist this is a relatively straightforward process and one which is carried out many times during the working day. Therefore the new engineering manager may wonder why so many management textbooks have been written on the problems associated with decision making (Bierman, *et al.*, 1973; Brown, *et al.*, 1974; MacArmman, 1974; Patz and Rowe, 1977; Radford, 1975; Sanford and Adelman, 1977).

Two case studies, 'A hammer job' and 'More dinshers', illustrate some of the differences between management and scientific decisions. Management decisions usually affect many more people and need to be made in a changing and uncertain environment.

Case study

A hammer job

Mike Smith, mechanical designer within the Engineering Systems Group, examined the new model of the automatic winder, which he had designed for the Floroll Assembly Shop. The prototype had been on a life run for the past two weeks and had unexpectedly stopped working. A decision was needed and Mike mentally ran through the steps in decision making, which he had learned on a recent course.

- **Step 1:** Identify the problem. That was clear enough; one of the cogs that drove the spring mechanism had become entangled in the spring.
- **Step 2:** Identify the alternative solutions. Mike realized that he could redesign the cog so that it was of a smaller size, he could eliminate the cog altogether, replacing it with a belt and pully system; or he could bend the spring mechanism fractionally outwards, so that it did not obstruct the cog.
- **Step 3:** Choose the best solution to meet the objectives. The objective was to get the machine back on test and to limit any redesign. Mike decided to bend the spring mechanism. He took a hammer out of his tool box and gave it a sharp tap. The mechanism moved outwards, and the machine purred back into action.

Mike Smith returned to his drawing board and recorded the change to the design, the spring mechanism being moved to its new position.

Case study

More dinshers

Adrian Elton, the Manufacturing Director, had come under considerable criticism at Jake Topper's Executive Staff Meeting for the low output from the Dinsher Shop. Peter Sell, the Sales Director, had been particularly scathing, claiming that his sales targets were being missed because of a shortage of florolls and long lead times.

Returning to his office Adrian asked John McCaully to come to see him.

'John', he said when John McCaully arrived, 'we have a problem. I took considerable stick at Jake's meeting this morning because we have missed our output target of florolls again this month. I realize that we have come close, but close is not good enough. Peter Sell is again making political capital, claiming that we are holding up his sales. The main problem is the poor output of dinshers. Can you look at how this can be increased, and let me have your recommendation by tomorrow?'

John McCaully set to work defining the problem. That was clear enough: to increase the output from the Dinsher Shop. He decided to draw up a decision matrix of the alternatives, along with the impact of each decision on key factors. This matrix is shown in Table 6.1.

The simplest change would be to introduce overtime working. John estimated that this would increase output by 8 per cent although costs would go up by 15 per cent in the short-term and by 20 per cent if overtime working continued, since payments would need to be increased according to a union agreement. Overtime could be introduced within a week, but could only be continued for about six months, since after this time efficiency would suffer. Overtime payments would be popular with the operators, especially with Christmas approaching.

The second alternative could be to change some of the work practices, which had been discussed on several occasions with the operators and unions. This would increase output by about 3 per cent only and put up costs by 5 per cent. It would require about six months before the changed practices became fully effective and once introduced they should be effective for about two years. John knew that he would face considerable union resistance to the changes.

Another alternative would be to increase the number of operators and to work shifts, so as to utilize the machines fully. This would give a substantial increase in output, up to 20 per cent, although the cost would go up by 40 per cent in the short-term and by 50 per cent longer term, due to increased maintenance of the machines. Shift working could be expected to last one year only, since after that time it would have a detrimental effect on the machines, which would need replacing. New operators would need to be recruited and trained, so shift working would take about three months before it could be introduced. It would be popular with the unions, since their membership would increase, but would be less popular with existing operators.

The fourth alternative would be to introduce further automation into the Dinsher Shop. This would eventually give a 40 per cent increase in output, but it would take five years before this was achieved. Initially costs would be high, about 70 per cent, which included the cost of new machines and operator redundancy costs. Running costs, however, would fall long-term by about 20 per cent due to the increase in efficiency. The effectiveness of the automation programme would be expected to last ten years. As expected there would be considerable resistance from operators and unions when it was realized that redundancies would be involved.

Having completed the decision matrix, John McCaully realized that he could not come up with a firm recommendation without further discussions with Adrian Elton, since his decision would affect many people. What worried him were the following points:

- The criteria for selection were not clear. How important were the political considerations: the need to please Jake Topper and Peter Sell?
- Would Adrian Screw back him up in any confrontation with the unions, or was this to be avoided?

Table 6.1 Decision matrix for increasing the output from the Dinsher Shop

	Impact of decision					
Alternative decisions	**Output increase (%)**	**Cost to implement (%)**	**Running costs (long term) (%)**	**Time to fully implement**	**Period of effectiveness**	**Other factors**
Introduce overtime	8	15	20	1 week	6 months	Popular with operators
Change work practice	3	5	5	6 months	2 years	Union resistance
Increase number of operators	20	40	50	3 months	1 year	Union choice
Introduce automation	40	70	(–20)	5 years	10 years	Union/operator resistance

- How much money was available, both short-term and long-term?
- What increase in productivity was being looked for, how soon did it have to be introduced, and what was the period over which it had to be effective?
- What were the market conditions like? If it took five years to reach peak output, would the demand for florolls have fallen by then, so that the increase in output was no longer needed?

6.1.1 Management decisions

As illustrated in the two case studies above, the process used for scientific decision making is similar to management decision making. In both the steps are the definition of the problem; gathering facts related to the problem; comparing these with right and wrong criteria based on knowledge and experience; and then taking the best course of action.

Management decision making, however, is often an art rather than a science. The conventional theories of decision making, such as information collection, information analysis, choice of best alternative and then full steam ahead on the favoured solution, do not always apply (Efzioni, 1989). The main characteristics of management decision making are:

- The problem is often ill defined or unpredictable. Engineers and scientists work to clearly defined laws of science, whereas most management decisions are related to human behaviour, which cannot be accurately defined or predicted.
- Management problems are usually much wider in scope than technical problems, and affect many more people and functions. All of these have their own objectives, requirements and interests, and see the problem differently. Conflicts of interest and political considerations need to be taken into account in the decision making process.
- Gathering information on which to base a management decision is often a hazardous process. A vast amount of information is usually available, and one cannot hope to collect it all, let alone only the relevant ones. There is also the difficulty of knowing what is relevant to the problem. The problem is continually changing, due to market or people considerations. Much of the information is usually based on guesswork, rumour, opinion or hearsay. The danger is that those taking part in the management decision-making process often forget this and consider the mass of data obtained as being exact and unchanging, and therefore expect decisions based on it to be accurate.
- The management decision, like the scientific one, is essentially one of choosing between several alternatives. This is easy to do if one can determine the consequences of each alternative and they do not change with time. Neither of these applies to management situations, which change with time

and the consequences of each alternative are often difficult to predict. There is also rarely a 'best' solution in management problems; they all involve compromises, and usually the manager is reduced to picking the best compromise and being aware that this may not be the best choice at some time in the future. It is highly probable that what is considered to be the best solution today may turn out to be the worst choice in the future. Managers usually act on part information and lots of gut feeling. They proceed with caution and continually check the consequences of their decision, being prepared to alter it if required by circumstances.

- Once a decision has been made it requires consensus and commitment from the people who are affected to implement the solutions. Political considerations then become paramount, since different factions effected may have different objectives.
- Following implementation the decision must be continually monitored to see whether it is still valid in a changing environment.
- The process of management decision making cannot be learned. Most of it is based on experience and judgement and must be acquired, usually in different management positions.

Figure 6.1 graphically illustrates the nature of management decision making. The problem has a set of alternative solutions, and within these there is probably the optimum choice. However, the problem itself changes with time and so do the solutions. The problem may get better or worse if left long enough. Even if the

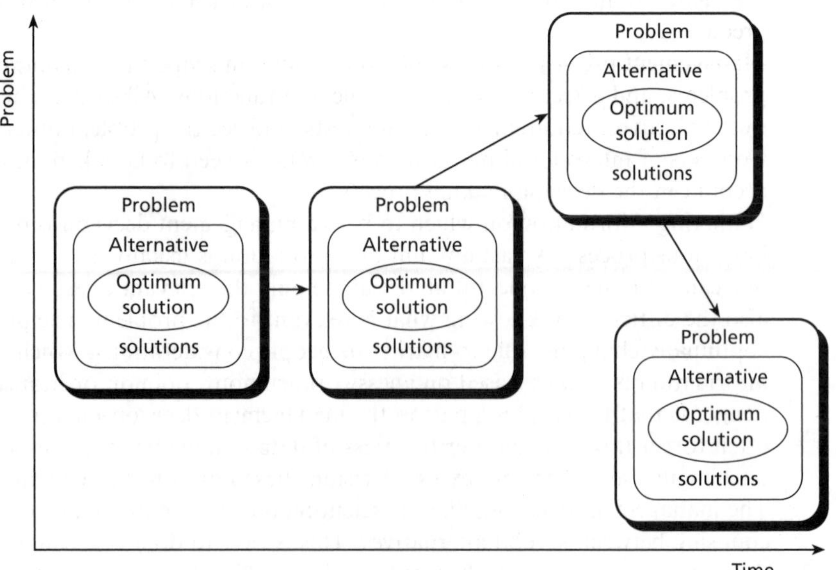

Figure 6.1 The problems of management decision analysis.

problem does not change the optimum solution probably will. For example, the problem may be defined as the need to sell more product, in order to increase market share. However, in time the market may become saturated, so this problem becomes difficult to solve. The problem itself may then change to one of increasing profitability by reducing product costs. Alternatively, the problem may initially be one of increasing profit margins, and the best solution for this may be defined as one of increasing price. In time a competitor may enter the market and make this a bad choice, so the solution may need to be redefined, for example to one of reducing price to maintain market share, but increasing advertising so as to grow the overall market size.

Decision-making styles are often affected by the organization in which a manager operates. For a strict mechanistic organization the roles and responsibilities of individuals are well defined and decision making is a routine process, where staff respond automatically in a prescribed way (Kay, 1982; Ashford and Fried, 1988). This results in efficient and quick decisions, with unexpected events being fed up the organization for decisions. The problem arises when events change, such as an unexpected change in process forced by circumstances. This would result in a considerable number of new decisions, which would all need to be handled by executives. The top-level decision-making structure would become overloaded (Burnes and Weeks, 1989) with senior management spending their time on small decisions (fire fighting) rather than on strategic ones.

Decisions can be divided into those which are routine and those which occur infrequently. Examples of routine decisions are the purchase of materials, with the decision on the quantity to order so as to obtain the best discounts. Routine decisions can usually be defined by a set of rules and are often automated, for example by computers. They can be taken at much lower levels in the organization, although in many cases they may have a major financial impact on the company. Decisions of an infrequent nature, which also usually have a major impact on the organization, need to be taken by senior management.

Figure 6.2 shows a simple flow diagram of the choices facing management when a problem arises. If the decision is not critical to the business, and it is of a routine nature, it can be delegated. If it is a critical decision, or if it is not of a routine nature, then it usually needs the manager's involvement. If managers do not feel that they have all the relevant information on which to base a decision, they would usually involve others. If they have all the relevant information then they might still wish to involve others, for example in order to gain commitment for implementing the solution, or as a training exercise for their staff.

It is often said that in management any decision, good or bad, is better than no decision at all. This is probably due to the fact that some managers, faced with the bewildering array of choices within an uncertain environment, are paralyzed and cannot act. However, the first question a manager needs to ask is: 'Is this decision necessary?'. If the decision can be deferred without detrimental effect then it should be.

Not all decisions need to be made on the spur of the moment, with managers throwing their full weight behind it to see that they are implemented immediately.

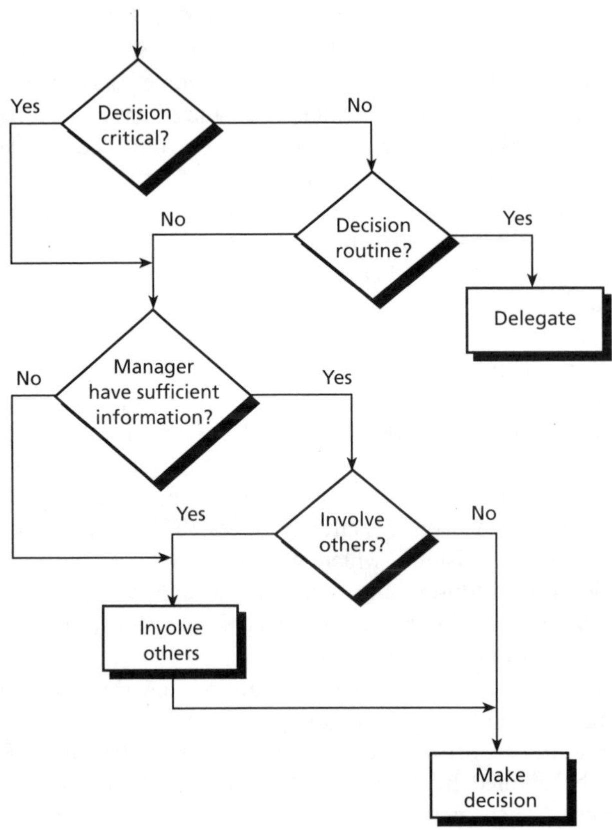

Figure 6.2 Management decision flow diagram.

Most decisions made in this manner are incorrect and only strive to maintain the *status quo*; that is, they correct the immediate problem and return to circumstances as they were before the problem occurred. The problem may, however, only be a symptom. What may be required is an entirely new strategy and direction for the organization, and managers need to buy time to ensure that they can reflect on the best way forward.

6.2 The decision-making process

The basic decision-making process has already been mentioned, and it is illustrated in Figure 6.3. The first task is to identify whether a decision is really required. Every decision introduces a change and, depending on the extent of the change, can result in a shock to the organization. Often this shock can be beneficial, but sometimes it can be disrupting.

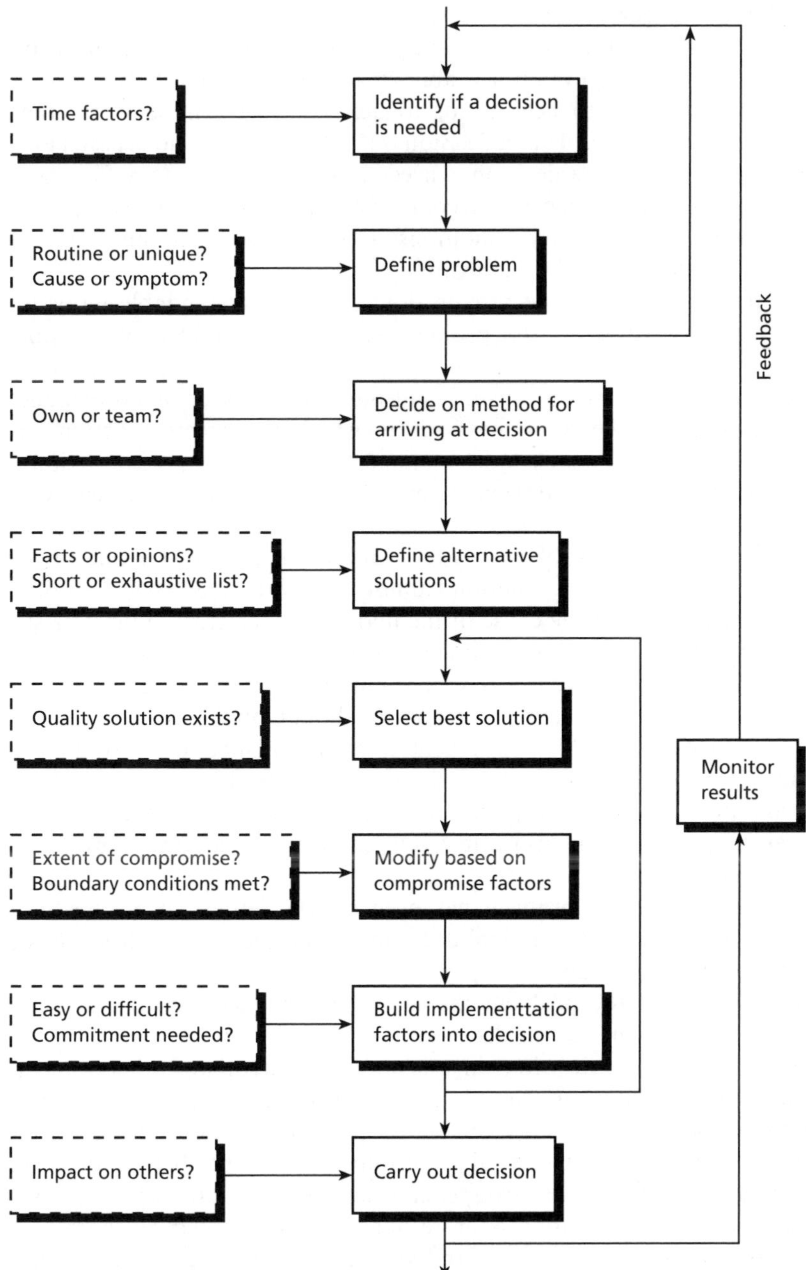

Figure 6.3 The decision-making process.

If things are going well then they should be left alone. No decision should normally be taken if the problem is likely to go away on its own, or if the problem is relatively insignificant, even if it is annoying. It is important, however, to make the decision on whether to act or not to act. Nothing can be worse than sitting indecisively on the fence. Half a decision is much worse than no decision at all.

Having decided that a decision is needed, the next step is to define the problem: what does the decision have to accomplish? This definition must included a clear statement of the boundary conditions: those factors that must be satisfied in order for the decision to be successful.

The problem needs to be broken down into understandable terms and small units. The true cause of the problem should be sought. For example, if the sales are low this may be due to the product costs being too high; the sales force being poorly trained; insufficient advertising used; or no demand for the product. One of the key questions is whether corrective action or a new strategy is needed. Usually most managers settle for corrective action, treating the symptoms of the problem, rather than getting at the cause, which may require a new strategy.

Problems are usually of two types, routine or unique. Unique problems are those for which a generic solution cannot be determined. They also often reveal the symptoms and not the cause of the underlying defect. (See the case study 'The case of the missing filters'.) Routine problems usually reveal the true cause of the defect, and for these there is a generic solution.

In defining the problem the manager should always first assume that the problem is routine and look for a generic solution. Once such a solution has been found it can be applied to many similar problems. Unique problems require a solution and a decision every time they occur; it is better if one is able to collect several unique problems into a few generic ones, so requiring fewer solutions to be implemented.

Note that, once the problem has been clearly defined, it is possible that the manager may reverse the original decision and decide that no decision is required in the present circumstances.

Having defined the problem the manager next needs to decide on the best method for arriving at a decision on its solution. Usually the choice is between making the decision on one's own or involving others, and a simple scheme for deciding on this was introduced in Figure 6.2. This will be elaborated on in the next section. It is important to note, however, that there is no evidence to suggest that groups arrive at better decisions than individuals in all cases. Groups can ensure that all the facts have been considered, and the solution is more likely to be accepted by those taking part in the decision-making process. Individual decisions are quicker to implement and may sometimes be better, since they do not need to involve a compromise between those taking part.

Alternative solutions to the problems now need to be defined. These may be as a result of fact collection or may be based on opinion (hypothesis). It is important,

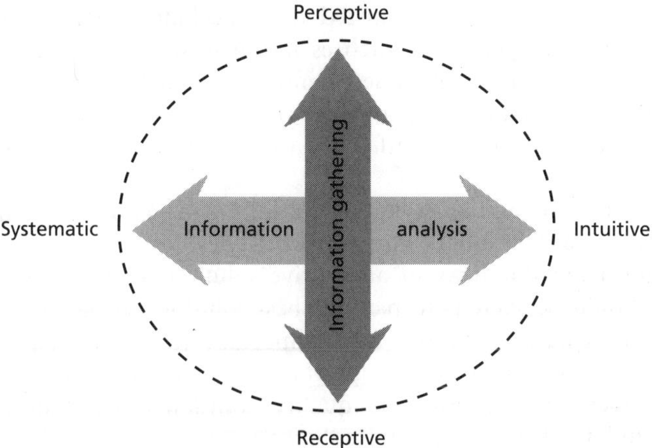

Figure 6.4 The cognitive map.

however, that these opinions are tested by facts, and those voicing the opinions should also be assigned the task of determining how best they can be tested.

Depending on the problem the manager may decide to produce an exhaustive list of solutions or only a few likely ones. Having several alternatives is usually useful since it provides a quick fall-back if the chosen solution cannot be used for any reason. It is important that 'organized conflict' exists during this stage of the process. Disagreement between the members taking part in the decision making is the most effective way of ensuring that all the facts are aired and considered, so avoiding the pitfall of a decision being made based on a preconceived idea of what the solution should be. Managers must see things from other's perspective, not just assume that their own solution is right.

McKenney and Keen (1974) have defined the cognitive style of a manager, which is how information is collected and analyzed in arriving at a decision. Most management styles can be plotted on a scale, as shown in Figure 6.4, and they vary between perceptive and receptive information gathering, and between systematic and intuitive information analysis.

In perceptive information gathering the manager builds up a picture of what information is being sought as the data are obtained, being guided by this perception. The analysis is partly done as the data collection progresses, so the decision-making process may be shorter. However, the danger is that much of the relevant data may be overlooked if it does not fit the manager's perception of what should be collected.

In receptive information gathering the manager first collects all the data and treats all this as relevant information. At this stage no attempt is made to see whether the data fit a perceived solution. After all the data have been collected they are analyzed before a solution is formulated.

In systematic information analysis the manager sets a plan for analysis of the data and sticks to it, being well organized and checking progress frequently against the plan. The manager who practises intuitive information analysis, on the other hand, prefers to start with one or more hypothesis and to match the data against these. Any data not matching the solution are discarded. The manager will often rely on hunches rather than on a strict interpretation of the information.

There is no one 'best' cognitive style, and usually if a team is working on a problem it is better to have members with a mix of styles.

The best solution from the array of alternative solutions is now selected. The selection process should be such as to pick the best solution and not be based on picking the most acceptable solution. It is important to know what the best solution is so that the impact of any compromises, introduced at the next stage, are known. Often there is no one unique 'quality' solution which is much better than any other, and in this case several solutions can go forward to the next stage.

The solution can now be modified to incorporate any compromises, such as the need to gain commitment from people to have it implemented. The extent of the compromise needs to be clearly defined. Although it is generally true that 'something is better than nothing', it is essential that the solution, after the compromise, still meets the boundary conditions set when defining the problem, since these were the minimum conditions that any acceptable solution needed to fulfil.

The implementation factors now need to be built into the decision. This is one of the most important stages in the decision-making process and often takes the longest time. A decision is worthless if it cannot be implemented, and the actions needed to carry it out must be part of the decision and not added later. One of the key considerations at this stage is the level of commitment needed from others in order to carry through the decision.

It is possible that considerations of implementation factors may require that the original selection of the best solution be re-examined, and this process will be repeated, as shown in Figure 6.3.

The final stage in the decision-making process is carrying out the actions. Often a manager will need to live a long time with the consequences of decisions, and it must not be assumed that changes can be introduced rapidly if a bad decision has been made. Therefore managers must ensure that they are happy with their decisions, and if they feel uncomfortable they should pause momentarily to reflect on the implications, and whether any factors were overlooked when arriving at the decision. However, this reflection period must be short, and if managers still believe that it is the right decision, they must ensure that it is followed through with commitment, even though it may not be a popular decision.

Once a decision is implemented the results from it must be constantly monitored. This will provide feedback on the original problem, which may need to be redefined.

Case study

The case of the missing filter

The new winding machine, designed by the Engineering Department, was extremely versatile. It was used within the Dinsher Shop and performed very well, except for the fact that it seemed to break down regularly, once a week.

'Seems like it has a clock inside it', joked Sam File, the maintenance engineer. 'Every week, come rain or shine, it stops. I could almost set my watch by it. Mind you, it doesn't take long to fix, just a quick clean out inside, but it is a nuisance.'

The problem remained a nuisance for some time. It was clearly a unique problem and the decision had been taken to treat it as such and to repair it whenever the need arose. The symptoms were being treated.

It was many months later that Sam met his counterpart in the Floroll Assembly Shop and discovered that they had the same problem with their new winding machine. 'Must be a time bomb in these new machines', muttered Sam, when he told 'Big' Joe Bull the story that afternoon.

All at once the unique problem was no longer unique. Joe contacted the Engineering Department. They surveyed all the ten users of their new machine and found identical symptoms. An investigation was started into the routine problem; the cause and not the symptom was about to be treated. It was discovered that dust was getting into the delicate balance mechanism and upsetting it. The generic problem was rectified; a filter was introduced into the design which cured the problem.

6.3 Decision-making techniques

Several mathematical tools are introduced in Chapters 7 and 8 for making decisions in specific situations. Examples of these are inventory control decisions; queue analysis for waiting time problems; and competitive strategy decisions in the theory of games.

Other techniques include decision trees and tables. Table 6.2, for example, shows a simple decision table which may be drawn up by the FloRoll Manufacturing Company to define the conditions under which a customer will be granted credit. The three considerations are: whether the customer has done business with the company before; whether it is a large organization, having a turnover above a prescribed value; and whether the customer has been recommended by an existing client which credit facilities. The accept or reject row indicates the combinations under which the customer's application for credit would be accepted. Routine problems, such as this one, can be programmed into a computer, so that the solution is rapidly obtained.

Table 6.2 Decision matrix on the criteria for credit

Question	Options 1	2	3	4	5	6	7	8
Done business before?	Y	N	Y	Y	Y	N	N	N
Turnover exceeds defined value?	Y	N	Y	N	N	Y	Y	N
Recommended by an existing client?	Y	N	N	Y	N	N	Y	N
Accept/reject?	A	R	A	A	R	R	A	R

Decision-making techniques usually adopt a two-stage process:

1. Determining the method for arriving at a decision, such as whether to involve others or not.
2. Arriving at the decision.

It was Maier (1963) who first defined the decision-making process in terms of the quality of the decision and the acceptability of the results to affected parties, who have to carry out the actions resulting from the decision.

Vroom and Yetton (1973) developed this quality–acceptability model further, and defined a method for deciding when to involve others in the decision-making process.

Kepner and Tregoe (1965) developed a method for the actual decision-making and problem analysis process.

6.3.1 The Vroom–Yetton model

Vroom and Yetton proposed a set of possible management decision-making styles, as shown in Table 6.3. These styles range from very autocratic (AI) to democratic (GII). Most managers use a variety of styles, depending on the circumstances and the Vroom–Yetton model proposes a set of seven questions, which provide a guide of the type of style to be used. These questions are given in Table 6.4.

Table 6.3 Management decision-making styles (Vroom and Yetton, 1973)

Style	Description
AI	The manager makes the decision entirely on his or her own
AII	The manager first collects information from others (possibly telling them why the information is needed)) and then makes the decision on his or her own
CI	The manager discusses the problem individually with others and gets their input before making the decision on his or her own
CII	The manager discusses the problem with a group of people (usually the most affected) and gets their collective input before making the decision on his or her own
GII	The manager discusses the problem with a group of people (usually the most affected) and lets the group make the decision. The manager does not try to influence the group to accept his or her solution

Table 6.4 Questions involved in the Vroom–Yetton method

Question	Description
1	Is one solution lilkely to be much better than any other; that is, is there a 'quality' solution?
2	Does the manager have enough information to make a quality decision?
3	Does the manager know what information is missing and where to find it?
4	Is commitment to the decision by others important for its implementation?
5	Will others accept the solution if they are not involved in the decision-making process?
6	Do all the affected people share in the same goals?
7	Will there be a conflict between the affected people regarding alternative solutions?

The questions are supposed to be asked for any situation in which the manager needs to decide on whether to involve others in the decision-making process. They provide a guide through a decision tree (a part of which is shown in Figure 6.5) which leads to the correct style.

If several alternatives are equally good then the choice is usually made on the basis of the least amount of effort (person-hours) involved.

A simple example of the use of the Vroom–Yetton technique is illustrated in the case study 'Feeding the line'.

Case study

Feeding the line

To ensure that the minimum of raw material is kept on the shop floor, Fred Shrew has recently been appointed Materials Handler for the dinsher line. His task is to keep the operators continually fed with raw material, so that operators have up to four hours of material at their stations. No further stock is to be held on the shop floor.

When Fred started his job one of the first decisions he had to make was how to determine the quantity of material to order, on a daily basis. He decided to use the Vroom–Yetton method, and started answering the seven questions given in Table 6.4. To the first question Fred answered 'yes', since obviously there was one quality solution: the one which gave minimum stockholding. He therefore progressed to point A in the decision tree of Figure 6.5. To the second question Fred answered 'yes', since he felt that he had all the relevant information. The men and machines were all working at a predictable rate, so he could tell exactly how much material was needed. This took Fred to point B on the decision model of Figure 6.5 and to question 4. To this question Fred answered 'no'. All the operators were committed to producing maximum output, since it affected their group bonus, and they would therefore accept Fred's decision if it kept the lines fully fed. This led Fred to adopt style AI, as indicated from Figure 6.5.

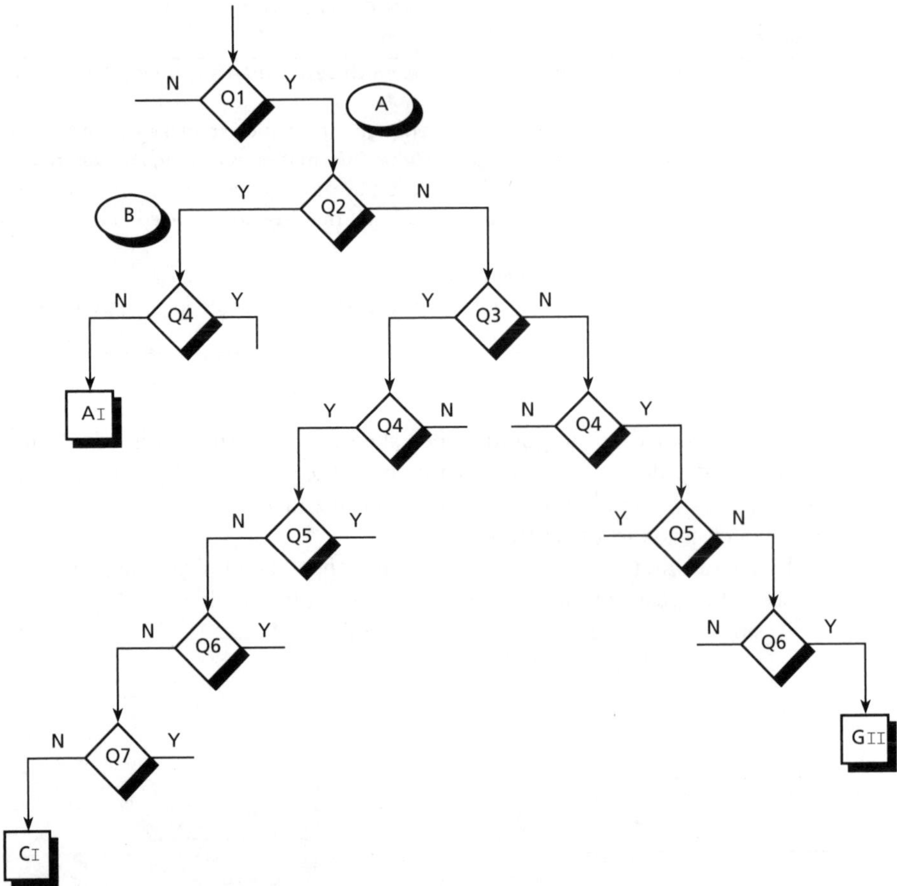

Figure 6.5 Example of part of the Vroom–Yetton decision model.

6.3.2 The Kepner–Tregoe method

The Kepner–Tregoe method for decision making and problem analysis specifies a series of steps to be followed when analysing a problem, as follows:

1. Specify the problem both in terms of what it is and what it is not. For example if the problem is only found to occur between the hours 10:00 p.m. and 2:00 a.m., then it does not occur at other times between 2:00 a.m. and 10:00 p.m.
2. Identify the differences (distinctions) between what the problem is and what it is not. For example, what happens between 10:00 p.m. and 2:00 a.m. that does not occur at other times?
3. Look for the causes (solutions) that explain these differences (distinctions).

4. Test the cause (solution). Compare what actually happened with what would have happened if the proposed cause (solution) had been in action. If the proposed cause (solution) explains what actually happened and did not happen, without requiring too many assumptions, then it is correct.

The following case study 'A hot stand', illustrates the use of the Kepner–Tregoe method.

Case study

A hot stand

A year after the new filters were fitted to the winding machines another problem was discovered: the machine in the Dinsher Shop kept breaking down at regular times. Sam File, the maintenance engineer, checked on the other winding machines being used within the plant and discovered that his problem was unique this time; all the other machines showed no problems.

Sam called in the Engineering Department. After extensive testing they could not find any cause for the regular breakdowns. The problem got to John McCaully, manager of the Dinsher Shop, who decided to try the Kepner–Tregoe method for tracking down the problem.

First John specified the problem in terms of what it was and what it was not, as in Table 6.5.

Table 6.5 Specification of the problem for the case study 'A hot stand'

What is it?	What is it not?
Machine breaks down between 2:00 p.m. and 3:00 p.m.	Machine does not break down at other times
Machine breaks down in the Dinsher Shop	Machine does not break down in any other area
Machine breaks down on Friday	Machine does not break down on any other day
Machine breaks down only after it has been making 'shewers'	Machine does not break down when making any other item

John then defined the differences between what the problem was and was not, taking the items from Table 6.5 and putting the results into Table 6.6.

This looked promising. John looked for causes that explained the differences in Table 6.6. It was clear that apprentice Mike Fly played a contributing rôle in the problem. To test this John had a video camera installed, which recorded Mike's movements while he was at the machine. On analysis it was discovered that, because of the shorter shift time available to apprentices on Friday, Mike Fly had fallen into the habit of eating his lunch (sandwiches) while still standing by the winding machine. One corner of the machine had a convenient dip in it for holding Mike's hot tea mug.

Table 6.6 Differences between 'is' and 'is not' (from Table 6.5)

Item	Difference
What is the difference between 2:00 p.m. and 3:00 p.m. and other times?	It follows the start of a new shift (2:00 p.m.)
What is the difference between the Dinsher Shop and other areas?	The Dinsher Shop is the only one to use apprentices on the assembly line
What is the difference between Friday and other days?	Apprentices work a short shift, leaving for college at 2:00 p.m.
What is the difference between the manufacture of 'shewers' and other products?	Only a few 'shewers' are needed and they are all made by apprentice Mike Fly.

To test the cause John went out onto the shop floor and placed a hot tea mug in the same position. Sure enough, the machine broke down half an hour later. Engineering were informed and traced the problem to the heat from the mug vaporizing the drive fuel, which was stored in a small tank under the cover. The fuel vapour affected the electronics within the machine and caused it to stop. It took about half an hour for the vapour to affect the electronics.

John decided to ban the consumption of food and drink from the shop floor, and provided a special area for this, a practice which was soon followed throughout the plant.

Summary

The basic steps used for engineering and management decision making are similar, although management decision making is much more complex because:

- Management problems are often ill defined or unpredictable, since they relate to human behaviour, which cannot be accurately predicted.
- Management problems are wide in scope and affect many people, all of whom have their own objectives, requirements and interests, and so see the problem differently.
- Gathering information on which to base a solution to a management problem is difficult, since the amount of data is usually large and is continually changing as market conditions change. Therefore much of the information is based on guesswork, rumour, opinion or hearsay.
- The consequences of management decisions are often difficult to predict and the 'best' solution will change as circumstances change.
- Management decisions require consensus and commitment from the large number of people affected, for successful implementation. Different political agendas may make this difficult to do.

Under these circumstances the decisions taken must be continually monitored and changed, if circumstances require it.

The first step in the decision-making process is identifying whether a decision is required; if not it should be postponed until later, when more information will be available to help in decision making. If a decision is needed it must be clearly defined. Managers need to decide whether to take decisions on their own or involve others. If it is a critical, non-routine decision, and managers have sufficient information to make the decision, and do not need 'buy-in' to get it implemented, then they should make the decisions on their own. Alternative solutions to the problem then need to be determined and the best solution chosen. This may subsequently be modified based on other factors, such as the need to meet political considerations, but the boundary conditions needed to satisfy the original problem should still be met. The implementation factors then need to be built into the decision and these acted upon. The results from these actions should be continually monitored to ensure that they meet the original problem, which may change with time.

Several techniques have been developed to assist in making management decisions. A simple technique is to draw up a decision matrix of the critical factors and to weight these to arrive at a solution. The method asks a series of questions and based on the replies to these it enables the manager to decide on the style to be used in solving the problem, the style varying from autocratic to democratic.

The Kepner–Tregoe method for decision making and problem analysis specifies a series of steps to be followed: specify the problem in terms of what it is and what it is not; identify the differences between these two; look for the causes that explain the differences; and test the cause (solution).

Case study exercises

6.1 Jane Furley, manager of the hardware design section, realized that her team would be needing a second workstation within the next twelve weeks if they were to complete the new design on schedule. The key decision required was whether the workstation should be rented or bought outright. Should Jane make the decision on her own or ask the design engineers? Jane believes she knows what type of equipment is best for the team and also has a good view of the team's forward workload.

6.2 Micro Dowling has been asked by Jake Topper to send one of his engineers to Australia for a month, to help with a floroll installation problem. Micro shortlists three members of his staff who he knows could do the job. He is certain that all three would jump at the chance of going abroad. However, they work for three different managers within his organization and Micro knows that these managers would be loath to lose their staff, especially as the projects are all at a critical phase. How should Micro Dowling make the choice?

6.3 On the recommendations of Jake Topper, Peter Sell has called in a group of management consultants, to advise him on the best way to organize his sales team for maximum effectiveness. The consultants have made their recommendations and Peter realizes that it affects several of his staff, who may oppose the changes. Peter is not completely convinced that the proposals which have been made are the most effective, but he feels that they should be implemented, at least in part. How should he proceed?

6.4 James Silver, Materials Manager, has received several complaints from his staff regarding the poor conditions within the procurement office. He has obtained Adrian Elton's approval to spend a limited amount of money on refurbishing the area. He has asked his staff for written recommendations on the changes which they consider to be high priority, and has received conflicting inputs. James believes he knows which changes are most urgently required and he also thinks that he can get them done at a very economic price by a local firm. How should he proceed?

References

Ashford, B.E. and Fried, Y. (1988) The mindlessness of organizational behaviours, *Human Relations,* **41**(4), 305–29.

Bierman, H.J. *et al.* (1973) *Quantitative Analysis for Business Decisions*, Richard D. Irwin Inc., Homewood, IL.

Brown, R.V. *et al.* (1974) *Decision Analysis for the Manager*, Holt, Reinhart & Winston, New York.

Burnes, B. and Weeks, D. (1989) *AMT: A Strategy for Success?*, NEDO, London.

Etzioni, A. (1989) Humble decision making, *Harvard Business Review*, July–August, pp. 122–6.

Kay, N. (1982) *The Evolving Firm: Strategy and Structure in Industrial Organization*, Macmillan, London.

Kepner, C.H. and Tregoe, B.B. (1965) *The Rational Manager*, McGraw-Hill, New York.

MacArmman, K. (1974) Managerial decision making. In *Contemporary Management*, Joseph McGuire (ed.), Prentice-Hall, Englewood Cliffs NJ.

Maier, N. (1963) *Problem-Solving Discussions and Conferences*, McGraw-Hill, New York.

McKenney, J. and Keen, P. (1974) How manager's minds work, *Harvard Business Review*, May–June, pp. 79–90.

Patz, A.L. and Rowe, A.J. (1977) *Management Control and Decision Systems*, John Wiley & Sons, New York.

Radford, K.J. (1975) *Managerial Decision Making*, Reston Publishing Co., Reston VA.

Sanford, E. and Adelman, H. (1977) *Management Decisions: A Behavioural Approach*, Winthrop Publishers, Cambridge MA.

Vroom, V. and Yetton, P. (1973) *Leadership and Decision Making*, University of Pittsburgh Press, Pittsburgh.

7 Information presentation

Introduction

As seen in Chapter 6, a large amount of information needs to be gathered and analysed during the decision making process. This chapter describes techniques which are available for presenting this information in order to make analysis easier.

7.1 Statistical analysis

The British Prime Minister Benjamin Disraeli is supposed to have said, 'There are lies, damned lies and statistics'. The fact that this has been so frequently quoted is evidence that statistics, misapplied, has gained a poor reputation.

Yet statistics is widely used in every branch of government and industry. So what is statistics and what is meant by the term 'statistical analysis'?

Statistics is the technique for comparing numbers and drawing conclusions from them. Two important factors are involved: using numbers to arrive at a solution, and comparing numbers. Usually a number in isolation provides very little information.

For example, suppose a Managing Director is told that his company made a profit of £200M in the previous year. By itself this number does not provide sufficient information for him to judge whether the company has been successful. If he is told that his company made profits of £50M, £100M and £200M over three consecutive years, then he can see that it is doubling its profits every year and so appears to be in good shape. However, if he is then told that a close competitor made profits of £30M, £90M, and £270M over the same three years (thus trebling its profits every year), then the MD can see that his company was underperforming in a climate which was clearly conducive to profit growth.

To be able to compare numbers they must be measured in the same way, be in the same units and must refer to the same items. For example, one cannot meaningfully compare the profits of one company with the turnover of another. If Company A is stated to have had a turnover of £2000M and Company B made a profit of £100M, then it is not possible to deduce which company was more profitable over this period.

Statistics was probably first used to record population details (called vital statistics), as required by rulers for raising taxes and armies. The Domesday Book of 1086, for example, was compiled for this purpose. Statistics is now used for a variety of reasons. Governments use it to plan social services, such as the number of schools needed, based on the birth rate. Within industry, statistics is used for competitive analysis; to carry out market research prior to the launch of a new product; and to gauge the quality of products.

It is important to appreciate that statistical analysis is a management tool, one of many tools available in decision making. The numbers which the technique provides cannot lie, but they are open to misinterpretation. For example, statistical analysis shows that more people die in bed than anywhere else. The conclusion can therefore be drawn that bed is the least safe place to be in!

Many errors, including bias, can occur in collecting and compiling data. A production manager in a large organization, for example, claimed that he was now producing 10 per cent more output, in monetary terms, with 20 per cent fewer staff, so that the productivity of his department had increased. What he neglected to say was that there had been a world shortage of the product, enabling the selling price to be doubled over this period, so in reality his productivity had fallen.

Companies obtain data from several sources. Government departments carry out surveys, such as into economic trends, the retail price index and the index of unemployment, and these are usually available for a modest fee. Several professional and commercial institutes also publish data relating to specific industries. Examples are share price indices, the salary range for members of a professional institute, and their qualification spread.

Companies also generate data, mainly for their own use. Examples are data on orders received, advertising effectiveness, labour turnover and quality measures such as project completion times and defect rates.

7.2 Presentation of data

The first step in information analysis is to place the numbers that have been collected into some sort of order, so that they can be more easily compared and conclusions obtained. Often these numbers can be presented so as to make a visual impact, by displaying the data in graphical format.

The items which are measured are referred to as variables. There can be two types of variable, discrete and continuous. Discrete variables are measured as single units, for example the number of people working in a factory, or the number of companies in a particular industry. Continuous variables, on the other hand, can have an infinite number of variations. For example the length of a piece of string can vary continuously between any two limits.

7.2.1 Tables

The use of tables is probably the most common technique for ordering numbers so that they are easy to see and compare. There are no hard and fast rules on how to present information in tabular form. Considerable imagination is needed to ensure that the tables are easy to read and achieve their purpose. The aim must be to keep the table as simple as possible and to ensure that it presents information such that the item on which a decision is needed can be easily identified. Often the position of columns relative to each other is important, since comparisons are then easier to make.

It is easy to read off the precise number from a table, which is not possible from other graphical methods of number presentation, as described in the following sections. However, the table is not as good for creating immediate visual impact. An example of a table is given in the case study 'Automating the Dinsher Shop'.

Case study

Automating the Dinsher Shop

A critical component used in the production of florolls is a dinsher. Because of its importance the FloRoll Manufacturing Company produces all its own supply of dinshers, rather than buy them in. The Dinsher Shop is now planning a four-year automation programme. It currently has 820 staff in total, made up of 500 operators, 200 production engineers and 120 supervisors.

The large number of supervisors and production engineers are needed because of the low level of automation and the need to set up each run individually. These support staff will be required right up to the end of the automation phase, since they will be involved in introducing the specialized equipment.

The manager of the Dinsher Shop estimates that he can automate the line in stages over a four-year period. At the end of the first year (the start of the second year) the number of staff will fall to 720, of which 420 will be operators, 190 will be production engineers and 110 will be supervisors.

The start of the third year will see a fall to 560, with operators, engineers and supervisors being 280, 180 and 100 respectively. At the beginning of the fourth year these will be 190, 160 and 80, giving a total of 430.

Automation will be complete by the end of the fourth year, so the production engineers and the supervisors can now be sharply reduced. The total number of people in the Dinsher Shop will now be 200, made up of 140 operators, 40 production engineers and 20 supervisors.

The manager of the Dinsher Shop has been asked to present these figures to the executive committee. He decides to prepare a table as in Table 7.1. This shows clearly the steady decline in the total number of staff, the fall in operators throughout the phase, and the sharp decline of production engineers and supervisors once automation has been completed. It also illustrates the value of presenting information in a much more readable tabular form, compared with the text version given above.

Table 7.1 Effect of automation in the Dinsher Shop

	Number of people			
Start of year	**Total**	**Supervisors**	**Production engineers**	**Operators**
1	820	120	200	500
2	720	110	190	420
3	560	100	180	280
4	430	80	160	190
5	200	20	40	140

7.2.2 Pictorial presentation

The advantages of pictorial presentation of data are that it gets over the essential facts very quickly and creates an impact. For large amounts of data it is also a good way for indicating trends, which may not be easy to deduce from the table.

Pictograms

Figure 7.1 gives a pictogram of the trend in total employment in the Dinsher Shop. This shows at a glance the reduction in staff over a five year period. The picture elements used in this representation can vary depending on the items being compared, for example people for employment (as here), machine tools for automation or a pound sign for profit trends.

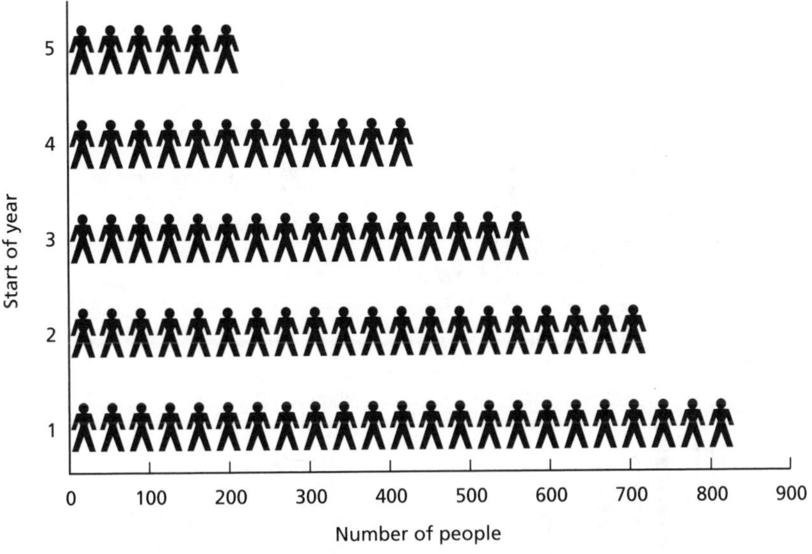

Figure 7.1 Pictogram of total employment in the Dinsher Shop.

Although creating effective visual impact, and putting over the essential facts quickly, a pictogram is very approximate. It is best used when whole items are being compared, each symbol representing a unit, although parts of units can be shown by a proportion of a drawing, such as half a person.

Bar charts

The bar chart is a pictorial representation technique which creates an impact as well as enabling information to be read with reasonable accuracy. It may consist of single bars, multiple bars or component bars. Generally the width of the individual bars is the same, their lengths varying to show the size of the item being compared.

Figure 7.2(a) shows a multiple bar chart of the employment in the Dinsher Shop. This indicates the decreasing trend in manpower, and the actual values can be read off from the scale.

An alternative presentation technique is the component bar chart of Figure 7.2(b). This is less cluttered than the multiple bar chart and therefore often more pleasing to see. It also shows the variation of the elements which make up the total for each bar, but now the actual values are more difficult to calculate since they do not start from zero.

Bar charts can be drawn in two dimensions or three dimensions. Generally it is easier to read the scale from two-dimensional bar charts (as shown in Figure 7.2(c)) although they are pictorially less effective.

Legends can be added to bar charts; that is, the actual numbers which make up

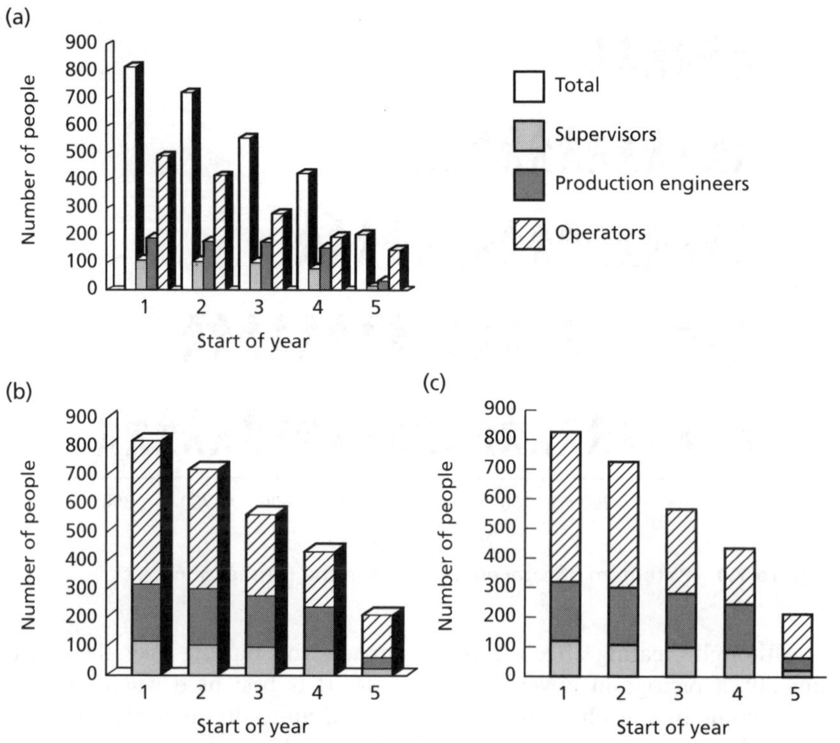

Figure 7.2 Bar charts of employment in the Dinsher Shop (from Table 7.1): (a) multiple bar chart; (b) component bar chart; (c) two-dimensional representation.

the bars can be placed next to them, so that they do not need to be read off the scale. However, this clutters the bar chart and some impact is lost.

Bar charts are useful for comparing information which is available at discrete points, such as year end sales. They are not very good when many items are involved.

The bar charts of Figure 7.2 could have been drawn horizontally, the x- and y-axes being interchanged, so that the bars appear horizontally. Vertical bar charts are, however, more common.

Histograms

An histogram is a special form of bar chart in which the areas under the rectangles that make up the bars represent the relative frequency of occurrence of the items. Usually the histogram is drawn with each rectangle having the same width (the same class interval), so the height of the bars determines the frequency. If the class intervals are unequal, then the heights must be adjusted accordingly. The use of a histogram is illustrated in the case study 'The problem of age distribution'.

Case study

The problem of age distribution

John McCaully would like to achieve staff reduction within the Dinsher Shop without introducing a programme of redundancy. Since the early staff reductions are mainly concentrated among the operators, he asks the HR Director, Jane Wilmore, to give him a breakdown of the ages of the operators within his Shop.

Jane gives him a list of 500 names, in alphabetical order, with their ages. To draw some conclusions from these figures John constructs a frequency table, as in Table 7.2, the ages of operators within the Dinsher Shop being grouped into ten year bands. To have constructed a table of individual ages for the 500 people involved would have produced a large number of figures. A frequency table allows these to be simplified by grouping ages into class intervals. The two extremes within this table (below 20 years and above 61 years) are assumed to have the same class interval (ten years) as those closest to them.

Table 7.2 Frequency table of ages in the Dinsher Shop

Age (years)	**20 and below**	**21–30**	**31–40**	**41–50**	**51–60**	**61 and over**
Number	90	120	180	70	30	10

John then plots the frequency table as a histogram, as in Figure 7.3. Note that there are no spaces between the bars, as exists for a bar chart. This histogram shows at a glance the spread of ages and how they are biased towards the young end. This means that the operators within the Dinsher Shop cannot be reduced without some form of redundancy programme, but that voluntary redundancy terms could be made attractive for the younger employees.

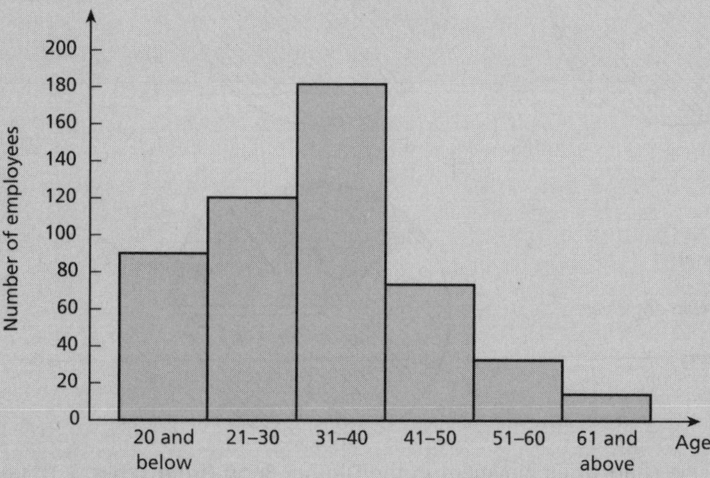

Figure 7.3 Histogram of ages in the Dinsher Shop (from Table 7.2).

Pie charts

The pie chart is a good method for illustrating subdivisions of the whole, although it cannot show total values. It is also not possible to read off absolute figures, unless these are added to the chart as numbers.

Figure 7.4(a) shows a pie chart of the employment in the Dinsher Shop, drawn from Table 7.1. This shows clearly the proportion of employees in the three categories, but not their actual numbers.

A separate pie chart must be used if comparisons over time are to be made. For example, Figure 7.4(b) shows the employment in year 4, and this indicates the proportionate increase of production engineers, since these are required to carry out the automation programme even though the number of operators is decreasing. If a pie chart for year 5 is drawn it will have proportions similar to Figure 7.4(a), since the numbers of production engineers and supervisors are now sharply reduced.

Note that these pie charts are unable to show the reduction in the total number of employees between years 1 and 4. Sometimes an attempt is made to indicate this by varying the size of the actual pie, but this is never successful.

Figure 7.4(c) illustrates an exploded pie chart, where a segment is shown separated from the total pie for greater impact or clarity.

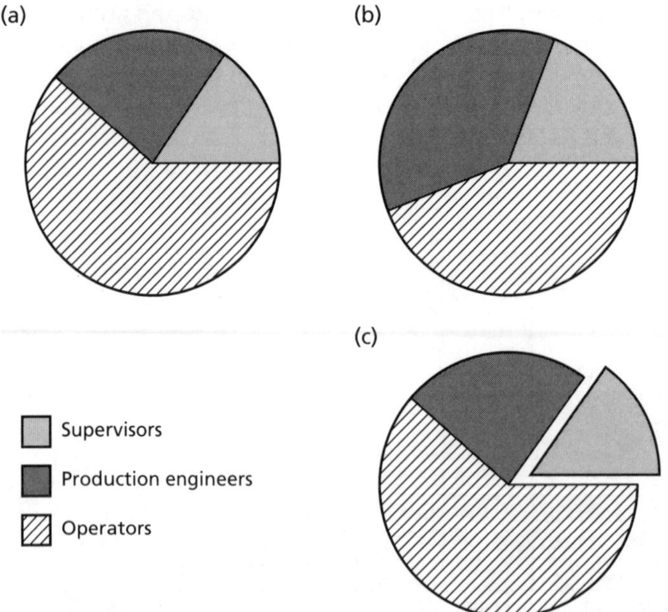

Figure 7.4 Pie chart of employment in the Dinsher Shop (from Table 7.1): (a) in Year 1; (b) in Year 4; (c) exploded pie chart.

7.2.3 Graphs

Graphs, or line charts, are a form of pictorial representation, although they usually provide less impact but more data. Graphs have x- and y-axes. Conventionally, the x-axis represents the independent variable and the y-axis the dependent variable. The variables may be discrete, occupying certain values only, or continuous, where the values vary infinitely between two limits.

There are very few rules in constructing a graph. Generally the scale should start from zero, unless one wishes to distort the curve. Two scales can also be used if it is necessary to show the interdependencies of two or more curves having very different values; for example, if the turnover and profits of a company are plotted on the same graph, the figures between them can vary by a factor of ten or more.

Honesty must also be used in drawing graphs, as illustrated in the case study 'Instant productivity increase'.

Case study

Instant productivity increase

When he joined the FloRoll Manufacturing Company from General Motors, Fred Steel, the Quality Director, introduced a system for measuring the productivity of all the departments. Some said that his measure of 'productivity factor' was too complex and did not give a true picture of productivity, but Fred persisted and, with the approval of the Chief Executive, ensured that productivity factors were published each year for every department. This factor was presented by each department head in his or her annual review with the Chief Executive.

At the end of the fifth year of automation within the Dinsher Shop the Chief Executive asked for a special review of achievements, including productivity factors. John McCaully, the Manager of the Dinsher Shop sat down to prepare his presentation and puzzled over how best to present the productivity factors, shown in Table 7.3. Fred Steel had insisted that the productivity of the Dinsher Shop be compared with that of the company average.

Table 7.3 Productivity factor comparison between the Dinsher Shop and the company average.

	Productivity factor	
Year	**Company average**	**Dinsher Shop**
1	25	5
2	29	8
3	34	10
4	41	12
5	46	15

John first tried to plot the curves on a single scale, as in Figure 7.5(a), but this only illustrated the apparent failure of automation. Productivity within the Dinsher Shop, although rising, was increasing at a slower rate than that of the company average and was also well below it.

(a)

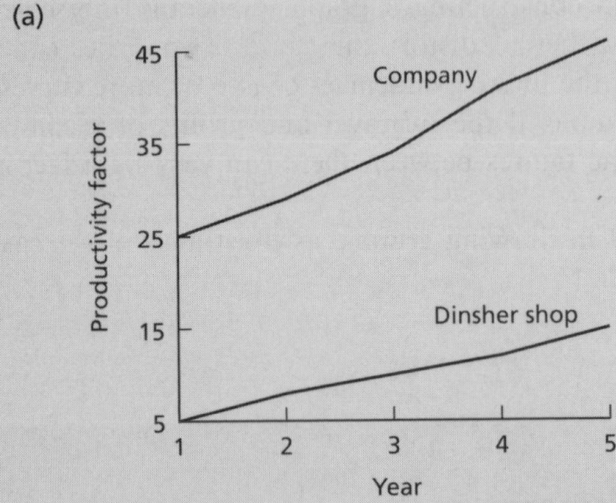

(b)

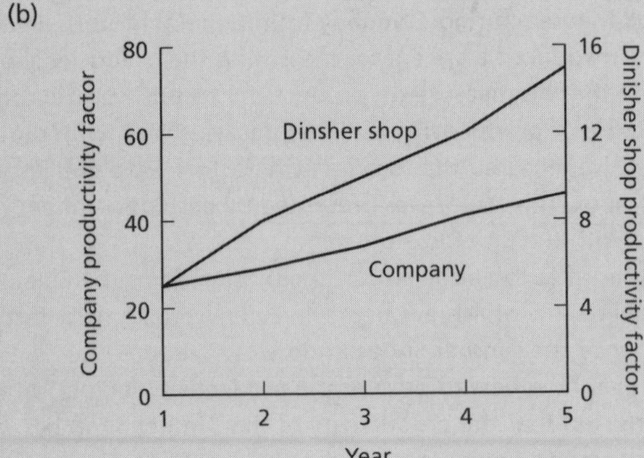

Figure 7.5 Productivity factor comparisons between the Dinsher Shop and company average (from Table 7.3): (a) equal scales; (b) unequal scales.

In his presentation to the Chief Executive John decided to plot the two curves to different scales, as in Figure 7.5(b). This looked much better! However, although this graph is technically correct, it is also dishonest. At a quick glance it gives the incorrect impression that the productivity of the Dinsher Shop is above the company average and is increasing at a faster rate. (It also gives credence to Disraeli's comments, quoted at the start of this chapter!)

Logarithmic scale graphs

The most common type of graph is the normal scale graph where the gradations on the two axes are linear. The logarithmic scale graph, or ratio graph, is an alternative were the gradations vary as a logarithmic or ratio. These graphs are used to show relative changes in data.

If only one of the axes, usually the *y*-axis, is logarithmic, the other being linear, then the graph is known as a semi-logarithmic scale graph. The use of a ratio graph is shown in the case study 'Linear growth'.

Case study

Linear growth

John McCaully's presentation to the Chief Executive was not as successful as he had hoped. Fred Steel referred to the data given in Table 7.3 to show that the overall productivity of the Dinsher Shop was poor compared with the rest of the company. Smarting with anger, John returned to his office and decided to carry out his own survey on the productivity of his Shop, using the production output figures as a measure.

Table 7.4 shows the output from the Dinsher Shop during the five years leading to full automation. This indicates that the output has been steadily increasing in spite of falling manpower. Plotted on a linear scale the graph is as in Figure 7.6 and indicates a steady increase in output.

Table 7.4 Production output for the Dinsher Shop

Year	**1**	**2**	**3**	**4**	**5**
Number (thousands)	1.0	1.5	2.0	2.5	3.0

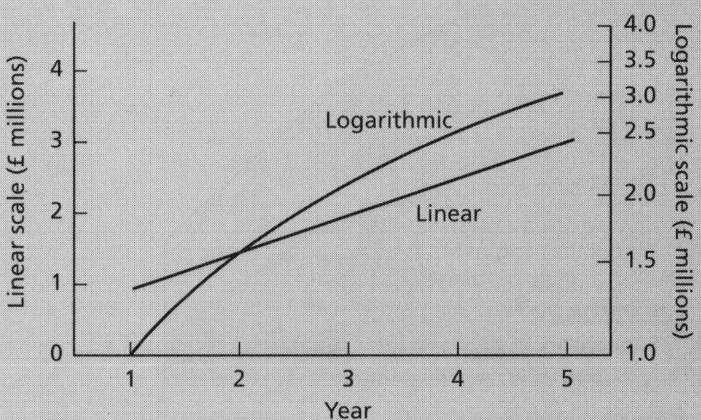

Figure 7.6 Productivity increase in the Dinsher Shop (from Table 7.4) plotted on a linear scale and a logarithmic scale.

However, John had learned his lesson and decided to present as true a picture as he could this time. Since he was comparing changes over several years, he then plotted the output figures on a logarithmic scale, as also shown in Figure 7.6. It is seen from this that the curve is bending over: the rate of increase in output is falling, and the trend is clearly towards zero increase over time. In the present example this is to be expected, since most of the productivity gains are obtained by automation and this will tail off once automation has been completed.

As is seen from Figure 7.6, a logarithmic scale graph is used when one is interested in the change in the ratio between numbers rather than in the absolute values of the numbers. On such a graph equal ratios are represented by equal distances, and when going up by equal ratios a straight line is obtained.

A ratio graph never starts from zero, but from the smallest number required. An advantage of such a graph is that it can cover a wide band or range of numbers, each division going up by a factor of ten, whereas in a linear scale graph the numbers go up in steps of one.

Strata or band graphs

In a strata graph, also called a band or layer graph, several curves are drawn on the same paper, the distances between them representing the actual values.

For example, Figure 7.7 shows the employment within the Dinsher Shop

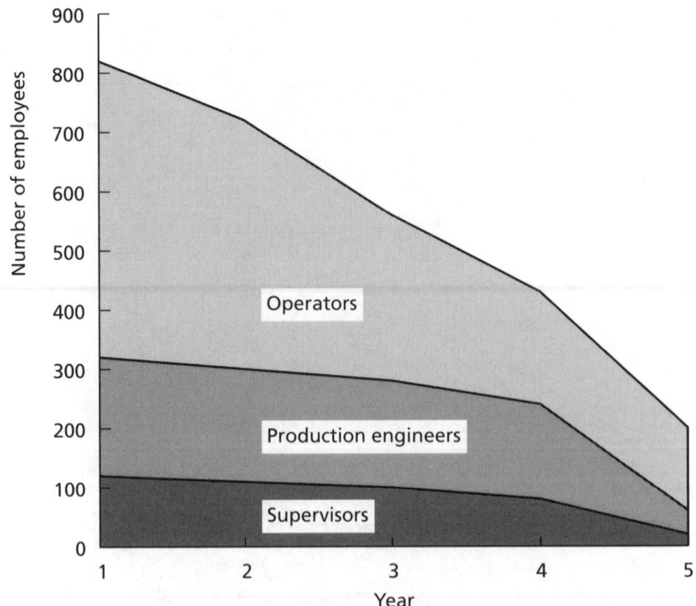

Figure 7.7 Employment in the Dinsher Shop plotted as a strata graph (from Table 7.1).

(from Table 7.1) plotted as a strata graph. In this graph the value for each category is equal to the distance between two of the graphs and cannot be read off from the *y*-axis directly. Only the total is given directly, as the sum of the three curves.

The strata graph gives a good pictorial representation of the variations of the various elements, but the actual values are more difficult to read compared with a conventional graph.

Ogives

Ogives are a graphical method for representing cumulative data. For example, the ages of employees in the Dinsher Shop, as shown in Table 7.2, can be presented in Table 7.5 as 'more than' and 'less than'.

The data from Table 7.5 can now be plotted in a graph as in Figure 7.8, resulting in an ogive or cumulative frequency curve. This curve can be used to estimate the value of the distribution at any given point. It is also usual to smooth out the ogive curves (not done in Figure 7.8) since the occurrences are not spread evenly across a class interval.

Table 7.5 Ages within the Dinsher Shop (reproduced from Table 7.2)

Age	Number of employees more than age	Number of employees less than age
0	500	0
20	410	90
30	290	210
40	110	390
50	40	460
60	10	490
65	0	500

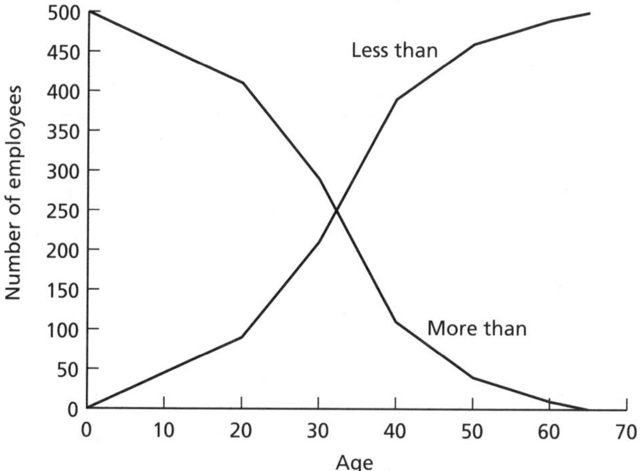

Figure 7.8 Ogives of operator ages within the Dinsher Shop (from Table 7.5).

Frequency polygons

A frequency polygon may be considered to be a graphical version of a histogram. It is usually shown as being derived from a histogram, although a histogram does not need to be drawn first.

Figure 7.9 shows the frequency polygon for the data in Table 7.2. Also shown (broken lines) is the histogram for this data, reproduced from Figure 7.3. The frequency polygon has been obtained by joining the midpoints of the class intervals. It is also conventional to extend the frequency polygon to meet the y-axis at the half class interval points.

As seen from Figure 7.9, the frequency polygon cuts off the corners of each histogram rectangle and adds a new corner of equal area, so the total area under the frequency polygon is equal to that under the histogram. This is also the reason why the frequency polygon is extended a half class interval at either end. Therefore the area under the frequency polygon is also equal to the relative frequency of occurrence of the items (as for an histogram).

If the x-axis is made up of very small subdivisions, then the frequency polygon will resemble a smooth curve.

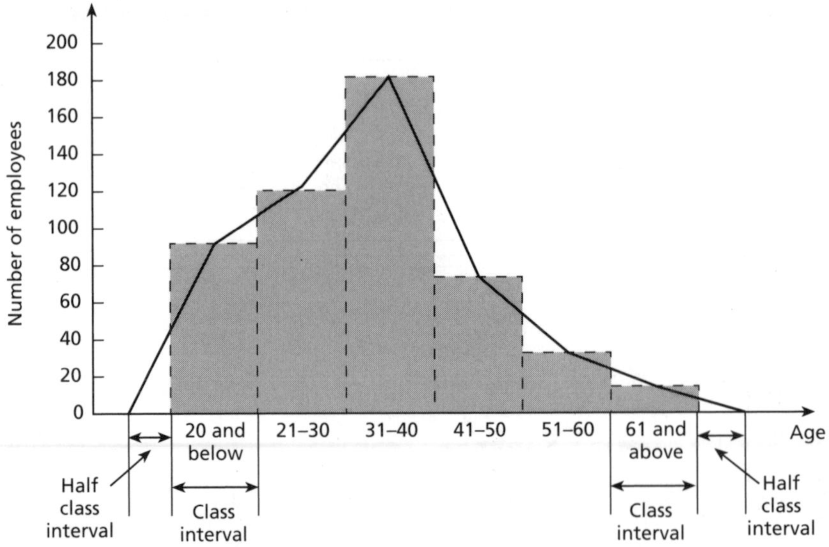

Figure 7.9 Frequency polygon of the histogram shown in Figure 7.3 (from Table 7.2).

Lorenz curves

In statistical analysis it is often found that a small proportion of items have the greatest influence. For example, a small percentage of the population have the highest income in a country; or a small number of large organizations employ the largest number of people, their total usually exceeding the sum of people employed in all the other smaller organizations.

This law of inequality can be shown graphically by a Lorenz curve, and it allows management attention to be focused on the few critical elements that have the greatest influence, rather than being diluted by the mass of less important items. An example in drawing and evaluating a Lorenz curve is given by the case study 'The gold nugget floroll'.

Case study

The gold nugget floroll

Johnny Bacon, the Financial Director, was becoming concerned at the steady profit margin erosion which the company was facing in most of its markets. This was partly due to competition and customer resistance forcing down the selling price, but it was also caused by the fact that the labour and raw material costs of florolls were escalating at a rate well in excess of the retail price index.

Production costs were so high that florolls were now jokingly referred to, within the company, as being worth their weight in gold. Something was being done about labour costs by introducing greater automation into the plant, starting with the Dinsher Shop, but action was clearly needed to control material costs.

Johnny Bacon tabled his concern at Jake Topper's Executive Staff Meeting, and the Manufacturing Director, Adrian Elton, received an action to present the cost breakdown of material used within a floroll at the next meeting.

Adrian had no problem collecting the data. This was presented to him by his procurement manager as a long list of items, and as there were 665 components in all Adrian knew that he had to classify them in some way before he could draw any meaningful conclusions from the figures.

Scanning down the list Adrian realised that there were a very large number of low-cost items (below £5) and a few items which were very expensive. He decided to group the items into bands, as shown in Table 7.6. This confirmed his hunch that the cost breakdown would be best presented and dealt with as a Lorenz curve.

Table 7.6 Cost breakdown of a floroll

Unit cost (£)	Number of items	Total contribution to cost (£)
Under 5	300	1 100
5 to under 10	230	1 800
10 to under 50	75	1 900
50 to under 100	35	2 800
100 and above	25	4 100
	665	11 700

Table 7.7 was then produced from Table 7.6. The main aim of this table is to obtain the columns for cumulative costs and cumulative number of items, which are to be plotted in the Lorenz curve. From this table it is seen that 45.1 per cent of the items that make up a floroll contribute only 9.4 per cent towards its material costs, or conversely, 54.9 per cent of the items contribute over 90.6 per cent of the costs, so that half the items can in effect be ignored.

Table 7.7 Cumulative cost breakdown of a floroll (from Table 7.6)

Number of items			**Total cost**		
Number	**Percentage of total**	**Cumulative percentage**	**Value**	**Percentage of total**	**Cumulative percentage**
300	45.1	45.1	1 100	9.4	9.4
230	34.6	79.7	1 800	15.4	24.8
75	11.3	91.0	1 900	16.2	41.0
35	5.3	96.3	2 800	23.9	64.9
25	3.7	100	4 100	35.1	100
665	100	–	11 700	100	–

Adrian then plotted the Lorenz curve from this table, as in Figure 7.10. (The Lorenz curve is also often drawn as a smooth curve.) He also plotted, on the same graph, the line of equal distribution. This line assumes that each item makes the same contribution to costs; that is 10 per cent of items have 10 per cent of costs, 20 per cent of items have 20 per cent of costs and so on. The distance between the line of equal distribution and the Lorenz curve therefore represents the extent of inequality in the Lorenz curve. Visually the greatest distance between these two curves occurs at approximately the 87 per cent items and 35 per cent cost point; in other words, 13 per cent of the items contribute 65 per cent of the costs.

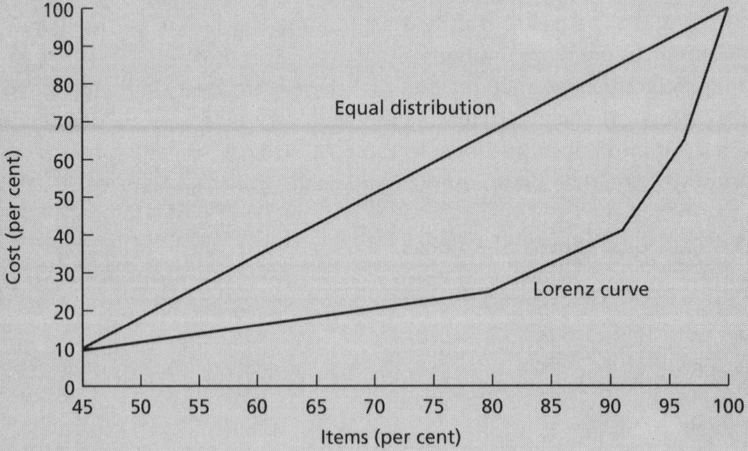

Figure 7.10 Lorenz curve.

Adrian Elton smiled in satisfaction. He decide to investigate these 13 per cent of the items (a manageable figure) in more detail, so that by the next Executive Staff Meeting he would not only show his analysis to the board, but would also present an explanation of how costs in these key 13 per cent of items had changed over the years, and a plan for reducing their cost, such as by a change in the design so that cheaper material could be used. If time permitted he could even consider some more items, using the Lorenz curve to determine what percentage of costs he was covering for any percentage of items being investigated.

7.3 Averages

Earlier sections of this chapter introduced techniques for condensing and presenting data so that meaningful conclusions could be drawn from them. The average takes this concept even further, presenting a single number, referred to as the central tendency, to represent many numbers.

There are several types of average, and using the wrong type for an application will provide erroneous results.

7.3.1 The arithmetic mean

The arithmetic mean is the most common form of average and is often simply known as the mean.

As an example consider that Elites, a component used in the manufacture of a floroll, come in boxes of 100, and a sample of five boxes have been taken at a goods incoming inspection and their contents counted. The results of this are as follows: 100, 99, 350, 101, 99.

The arithmetic mean of these is then $(100 + 99 + 350 + 101 + 99)/5 = 149.8$.

The advantage of the arithmetic mean is that it is easy to calculate and that it takes account of all the numbers. Its disadvantage is that it is affected by extreme values. In the example of Elites the number 350 is clearly an error and should be discarded.

Another disadvantage of the arithmetic mean is that the final result may not be a whole number even though each item is a whole number, as in the example of Elites, above.

Mathematically the arithmetic mean of n numbers, x_1 to x_n can be written as in Equation (7.1).

$$A_a = \frac{\text{Total of all items}}{\text{Number of items}}$$
$$= \frac{x_1 + x_2 + \ldots + x_n}{n} \qquad (7.1)$$
$$= \frac{\sum_{r=1}^{n} x_r}{n}$$

To obtain the arithmetic mean from a frequency distribution requires the total of the values to be found and divided by the total number of items.

For example, suppose that it is required to find the average age in the Dinsher Shop from the frequency numbers of Table 7.2. This is reproduced in the first two columns of Table 7.8, but the upper and lower limits have been added since no one below the age of 16 or above the age of 65 is employed by the company.

Table 7.8 Ages in the Dinsher Shop (from Table 7.2)

Age	Number	Mid-point age	Mid-point age × number	Cumulative number of employees
16–20	90	18	1 620	90
21–30	120	25.5	3 060	210
31–40	180	35.5	6 390	390
41–50	70	45.5	3 185	460
51–60	30	55.5	1 665	490
61–65	10	63	630	500
	500		16 550	

Also shown here are the mid-point values and the products of these with the number of employees. Use of mid-points results in error, since it is assumed that all the employees have an age equal to the mid-point of the class concerned. The average age is obtained by dividing the total age by the total number of employees: $16550/500 = 33.1$ years.

When calculating the arithmetic mean of percentage figures the weighted average must be obtained, as illustrated by the case study 'Consolation for operators'.

Arithmetic means are very well known and commonly used for most averaging calculations. This is one of its problems, since it is often misapplied, as in the case study 'Ted Court's puzzle'.

Case study

Consolation for operators

The earnings increase in the Dinsher Shop over two successive years is as shown in Table 7.9 for the three categories of employees: supervisors, production engineers and operators. The percentage increase between the two years is seen to vary from 2 per cent for supervisors to 10 per cent for operators. These figures include overtime working.

Table 7.9 Earnings in the Dinsher Shop

Category	Year 1 (£)	Year 2 (£)	Percentage increase
Supervisors	50 000	51 000	2
Production engineers	45 000	46 800	4
Operators	40 000	44 000	10
	135 000	141 800	

John McCaully fought hard to set rates such that the operators could substantially increase their earnings, as a cushion for the erosion of jobs in this category. However, he was told by Jane Wilmore that the average increase in the Dinsher Shop was not to exceed 5 per cent, in line with the company average.

John calculated the average increase in the Dinsher Shop as $(2 + 4 + 10)/3 = 5.3$ per cent and realized that he had exceeded his limit. Or had he?

If the arithmetic mean is calculated by using total values, then it is given by $(141\,800 - 135\,000)/135\,000 = 5.0$ per cent. So why the difference?

When calculating averages for percentages, the weighted average must be taken, as stated earlier, each percentage being weighted by its corresponding base year figure. In the present example this would give an arithmetic mean of $(500\,000 \times 2 + 45\,000 \times 4 + 40\,000 \times 10)/135\,000 = 5.0$ per cent, which is the correct figure. Therefore John McCaully has not exceeded the average increase limit set for the Dinsher Shop.

Case study

Ted Court's puzzle

Ted Court, the stock control supervisor, has been asked to evaluate the stock for accounting purposes. He knows that for one of the materials used in the manufacture of florolls, called Elite, he has on his shelves three batches which were bought over three periods of time. The orders were all for £2000 worth of Elite, but the prices over the three periods were £20 per item, £40 per item and £50 per item. Ted is asked to provide the average unit price of the stock, and he calculates this as $(20 + 40 + 50)/3 =$ £36.67.

Being a cautious man he decides to check this number. If £36.67 is correct then the total number of items he should have in stock, considering that a total of £6000 was spent, is $6000/36.67 = 163.63$. However, Ted knows that he has 190 items in all. This is confirmed if the numbers from the individual orders are calculated: 2000/20 + 2000/40 + 2000/50 = 190.

Ted scratched his head in bewilderment. What had he done wrong? (See Section 7.3.5.)

7.3.2 The median

The median, or middle one, is found by placing the figures in an ascending or descending order and then taking the middle figure. If there are an even number of items then the arithmetic mean of the two central numbers is taken as being the median.

For example, in the count of the box of Elites mentioned earlier, the median is found by arranging the numbers in ascending order, such as 99, 99, 100, 101, 350, and selecting the middle number, here 100, as the median.

The advantage of the median as an averaging technique is that it can be used even when the items cannot be expressed as a number. For example, shades of a colour can be placed in order, say from light to dark, and the median chosen. A further advantage of the median is that it is not affected by extreme values.

The disadvantage of the median is that, since it does not consider all the figures, the extreme numbers can change without having any effect on the median. It is therefore not very useful for further mathematical calculations. The median is also not representative of the result if the numbers are erratic and widely spread. For example the median of a range of numbers 1, 2, 3, 1450, 2820 is 3, which takes no account of the very large numbers.

In a frequency distribution of n items, the position of the median is found as the value located at the $n/2$th item. For example the median age in the Dinsher Shop is found from the frequency table of Table 7.8 as the $500/2 = 250$th item. This is clearly in the range 31–40 years. The problem is now of finding the precise year, since it will be closer to 31 than to 40 (since 250 is closer to 210 than to 460, the adjoining classes).

Although the exact age can be found mathematically it is just as easy to read it off the ogive, as in Figure 7.8. Only one of the curves (more than or less than) is required and the age is read off at the 250 employee point.

7.3.3 The mode

The mode, or most fashionable one, is the item that appears most often. For example in the count of Elites given earlier, the mode from the numbers 100, 99, 350, 101, 99 is equal to 99, since this is the only number that appears more than once.

The advantage of the mode as an averaging technique is that it is not affected by extreme values and one does not need to know the actual number to find it.

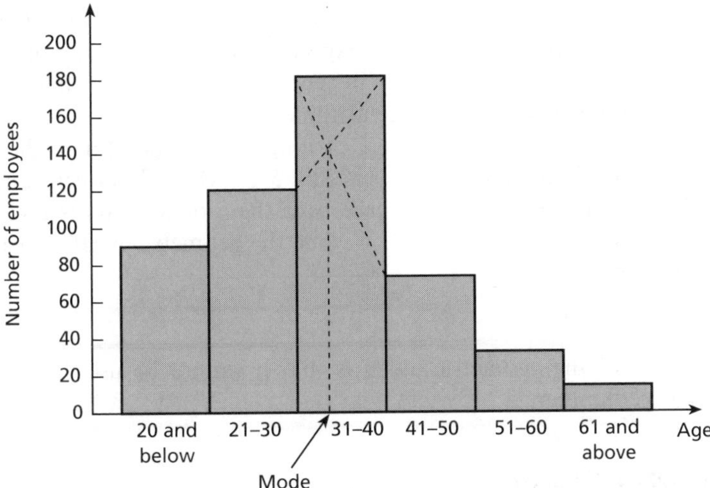

Figure 7.11 Obtaining the mode from a histogram (see Figure 7.3).

For example, one could compare the colours of an item and the mode would be that colour which appears most often.

The disadvantage of the mode is that there can be several modes (referred to as multimodes) when different items appear the same number of times, or there may be no mode if no item is repeated. Since the mode does not take numeric values into account it is also not suitable for calculations. Modes can give large errors when dealing with widely spread and erratic numbers. For example the mode of the numbers 1, 20, 1, 2900, 750 is equal to 1.

In the case of a frequency distribution, such as that shown in Table 7.8, the mode is within the class interval which occurs most often: age 31–40. The actual age is found conveniently from the histogram, as shown in Figure 7.11. As expected, the mode is closer to 31 than to 40 since the histogram is biased towards the lower age.

7.3.4 The geometric mean

In the arithmetic mean the items are added together and divided by the number of items, as in Equation (7.1). In the geometric mean the items are multiplied together and the root of the result is taken, the base of the root being equal to that of the number of items, as in Equation (7.2).

$$A_g = (x_1 \times x_2 \times \ldots \times x_n)^{1/n} \tag{7.2}$$

For example the geometric mean of the numbers 100, 99, 350, 101, 99 is given by:

$$(100 \times 99 \times 350 \times 101 \times 99)^{1/5} = 128.2.$$

The geometric mean is mainly used to find the averages of quantities which follow a geometric progression or an exponential law, for example rates of change, where the value of the quantity depends on its previous value. For these applications it is more accurate than the arithmetic mean.

For example, suppose that the average earnings of an operator in the Dinsher Shop increased over three consecutive years by 5 per cent, 7 per cent and 10 per cent. Then the average percentage increase over these three years is given by the arithmetic mean as $(5 + 7 + 10)/3 = 7.33$, while the geometric mean gives 7.04, as follows, which is more correct:

$$(5 \times 7 \times 10)^{1/3} = 7.04$$

The disadvantage of the geometric mean is that it cannot be used if any item is zero or a negative number.

7.3.5 The harmonic mean

The harmonic mean is calculated as the inverse of the mean of the reciprocals of each individual item. For example, for n items of value x_1 to x_n, the harmonic mean is given by Equations (7.3) and (7.4).

$$\frac{1}{A_{\mathrm{h}}} = \frac{1}{n}\left(\frac{1}{x_1} + \frac{1}{x_2} + \ldots + \frac{1}{x_n}\right) \tag{7.3}$$

$$A_{\mathrm{h}} = \frac{n}{\sum_{r=1}^{n} \frac{1}{x_r}} \tag{7.4}$$

In considering items such as A per B (metres per second, or dollars per gram), if the items are for equal A then the harmonic mean should be used to find the average, but if it is for equal B then the arithmetic mean should be used.

For example if a car travels three equal distances at speeds of 60 km per hour, 80 km per hour and 120 km per hour, then the average speed is found as the harmonic mean, but if the car travelled for equal intervals of time (say twenty minutes) at these speeds, then the average speed is found as the arithmetic mean.

Consider now the example given earlier, the case study 'Ted Court's puzzle'. Since the problem consists in determining the cost per item and the items were bought for equal cost, the harmonic mean must be used in averaging them. Therefore the average item cost is:

$$\frac{3}{\frac{1}{20} + \frac{1}{40} + \frac{1}{50}} = 31.58$$

That this is correct can be found by dividing 31.58 into the total expenditure to give the total number of items as $6000/31.58 = 190$, which was what Ted had on his shelves.

Note that if the buyer had bought 100 items over three periods of time and had paid £20 per item, £40 per item, and £50 per item, then the arithmetic mean would be used to find the average cost: $(20 + 40 + 50)/3 = 36.67$.

7.4 Index numbers

An index number is a special type of average which is used when comparing different types of item, for example industrial output, which is made up from TV sets, motor cars, chemicals and so on. It is an average of a group of items, measured over time, to show the change in the items. Although it can be used to measure single items, such as the output of one factory and one product over time, index numbers are more useful when they represent a collection of items, such as the output from many factories and many products.

Index numbers are widely used for government-produced statistics, such as the index of industrial output, the retail price index and the index of unemployment.

In compiling an index number it is important to decide on the type of items to be compared, the number of items and the base year. The base year, to which all the other numbers are compared, must be as 'normal' as possible, since if it is unrepresentative of subsequent years in any way (for example because a hurricane occurred in that year which affected all the items) then the subsequent comparisons will be biased.

The method for calculating an index number is illustrated in the case study 'What makes a floroll flow?'. Because one is dealing with different types of item a simple average cannot be taken to find the index number, and a weighted average is used. Also, many different types of index numbers can be calculated, depending on the type of average used (such as a geometric index using geometric averaging), although in the present case study the arithmetic mean will be used.

Calculations can also be made using a variable base, as in the chain base method, where changes are based on the previous year as a base. This technique is used when weights are changing rapidly, but is not used in the case study 'What makes a floroll flow?'.

Case study

What makes a floroll flow?

In a further study into the main components which make up a floroll, Adrian Elton was told that there were four critical components, Elite, Hector, Zing and Peddle. The prices and usage of these components varied over time, since the engineers modified the design of the product depending on the anticipated price of the key components for the following year.

For example, the prices of Zing over the previous three years were £25, £30, £45 and the quantities used over these same years were 10 litres, 5 litres and 3 litres.

To compare these figures Adrian decided to calculate index numbers. Since prices and quantities are involved he realized that he would need two index numbers: the price index and the quantity index. These are shown in Table 7.10.

Table 7.10 Variation of price and quantity of Zing used over three years

Year	Price (£)	Quantity (litre)	Price index	Quantity index
1	25	10	100	100
2	30	5	120	50
3	45	3	180	30

Year 1 is taken as the base year and prices in subsequent years are expressed as a percentage of the price in the base year to calculate the price index. Similarly, the quantity is compared with the quantity in the base year to find the quantity index. This shows that although the price of Zing has almost doubled over the three years, the quantity used has been reduced to a third, so that overall a cost saving has been made.

Adrian Elton now decided to determine the price index and quantity index for all four of the components, Elite, Hector, Zing and Peddle. Their usage and price is shown in Table 7.11 for two years. It is important to note that the four components are measured in very different units, so that a simple averaging technique could not be used. The price index and quantity index in Year 2, taking Year 1 as the base year, are found by weighting the items, as in Equations (7.5) and (7.6) respectively:

$$\text{Price index} = \frac{\text{Price Year 2 } (P_2) \times \text{Quantity Year 1 } (Q_1)}{\text{Price Year 1 } (P_1) \times \text{Quantity Year 1 } (Q_1)} \times 100 \qquad (7.5)$$

$$\text{Quantity index} = \frac{\text{Quantity Year 2 } (Q_2) \times \text{Price Year 1 } (P_1)}{\text{Quantity Year 1 } (Q_1) \times \text{Price Year 1 } (P_1)} \times 100 \qquad (7.6)$$

From these equations the price index and quantity index in Year 2 are found as:

$$\text{Price index} = \frac{30 \times 30 + 20 \times 20 + 30 \times 10 + 35 \times 25}{30 \times 35 + 20 \times 20 + 10 \times 25 + 25 \times 15} \times 100 = 119$$

$$\text{Quantity index} = \frac{30 \times 35 + 25 \times 20 + 5 \times 25 + 15 \times 15}{30 \times 35 + 20 \times 20 + 10 \times 25 + 25 \times 15} \times 100 = 92$$

These indices show that, although the quantity index has gone down by 8 per cent the price index has gone up by 19, so that overall there has been a detrimental effect on the cost of the floroll. This is different to the result obtained when Zing alone was considered, as in Table 7.10, where between years 1 and 2 the price index went up by 20 per cent but usage was halved. Clearly the engineering department was focusing on the wrong component (Zing) in its cost reduction efforts. Adrian Elton decided to call a meeting with Micro Dowling, the Engineering Director, to discuss this further.

Table 7.11 Variation of usage and price for four of the main ingredients used to make florolls

		Year 1		**Year 2**	
Item	**Unit of measure**	**Number of units (Q_1)**	**Price per unit (£) (P_1)**	**Number of units (Q_2)**	**Price per unit (£) (P_2)**
Eliite	Boxes	30	35	30	30
Hector	1 kg	20	20	25	20
Zing	1 litre	10	25	5	30
Peddle	500 g	25	15	15	35

7.5 Dispersion from the average

The average provides useful information about the numbers which it represents, but often more information is needed. For example, if John McCaully is told that the average output per operator from his Shop is 20 dinshers per day, he can work out the average daily output of the shop as 1000 dinshers, since he knows that he has 500 operators.

However, John wants to know more; for example, are some operators performing better than other, and if so by how much? This is a measure of the spread or dispersion of the figures from the average.

Dispersion can be seen from tables or graphs, but they are often required to be represented by one or two numbers, and the techniques by which this can be done are discussed in this section. Generally, the dispersion has the same units as the quantity being measured.

7.5.1 The range

The range is found very simply as the difference between the largest and smallest numbers being considered, as in the case study 'Is everyone pulling their weight?'.

Case study

Is everyone pulling their weight?

To get an idea of the spread in output between the operators in the Dinsher Shop, John McCaully took a sample of ten operators and measured their output per day over a month. The average figures obtained for these ten operators were calculated to be as follows:

Daily average output = 13, 14, 17, 19, 22, 22, 22, 23, 23, 25

The average for the ten operators is therefore 20 and the range is equal to (25 – 13) = 12, which is quite a large dispersion from the average, indicating to John McCaully that he has at least two operators whose output varies by 50 per cent of the average.

The advantage of the range as a measure of dispersion is that it is easy to calculate and understand. Its disadvantage is that it does not effectively indicate the degree of clustering and does not take into account all the numbers.

For example the numbers between 13 and 25 could change to any value without affecting the result so long as they remained within the range 13 to 25. Therefore nine operators could produce 25 units per day and only one produce 13, but the range would still be 12.

The units of dispersion are those of the series of item being measured, for example dinshers, so they cannot be used to compare different series.

7.5.2 Quartile deviation

The median was seen to divide the series of figures into two equal parts. A quartile divides it into four equal parts.

For the example above (the case study 'Is everyone pulling their weight?') the first quartile of the series, representing the output from the operators in the Dinsher Shop, is 17 and the third quartile is 23. The interquartile range is the difference between the first and third quartile numbers, in this case 23 – 17 = 6, and the quartile deviation is half the interquartile range, here 3.

Therefore the dispersion from the average output of 20 dinshers can be stated to have a quartile deviation of 3.

The advantages of the quartile deviation as a measure of dispersion are that it is easy to calculate and is not affected by extreme values. However, its disadvantages are that it in effect ignores half the values in the series and gives no indication of clustering. The extreme values can change to any number without affecting the result.

The units of measure for the quartile deviation are the same as the series of numbers from which they are derived; dinshers in the above example. The quartile coefficient of dispersion is used when series of different items are to be compared, since it is unitless. It is equal to the ratio of the quartile deviation to the median, as in Equation (7.7).

$$\text{Quartile coefficient} = \frac{\text{Quartile deviation}}{\text{Median}}$$

$$= \frac{\text{Difference between quartiles}}{\text{Sum of quartiles}} = \frac{23 - 17}{23 + 17} = 0.15 \qquad (7.7)$$

7.5.3 Mean deviation

The mean deviation is the average of all deviations. It is found by taking the mean of the difference between each figure and the arithmetic mean or the median of the series, signs being ignored. This is given by Equation (7.8).

$$\text{Mean deviation} = \frac{\sum_{r=1}^{n} |A_a - x_r|}{n} \qquad (7.8)$$

For example, Table 7.12 shows the output of the ten operators sampled in the Dinsher Shop. The arithmetic mean is found from the second column as 20. The third column gives their deviation from this mean, and the sum, ignoring signs, is 34. (Note that if the signs are not ignored the result will always be zero, since the deviations lie equally on either side of the mean.)

The mean deviation is therefore $34/10 = 3.4$. This is the dispersion from the average value of 20.

The advantage of the mean deviation as a measure of dispersion is that it takes account of all the items in the series. However, since signs are ignored, it is not suitable for use in further mathematical analysis.

7.5.4 Standard deviation

The standard deviation is the most frequently used measure of dispersion from the average. The arithmetic mean is always used to calculate standard deviation, never the median.

The standard deviation (σ) is found by squaring the deviation from the mean (so eliminating signs), adding all the results, finding their arithmetic mean, and then taking the square root of the result. This is illustrated in Equation (7.9).

$$\sigma = \frac{\left((A_a - x_1)^2 + (A_a - x_2)^2 + \ldots + (A_a - x_n)^2\right)^{1/2}}{n}$$

$$= \left(\frac{\sum_{r=1}^{n} (A_a - x_r)^2}{n}\right)^{1/2} \qquad (7.9)$$

For the example of the output from the Dinsher Shop, the fourth column in Table 7.12, derived by squaring the values in the third column, shows that the

Table 7.12 Sample of output from the Dinsher Shop

Sample number	Output per day	Deviation from mean (20)	Square of deviation from mean (20)
1	13	−7	49
2	14	−6	36
3	17	−3	9
4	19	−1	1
5	22	2	1
6	22	2	4
7	22	2	4
8	23	3	4
9	23	3	9
10	25	5	25
Total (ignoring sign)		34	150

sum of the square of the deviations from the arithmetic mean is equal to 150. The standard deviation is therefore:

$$\sigma = \left(\frac{150}{10}\right)^{1/2} = 3.873$$

The advantages of the standard deviation are that it includes every value in the series and is mathematically correct, so it can be used for further calculations. Its disadvantage is that it is more difficult to understand and calculate. It can also give greater weight to extreme values, since the deviations are squared.

Several other measures of dispersion are derived from the standard deviation. For example, the variance is equal to the square of the standard deviation, and is therefore found from Equation (7.9) before the final square root is taken.

The coefficient of variance is found by dividing the standard deviation by the arithmetic mean, as in Equation (7.10).

$$\text{Coefficient of variance} = \frac{\sigma}{A_a} \times 100 \tag{7.10}$$

The coefficient of variance gives a dimensionless ratio, unlike the standard deviation, which is in the same units as the series from which it has been derived. Therefore the coefficient of variance can be used to compare two sets of numbers which have very different units or are widely different in value.

When calculating the standard deviation from a frequency table the mid-points must be used. This is shown in Table 7.13 for the ages in the Dinsher Shop, derived from Table 7.8.

The arithmetic mean is equal to 33.1, as found in an earlier example, and the final column of Table 7.13 gives the total sum of the squares of the deviations from this mean as 63 245. The standard deviation is therefore equal to:

$$\text{Standard deviation } (\sigma) = \left(\frac{63\,245}{500}\right)^{1/2} = 11.25$$

Table 7.13 Frequency table of ages in the Dinsher Shop

Number (q)	Mid-point age (x_r)	($A_a - x_r$)	$(A_a - x_r)^2$	$(A_a - x_r)^2 q$
90	18.0	15.1	228.01	20 520.9
120	25.5	7.6	57.76	6 931.2
180	35.5	−2.4	5.76	1 036.8
70	45.5	−12.4	15.76	10 763.2
30	55.5	−22.4	501.76	15 052.8
10	63.0	−29.9	894.01	8 940.1
500				63 245

7.5.5 Skewness

The distribution of numbers within a series usually lies unequally on either side of the middle, as illustrated in the frequency distribution of Figure 7.12. The mode is always the point of most frequent occurrence and is the peak of the distribution. The median divides the area under the curve into two equal parts. The arithmetic mean cannot be found from the curve; it must be determined mathematically and then added to the curve.

The distribution is said to be positively skewed if it is biased to low values, as in Figure 7.12(a) and it has a negative skew if biased the other way, as in Figure 7.12(b). If the distribution is symmetrical (has no skew) then the arithmetic mean, median and mode will all coincide.

(a)

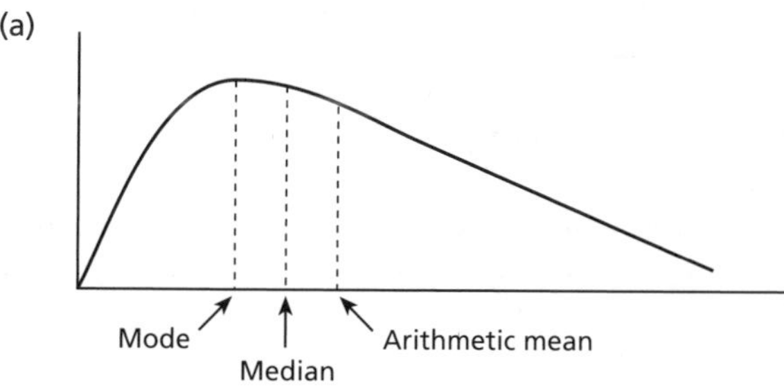

(b)

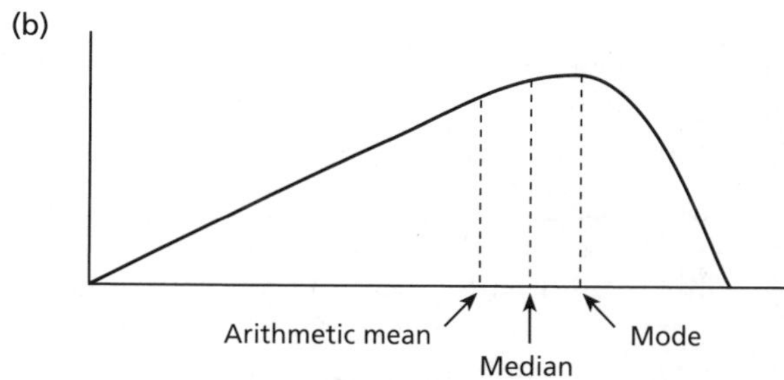

Figure 7.12 Skewed distributions: (a) positive skew; (b) negative skew.

Several mathematical techniques exist for expressing the skewness of the distribution, and they all give a measure of the deviation between the mean, median and mode. Usually this skewness is stated in relative terms, to make comparisons between different series easier. The Pearson coefficient of skewness is given by Equation (7.11) or by Equation (7.12).

$$\text{Pearson coefficient } (P_{\text{k}}) = \frac{\text{mean} - \text{mode}}{\text{standard deviation}} \tag{7.11}$$

or

$$\text{Pearson coefficient } (P_{\text{k}}) = \frac{3\,(\text{mean} - \text{median})}{\text{standard deviation}} \tag{7.12}$$

If the results from Equations (7.11) or (7.12) are positive then the distribution has a positive skew, and if negative it has a negative skew. The higher the absolute value of the number the greater the skew. If there is no skew then the Pearson coefficient will be zero.

Summary

The first step in information analysis is to place the data into an ordered form. Tables are frequently used, although pictorial representation is better for indicating trends and for visual impact.

Pictograms give good visual impact, but are only suitable for relatively simple amounts of data. Bar charts combine the visual impact of pictograms while at the same time providing more detailed information on several simultaneous items of information. There are different forms of bar chart, such as multiple bar charts and component bar charts, as in Figure 7.2.

Histograms are a special form of bar chart, which are used when a large amount of information is to be presented. Data is classified into groups or classes, and a bar is drawn for each class rather than for each item within the class. A pie chart is used when subdivisions of a whole are to be illustrated and compared. It is not very good for showing absolute values and how these change.

Most engineers are familiar with graphs, both linear and logarithmic, as these are frequently used to analyze technical problems. Graphs are also useful for management problem analysis. A linear or normal scale graph should be used when absolute values are important, while a logarithmic graph is used when relative changes of the data are to be considered. Strata or band graphs are also used since they combine information presentation with visual impact, especially when several different items are being compared. Cumulative data is best represented on another type of graph, the ogive, which can be plotted as 'less than' or 'greater than' curves.

The frequency polygon is a graphical version of an histogram and is obtained by joining the mid-point of the class intervals of the histogram. Lorenz curves are used to determine instances when a small number of items have the greatest influence on the outcome. They allow management attention to be focused on the few critical elements, rather than being diluted by the mass of less critical data.

Often it is useful to reduce the large amount of data obtained into a single number for comparison with other data. This can be done by taking an average of the numbers. The most common average is the arithmetic mean. It is the easiest to apply, but can be affected by extreme values. The median is simpler to obtain and is found by placing the numbers in order and then choosing the middle one. It is not very accurate, but it can be used in cases where the information cannot be expressed as a number, for example when comparing shades of colour.

The mode is also a simple averaging technique which selects the item or number which appears most often. Like the median it can be used in instances where the data cannot be expressed as numbers. Its disadvantage is that it can give large errors when dealing with widely spread and erratic numbers.

The geometric mean is obtained by multiplying the items of data together and taking the root of the result, the base of the root being equal to that of the number of items. This averaging technique is primarily used to find averages of quantities which follow a geometric or exponential law, such as rates of change, where it is more accurate than the arithmetic mean.

The harmonic mean is calculated as the inverse of the mean of the reciprocals of each individual item. When considering items such as A per B the harmonic mean is used to find the average when the items are for equal A; if they are for equal B then the arithmetic mean should be used.

Index numbers are a special form of average, used for comparing different types of items, such as industrial output made up from cars, TV sets, chemicals and so on. It is an average of a group of items, measured over time, to show the change in the items.

Apart from the average of the data, it is often useful to know the spread or dispersion of the data from this average. Several measures exist for illustrating this spread. The range is the simplest, consisting of the difference between the largest and the smallest number. It is easy to implement but does not take all the data into account. Quartile deviation is the difference between the quartiles, which divide the series of numbers into four equal parts when placed in order. Like the range, the quartile deviation is easy to implement, but in effect ignores half the values in the series.

The mean deviation is the average of all deviations and is found by taking the mean of the difference between each figure and the arithmetic mean or the median of the series, ignoring signs. It is a useful measure of dispersion in that it takes account of all the data in the series, but since signs are ignored it is not suitable for further mathematical analysis. The standard deviation is the most frequently used measure of dispersion and overcomes this problem. It is found by squaring the deviation from the mean (so eliminating signs) adding the results, finding their arithmetic mean, and then taking the square root of the result.

Case study exercises

7.1 There are three types of floroll: floroll LC, floroll S and the top of the range floroll GX. The profit margin for floroll LC is 20 per cent, for floroll S it is 40 per cent and for floroll GX it is 60 per cent. The sales of the three products over the past three years have been as below. Present these figures so as to show clearly the variation of turnover and profits between the three lines. What conclusions can be drawn regarding product mix?

floroll LC: £950M, £1430M, and £1910M
floroll S: £520M, £1100M, and £1510M
floroll GX: £420M, £610M, and £730M

7.2 Two measures of productivity have been introduced: the productivity factor (Figure 7.5) and the output (Figure 7.6). What other measures could be used and how would you present them to help John McCaully show productivity in his Shop in the most favourable light?

7.3 James Silver, the Materials Manager, has been given the task of rationalizing the supplier base, to see if more favourable terms can be obtained if the materials needed were bought from fewer sources. James found that 220 suppliers had been used in the previous year and £1936M had been spent with them. He grouped the suppliers and spend into classes, as in Table 7.14.

Present these figures in a form that will enable decisions to be taken on which suppliers should be tackled first.

Table 7.14 The supplier base

Number of suppliers	**50**	**45**	**36**	**30**	**26**	**18**	**15**
Level of spend (£M)	76	130	190	220	290	380	650

7.4 As part of the manufacturing process florolls are placed in an 800 litre tank which is then gradually filled with a special fluid, a process known as 'drowning a floroll'. The pump used to fill the tank is old and the maintenance engineer decides to check its speed. He finds that, to fill the tank on three occasions, it worked at speeds of 10 litres per minute, 20 litres per minute and 40 litres per minute. What was the average operating speed of the pump? The engineer then carries out maintenance on the pump and measures its speed again, but this time over three equal intervals of time (5 minutes). The speeds measured were 26 litres per minute, 28 litres per minute, and 31 litres per minute. What is now the average operating speed of the pump?

7.5 John McCaully was concerned that redundancies in the Dinsher Shop would result in strikes. He decided to determine the amount of time lost due to strikes in the past three years. This is shown in Table 7.15.

Table 7.15 Time lost due to strikes in the Dinsher Shop

Man hours lost through strikes	Number of days concerned
<5	5
6–10	15
11–15	22
16–20	190
21–25	350
26–30	180
31–35	80
36–40	55
41–45	20
46–50	3
>50	1

From this table, calculate the average number of days lost and the dispersion from the average in as many different ways as may be appropriate. Is there a 'best' method?

Mathematical models in decision making

Introduction

As seen in Chapter 6, the decision-making process requires the collection of a large amount of data, followed by the analysis of these data. Chapter 7 introduced techniques for presentation of data, to aid in their analysis. The present chapter will describe a few examples of mathematical models that can be used to carry out the analysis of the information collected, to help in the decision-making process.

8.1 Modelling

A model usually refers to a representation of an actual object to a reduced scale, such as a wooden model of a building. These models are often used to study the behaviour of the actual object. For example, a scale model of an aeroplane can be used in a wind tunnel to simulate the behaviour of the actual aeroplane in flight.

Mathematical models perform a similar function. They represent the actual object or event as a series of mathematical expressions, which can then be studied further. The key is that if one can define the problem in mathematical terms, then this can be solved to find answers.

Mathematical models are often programmed into computers, so the manager does not need to understand the theory behind them, just how to use the computer program. Sometimes a mathematical model is so complex that a computer is essential if it is to be solved.

The present chapter will look at a few examples of mathematical modelling used as an aid to decision making, including the theory behind them. Four main types of problem are considered: those solved by linear programming; inventory control problems; the problems of queues; and the formulation of competitive strategy, also known as game theory.

8.2 Linear programming

Several management situations are concerned with the problem of utilizing limited resources to best advantage. Linear programming is the simplest and most widely used technique for solving these problems. The aim of this method is to determine how to meet the desired goals while taking constraints into account.

The term linear implies proportionality; for example, if one machine produces 100 items per hour, then two machines will produce 200 items per hour. Programming implies some form of systematic analysis of solutions until the optimum one has been found.

Two main techniques are used for solving the mathematical models in linear programming: the graphical method and the Simplex matrix method. The graphical solution is easy to follow and illustrates the principles involved. It will be used in this chapter, although it is restricted to equations with about three variables, above which it becomes cumbersome to apply. The Simplex matrix method is not restricted to any number of variables. It is, however, tedious and is best solved by a computer. The principle is the same: all feasible solutions are examined in turn to find the optimum solution.

Transportation problems are a good example of solution by the Simplex matrix method. They concern the problem of moving goods from several warehouses to several customers in the most economical way.

Linear programming assumes that all the units under consideration can be infinitely divided, such as machine hours or kilograms of weight. If discrete items are being considered, such as number of people or machines, then an extension of linear programming, called integer programming, is used, since it gives answers in whole numbers. A further extension of linear programming is dynamic programming. It is also used to select the optimum allocation of resources, but is dynamic in that decisions are taken at several stages *en route* to the final solution.

Case study

The secret formula

The two main items which go into the manufacture of Zing (a component of a floroll), are known simply as Formula X and Formula Y. The composition of these two ingredients is a closely guarded secret, but both are produced by two processes, Upper and Lower. To produce one litre of Formula X requires 4 hours in the Upper process and 2 hours in the Lower process. To produce one litre of Formula Y requires 1 hour in the Upper process and 3 hours in the Lower process. The total amount of time available, in a day, in the Upper process is 80 hours, and in the Lower process it is 120 hours. The manufacturer makes a profit of £10 by selling one litre of Formula X and £15 by selling one litre of Formula Y.

Table 8.1 Determining the optimum production of Formula X and Formula Y

	Formula X	**Formula Y**	**Total process hours available**
Hours required in upper process	4	1	80
Hours required in lower process	2	3	120
Profit per unit (£)	10	15	

Table 8.1 summarizes this information. The objective is to maximize profits, while working within the constraints of the processes, and this problem is best solved by first determining mathematical models. If the company makes x litres of Formula X and y litres of Formula Y, then the profit equation is given by Equation 8.1, and this must be maximized. The constraints are specified by Equations (8.2)–(8.5), the last two equations being required to indicate that negative quantities cannot be made.

$$\text{Profit} = 10x + 15y \tag{8.1}$$

$$4x + y \leq 80 \tag{8.2}$$

$$2x + 3y \leq 120 \tag{8.3}$$

$$x \geq 0 \tag{8.4}$$

$$y \geq 0 \tag{8.5}$$

The problem can be solved graphically as in Figure 8.1. If the lines given by $4x + y = 80$ and $2x + 3y = 120$ are plotted, then, since the values 80 and 120 must not be exceeded, the feasible region must lie between these lines and is shown shaded. If a profit of £150 is assumed then the line $10x + 15y = 150$ can also be drawn. This gives a feasible solution, but not the optimum one. A profit of £300 is also feasible for those portions of the line which lie in the shaded region. Therefore the optimum solution is on a line parallel to these two lines and passing through point A. This can be read off from the graph as $x = 12$ and $y = 32$, so that to maximize profits the company should produce 12 litres of Formula X and 32 litres of Formula Y, and its profits will be $10 \times 12 + 15 \times 32 = £600$.

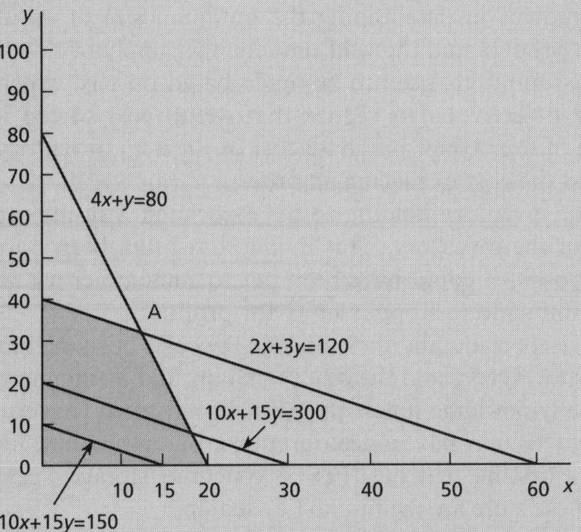

Figure 8.1 A graphical solution to a linear programming problem.

Suppose now that it is possible to increase the number of hours available in the lower process to 180, but at an increased cost. The problem is to determine the maximum cost that one can pay and still make this worthwhile.

The mathematical models, denoted by Equations (8.1), (8.2), (8.4) and (8.5) are unchanged, but Equation (8.3) changes to that in Equation (8.6).

$$2x + 3y \leq 180 \quad (8.6)$$

Solving these equations graphically, as before, will give the new optimum production quantities as $x = 6$ litres, and $y = 56$ litres, the total profit being £900. Therefore there is an increase of £300 over the previous profit figure and the company has bought another 60 hours in the Lower process. If this extra time can be obtained at less than $300/60 = £5$ per hour, it is worth having. The value of £5 is called the shadow price of the process.

8.3 Inventory control

An inventory, or stock of goods, is required for two principal reasons, namely, to enable items to be built or to be sold and to act as a buffer against an unpredictable high usage. The commonest occurrences of inventories are as piece parts in stores, from which items are drawn for manufacture, or as a store for finished goods from which items are shipped to customers. The two questions to be answered in both these cases is: how large should the inventory be and how often should it be replenished?

There are many unknowns in determining the optimum size of an inventory, such as demand for the product and the lead time for the supply of raw materials. In most cases several assumptions need to be made based on past experience. In all cases, however, the objective is to reduce the overall cost of the inventory system. These costs are of four types: purchase cost of the item; inventory holding cost; stock-out cost; and the cost of placing an order.

The largest part of the inventory holding cost is associated with the capital cost of the money tied up in the inventory. This money has often been borrowed at interest, and if not borrowed it could have been put to some other use within the organization and so amounts to lost opportunity for profits.

Storage costs are also important, and these include the costs of space, heating and lighting, the cost of the storekeeper and the cost of making and maintaining records. Insurance costs are usually not large unless the items being stored constitute a high risk. In some cases the goods may have a deterioration cost, and in these instances it is essential to institute a first in, first out (FIFO) system of storage, such that the goods that arrive first at the store are the first to be used up.

Order costs are associated with the clerical and managerial time required to place and process an order; the shipping and billing cost; the cost of receiving the goods and putting them into inventory; and the cost of using the computer (for clerical effort) to update the stock records. If the order is for more parts to be made internally in the company for sending to the finished goods store then the order costs include the set-up cost of machines and the scheduling costs.

Stock-out costs are difficult to quantify, since they depend on the type of business involved. They include items such as lost profits; the cost of borrowing the stock or buying it at a premium; the loss of goodwill and of customers; the loss of market share; line stop cost and poor staff relationships; and a halt in the production process, which may be long and expensive to restart.

In the sections which follow, analytical techniques are illustrated for determining inventory sizes that minimize overall costs within the company. In most companies about 10 per cent of the inventory items account for over 80 per cent of the inventory costs, so that complex control techniques are usually applied to these items and simpler methods are used for the remainder.

There are two main types of inventory situations, namely those in which the demand or usage is known or predictable and those in which it is unknown.

8.3.1 Fixed demand

Four different types of situation can occur: the basic case of known demand; quantity discounts; cases in which stock-outs are permitted; and the case where the receipts into inventory are not instantaneous.

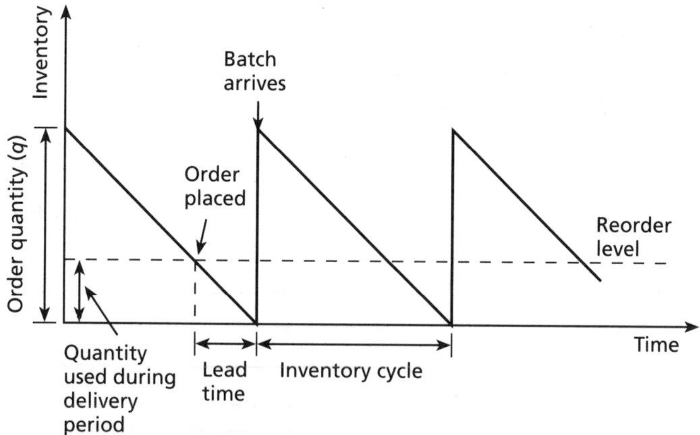

Figure 8.2 Inventory under conditions of constant demand, fixed lead time and instantaneous batch receipts.

The assumptions made in the base case are that the demand for the product is known over a period and is fairly uniform, the time between the placing of an order and its arrival into stores (lead time) is zero or a known fixed value, and that all items in a delivery (the batch quantity) arrive at the same time. Under these conditions Figure 8.2 shows how the inventory will vary with time. Whenever a batch of q items arrives the inventory increases to this value. The inventory is then gradually decreased over a period until the reorder level is reached, when a new order is placed for q items. Since demand is assumed to be known, the inventory is allowed to fall to zero at the end of each inventory cycle without risk of shortages.

The lead time shown in Figure 8.2 can sometimes be taken to be zero when, for example, the vendor is prepared to make an express delivery.

The problem in all inventory management systems is to find the most economical order quantity (EOQ), which the user should order so as to minimize the overall costs. If d is the demand per annum, q is the batch size for each order, o is the cost per order, i is the cost per unit item and s is the stockholding cost per item as a fraction of the inventory, then the total order cost is given by Equation (8.7), the total stockholding cost by Equation (8.8) and the total unit cost by Equation (8.9).

$$\text{Total order cost} = \frac{o\,d}{q} \tag{8.7}$$

$$\text{Total stockholding cost} = \tfrac{1}{2}q\,i\,s \tag{8.8}$$

$$\text{Total cost of units} = i\,d \tag{8.9}$$

The cost of the units is stated to be unaffected by the batch size q, so that if the total order cost and total stockholding cost are plotted against q the graphs

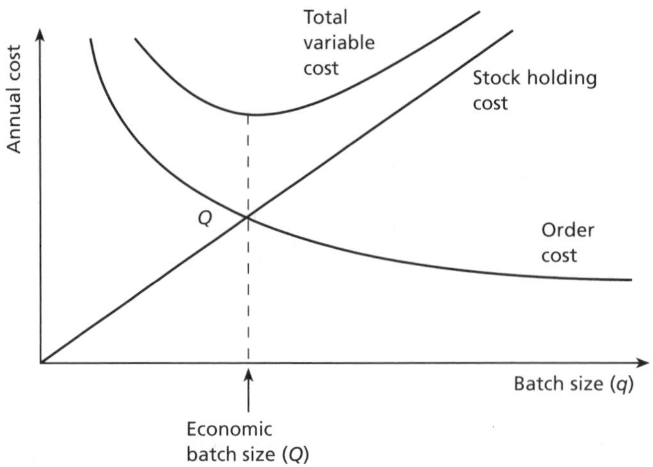

Figure 8.3 The dependence of stockholding cost and order cost on batch size.

shown in Figure 8.3 are obtained. The total variable cost curve has a minimum at the EOQ (denoted by Q) and this is also the point at which the cost of stockholding and ordering are equal. Therefore it is given by Equation 8.10.

$$\text{Economic order quantity } (Q) = \left(\frac{2\,o\,d}{i\,s}\right)^{1/2} \tag{8.10}$$

This model gives the basic EOQ formula and specifies the batch size which will minimize total inventory costs. The number of times orders need to be placed in a year is equal to q/Q.

Case study

Antistatic dinshers

As part of the production process, dinshers need to be covered by a thin film of special antistatic fluid. James Silver, the Materials Manager, has been informed that during the next year it is estimated that 500 kg of this antistatic fluid will be required, and he assumes that the demand over an inventory cycle will be constant and that lead times from the supplier are fixed and known. James wishes to estimate the Economic Order Quantity and the number of orders to be placed in the year.

The company's internal costs for placing an order are estimated at £100 per order, and the vendor charges £50 per kilogram of the antistatic fluid. Stockholding costs are estimated at 20 per cent of the inventory cost, the bulk of these costs being due to the interest on money tied up in the inventory.

Therefore, in Equation (8.10), $d = 500$ kg; $o = £100$; $i = £50$ per kilogram; and $s = 20$ per cent or 0.2, giving the EOQ as:

$$Q = \left(\frac{2 \times 100 \times 500}{50 \times 0.2}\right)^{1/2} = 100 \text{ kg}$$

The number of orders per year is 500/100 = 5. Therefore James Silver needs to place 5 orders per year, at equal intervals of time, for 100 kg at each order.

It should be noted that the total cost curve of Figure 8.3 is relatively flat in the region of the EOQ, so that it is not very sensitive to errors made when estimating the various individual costs. For example, if the price of the antistatic fluid changes from £50 per kilogram to £100 per kilogram, an increase of 100 per cent, the EOQ will change from 100 kg to 70.7 kg, a decrease of 30 per cent. Examination of Equation (8.10) shows that this is due to the fact that the EOQ varies as the square root of the individual items.

Case study

The more the cheaper

James Silver has negotiated quantity prices for antistatic fluids: £50 per kilogram for a batch size of 100 kg; £48 per kilogram for a batch size of 250 kg; and £47 per kilogram for a batch size of 500 kg. He now wishes to calculate the EOQ, and determines the total costs at the various batch sizes, as in Table 8.2. Once the batch size becomes bigger than the EOQ without discounts the cost increases between the discount points; therefore only these points need to be considered in any EOQ calculations.

Table 8.2 Calculation of EOQ with quantity discounts

		$d = 500, o = 100, s = 0.2$		
Item	**Model**	$q = 100$, $i = 50$	$q = 250$, $i = 48$	$q = 500$, $i = 47$
Total unit cost	$i\,d$	25 000	24 000	23 500
Total order cost	$\frac{o\,d}{q}$	500	200	100
Total stockholding cost	$\frac{1}{2}q\,i\,s$	500	1 200	2 350
Total		26 000	25 400	25 950

From Table 8.2 James Silver can see that the EOQ has changed to 250 kg, and it is not worth accepting the highest quantity discount offered.

The advantages of buying in quantity are lower unit prices; lower ordering costs; lower transportation costs; fewer stock-outs; and preferential treatment by the vendor. Some of the disadvantages are larger stocks and more capital tied up in inventory; less flexibility to be able to change the usage pattern (for example because of a design change in the product) and greater risk of deterioration of the stock. The non-economic considerations do not come into the model of the EOQ, so James Silver realizes that the model must not be applied mechanistically for every application. He checks with Engineering that the longer stockholding time will not adversely affect the life of the antistatic fluid.

Shortages and non-instantaneous receipts

If the demand over a period is known and is constant, with fixed lead time, the danger of running out of stock is low provided sufficiently large batch sizes are ordered at the correct time. However, even under these conditions, a company may deliberately plan to let the inventory run into shortages if, for example, it can cover these shortages by some other means, such as borrowing.

If c is the stock-out cost per unit, q is the batch size, r is the starting inventory; and p is the stock-out or shortage, then if the inventory is allow to run into shortages the models given in Equations (8.11)–(8.15) apply. Every time a new batch arrives the original shortage is made up (for example any borrowings are repaid), so the inventory is less than the batch size.

The total inventory is equal to the average inventory times the unit cost of inventory and the total stock-out cost is the average level of stock-out times the cost of borrowing per unit.

$$\text{Average inventory} = \tfrac{1}{2}\frac{r^2}{q} \tag{8.11}$$

$$\text{Average level of shortage} = \tfrac{1}{2}\frac{p^2}{q} \tag{8.12}$$

$$\text{New EOQ } (Q_1) = \left(\frac{2od}{is}\right)^{1/2}\left(\frac{is+c}{c}\right)^{1/2} \tag{8.13}$$

$$\text{Starting inventory } (r) = \left(\frac{2od}{is}\right)^{1/2}\left(\frac{c}{is+c}\right)^{1/2} \tag{8.14}$$

$$\text{Stock-out } (p) = q - r \tag{8.15}$$

It has also been assumed so far that the total batch quantity is received into inventory all together. This is not always the case, since vendors often deliver the order quantity in groups, and when parts are being added to the finished goods store the manufacturer is selling the items at the same time that it is making them.

If g is the receipt rate in items per unit of time and u is the usage rate of the same units, then the new EOQ (Q_2) for the inventory system is given by Equation (8.16), and the average stock by Equation (8.18).

$$Q_2 = \left(\frac{2\,o\,d}{i\,s[1-(u/g)]}\right)^{1/2} \tag{8.16}$$

$$\text{Average stock} = \frac{q}{2}\left(1-\frac{u}{g}\right) \tag{8.17}$$

For very small values of u compared with g, Equation (8.16) becomes equivalent to Equation (8.10), since the receipts are now in effect instantaneous.

Case study

Dinsher shortages

James Silver has been told that the company is able to borrow antistatic fluid from another division of the company at £20 per kilogram, if required. This then represents the stock-out cost (c) and, from the earlier example (see the case study 'Antistatic dinshers'), the value of the Economic Order Quantity and various costs can be calculated from the mathematical models, as in Table 8.3.

Table 8.3 Calculation of EOQ with allowable shortages

Item	**Symbol**	**Model**	$d = 500$, $i = 50$, $o = 100$, $s = 0.2$, $c = 20$
Economic batch quantity	Q_1	$\left(\frac{2od}{is}\right)^{1/2}\left(\frac{is+c}{c}\right)^{1/2}$	122
Starting inventory	r	$\left(\frac{2od}{is}\right)^{1/2}\left(\frac{c}{is+c}\right)^{1/2}$	82
Stock-out	p	$Q_1 - r$	40
Total stockholding cost		$\frac{1}{2}\frac{r^2si}{Q_1}$	276
Total order cost		$\frac{od}{Q_1}$	410
Total unit cost		id	25 000
Total stock-out cost		$\frac{1}{2}\frac{p^2c}{Q_1}$	131
Total cost			25 817

Table 8.3 also gives the total cost, which is seen to be less than the total cost when shortages are not allowed (Table 8.2, third column). Therefore James Silver decides to order in batches of 122 kg, and to place orders four times a year (= 500/122). There will be a stockout of 40 kg at the end of each inventory period which will need to be made up by borrowing.

Returning to the original problem (case study 'Antistatic dinshers'), the supplier has informed James Silver that, due to a change in policy, he will be delivering goods at the rate of 4 kg per day (g). James knows that the rate of usage from inventory is 3 kg per day (u); therefore the new EOQ can be found from Equation (8.16) as:

$$Q_2 = \left(\frac{2 \times 100 \times 500}{50 \times 0.2(1 - 0.75)}\right)^{1/2} = 200 \text{ kg per day}$$

The costs under these conditions are found as in Table 8.4.

Table 8.4 Costs under conditions of non-instantaneous receipt into inventory

Item	Model	$d = 500$, $i = 50$, $o = 100$, $q = 200$, $s = 0.2$, $u = 3$, $g = 4$
Total unit cost	$i\,d$	25 000
Total order cost	$\frac{o\,d}{q}$	250
Total stockholding cost	$\frac{i\,s\,q}{2}\left(\frac{1-u}{g}\right)$	250
Total		25 550

8.3.2 Variable demand

When the demand over a period is variable it must be estimated in terms of its mean and standard deviation before calculations of economic batch quantity can occur. If M_x and S_x are the mean and standard deviation of values $x_1, x_2, x_3, \ldots x_n$ and $d = x_1 + x_2 + \ldots + x_n$ then Equations (8.18) and (8.19) are obtained.

$$M_{\mathrm{d}} = M_x n \tag{8.18}$$

$$S_{\mathrm{d}} = S_x\,(n)^{1/2} \tag{8.19}$$

If the mean demand M_{d} is substituted for the actual demand in the EOQ model of Equation (8.10), then a good approximation can be obtained, as in Equation (8.20).

$$\text{EOQ }(Q_3) = \left(\frac{2\,o\,M_{\mathrm{d}}}{i\,s}\right)^{1/2} \tag{8.20}$$

The actual demand is unknown, so that even though the EOQ is known it is not possible to calculate the order frequency accurately. This is given approximately by Equation (8.21), but if the demand exceeds M_{d} then more orders will need to be placed to avoid stock-out.

$$\text{Order frequency} = \frac{M_{\mathrm{d}}}{Q_3} \tag{8.21}$$

There are two systems that may be used when placing orders:

1. The fixed quantity or continuous review system. This continuously reviews the inventory and places an order, for a fixed amount, when this falls below a critical reorder level. A high value of reorder level will result in large residual stock and increase stockholding costs, whereas a low level will cause frequent stockouts and so increase stock-out costs. The reorder level is generally determined by considering a probability of stock-out that the company is willing to accept.
2. The fixed frequency or fixed period system. In this the time at which the order is placed is fixed, but the order quantity is varied according to the inventory level at that time. Since the order will need to cover the lead time plus one inventory period, the batch size is determined by the maximum probable demand during this period and the inventory level at that time is subtracted from this.

In both of the above systems the prime objective is to determine when to order and how much to order. Since the demand is variable most companies adopt a service level policy in answering these questions; that is, they hold sufficient buffer stock that the probability of being out of stock is low.

Generally a fixed period system involves a larger safety stock and higher holding cost, so it is less desirable than a fixed quantity system when unit cost is low and unit inventory cost is high. However, a fixed quantity system needs continuous inventory monitoring and this adds to overall inventory costs. A fixed period system is also more convenient to use when different types of goods are obtained from one supplier, since costs can often be reduced by placing several orders together.

Case study

Non-constant usage

Suppose that the demand for antistatic fluid (see the case study 'Antistatic dinshers') is not constant but varies with a mean of 50 kg per month and a standard deviation of 5 kg. Then the mean demand over the year, the standard deviation and the new EOQ can be obtained from Equations (8.18)–(8.20) as:

$$M_d = 50 \times 12 = 600 \text{ kg}$$

$$S_d = 5 \times (12)^{1/2} = 17.3 \text{ kg}$$

$$\text{EOQ } (Q_3) = \left(\frac{2 \times 100 \times 600}{50 \times 0.2}\right)^{1/2} = 110 \text{ kg}$$

Suppose that a fixed quantity system is used for reorder of goods. If the lead time for an order is 1.5 months, then Equations (8.18) and (8.19) give the demand (M_L) and standard deviation (S_L) of the demand during the lead time as:

$$\textbf{Demand } (M_L) = 50 \times 1.5 = 75\,\text{kg}$$

$$\textbf{Standard deviation } (S_L) = 5(1.5)^{1/2} = 6.1\,\text{kg}$$

Suppose that the demand is distributed normally and that James Silver is prepared to accept only a 5 per cent probability of stock-out (referred to as a 95 per cent stock-out level). Assume that to obtain this the reorder level must be set to L.

For probability 0.05 (5 per cent), $1 - 0.05 = 0.95$, so that, from probability theory, $\omega = 1.64$, and

$$1.64 = \frac{L - 75}{6.1}$$

or

$$L = 85\,\text{kg}$$

Therefore every time the inventory falls below 85 kg, James Silver should place a new order for a batch size of 110 kg. The difference between the reorder level and the mean demand during the lead time is called the safety stock. In this case it is $85 - 75 = 10$ kg. It does not mean that this stock is always present at the end of the inventory period since its residual value is determined by the demand during the lead time. The high reorder level has been caused by the acceptable low stock-out probability and the relatively long lead time. The inventory period will vary depending on the rate of usage. The stockholding cost increases with the service level, and above 95 per cent the increase is very rapid. Therefore very few companies attempt to obtain a service level above 95 per cent to 97 per cent.

Suppose now that James Silver decides to use a fixed period system for reordering stock. The mean demand per year is 600 kg and the EOQ was found to be 110 kg, so the mean number of orders placed per year is equal to $600/110 = 5.5$ and the order period is $12/5.5 = 2.2$ months. Since the lead time is 1.5 months, the order placed will need to cover a demand for 3.7 months. The mean demand is 50 kg per month with a standard deviation of 5 kg, which gives the mean and standard deviation, from Equations (8.18) and (8.19) as:

$$M_d = 50 \times 3.7 = 185\,\text{kg}$$

$$S_d = 5 \times (3.7)^{1/2} = 9.6\,\text{kg}$$

Therefore, assuming a normal distribution, the order quantity L needed to obtain less than a 5 per cent probability of stock-out can be found as before, from probability theory, as:

$$1.64 = \frac{L - 185}{9.6} \quad \text{or} \quad L = 201\text{ kg}$$

Therefore every 2.2 months James Silver should place an order for 201 kg minus the stock at that time, so as to have a stock-out probability of below 5 per cent.

8.4 Queues

8.4.1 Queuing theory models

In a simple waiting time situation suppose that the customers arrive at a rate denoted by λ and that these customers are served at a rate of μ. Then the traffic intensity (ρ) is given by Equation (8.22).

$$\rho = \frac{\lambda}{\mu} \tag{8.22}$$

Traffic intensity is the probability that an arrival will have to queue and in a stable situation the intensity must be less than one or the queue will grow to an infinite length.

If one assumes that the arrival rate and the service times vary randomly, then with μ greater than λ it does not mean that there is no queue. This is because when there are no customers the service mechanism is idle, and this idle time cannot be stored and used when several customers arrive within a short interval of each other.

The average number of units in a queue, and the average waiting time, are given by Equations (8.23) and (8.24). These represent the two basic queuing theory models and are derived from several assumptions:

- There is a single service channel. When several parallel channels exist the models are much more complex.
- The arrival rates are random and follow a Poisson distribution. The service times are also random and are exponentially distributed. This means that a large number of customers need a very low service time, whereas a few customers require very long service times.
- There is no limit to the maximum size of the queue and items in the queue do not leave the queue no matter how long it gets.
- The population from which the arrivals are drawn is infinite, so that the arrival rate is not influenced by the number in the queue.
- The service rate is independent of the arrival rate. In practice, in the case of human servers, the tendency is to increase the service rate as the queue gets longer.
- There are no simultaneous arrivals.
- Arrivals are served on a FIFO system and there are no priorities.

- All units are discrete, for example telephone callers, customers in a supermarket and machine failures. Continuous variables, such as a liquid flowing into a tank which is served by outlet valves, are not considered here.

$$\text{Average number in queue} = \frac{\rho^2}{1-\rho} \tag{8.23}$$

$$\text{Average waiting time in queue} = \frac{\rho}{\mu-\lambda} \tag{8.24}$$

Case study

Slurpers

One of the processes used in the manufacture of a dinsher is known as slurping, which is done by use of a large number of slurping machines called slurpers. These machines break down on an average of three per month and John McCaully, the manager of the Dinsher Shop, has determined that the breakdown pattern is random, with a Poisson distribution. The time taken to repair the machines depends on the extent of the damage and varies randomly with an exponential distribution, but on average the repair rate is four per month.

It costs £15 000 per month in lost production when a machine is out of action and the cost of repair, in terms of the overhead of the repair department, is £6000 per month. John McCaully is considering further mechanization in the repair department, which can be done in two stages. The first stage would allow it to increase its repair rate to 5 machines per month, although the operating costs would then rise to £16 000 per month. The second stage would increase the repair rate to 8 per month but the operating costs would also increase to £32 000 per month. John McCaully needs to know whether he should go ahead and modernize the repair facility and to what stage.

The problem can be solved using the queuing models of Equations (8.22) and (8.23), as shown in Table 8.5 where stage 0 represents the current situation. From this table it is clear that John McCaully should carry out the first stage of modernization but not the second stage.

Table 8.5 Increasing the capability of the repair centre in the Dinsher Shop

Stage	Arrival rate λ	Service rate μ	Traffic intensity $\rho = \lambda/\mu$	Average number of machines out of action $\rho^2/(1-\rho)$	Total cost per month £
0	3	4	0.75	2.25	$2.25 \times 15\,000 + 6000 = 39\,750$
1	3	5	0.6	0.9	$0.9 \times 15\,000 + 16\,000 = 29\,500$
2	3	8	0.375	0.225	$0.225 \times 15\,000 + 32\,000 = 35\,375$

Case study

Piece parts

There is a single storeman in the piece parts store in the Dinsher Shop and he serves collectors who come to the store for different types of production piece parts. The arrival rate is random (Poisson distribution) with an average of 18 per hour. The service time depends on the parts required and is also random (exponential distribution). A collector is paid £10 per hour.

Currently the storeman can serve on average 19 customers per hour and his cost is £48 per day. John McCaully is considering increasing the service rate and has been given three options:

1. Provide the storeman with a sophisticated index system. This will enable him to increase the service rate to 20 per hour, but his cost will increase to £60 per day.
2. Employ two storemen with a larger store. The service rate will now increase to 22 per hour and the cost will increase to £150 per day.
3. Employ three storemen with a larger store. The service rate and costs will now be 24 per hour and £300 per day, respectively.

John McCaully wishes to know which option to follow. Table 8.6 (see page 218) gives the solution, where option 0 is the current system of a single storeman. The cost of collectors per day assumes an eight hour day. From this table it is seen that option 2 has the lowest overall cost. However, John McCaully also notes that option 3 has only a slight cost increase but the waiting time is almost halved, so he decides to adopt this option since it would considerably reduce the frustrations which collectors have experienced when waiting for service.

8.4.2 Simulation techniques

Simulation is a method for obtaining information about the system that is being studied and then working out, step by step, what happens when customers arrive and go through the service mechanism. The best known method is called Monte Carlo simulation, and, as its name implies, it is based on the assumption of randomness or chance.

Simulation methods are tedious to implement and require the use of a computer programme. However, they are based on very few assumptions and can therefore be applied to a wide range of systems. The simplified case study 'Trucks in queue' shows how simulation techniques can be used.

Table 8.6 Modernizing the piece part store in the Dinsher Shop (see the 'Piece parts' case study on page 217)

Option (1)	**λ (2)**	**μ (3)**	**ρ (4) = (2)/(3)**	**$\frac{\rho^2}{1-\rho}$ (5)**	**Cost of collectors per day (£) (6) = (5) × 8 × 10**	**Cost of storemen per day (7)**	**$\frac{\rho^2}{\mu-\lambda}$ hours (8)**	**Total cost per day (9) = (6) + (7)**
0	18	19	0.95	18.0	1440	48	0.95	1488
1	18	20	0.90	8.1	648	60	0.45	708
2	18	22	0.82	3.74	299	150	0.21	449
3	18	24	0.75	2.25	180	300	0.125	480

Case study

Trucks in queue

In the goods-in bay of the FloRoll Manufacturing Company trucks are constantly arriving, with deliveries of components used in the manufacture of florolls. The trucks need to be unloaded and the goods stored until required. The times between truck arrivals and the unloading times both vary randomly. By observing the system for a few days, Stewart Jones, the Site Services Manager, has obtained the data given in the first two columns of Tables 8.7 and 8.8 for the time between truck arrivals and the service time for these trucks.

Table 8.7 Truck arrival times

Time between arrivals (minutes)	Frequency	Frequency (%)	Random number allocated
5	80	21	00–20
10	110	29	21–49
15	140	37	50–86
20	50	13	87–99

Table 8.8 Truck service times

Service time (minutes)	Frequency	Frequency (%)	Random number allocated
5	50	13	00–12
10	140	37	13–49
15	120	32	50–81
20	70	18	82–99

Therefore, during the observation period, trucks arrived at five minute intervals 80 times and at ten minute intervals 110 times, and so on. Similarly, trucks were emptied in five minutes on 50 occasions and in ten minutes on 140 occasions, and so on. In both tables the frequency can be calculated as a percentage of the total number of observations and random numbers between 0 and 99 allocated to this, as in the fourth columns of Tables 8.7 and 8.8. Stewart Jones wants to determine parameters for the service system in order to decide whether it needs to be improved.

Table 8.9 shows the simulation over an hour and ten minutes of the data given in Tables 8.7 and 8.8. Columns 5 and 6 represent a series of random numbers between 0 and 99. At the start of the day, at 9:00, the first truck (A) is assumed to arrive. The service random number is 65, so that from Table 8.8 the service time will be 15 minutes and column 4 shows that service will therefore be completed at 9:15. The arrival random number at 9:00 is 09, so that from Table 8.7 the arrival time between trucks A and B is 5 minutes. Therefore truck B arrives at 9:05, but it cannot be served until 9:15 so it has to wait. The arrival random number at 9:05 is 18, so that truck C arrives 5 minutes after B, and it also has to wait. At 9:15 truck A has been served and leaves the queue, so that truck B can be served. The service random number is 48 giving a service time of 10 minutes, so that service is completed at 9:25. The rest of the table can be worked through as shown. For the portion of the simulation in Table 8.9 it is seen that the maximum queue is two, the maximum waiting time is 15 minutes and there is a service idle time of 5 minutes.

Table 8.9 Simulation of truck arrival and service (from Table 8.7 and Table 8.8)

Time (1)	Truck arriving (2)	Truck awaiting service (3)	Truck being served (4)	Service random number (5)	Arrival random number (6)
9:00	A		A	65	09
9:05	B	B	A		18
9:10	C	C, B	A		82
9:15		C	B	48	
9:20		C	B		
9:25	D	D	C	11	00
9:30	E	E	D	76	97
9:35		E	D		
9:40		E	D		
9:45			E	06	
9:50	F		F	74	90
9:55			F		
10:00			F		
10:05					
10:10	G		G	80	04

8.5 Competitive strategy

Managers are constantly making decisions in a competitive environment, in which the outcome depends not only on their own actions but also on those of competitors. The scientific technique that has been developed to help managers in making decisions under these circumstances is called the theory of games.

8.5.1 Zero-sum games

The simplest type of situation is known as the two-person zero-sum game, in which the gain of one competitor is exactly equal to the loss of the other. Situations such as this exist when two companies are operating in a fixed market, where one company can increase its share only at the expense of the other.

In the theory of games it is assumed that each competitor knows of all the alternative moves its rival can make, although it does not know which of these alternatives is to be adopted at any time.

Case study

Competing florolls

Jake R. Topper II, Chief Executive of the FloRoll Manufacturing Company, sat in his luxurious office and gazed absent-mindedly out of the window. He had been informed by Harry Dean, the Special Projects Director, that their nearest rival in the German market, Winger Inc., was planning an aggressive campaign to gain market share.

There were three options, which both companies could adopt: do nothing, increase advertising, or reduce price. Jake decides to write down the effects of these alternatives. This is shown in Table 8.10, where the payoff matrix for all the options has been shown, the loss or gain in market share being from the perspective of the FloRoll Manufacturing Company.

From Table 8.10 it is seen that if both companies do nothing their share of the market will remain unchanged. If the FloRoll Manufacturing Company goes in for increased advertising but Winger Inc. does nothing it will increase its share by 6 per cent. If Winger Inc. also increases advertising then the FloRoll Manufacturing Company will see its share of the market decline by 2 per cent, while Winger Inc. will increase its share by 2 per cent. This is because the FloRoll Manufacturing Company is already carrying out a considerable amount of advertising, so an increase will have less impact.

Table 8.10 Payoff matrix for floroll market share

FloRoll Manufacturing Company strategy	Winger Inc. strategy: Do nothing	Winger Inc. strategy: Increase advertising	Winger Inc. strategy: Reduce price
Do nothing	0	−5	−10
Increase advertising	+6	−2	+5
Reduce price	+5	+1	+4

Game theory states that each player will choose a strategy which results in least loss. This is based on the assumption by each player that the opponent is very shrewd and will always play so as to minimize its opponent's gain. In Table 8.10, if the FloRoll Manufacturing Company does nothing it could lose up to 10 per cent market share, depending on the strategy adopted by Winger Inc. If the FloRoll Manufacturing Company increases advertising, it could lose up to 2 per cent market share. If the strategy of reducing price is adopted, then it can be certain of gaining at least 1 per cent market share. Jake Topper therefore decides to adopt this policy.

Similarly Winger Inc. will go in for increased advertising, since under these circumstances it only loses 1 per cent of its market share, whereas for the other two alternatives it can lose up to 6 per cent and 5 per cent. This is called the minimax strategy, since the company is minimizing its maximum possible loss.

The end result of the above competitive position is that the FloRoll Manufacturing Company carries out a price reduction while Winger Inc. increases advertising and loses 1 per cent market share. This is a stable situation, since even when both players realize each other's strategy they do not change their own strategies, since they are the best ones considering what the opponent is playing.

Note that the point +1 per cent is the smallest element in its row and the largest in its column. It also represents the point where the maximum loss to Winger Inc. for any strategy is also the minimum gain to the FloRoll Manufacturing Company. This point is known as the saddle point of the payoff matrix, and the value of the game to the FloRoll Manufacturing Company is +1. Both players are also said to adopt a pure strategy, since they never change their strategy, and this is always done where the matrix has a saddle point.

On examining Table 8.10 it is seen that for every move that Winger Inc. makes, the FloRoll Manufacturing Company is better off if it increases advertising rather than doing nothing. The same considerations hold good for Winger Inc. for the three alternative moves made by the FloRoll Manufacturing Company. Therefore neither company will adopt the strategy of doing nothing, and this can be removed from the pay-off matrix, as in Table 8.11. The previous table was said to have dominance, since one strategy was superior to another strategy irrespective of the opponent's move. (Note that Table 8.11 also contains dominance, since Winger Inc. will always be better off with increased advertising.)

Table 8.11 Modification to Table 8.10 to remove some dominance

	Winger Inc. strategy	
FloRoll Manufacturing Company strategy	**Increase advertising**	**Reduce price**
Increase advertising	−2	+5
Reduce price	+1	+4

Suppose that the payoff matrix was as illustrated in Table 8.12. There is now no saddle point and both players will adopt a mixed strategy, rather than a pure strategy, since they will vary their moves depending on what they think their opponent is playing. Mathematical techniques exist for determining the best mix of moves in order to maximize or minimize loss, but these will not be considered here.

The game shown in Table 8.12 is also unstable, since once a player discovers what the opponent is playing that player can improve his or her gain by changing the move. For example, if Winger Inc. is playing 'increase advertising' and the FloRoll

Manufacturing Company is playing 'reduce price', then when Winger Inc. discovers its opponent's move it can change its own play to 'reduce price' and so increase its gain. When the FloRoll Manufacturing Company learns of this change it will move to 'increase advertising' to increase its gain, and so on.

Table 8.12 Alternative payoff matrix for floroll market share

FloRoll Manufacturing Company strategy	**Winger Inc. strategy**	
	Increase advertising	**Reduce price**
Increase advertising	−2	+5
Reduce price	+1	−1

8.5.2 Games against nature

Very often a manager needs to work in a competitive situation against nature. Nature is not trying to outwit the manager, but in all other respects the opponent's move is unpredictable. In these instances the manager needs to obtain as much information as possible and then adopt a strategy which maximizes gain.

Case study

Rain, rain, go away

The Social Club at the FloRoll Manufacturing Company is organizing a special outdoor event and the club secretary, Clare Jolly, is trying to decide whether to insure against rain or not. The probability of rain on the day is 0.2 and the cost of the insurance policy will reduce the club's profits on the day by 2 per cent. If no insurance is taken out and it rains then profits are expected to be 5 per cent lower. If no insurance is taken out and it does not rain, or if insurance is taken out and it does rain, the profit is expected to be unaffected. The payoff matrix is as in Table 8.13.

Table 8.13 Payoff table for insurance against rain

Club secretary	**Nature**	
	Rain	**No rain**
Insure	0	−2
Do not insure	−5	0

The probability of rain is 0.2, and so the probability of no rain is 0.8. Therefore:

Gain if insured $= 0.2 \times 0 + 0.8 \times (-2) = -1.6$

Gain if not insured $= 0.2 \times (-5) + 0.8 \times 0 = -1$

Clare Jolly therefore decides to maximize her gains (minimize her losses) by not insuring. This will be the case until the probability of rain increases to about 0.29.

Summary

Mathematical models are similar to physical models; they represent the actual object or event as a series of mathematical expressions which can be analyzed in order to predict the behaviour of the original object or event. Four examples of mathematical modelling were considered in this chapter: those solved by linear programming; inventory control problems; the problems of queues; and models used for developing competitive strategy.

Linear programming is used for solving problems concerning the utilization of limited resources for maximum gain. It assumes proportionality: that is, if one machine produces 10 units in one hour then two machines will produce 20 units. In linear programming operating equations are developed to describe the situation, and then all feasible solutions are examined in turn to find the optimum one. This can be done graphically, which is usually limited to about three variables, or by the Simplex matrix method, which is not limited to any number of variables but is tedious and requires the use of a computer.

In inventory control situations the problem is to determine how large the inventory should be and how often orders should be placed to replenish it. The aim is to ensure that inventory costs, which include order costs, are minimized. There are two main types of inventory situation: those in which the demand or usage is known or predictable and those in which it is unknown or variable.

The assumptions made in the case of known demand are that the demand for the product over a period is fairly uniform, the lead time between placing an order and its arrival into stores is also fixed, and that all items in a delivery arrive at the same time. Other variables that were considered are: when quantity discounts are applicable; when stock-outs are permitted; and when receipts into inventory are not instantaneous. For the case in which demand is unknown or variable it must be estimated in terms of its mean and standard deviation before calculations can occur. One of two systems is then used when placing orders for new stock: the inventory is continuously reviewed and orders placed for a fixed amount when this falls below a critical reorder level; or orders are placed at fixed intervals, but the quantity ordered is varied according to the inventory level at that time.

Queuing-related problems aim to determine the optimum service level to meet a given requirement. They occur frequently within companies, such as the queue of trucks waiting to be unloaded or filled with goods, the queue of operators waiting to draw goods out of stores, and the queue in the works canteen. Mathematical equations can be derived to cover the instances where the arrival rate of customers, and the service rate with which they can be provided, are random but follow known distributions, such as Poisson and exponential distributions. Simulation techniques, such as the Monte Carlo simulation method, are based on very few assumptions and are therefore applicable to a much wider range of applications. In this method the arrival and service rates are simulated so as to determine parameters such as the maximum queue length, the maximum service time and the service idle time.

One of the key aims of strategic decision making is to gain competitive advantage, and mathematical techniques, such as the theory of games, have been developed to assist in this area. The simplest type of situation is the two-person zero-sum game in which the gain of one of the competitors exactly equals the loss of the other. Games also exist against nature, where nature is not deliberately trying to outwit the manager, but its intentions are unpredictable, and the manager is working to maximize gains under these circumstances.

Case study exercises

8.1 Medlay Inc., located adjacent to the FloRoll Manufacturing Company, manufactures two products, P1 and P2. Each P1 requires 5 kg of material M and 3 kg of material N. Each P2 requires 3 kg of M and 3 kg of N. In total there are 350 kg of M available and 250 kg of N. The profit on each P1 is £50 and on each P2 it is £40. What mix of P1 and P2 should be manufactured in order to maximize total profit?

8.2 Medlay Inc. uses 10 000 kg of material N per year. A kilogram of material N costs £5, the cost of placing an order is £200 and the stockholding cost, expressed as a fraction of the inventory, is 0.4. Calculate the economic order quantity (EOQ) for material N.

8.3 The Inventory Stores in Medlay Inc. stocks product P1, which is then shipped to customers on demand. It is estimated that during the next period the demand will be 100 000 units. The overhead cost involved in setting up the machines to commence making a batch of product P1 is £1000. The cost of each P1 is £10 and the inventory holding cost, as a fraction of inventory, is 0.6. The rate of receipt of product P1 into inventory is 2000 per day when production commences, and the rate of withdrawal is an average of 300 per day through the period. Find the EOQ.

8.4 The service rate in the canteen of Medlay Inc. is 20 per hour and the arrival rate of customers is 18 per hour. Calculate the average number of people in the queue and the average waiting time per person. If the arrival rate is reduced to 15 per hour, for example by staggering the lunch break, what is the new value for average queue size and waiting time? What other parameters could be changed to reduce the length of queue and the waiting time?

8.5 Trucks arrive at the goods-in bay of Medlay Inc. at an average rate of 6 per hour and they are unloaded at an average rate of 8 per hour. Calculate the average number of trucks in the queue and the average time spent in the queue. State all assumptions made in the calculations.

8.6 The year's warranty on the new television set for the Medlay Inc. Social Club has run out, and the club secretary needs to decide whether or not to insure it for the next year. She estimates that there is a 0.1 probability of the set going wrong during the year. The cost of the insurance policy is £50 and the cost of repair, if not insured, is likely to be £200. Construct a payoff matrix and advise the club secretary whether she should insure the set.

9 Forecasting

Introduction

This chapter describes some of the techniques that may be used to forecast future data, such as the size of a market or the technology available to meet a given market requirement.

9.1 Forecasting the future

Forecasting can be likened to looking into a crystal ball, except that in most cases managers would replace the crystal ball with a computer. Often, unfortunately, the end result is not much better!

Forecasting is an art and a science – it uses scientific tools to help the manager get to an informed guess. Its aim is to provide information for planning and decision making. A forecast tries to define what one believes will happen in the future. A plan defines what one would like to happen in the future, and maps out the activities needed to meet this aim. Often a forecast feeds into a plan – the plan aims to change the results that the forecast predicts will occur if no action is taken.

There are many applications for forecasting within industry, such as determining the factory output needed; forecasting the volume of sales of a new or existing product; Q forecasting the turnover or profit of a company; and predicting the price trends on competitor products. Forecasts have many knock-on effects. For example, a high sales forecast may mean that the company needs to embark on an automation programme in its plant, so as to ensure that it can meet the level of demand anticipated.

Forecasts have different time frames. For example, forecasting the level of goods needed on a supermarket shelf would have a time span of one to two days, while forecasts to determine whether automation is required within a plant would have a time span of two to five years. Computerized forecasting systems are available for handling large numbers of items quickly and carrying out repetitive tasks, such as predicting the demand for the many thousands of items normally stocked by a supermarket. However, in many cases the actual forecast requires management judgement and cannot be automated.

There are many techniques for forecasting and no one method is best in all situations (Georgoff and Murdick, 1986; Makridakis, 1987; Makridakis and McGee, 1983). Generally, forecasting methods can be classified into three broad groups:

- Qualitative forecasting, which is based on judgement of past experience and future trends. This is generally the method used for forecasting events with long time-scales. Because it is based on management judgement different results can be obtained by different forecasters, although most methods take an average from several individuals. This method is used when no data are available or when past data are unreliable for predicting the future.
- Quantitative techniques that use time series for forecasting. Past data is extrapolated into the future using a time series.
- Quantitative techniques that use the causal model. Again, past data are used to predict the future, but the technique for this is based on the causal model of cause and effect analysis.

In forecasting it is important to recognize that:

- Forecasting uses the best available data and makes the best informed guess. There will be errors involved, so that all forecasts should state a plus or minus margin of error. The use to which the forecast is to be put determines the resources that are used to collect the data.
- The further away the forecast the less accurate it is likely to be. In all forecasts it is also important to take account of discontinuities, which can affect the data trend. For example, high exports of a product into a country in previous years may suddenly come to an end due to government action banning all imports of the product.
- All forecasting systems should include a feedback mechanism where the actual results, when they are available, are compared with the forecast values and the forecasting system is modified to make it more accurate in future.

There are many forecasts, produced by governments, which are freely available to managers. These can be used to supplement internal forecasts. Examples of these are, the Gross National Product (GNP) and forecast of employment trends: inflation forecasts. Related statistics can also be used. For example, unemployment trends may be an indicator of demand for consumer luxury goods.

9.2 Qualitative methods

Qualitative techniques are used in two common circumstances:

1. When past data cannot be used reliably to predict the future. In these circumstances past data may still be used but they will be modified based on judgement.
2. When no past data are available, usually because the situation is very new. In these circumstances predictions can be made either by doing research to gather information on which to base a judgemental prediction, or else by making the prediction by analogy to a similar situation on which past data are available.

It is important in qualitative methods to use a symmetrical approach. Several techniques are available, only four being described in the following sections.

9.2.1 Delphi method

This technique is based on using the judgement of a panel of experts to arrive at a convergence regarding the forecast. The panel is presented with a series of questions and their replies are treated anonymously. The answers are fed back to the group along with other information, such as the mean, median, interquartile range and standard deviation. It is also possible to ask panel members to provide reasons for their forecasts, and these may also be fed back anonymously to the panel. Strict anonymity is observed to prevent members of the panel from being influenced by each other, especially if one of them is supposed to be an authority on the subject, or to avoid the tendency for a member to go along with the majority.

The members of the panel are then asked to reconsider their answers and to provide a second answer. Once again these are fed back anonymously to the panel. The process is repeated four to six times, until there is sufficient convergence between the members of the panel. The results from the last round are then treated as the final forecasts.

The Delphi forecast can also be conducted by post, using an anonymous panel of experts. In another variant, if the experts do not have equal knowledge of the subject they may be asked to weight their own knowledge on each question, and their replies are then weighted by this amount.

The Delphi method is illustrated in the case study ‘Years to launch’.

Case study

Years to launch

Jack Fox, Director of Marketing for the FloRoll Manufacturing Company, wanted to obtain a forecast for the launch of the next generation of floroll. He wanted to target this product at the leisure market, especially for use in water sports. Jack knew that the availability of the product was closely linked to the development of technology in several areas, and that technology was difficult to forecast.

Cost was also a crucial factor if high market penetration was to be achieved. In order to formulate an effective strategy for the company's product portfolio, Jack decided to use the Delphi method to obtain a forecast. A panel of five experts was invited to participate, many of them eminent in the scientific field. The panel were kept anonymous, so they were unaware of who the other members were.

Two questions were put to the panel:

1. In how many years will technology have reached a sufficient state of development to enable a floroll to be launched as a water-based product?
2. What will be the unit cost of such a product in proportion to its current cost?

Table 9.1 shows the answers from the panel of experts after the first round. So, for example, Expert A thought that it would be five years before the technology would have developed sufficiently to make such a product feasible, and the product would then cost four times as much as a present day floroll.

Table 9.1 A Delphi forecast

	1st round		2nd round		3rd round		4th round	
Expert	**Years**	**Cost factor**	**Years**	**Cost factor**	**Years**	**Cost factor**	**Years**	**Cost factor**
A	5	4.0	4	2.0	3	2.0	3	1.0
B	2	0.3	3	0.5	4	0.7	3	0.7
C	6	0.7	6	1.1	4	1.0	3	0.8
D	5	2.2	4	2.0	4	1.5	4	0.7
E	8	3.0	6	2.5	3	1.0	3	1.0
Mean	5.2	2.04	4.6	1.62	3.6	1.24	3.2	0.84

As can be seen from the table there is a wide spread in the estimates, and the mean is 5.2 years, with a cost a little over twice the current cost.

These results were then fed back to the experts and they were asked to make a second forecast. This is shown in the second set of columns of Table 9.1. There is greater convergence between the figures. After round three the convergence has increased further, while the figures after round four were considered to be close enough not to warrant a fifth round.

Jack Fox therefore assumed that technology would enable the new florolls to be developed in about three years time and the cost of the product would be 16 per cent less than present-day florolls.

9.2.2 Scenario building

In this forecasting technique the parameters of importance to the company are first recorded. A number of assumptions are then followed through on how these parameters may change. Several scenarios are developed based on these assumptions. It is then assumed that each scenario happens, and the implications for the organization are analyzed. The probability of occurrence of each scenario is also defined. Action may then be taken on the high-probability events.

Research is often carried out on future developments and these may be based on broad industry reports, which are procured commercially and then tailored so as to be specific to the organization.

9.2.3 Normative relevance analysis

In this method the future situation that is being forecasted is first assumed. It is then broken down into a series of steps or items that are needed to make the forecast happen. These steps may be further and further subdivided, leading to a relevance tree diagram, as illustrated in the case study 'Going faster'.

Probabilities, or weighting factors, can then be applied to each step, so as to find the probability for the total event.

Several routes may exist to the final situation, so alternative probabilities, with dates, may be obtained. Analysis can also lead to alternatives not originally considered.

Case study

Going faster

It was technology forecasting time again at the FloRoll Manufacturing Company. Jack Fox, the Director of Marketing, wanted to know whether, in two years time, florolls could be developed to go three times faster than they did at present.

A team of specialists were consulted from Engineering, and they drew up the relevance tree diagram of Figure 9.1. For florolls to travel at three times their present speed required improvements in the guidance control mechanism, the air shield and the power drivers. These could then be further subdivided, as in Figure 9.1.

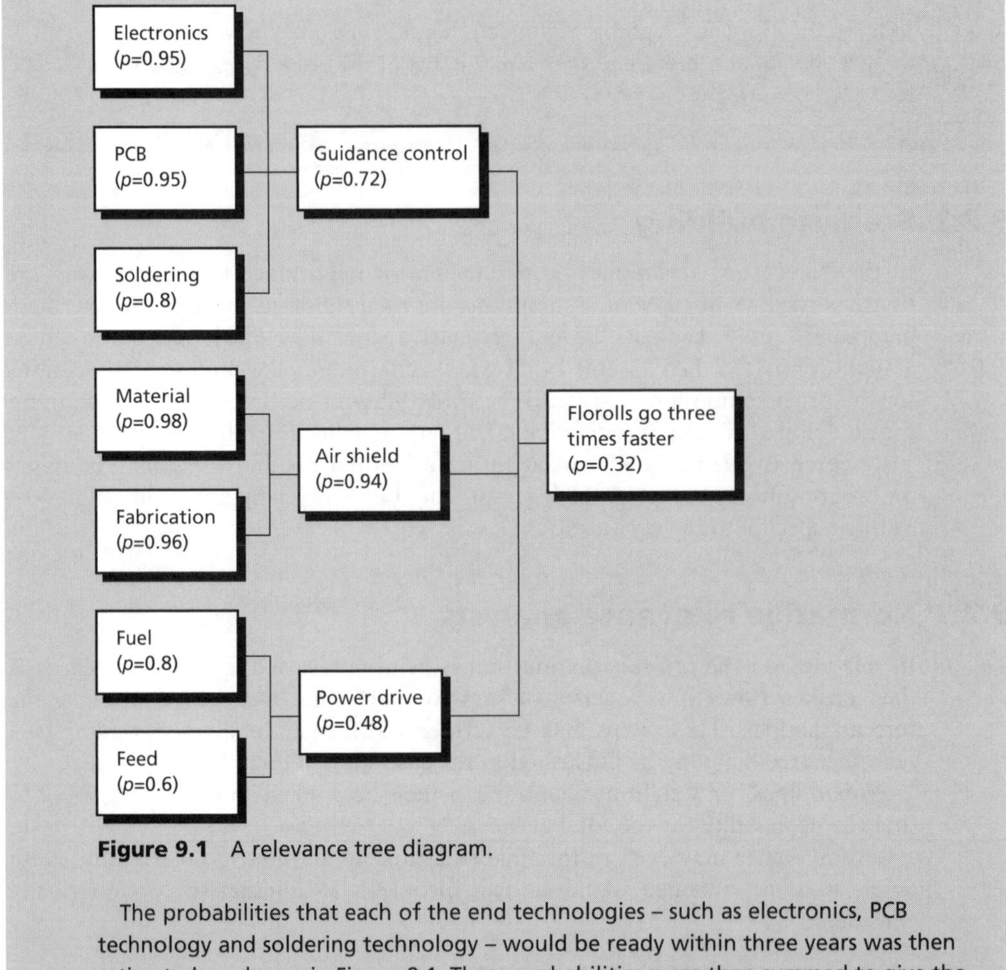

Figure 9.1 A relevance tree diagram.

The probabilities that each of the end technologies – such as electronics, PCB technology and soldering technology – would be ready within three years was then estimated, as shown in Figure 9.1. These probabilities were then summed to give the probability of obtaining super-fast florolls within two years. As can be seen, the probability that this could occur is only 32 per cent, primarily due to the low probability that the power driver technology would be available then.

9.2.4 Informed judgement

This is also known as the consensus of opinion method. The forecast is done by a group and is based on experience. No formal method is used and it could include the opinions of executives and outside experts.

Informed judgement is often used by executives and is done in their minds. It can become a very introspective method and may result in unreliable forecasts.

9.3 The time series

9.3.1 Defining a series

The time series technique for forecasting is essentially an averaging method where the average over several past periods is used to predict the future for the next period (Wheatley, 1991). Statistical techniques are often needed to forecast the future, especially if the past data are variable. For example, if the sales of a product over ten previous periods were 2, 3, 3, 2, 4, 3, 3, 4, 2, 3, then it is not immediately evident what the number for the eleventh period is likely to be.

Two time series techniques are described here: forecasting by moving averages and by exponential smoothing. These techniques analyze past data and project the information into the future. They assume that the past data, which are broken down into their various components, can be reliably carried forward to the future. Figure 9.2(a), for example, shows how information was found to vary over time, and Figure 9.2(b) gives its components.

The level is the amount that is relatively stable with time and can be extrapolated to be the amount at 'time zero'. Random variations are an

(a)

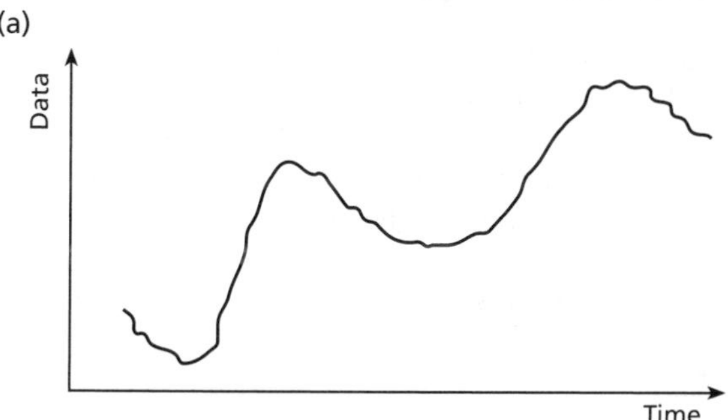

(b)

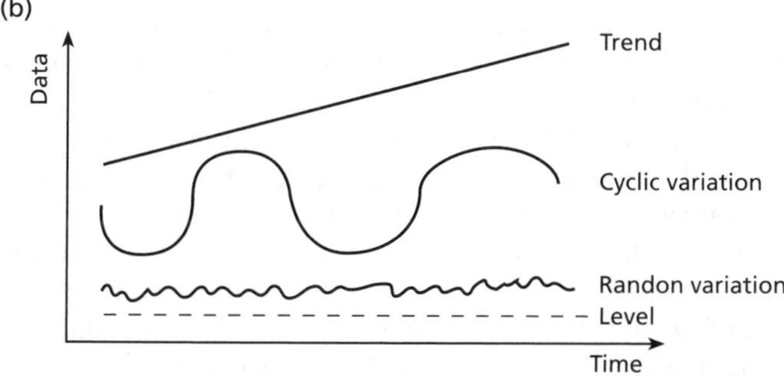

Figure 9.2 The elements of a series: (a) actual data; (b) components of the data.

unpredictable component which result in errors in the forecasts. Cyclic variations are regular variations in the data, but the cycle period is difficult to predict, such as, for example, the trade cycle. By contrast, seasonal variations, which are not shown in Figure 9.2, are also regular variations, but much more predictable, such as the increase in sales of toys at Christmas. Finally, the trend is the variation, either increase or decrease, in the overall level of the data with time.

The forecast value in relation to time t, $F(t)$, can be defined by Equation (9.1), where $C(t)$ is the cyclical variation, L is the level, $T(t)$ is the trend and $R(t)$ the random variation.

$$F(t) = C(t)[L + T(t)] + R(t) \tag{9.1}$$

Some forecasts cannot be reliably made on the basis of past data, for example the weather. It is also obviously important, in carrying forward past data, to use common sense. The output from a factory over the past five periods may have doubled in each period (equal to 2, 4, 8, 16, 32), but it would be foolish to forecast a value of 64 for the next period if a fire has just occurred that has burned down half the factory.

9.3.2 Forecasting by moving averages

This method of forecasting usually assumes that the time series has a level and a random component only; that is, there are no seasonal or cyclic variations, nor any trend. More advanced methods are available which include these.

The basis principle consists of selecting a number of periods from the past data and calculating the average of the figures to give an average past value (v_a). This is shown in Equation (9.2), where v_1, v_2, v_3 and so on are the values at the various periods, and n is the number of periods over which the average is taken.

$$v_a = \frac{v_1 + v_2 + v_3 + \ldots + v_n}{n} \tag{9.2}$$

Since a flat series is assumed this average value is then taken to be the new value during the next period. In determining the average the same number of periods n are always considered, but as new actual figures are obtained the oldest one is dropped, so the average 'moves' in time.

Forecasting on a moving average will give zero bias when the data has no trend, but when trend exists the forecast will always be too high or too low, depending on the direction of the trend. This is illustrated in Figure 9.3. The average value represents the moving average of a, b, c and d, and this is carried forward to give the forecast value. To find the trend that exists, the averages of a and b, and of c and d may be found separately and then the smaller average taken from the larger one. This is then divided by two to give the trend T per period.

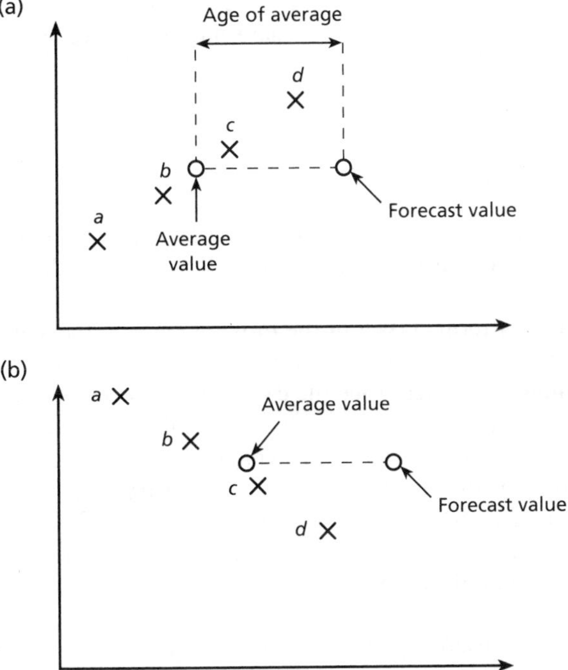

Figure 9.3 The introduction of bias due to a trend when using moving average forecasting: (a) positive bias (forecast too low); (b) negative bias (forecast too high).

The age of the moving average is the number of periods by which it lags behind the forecast value. For the four-term average of Figure 9.3 this is 1.5 periods, so to correct for trend it is necessary to add $2.5T$ to the forecast value. In general, an n term moving average's age is given by Equation (9.3) and the correction value for trend by Equation (9.4).

$$\text{Age of average} = \frac{n-1}{2} \tag{9.3}$$

$$\text{Correction value for trend} = \frac{(n+1)T}{2} \tag{9.4}$$

The technique for forecasting using moving averages, where no trend exists, is illustrated in the case study 'The moving average'. The effect of trend is shown in the case study 'The problem with trend'.

The larger the averaging period n the smaller the forecast variations will be, but the less the forecast now responds to recent events. Therefore a large averaging period gives stability but has less connection with real recent changes. Weighting techniques can be used to ensure that recent events exert a greater influence on the forecast. If the values of the past data are given by v_1, v_2, v_3 and so on, and weights w_1, w_2, w_3 and so on are allocated to these data values, then the forecast value $F(t)$ in time t is given by Equation (9.5) where the weights all add to one, as in Equation (9.6).

$$F(t) = w_1 v_1 + w_2 v_2 + \ldots + w_n v_n \tag{9.5}$$

$$w_1 + w_2 + w_3 + \ldots + w_n = 1 \tag{9.6}$$

In the simple moving average, as given by Equation (9.2), each component has an equal weight equal to $1/n$. The use of weighting is illustrated in the case study 'Weighting data'.

The greater the number of terms taken in the moving average the smoother the forecast series. Too many terms should not be taken or else cyclical or seasonal variations which exist may be masked. (See the case study 'The cycle'.)

The age of the average also increases with the number of terms, and this has the danger that the circumstances in which the series of data were obtained may have changed during that period. If necessary one can overcome this by giving more recent data values a greater weight in the average, although the choice of these weights is now difficult, and other forecasting techniques, such as exponential smoothing, described in the next section, are then preferred.

Smoothing techniques are useful in ironing out any random variations which occur in the data, and to emphasize any past trends, which may be used by the forecaster to extrapolate into the future. The choice of averaging period is now important, and to discover seasonal or cyclical variations it should equal the expected length of these variations. The seasonal or cyclical factor at any period may then be found by taking the ratio of the actual data at that period to the smoothed variation. These factors should be explainable and should be studied over several cycles before they are used.

In forecasting it is the usual procedure to smooth out any seasonal variations and then to reintroduce the variations by the use of expected seasonal factors for that period. Cyclical effects are much more difficult to forecast, since they may be longer term and are influenced by many unpredictable factors. It is necessary to determine these effects, as far as possible, and to incorporate them into the forecast; that is, to change the forecast into a prediction.

Case study

The moving average

Sales of a new version of florolls over the past ten periods are shown in the first two columns of Table 9.2. Peter Sell, the Sales Director, wants to forecast sales over the next period and has decided to use the moving average technique.

Table 9.2 Illustration of the moving average forecasting method

		Moving average			
		Two periods		**Five periods**	
Period	**Sales (£00 000)**	**Forecast (£00 000)**	**Error**	**Forecast (£00 000)**	**Error**
1	2				
2	3				
3	3	2.5	+0.5		
4	2	3.0	−1.0		
5	4	2.5	+1.5		
6	3	3.0	0	2.8	+0.2
7	3	3.5	−0.5	3.0	0
8	4	3.0	+1.0	3.0	+1.0
9	2	3.5	−1.5	3.2	−1.2
10	3	3.0	0	3.2	−0.2
11		2.5		3.0	
			0		−0.2

The third column of Table 9.2 gives a running average of the previous two sales figures: $(2+3)/2$, $(3+3)/2$, $(3+2)/2$, $(2+4)/2$, and so on. This constitutes the forecast for that period based on the previous two periods. On this basis the forecast for period 11 is $(2+3)/2$, or 2.5. The error between the actual and forecast sales is shown in column 4. It is seen that the total is zero, so that the forecast is unbiased. The forecast also lags the data, since when sales go up from one period to the next the forecast goes down, and vice versa.

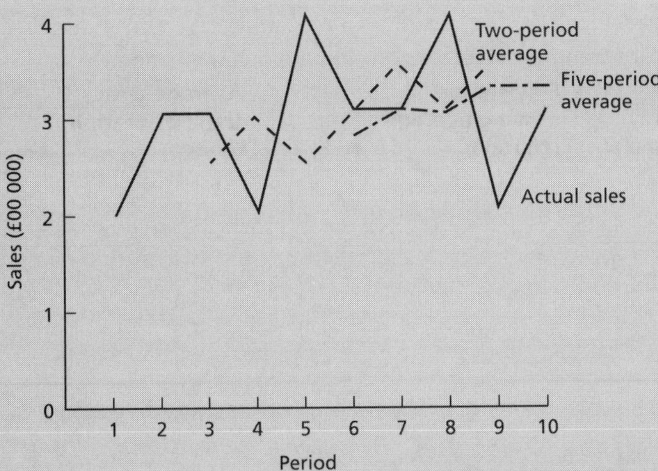

Figure 9.4 The effect of averaging period on the fluctuation in forecast data (from Table 8.2).

The last two columns in Table 9.2 show the forecast and corresponding error obtained by having a moving average based on five previous periods. The sum of the errors is low enough not to indicate any bias. By comparing with the forecast in the third column it is seen that the greater the number of terms that are averaged the smoother the forecast, as illustrated in Figure 9.4. However, the larger averaging period also means that the forecast is less able to respond quickly to changing circumstances.

Peter Sell decided to use the five period average to obtain the forecast for the eleventh period.

Case study

The problem with trend

Sales of florolls over another ten periods are illustrated in the first two columns of Table 9.3, and Peter Sell again decided to use the moving average technique to determine the sales in period 11, using four periods for the moving average. The result is given in the third column of Table 9.3 with the error in the fourth column. The forecast in period 11 was found to be 9.5, but it was obvious to Peter Sell that there was significant bias in the prediction, indicating trend, as was illustrated in Figure 9.3.

Peter Sell decided to calculate the trend T by taking the average of adjacent data from each other, as in Equation (9.7).

$$T = \frac{(5+4)/2 - (3+2)/2}{2} = 1 \tag{9.7}$$

Table 9.3 Effect of trend on forecasting using a moving average

Period	Sales (£00 000)	Four-period moving average (£00 000)	Error	Average with trend correction (00 000)	Error
1	2				
2	3				
3	4				
4	5				
5	6	3.5	+2.5	6.0	0
6	7	4.5	+2.5	7.0	0
7	8	5.5	+2.5	8.0	0
8	9	6.5	+2.5	9.0	0
9	10	7.5	+2.5	10.0	0
10	11	8.5	+2.5	11.0	0
11		9.5		12.0	
			+15.0		0

He then obtained a correction factor as given by Equation (9.4) and shown in Equation (9.8) for his prediction.

$$\text{Correction factor} = \frac{(4+1)\times 1}{2} = 2.5 \quad (9.8)$$

Therefore 2.5 needs to be added to the forecast in each period to eliminate bias, and this gave zero forecasting error, as in the last two columns of Table 9.3. The corrected forecast for period 11 was therefore taken to be 12.0.

Case study

Weighting data

In a separate study Peter Sell found the sales of a competitor's florolls over three periods to be as in the first two columns of Table 9.4.

Table 9.4 Moving averages and weights

Period	Sales (£00 000)	Three-period moving average (£00 000)	Weight	Weighted sales (£00 000)	Weighted moving average (£00 000)
1	2		0.1	0.2	
2	10		0.2	2.0	
3	3		0.7	2.1	
		5.0			1.4

Using a three-period moving average he forecast the sales in the fourth period to be 5.0. However, by applying a weight which gave considerable significance to the sales in the most recent period, as in the fourth column, he obtained a forecast of sales in the fourth period of only 1.4.

Case study

The cycle

Peter Sell was again busy analyzing the sales of florolls in order to determine any trends or periodic variations. He obtained the sales over twenty periods, as given in the first two columns of Table 9.5 and plotted this as in Figure 9.5.

Table 9.5 Smoothing to remove cyclical or seasonal variations

		Moving average	
Period	**Sales (£00 000)**	**Four-period (£00 000)**	**Eight-period (£00 000)**
1	10		
2	9		
3	6		
4	8		
5	6	8.3	
6	5	7.3	
7	5	6.3	
8	4	6.0	
9	8	5.9	6.6
10	7	5.5	6.4
11	7	6.0	6.1
12	6	6.5	6.3
13	4	7.0	6.0
14	3	6.0	5.8
15	6	5.0	5.5
16	3	4.8	5.6
17	5	4.0	5.5
18	5	4.3	5.1
19	6	4.8	4.9
20	5	4.8	4.8

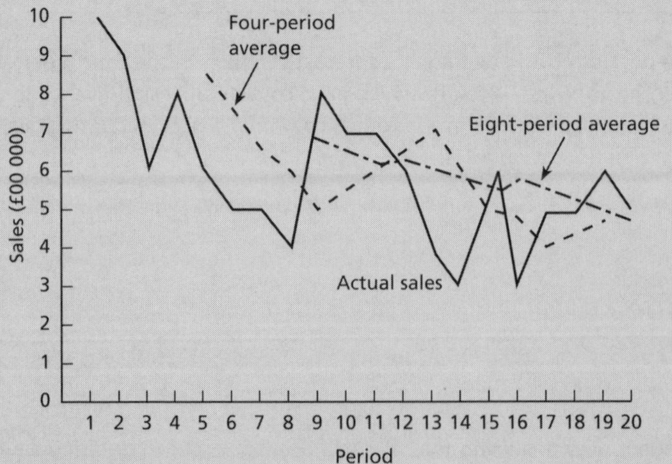

Figure 9.5 The effect of smoothing on cyclical or seasonal variations (from Table 9.5).

This graph shows that there is a clear trend, downwards, and that periodic variations exist. These variations are approximately over four periods, so Peter then calculated the four-period moving average, as in the third column of Table 9.5.

When this was plotted it showed the periodic variation much more clearly. Peter then increased the averaging period to eight, as in the final column of Table 9.5. This showed the downward trend very clearly, and it was also plotted in Figure 9.5.

Peter Sell suspected that he knew the reason for both the cyclical variations and the steady downward trend. The troughs in the variations coincided with the periods when the marketing budget was reduced, and even though sales recovered when marketing was increased, it never reached its original value. Since these cuts were imposed by the Chief Executive, Jake Topper, Peter decided to approach Jake with this information in order to ensure that there were no further cuts in the marketing budget.

9.3.3 Exponential smoothing

The simple moving average method of forecasting has some disadvantages:

- Several periods have to pass before the first average can be taken, so a trend for the period of the whole series cannot be found.
- A large amount of data, equal to the period over which the average is being taken, has to be retained.
- The trend is based on arithmetic averages, so it is influenced by extreme values. If the cyclical or seasonal variations are large this means that it may not be possible to smooth them out effectively and arrive at the underlying trend.
- All data (in the simple moving average technique) are weighted equally and data which are too old to be included are weighted by zero. Generally it is probable that recent data are more important and should have higher weights.

The exponential smoothing technique overcomes the above disadvantages. In this system the forecast for the present period is found as the forecast for the last period plus a proportion of the error made in the last forecast. This is given by Equation (9.9).

$$\text{This period forecast} = \text{Last period forecast} + \alpha \\ (\text{last period actual} - \text{last period forecast}) \tag{9.9}$$

This technique of forecasting therefore corrects for past errors, and it is also called adaptive forecasting. It can be shown that the exponential smoothing method gives a forecast which is based on an unlimited number of past actual data terms in which the terms are weighted exponentially, the most recent data having the highest weight. This is therefore an exponentially weighted moving average method using an unlimited number of terms.

In Equation (9.9) α is known as the smoothing constant, and its value varies between zero and one. The age of the average is given by Equation (9.10).

$$\text{Age of average} = \frac{(1-\alpha)}{\alpha} \qquad (9.10)$$

For $\alpha = 1$ the age is zero, which means that the present period forecast equals the last period actual and there is no smoothing. For $\alpha = 0$ the age is infinite, indicating that the present period forecast equals the last period forecast and is not influenced by the actual values; the forecast once set never changes, so that there is infinite smoothing. The value of α is usually chosen such as to give the lowest bias (error) and deviation in the forecasts. For most industrial applications this is usually between 0.1 and 0.2.

Simple exponential smoothing is not suitable for use when trends or seasonal variations are present. Other techniques are better in these circumstances (Wheelwright and Makridakis, 1985).

The advantages of the exponential smoothing technique are:

- There is no waiting period of several sets of data before reliable forecasts can be made, since further estimates can be obtained based on the initial estimate and on fresh data.
- The values of the weights decrease with time and there is no cut-off period after which data are not considered.
- It is only required to retain three figures for any estimate: the past forecast, the current actual and the smoothing constant.
- The value of α can be made to change or adapt to changed circumstances, such as for example to make the series more sensitive to rapidly changing data.

The use of the exponential smoothing method is illustrated in the case study 'Smoothing the results'.

Case study

Smoothing the results

Peter Sell decided to use the exponential smoothing technique to compare the forecasts obtained with the moving average method. The sales over ten periods, given in Table 9.2, are reproduced in the first two columns of Table 9.6.

Table 9.6 The exponential smoothing forecasting technique

		$\alpha = 0.1$				$\alpha = 0.5$			
Period	**Sales (£00 000)**	**Forecast for period (£00 000)**	**Error**	**0.1 error**	**Forecast for next period (£00 000)**	**Forecast for period (£00 000)**	**Error**	**0.5 error**	**Forecast for next period (£00 000)**
1	2	3.0	−1.0	−0.1	2.9	3.0	−1.0	−0.5	2.5
2	3	2.9	+0.1	0	2.9	2.5	+0.5	+0.3	2.8
3	3	2.9	+0.1	0	2.9	2.8	+0.2	+0.1	2.9
4	2	2.9	−0.9	−0.1	2.8	2.9	−0.9	−0.5	2.4
5	4	2.8	+1.2	+0.1	2.9	2.4	+1.6	+0.8	3.2
6	3	2.9	+0.1	0	2.9	3.2	−0.2	−0.1	3.1
7	3	2.9	+0.1	0	2.9	3.1	−0.1	−0.1	3.0
8	4	2.9	+1.1	+0.1	3.0	3.0	+1.0	+0.5	3.5
9	2	3.0	−1.0	−0.1	2.9	3.5	−1.5	−0.8	2.7
10	3	2.9	+0.1	0	2.9	2.7	+0.3	+0.2	2.9
11		2.9				2.9			
			−0.1				−0.1		

Peter used two smoothing constants of 0.1 and 0.5 and made an initial forecast for the first period of £300 000.

Table 9.6 illustrates the calculations for the two smoothing constants. In each case the forecast for the next period is found as the present forecast plus α times the forecasting error for the present period, as in Equation (9.9). In both cases the forecast for period 11 is 2.9, although the amount of smoothing varies significantly, as illustrated by Figure 9.6. There is much less smoothing with $\alpha = 0.5$ than with $\alpha = 0.1$.

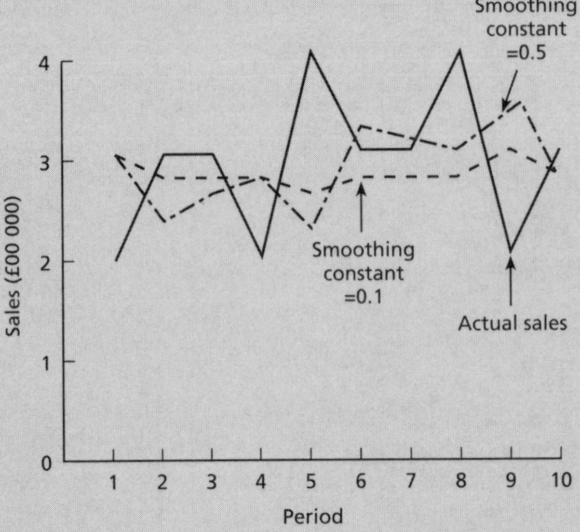

Figure 9.6 The effect of an exponential smoothing constant.

From Table 9.6 it can also be seen that both smoothing constants give negligible forecasting error. The standard deviation in the error for the 0.1 case can be calculated as 0.75 and the distribution is approximately normal, since 50 per cent of the terms lie within plus and minus one standard deviation and 100 per cent lie between plus and minus two standard deviations. The standard deviation of the error for the 0.5 case can be found as 0.9, and 60 per cent and 100 per cent of the terms lie between the plus and minus one and two standard deviation limits.

9.4 Causal models

The causal model forecasting method consists of developing a model of cause and effect, based on collected data, and then using this model to forecast the future. The cause behind the figures needs to be understood, and this is then used to determine how the various factors involved will change to affect the final result. For example, a company wishing to set up an ice cream stall on a public beach may want to determine the sales level it could expect. This will be determined by several factors, such as the number of people at any time on the beach; the composition of the people involved, such as the percentage of children; the time of day, since ice cream sales will probably be low very early in the morning and late at night; the weather; and the quality of the ice cream. These are the causes, and each needs to be analyzed to determine the effect it would have on the sales of ice cream before a forecast can be made.

The most commonly used form of causal model is regression, and the simplest example of this is a single-variable linear model. A straight line is assumed to be able to pass through the data such that the sum of the squares of variation from this line is equal to a minimum. This gives an equation for the line as in Equation (9.11), where y is the dependent variable, t is the independent variable, m is the slope of the line, given by Equation (9.12), and c is the intercept of the line on the y-axis, given by Equation (9.13).

$$y = mt + c \tag{9.11}$$

$$m = \frac{\Sigma yt - (\Sigma t\,\Sigma y/n)}{\Sigma t^2 - \Sigma t^2/n} \tag{9.12}$$

$$c = \frac{\Sigma t \Sigma yt - \Sigma y\,\Sigma t^2}{(\Sigma t)^2 - n\Sigma t^2} \tag{9.13}$$

Generally linear regression forecasting techniques are used in 'once-and-for-all' forecasts. They may be used in applications which require continuous forecasts over several periods by including the latest data or by using the latest n figures of the data, but usually it is much better to use other forecasting techniques, such as moving average or exponential smoothing, which make it easier to continually update forecasts.

The use of this single-variable linear model for forecasting is illustrated by the case study 'Fitting a curve'.

Multiple regression models are more useful than single-variable ones, since forecasts often relate to more than one factor, as for example in the case of the sales of ice cream described above. These models are more complex and require a considerable amount of data in order to develop them. They are usually based on computer models.

Other forms of causal models are also common, such as simulation, and some of these were described in Chapter 8.

Case study

Fitting a curve

Peter Sell decided to use the regression method for forecasting the sales of florolls in period 11. The sales for the first ten periods were as described before and are repeated in the first two columns of Table 9.7.

Table 9.7 Using regression (causal model) for forecasting

Period (*t*)	**Sales (£00 000) (*y*)**	**Forecast $y = .055t + 2.6$ (£00 000)**	**Error**
1	2	2.65	−0.65
2	3	2.71	+0.29
3	3	2.76	+0.24
4	2	2.81	−0.81
5	4	2.87	+1.13
6	3	2.93	+0.07
7	3	2.99	+0.01
8	4	3.04	+0.96
9	2	3.10	−1.10
10	3	3.14	−0.14
55	29	–	0

From these data, and Equations (9.11)–(9.13), the best straight line through the points can be found as in Equation (9.14).

$$y = 0.055t + 2.60 \tag{9.14}$$

This equation has been used to calculate the forecast values given in the third column of Table 9.7 and the resulting error is seen to give an unbiased forecast. The forecast for the eleventh period is therefore, from Equation (9.14), equal to:

$$y = 0.055 \times 11 + 2.6 = 3.21$$

9.4.1 Choice of forecasting method

Several forecasting techniques have been described in this chapter and there are many more which have not even been mentioned. As illustrated by some of the case studies, most of the methods give different, but related, results even with the same input data. The choice of system is usually determined by three major considerations:

1. The use to which the forecasts are to be put. This covers the accuracy required and the time horizon for the forecast. High-accuracy, short-term forecasts are suited to time series techniques, whereas longer term forecasts, which can only be approximate, require qualitative or causal model methods. Mid-time-range forecasts can be covered by time series or causal models.
2. The time and resources available for forecasting. Time series forecasting methods, such as exponential smoothing, usually require less data and

resources than developing complex causal models for forecasting. The resources needed to obtain good accuracy with qualitative methods can also be high.

3. The quantity and pattern of data available for the forecast. Some forecasting methods require a considerable amount of data for reliable forecasting. Simple time series methods require the least amount of data, but if the data pattern contains trends or cyclical variations then second- or third-order models must be used, which increase the amount of data needed. Generally the data should first be plotted in order to determine whether it has any trends, seasonal or cyclical variations. Where numeric data is not available, qualitative methods must be used.

Summary

Forecasting uses scientific techniques to help the manager come to an informed guess; it is therefore an art and a science. Forecasts generally aim to determine what is likely to happen in the future if no major changes occur. A forecast normally feeds into a plan, which may set out to put in place actions to change the forecast values.

There are three groups of techniques for forecasting: qualitative, quantitative using a time series, and quantitative using the causal model. Qualitative techniques are used when no past data or only unreliable past data are available. Several qualitative methods exist, such as Delphi, scenario building, normative relevance analysis and informed judgement. The Delphi method uses the judgement of a panel of experts to arrive at a converged result on the forecast value. In scenario building a number of assumptions are made about the parameters of importance and scenarios are developed, based on these assumptions, including probability of occurrence. The implications for the organization of these are analyzed and actions taken on the high-probability events.

Normative relevance analysis starts with an assumption of the future forecast and then breaks this down into a series of steps needed to get to it. By assigning probabilities to each of these steps the probability of the eventual forecast can be determined. Informed judgement uses a group of people to come to a conclusion about the forecast. It is generally not a very reliable technique, but may be the only one available in some circumstances.

The time series technique for forecasting is essentially an averaging method in which the average over several past periods is used to make a forecast for the next period. Several time series techniques exist for forecasting, such as moving averages and exponential smoothing. The moving averages method assumes that the time series has a level and a random component only, with no seasonal or cyclic variation and no trend. It selects a number of periods from the past data and calculates the average of these. This is then taken to be the new value during the next period of the forecast. The larger the number of periods taken to determine the average the smaller the forecast variations will be, but the less the forecast will now respond to recent events. Weightings can, however, be used to ensure that recent events exert greater influence on the forecast.

Although relatively easy to apply, the simple moving averages method has several disadvantages: several periods need to pass before the first average can be taken, so a trend for the whole series cannot be found; a large amount of data, equal to the period over which the average is being taken, has to be retained; the trend is based on arithmetic averages, so it is influenced by extreme values; unless unequal weightings are used all data are weighted equally and old data are omitted.

The exponential smoothing technique aims to overcome these problems. In this method the forecast for the next period is found as the forecast for the last period plus a proportion of the error made in the last forecast. This method corrects for past errors and as such it is also known as adaptive forecasting. It determines forecasts based on an unlimited number of past actual data terms, in which the terms are weighted exponentially, the most recent data having the highest weight. There is also no waiting period of several sets of data before reliable forecasts can be made, and only three sets of figures need to be retained for a forecast: the past forecast, the current actual value and the exponential constant used to make the forecast. However, simple exponential smoothing is not suitable for use when trends or seasonal variations are present.

The causal model method of forecasting consists of developing a model of cause and effect, based on past collected data, and then using this model to forecast the future. Several types of model can be used, such as regression and simulation.

Case study exercises

9.1 The sales over 12 periods for the FloRoll Manufacturing Co. are as in Table 9.8. Plot these figures on a graph and prepare a report describing the trend and putting forward reasons for its occurrence. Use an appropriate forecasting method to forecast the sales for period 13.

Table 9.8 Twelve-period sales

Period	**1**	**2**	**3**	**4**	**5**	**6**	**7**	**8**	**9**	**10**	**11**	**12**
Sales (£00 000)	1	5	7	6	5	8	11	11	8	9	12	16

9.2 The output from the Dinsher Shop over five periods is as in Table 9.9. Fit a curve to these figures and use this to forecast the output in period 6.

Table 9.9 Five-period output

Period	**1**	**2**	**3**	**4**	**5**
Output (000)	5	7	15	12	18

9.3 What are the advantages and disadvantages of the moving average method of forecasting? If the sales of florolls over eight periods are as in Table 9.10 use a three-period moving average to determine a trend and forecast the sales in period nine.

Table 9.10 Eight-period sales

Period	**1**	**2**	**3**	**4**	**5**	**6**	**7**	**8**
Sales (£00 000)	2	10	6	8	16	12	14	22

9.4 What is meant by exponential smoothing? The floroll sales forecast for Period 5 was £20 000 and the actual sales for that period was found to be £15 000. Using a smoothing constant of 0.2 find the forecast for period 6.

References

Georgoff, D.M. and Murdick, R. (1986) Manager's guide to forecasting, *Harvard Business Review*, January–February, pp. 110–20.

Makridakis, S. (1987) *Handbook of Forecasting*, Wiley, New York.

Makridakis, S. and McGee, V.E. (1983) *Forecasting: Methods and Applications*, Wiley, New York.

Wheatley, M. (1991) Making a start with the future, *Management Today*, August, pp. 67–70.

Wheelwright, S. and Makridakis, S. (1985) *Forecasting Methods for Management*, Wiley, New York.

References

Part IV

Financial management

10 The financial environment

Introduction

This chapter introduces the basic financial environment in which an organization works. This includes financial accounting principles, the concept of company profitability, the importance of financial budgets and controls, the methods an organization uses to obtain finance for its operation, and the techniques used to value a company.

10.1 The basics

Money may be the 'root of all evil' but it is also the commodity that 'makes the world go round'! Companies are preoccupied with financing in several ways, such as capital acquisition to commence trading, profit generation to repay the capital providers and to ensure that operations are self-financing, and cost determination and control to improve profitability and competitiveness (Broadbent, 1993).

Organizations are complex structures with many facets to their operations, dealing with human and product aspects. Financial considerations for an organization therefore represent only one aspect of its operations, although a very important aspect. The accounts of the company attempt to present this financial aspect in a clear and concise manner to enable comparison and evaluation of the organization's performance. A vast amount of financial information is generated every minute of every day during the operations of a typical company and the accountant's job is primarily to be selective in condensing this information so that it can be readily comprehended and acted on.

A company is an investment; money is raised to finance this investment by equity and by debt. The equity-holders own the company, although the debtors have first call on the assets of the company and the equity-holders own what is left after the debt has been paid. The job of the accountant is to ensure that the financial interests of all the parties concerned are looked after.

Accounting practices within a company usually fall into two broad groups: financial accounting and management accounting. The financial accountant is primarily concerned with historical data, being tasked with preparing annual accounts for the organization. In preparing this the accountant is bound by the Companies Acts to follow principles which ensure standards and uniformity across companies and even across countries. Examples of these are the practices laid down by the Treaty of Rome, enhanced by the Maastricht Treaty, in order to establish harmonization across countries of the European Economic Community.

The financial accountant produces information in areas such as profit and loss and use of assets, with a duty to be fair to the shareholders and to work within legal guidelines to minimize the tax burden of the company. Although most large organizations have their own internal financial accountants, formal audits are also carried out by external auditors.

The management accountant's main aim is to provide feedback to management on the financial state of the company, in order to enable effective decision making. Examples of these could be performance against budgets and evaluating capital investment decisions from a financial aspect. An important part of management accounting is cost accounting.

The management accountant uses both past data and predicted future information, which is then presented to management to help in planning, control and decision making. To obtain this information management accountants must work closely with peers in other functions and for maximum effectiveness it is important that they understand the technology of the industry in which they work.

Although accounting deals with numbers it is not a precise science, such as the natural sciences. It is based on certain agreed ways of calculating and presenting information, which have been developed by usage, by various professional and accounting bodies, and by legislation. Considerable leeway, however, exists in interpretation of the financial data.

10.2 Financial accounts

The basic purpose of financial accounting is to present a 'fair and true view' of the financial position of the business at any instant in time (Holmes and Dunham, 1994). It makes no attempt to calculate the 'net worth' of the organization and therefore ignores intangibles, which determine the success of the company, such as business know-how, product portfolio, and market share.

Financial accounts are based on several accounting principles:

1. They treat the company as an 'ongoing concern': the assumption is made that the organization will continue to trade into the foreseeable future. Therefore assets are valued at cost, although if sold, for example when the company is being wound up, they are likely to fetch very little revenue. As stated earlier, no attempt is made to measure the worth of the business to a potential buyer. However, some assets may be depreciated, as discussed later, and other assets, such as property, may be revalued periodically.
2. All statements are based on the principle of 'accruals'. Profits are not calculated from the actual cash received by the business or from the cash spent. Instead, the value of work done, whether payment has been received for it or not, is used for receipts, and the value of assets used up in doing the work is taken as the cost. Provisions can be built into these calculations for future events which are expected to occur, as a result of trading, for example a provision for non-payment by some debtors (bad debt).
3. Only those factors are considered within a financial statement that can be related in monetary terms. Therefore important factors are ignored, such as the technical capability of the organization; a new product range which has just been developed and which would result in greatly increased sales in the following year; or the entry of a competitor with a better and cheaper product, which would result in a sharp reduction in sales next year. Therefore a financial statement is not necessarily a good measure of the strength or future potential of a company (Barfield, 1994).
4. Prudence is applied in preparing the financial statement. When in doubt the statement should err on the safe side. It should only include revenues that are certain to be obtained, and all losses should be allowed for if expected, even if the actual value is not known.
5. Consistency should be used in the accounts, especially between one period and the next. Items should be handled in the same way to enable comparison between periods; for example, the method used for depreciation should be the same.
6. The financial accounts are for the business, which is quite separate from that of its owners. It is assumed that the owners (shareholders) have entrusted the business to the directors of the company and the financial accounts show how well the directors have carried out their trust. Because the company is

considered to have an identity of its own, someone owns every part of it, such as the equity-holders and the loan providers.

Every company is required by law to deliver a copy of its accounts to the Registrar of Companies for filing, and these are available for public inspection. These accounts must contain the balance sheet, a profit and loss statement and an auditor's report. Small companies can present an abbreviated balance sheet only, where a small company is defined as an organization with a turnover below a certain amount, total assets below a stated limit, and with less than a specified number of employees. The levels for turnover, assets and employees are periodically reviewed.

Accounts for subsidiaries (companies that are owned by another larger company, called the holding company) are usually shown within the consolidated accounts of the holding company. Therefore the holding company and its subsidiaries are treated as if they are one single company and a single consolidated account is produced.

Three types of financial statement are described in this section: the balance sheet, the profit and loss statement, and the source and application of fund (or funds flow) statement (Holmes and Sugden, 1994; Oldcorn, 1980; Pocock and Taylor, 1981; Sizer, 1985).

10.2.1 The balance sheet

The basic equation of a balance sheet is given by Equation (10.1).

$$\text{Assets} = \text{Owner's equity} + \text{Liabilities} \tag{10.1}$$

The profits, which remain after liabilities are met, add to the owner's equity.

Figure 10.1 shows the main elements of a balance sheet. This is known as a double entry system, since entries on one side balance out those on the other; thus liabilities for the company equal its assets. Any difference between the actual assets owned by the company and that owed as liabilities to non-shareholders is made up by the shareholders (equity-holders) in terms of an increase or decrease in capital. Owners' equity includes their initial investments plus any earnings of the company which have not been distributed as dividends (called retained earnings).

Liabilities	Assets
Capital Long-term liabilities Current liabilities	Fixed assets Current assets

Figure 10.1 The main headings of a balance sheet.

Long-term liabilities are those which do not need to be met over the coming year, such as long-term loans and mortgages. Eventually even these loans will reach the end of their period and will have to be repaid, when they will become current liabilities.

Current liabilities are those which are of a short-term nature and have to be met within the coming year, such as money owed to suppliers, and some bank overdrafts. Provision can also be made within the balance sheet for liabilities and charges (expected liabilities within the coming year) even though the exact amount is not as yet known. There are many other sources of finance available to a company, long, medium and short-term (Johnston, 1991).

Fixed assets are those used in the long-term production of goods and which are not to be converted to cash in the near future. Examples of these are plant, machinery and buildings. These assets are usually valued in the balance sheet at the original cost value unless there is a permanent and significant change to their value, which is then reflected in the balance sheet. For example, if a new machine is available which has a price tag of one quarter that of the original machine and can produce as many or more goods in a given time, then it may be prudent to write down the value of the existing asset to one quarter of its original cost price. Similarly, properties are often revalued at intervals of three to ten years, and the new value, if significantly different from that stated on the balance sheet, is taken into the updated balance sheet. Property is often revalued at times of rapidly rising property prices in order to show the real worth of the company and so prevent a takeover by predators who are primarily interested in selling off its assets.

When assets are revalued the new value is included in the balance sheet, and this is balanced by a 'revaluation reserve' on the other side of the balance sheet. If the asset is sold and a profit realized then this is included in the profit and loss statement.

Current assets are those assets that are expected to be used up or converted into cash during the coming year. Examples of these are stocks, work in progress and money owed by trade debtors for goods or services already provided.

There are two other items which are sometimes seen on balance sheets:

1. Prepayments or advanced payments or deferred charges. Sometimes in any one year a company incurs an extraordinary expenditure, the benefits of which continue for several years, so only a part of the money spent is shown to have occurred for that year. Examples arise when a company takes out a three-year insurance policy and pays the premium for all three years during the first year.
2. Intangibles, such as goodwill. These represent an asset of the company, although it is very difficult to place a monetary value on them. It is undesirable to give them a high value in the balance sheet since they could give a false statement of the company's real position.

An example of a balance sheet is given in the case study 'The balancing act'. It is important to appreciate that the balance sheet is a statement of a company's position on a given day – it is not an account. The build-up of a balance sheet over several periods gives a picture of the company's operation, as illustrated in the case study 'Starting up'.

Case study

The balancing act

In the early days the FloRoll Manufacturing Company was a small start-up operation and had a relatively simple balance sheet. This is shown in Figure 10.2. It uses double entry book keeping, as described earlier, where every item is looked at from two different aspects and tabulated in two columns, which balance. As also stated earlier, this balance sheet is a snapshot of the company on a particular day, in this instance 31 December, and the document relates to the company and not to the owners of the company. Therefore the assets represent the items which the company (and not the owners) owns, and the liabilities are the items which the company owes other people, including the owners. Since all the assets that are not owed to debtors belong to the owners, as in Equation (10.1), the asset and liability columns balance.

Liabilities		Assets	
Capital		*Fixed assets*	
Ordinary shares of £1	80 000	Property	50 000
Retained profits	40 000	Plant and equipment	70 000
10% preference shares	30 000		
	150 000		120 000
Long-term liabilities			
10% secured mortgage	20 000		
Current liabilities		*Current assets*	
Accrued expenses	5 000	Stock and WIP	40 000
Trade creditors	10 000	Trade debtors	20 000
Compant tax	24 000	Investments	40 000
Bank overdraft	2 000	Cash	10 000
Dividends	19 000		
	60 000		110 000
	230 000		230 000

Figure 10.2 The balance sheet on 31 December for the FloRoll Manufacturing Company.

From Figure 10.2 it is seen that the ordinary shareholders of the company have put up £80 000 since the formation of the company in order to finance its operation. Although this finance is stated in units of £1, this does not mean that the current market price is £1, since the price can go up or down, depending on the demand for the shares on the stock market. These market fluctuations do not affect the day-to-day running of the company, the only effect being when the company wishes to raise more capital by the issue of new shares, since the price is influenced by the market value at that time.

The ordinary shareholders wish to obtain interest on their investment, called dividends, and hopefully to have their original investment maintained or increased in value.

The company has saved £40 000 from its operating profits over previous years, to help finance its operations. These profits were those remaining after paying debts and taxes and profit distribution to ordinary shareholders as dividends. By allowing the company to retain some of their dividends the shareholders would rightly expect the net worth of their company to increase, and this is evident by increased profits in subsequent years and an increase in the share price.

Since the capital of the company has gone from £80 000 to £120 000 with the retained profits, the shareholders could expect the share price to go from £1.00 to £1.50. However, the stock-market price does not exactly follow this trend, since it is influenced more by how investors view the future prosperity of the company. (Note, however, that usually a company with a small capitalization, such as the FloRoll Manufacturing Company has at present, would not be quoted on the Stock Exchange.)

In addition to the ordinary shares issued and retained profits, the company has also obtained £30 000 from the issue of preference shares, taking the total capital to £150 000. These shareholders get 10 per cent interest each year. They are called preference shareholders since this interest is paid before any of the ordinary dividends, and if the company has made insufficient profit in any year to pay this then it is usual to carry forward the amount owing from one year to the next. Preference shareholders are usually repaid before ordinary shareholders if the company goes into liquidation. Generally, preference shareholders have no voting rights in the company and even in a good year their dividends are limited to a fixed percentage, although no such restrictions apply to ordinary shareholders, since they carry the main risks of the company.

Apart from the capital of £150 000 the company obtained a loan of £20 000 at 10 per cent interest. This loan will have to be repaid at a specific time and in the event of default some of the assets of the company will be sold to meet the repayment.

The FloRoll Manufacturing Company therefore had £170 000 (£150 000 + £20 000) on 31 December to finance its operations. The assets column shows what it did with this money. First, the company bought £120 000 worth of plant and equipment, and buildings from which to conduct its business. These are fixed assets, since the company needs these to get into a trading position. The assets are usually quoted in the balance sheet at cost, although the plant and equipment are depreciated, as described later.

At the time of the balance sheet the current assets of the company contained stock and work in progress (WIP) of £40 000. These will eventually become finished products and will be sold for cash and the cash used to buy more stock and WIP, and so on. It should be noted that it is not easy to put a worth accurately on stock and WIP for balance sheet purposes. For example, stock and WIP can become obsolete before they are sold. Prices of raw material also change; should the original price paid or the current (and therefore replacement) price be considered? In calculating WIP should the raw material cost only be taken into account or also an element of the production labour and overheads? Generally the accounting practice of prudence requires that the costs used be conservative; for instance, one should use current cost or the realizable value, whichever is lower. The different methods used for costing stock and WIP are discussed in Chapter 11.

The company also has £20 000 owing to it from customers for goods supplied. It is expected that all of this will be obtained. If some of the debts were expected to be unpaid then a proportion of these would be subtracted as a provision for bad debts.

The investment of £40 000 is a short-term investment in 'safe' securities, since the company temporarily has a surplus of cash. The cash of £10 000 is unproductive and is being held since it is expected to be used shortly.

The company has £170 000 and it spent £120 000 to buy fixed assets and used £110 000 to finance its day to day operations. This meant that it overspent by £60 000 and obtained this money by borrowing, as shown under current liabilities. Accrued expenses are items such as salaries to employees which are owed by the company on 31 December. This does not mean that the company is not paying its staff. Salaries are not due until the middle of each month, so in effect the company has 'borrowed' two weeks' labour from its employees to finance some of its current assets.

Trade creditors are represented by invoices received from suppliers that have not yet been paid. Once again the company is not necessarily in default, since most suppliers allow 30 days for the settlement of bills. However, since the goods have been received by 31 December, and these are shown as company assets under stock, the assets have been financed by in effect borrowing £10 000 from the suppliers. Unfortunately, many companies use their suppliers as a major source of finance for their working assets. This is a dangerous policy, since if the company constantly delays payments it may find itself cut off from further supplies of the raw materials it needs.

Liabilities		**Assets**	
Capital	150 000	Fixed assets	120 000
Long-term liabilities	20 000	Current assets	110 000
Current liabilities	60 000		
	230 000		230 000

Figure 10.3 Abbreviated balance sheet of Figure 10.2.

The tax liability for the year is £24 000 and although this is not yet due it has arisen out of the trading assets for this year, so that again the company has used this as a source of finance. The bank overdraft is a direct form of bank borrowing. This is a short-term loan, and although many companies depend heavily on this form of finance it is not very reliable, since it can be affected by government monetary policies. The dividends declared for this year, for ordinary and preference shares, are £19 000 and these have not as yet been paid out, so they are shown as a liability.

Figure 10.3 shows an abbreviated version of the balance sheet of Figure 10.2, and this indicates clearly the principle that any excess of assets over capital must be made up by long- or short-term borrowings.

Case study

Starting up

The balance sheet of Figure 10.3 gives the position of the FloRoll Manufacturing Company on one particular day. The company went though several phases when it first started up, and its development can be traced by a study of its balance sheet over four periods. These are given in Figure 10.4.

The company was set up in Period 1 with an ordinary shareholder capital of £80 000. This was, at this stage, in the form of cash, as shown in the assets column for Period 1.

In Period 2 the company spent most of this capital. It first bought property for £50 000 and some plant and equipment, costing £20 000, to start manufacturing florolls. It still retained £10 000 as cash to buy raw materials.

During Period 3 more plant and equipment were added, making the total up to £50 000. At the same time stock was bought to start making florolls. £5000 remained in the bank. To finance all this increased activity the company issued preference shares, to the value of £30 000, and also built up trade creditors to £5000.

In Period 4 florolls were being produced and sold. The fixed assets did not need to be increased, since there was adequate capacity to meet anticipated demand. However, stock and work in progress increased to £20 000 and trade debtors to £10 000. Cash was also at £10 000. To finance these operations £10 000 of profits were retained, trade creditors increased to £10 000 and a bank overdraft facility for £5000 was utilized. In addition, tax of £5000 was due.

Period	Liabilities		Assets	
1	*Capital*		*Current assets*	
	Ordinary shares of £1	80 000	Cash	80 000
2	*Capital*		*Fixed assets*	
	Ordinary shares of £1	80 000	Property	50 000
			Plant and equipment	20 000
				70 000
			Current assets	
			Cash	10 000
				80 000
3	*Capital*		*Fixed assets*	
	Ordinary shares of £1	80 000	Property	50 000
	10% preference shares	30 000	Plant and equipment	50 000
		110 000		100 000
	Current liabilities		*Current assets*	
	Trade creditors	5 000	Stock	10 000
			Cash	5 000
		115 000		
				15 000
				115 000
4	*Capital*		*Fixed assets*	
	Ordinary shares £1	80 000	Property	50 000
	10% preference shares	30 000	Plant and equipment	50 000
	Retained profits	10 000		
				100 000
		120 000	*Current assets*	
	Current liabilities		Stock and WIP	20 000
	Trade creditors	10 000	Trade debtors	10 000
	Company tax	5 000	Cash	10 000
	Bank overdraft	5 000		
		20 000		40 000
		140 000		140 000

Figure 10.4 The balance sheet of the FloRoll Manufacturing Company over four periods.

10.2.2 Profit and loss

By law the profit and loss account is the second financial statement (with the balance sheet) that must be completed by a company. The only exception is if the company is not trading for profit (for example if it is a charitable trust), when it needs to present an income and expenditure account instead.

The basic equation for the profit and loss statement is given in Equation (10.2).

$$\text{Profit} = \text{Revenue} - \text{Expenses} \tag{10.2}$$

The revenue or sales should include all income-generating activities, such as goods sold, consultancy and services (for example maintenance and installation). From these, items such as discounts and returns need to be subtracted.

Expenses are generally of two types, direct and indirect. Direct expenses are items which can be clearly identified as going in to produce the various items of revenue. They vary in direct proportion to the amount of output. Examples are labour costs, raw material and maintenance of equipment used in producing the product. Indirect expenses cannot be related directly to the items of revenue. For example, the salaries of sales staff who sell a complete line of products cannot be easily split out to individual products. Some advertising expense, when it relates to a single product, can be treated as a direct expense, but advertising which promotes the company image, and indirectly benefits all the products it makes, is an indirect expense.

R&D costs are usually difficult to allocate to individual products since they benefit a range and are often technology- rather than individual product-related. Rates and building insurance fall into this category as well, unless the building is being used exclusively for the manufacture of a single product.

Companies sometimes convert indirect expenses (often treated as overheads) into direct expenses by allocating a proportion of the overheads to each product line. This can be done on several bases, such as the volume of each product line being produced. The assumption made is that the higher the output the greater the amount of indirect resources the product is using.

Sometimes adjustments are made on profit and loss account statements for extraordinary items, which are those that fall outside the ordinary activities of the company and so are not expected to occur frequently. Examples of this are sales of a part of the business or provision for a high level of redundancy.

Figure 10.5 shows the basic elements of a profit and loss statement. The item remaining after the removal of direct expenses from sales is termed the gross profit.

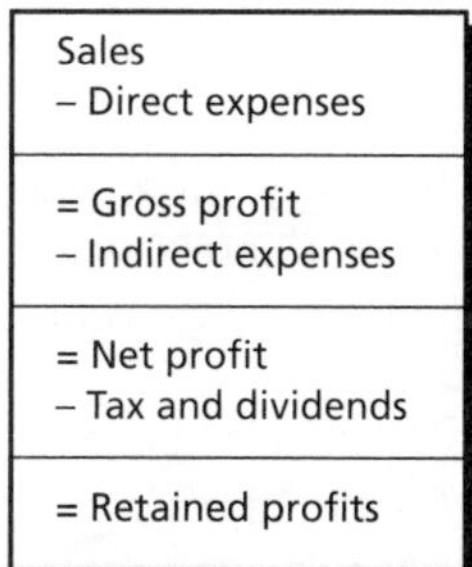

Figure 10.5 The basis of a profit and loss statement.

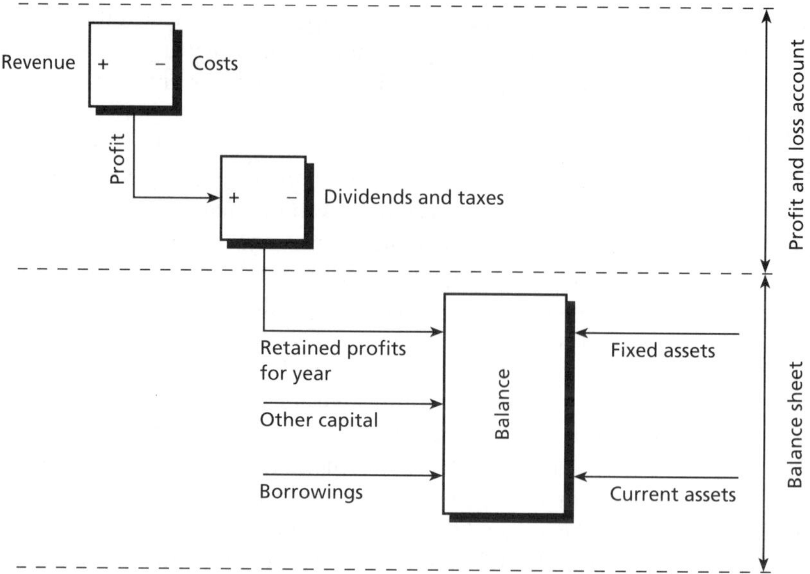

Figure 10.6 Linking the profit and loss account and the balance sheet.

Net profit remains after indirect expenses are also subtracted, and tax is paid on this amount. The item remaining after tax and dividends have been subtracted is the retained earnings for the period. (See the case study 'Making a profit'.)

Figure 10.6 shows the link between the profit and loss account and the balance sheet. The retained profits remaining after costs, dividends and taxes have been taken from the revenue for the period are an item in the liabilities column of the balance sheet. The liabilities consist of capital (which includes retained profits), and borrowings (current liabilities and long-term liabilities). This is balanced by the fixed and current assets.

Case study

Making a profit

The profit and loss account for the FloRoll Manufacturing Company, for the year ending 31 December, is shown in Figure 10.7. The company sold £700 000 worth of goods during the year. These are the actual sales for which invoices were issued and do not include finished goods not yet delivered.

The direct cost of these sales (referred to as 'cost of sales' in the profit and loss statement) was the labour and material used to make the florolls, plus a proportion of factory overheads. After removal of cost of sales the company made a gross profit of £100 000.

		£
Sales		700 000
	£	
Cost of sales		
Labour	300 000	
Materials	250 000	
Overheads	50 000	600 000
Gross profit		100 000
Administration expenses	50 000	
Intrest on mortgage	2 000	52 000
Net profit		48 000
Tax		24 000
Profit available for distribution		
Dividends		
Preference	3 000	
Ordinary	16 000	19 000
Balance		5 000
Profit retained from previous years		35 000
Total profit retained		40 000

Figure 10.7 Profit and loss account for the year ended 31 December for the FloRoll Manufacturing Company.

The indirect expenses consisted mainly of the R&D, selling and clerical expenses, which are shown as administration expenses of £50 000. In addition, there was interest payment on a mortgage loan, which was seen from the balance sheet of Figure 10.2 to be 10 per cent of £20 000 or £2000. The net profit, remaining after removal of indirect expenses, is £48 000.

Tax is paid on net profit. The rate of tax for the year is 50 per cent and this gives a tax bill of £24 000, leaving the balance from net profit, or £24 000 for the shareholders. Preference shareholders are paid 10 per cent on their investment of £30 000 (as in Figure 10.2), or £3000, and £16 000 is distributed to ordinary shareholders as dividends (a rate of 20 per cent on their equity of £80 000). The balance of £5000 is added to the retained earnings of previous years to make the total retained earnings up to £40 000, as shown in the bottom line of the profit and loss account and in the balance sheet of Figure 10.2.

10.2.3 Depreciation

There are two types of asset, as discussed before: revenue or current assets and capital or fixed assets. Revenue assets include stocks and work in progress, which are used up fairly quickly. Capital assets include buildings, plant and equipment, which have a much longer life.

The expenses involved in revenue assets are written off as an expense in the profit and loss account for that year, but the capital asset is kept on the balance sheet as a fixed asset and its original cost is written off slowly over several years. This is called depreciation. It appears on the balance sheet as a cumulative fund for each successive year and as an expense on the profit and loss account.

The aim of depreciation is to gradually remove the cost of an asset from the financial records of the company by moving it as a cost into the profit and loss account.

Depreciation is not a method for building up a fund to replace the asset, and neither can the method of depreciation be chosen to minimize the tax liability of the company. Inland Revenue rules exist for capital allowances against fixed assets. The main impact of depreciation is to reduce the profit for the year and so also reduce the amount available for distribution to shareholders. Indirectly, therefore, it does make funds available for use by the company for other purposes, such as the purchase of new fixed assets. Depreciation, however, bears no relation to the change in the market value of the asset nor to the cost of replacement of the asset at the end of its life.

It is very difficult to decide what should be expensed and what should be depreciated. For example, if a repair is carried out to an existing machine to maintain it in its current condition the work done is revenue expense and should be written off at once. If, however, during maintenance the machine is modified to improve its performance, then the work could be treated as a capital expense and added to the cost of the machine and depreciated with it.

Several techniques are used to depreciate an asset; only two of these, the straight line and the reducing balance methods, are discussed here.

Straight line method

In the straight line method of depreciation the asset is assumed to lose its value in a linear fashion during its useful life. Therefore, if the initial cost (C_i) of the asset, the number of years of useful life (n) and the residual value (C_r) at the end of this life are known, the depreciation per year (D_n) can be found as in Equation (10.3).

$$D_n = \frac{(C_i - C_r)}{n} \tag{10.3}$$

For example, suppose an asset is bought for £10 000 and is estimated to have a useful life of four years, with a residual value at the end of this period of £600. Using the straight line method of depreciation the rate of depreciation per year is given by Equation (10.3) as:

$$D_n = \frac{(10\,000 - 600)}{4} = £2350$$

The value of the asset at the end of the first, second, third and fourth years is found as in Table 10.1 by removing £2350 from the remaining cost in each year.

Although this method of depreciation is simple to implement, it is difficult to forecast the end of life value or the expected life of many assets. If the asset continues beyond its predicted life no further depreciation is applied to it. Another problem with the straight line method of depreciation is that it assumes a uniform loss of value throughout the life of the product. However, most assets (such as vehicles) tend to lose their value quicker in the early years rather than later on in their life. The reducing balance method of depreciation caters for this problem.

Table 10.1 Straight line depreciation method

Year	**0**	**1**	**2**	**3**	**4**
Asset value (£)	10 000	7650	5300	2950	600

Reducing balance depreciation

In this method of depreciation it is assumed that initially the asset depreciates quickly and then the rate of depreciation levels off. For example, if the value of an asset in any year is V_n and it is depreciated at a rate of x per cent per year, then the value of the asset in the following year (V_{n+1}) is given by Equation (10.4).

$$V_{n+1} = V_n - \frac{V_n x}{100} \tag{10.4}$$

Therefore if an asset is bought for £10 000 and it is depreciated at a rate of 50 per cent, its value at the end of the first, second, third and fourth years would be as in Table 10.2.

Table 10.2 Reducing balance depreciation method

Year	**0**	**1**	**2**	**3**	**4**
Asset value (£)	10 000	5000	2500	1250	625

The effects of the straight line and reducing balance methods of depreciation on the residual value of an asset, as given by Tables 10.1 and 10.2, are plotted in Figure 10.8, which shows that the reducing balance method gives much greater depreciation during the early life of the product, although this rate is less in the later years of the product's life.

Depreciation is considered to be a cost of sale in the profit and loss statement and an asset in the balance sheet, as shown in the case study 'The effect of depreciation'.

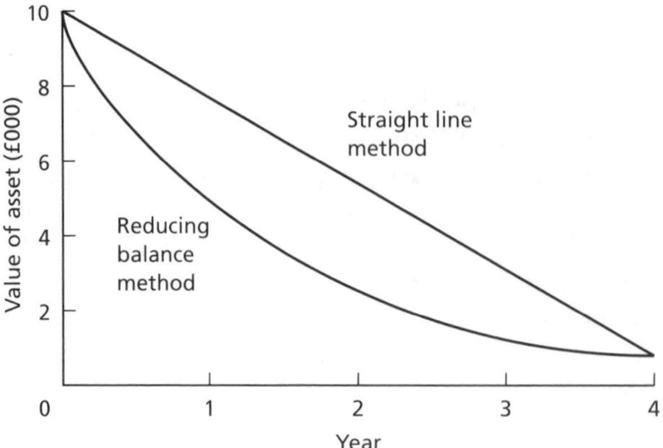

Figure 10.8 Comparison of the straight line and reducing balance depreciation methods.

Case study

The effect of depreciation

The plant and machinery shown in the balance sheet of the FloRoll Manufacturing Company (Figure 10.2) is to be depreciated using the straight line method. It has been determined that the £70 000 worth of assets will have a useful life of six years and would have a value at the end of this time of £10 000.

Using the straight line method the rate of depreciation per year is given by Equation (10.3) to be:

$$D_n = \frac{(70\,000 - 10\,000)}{6} = £10\,000$$

If the asset was bought five years earlier, then by 31 December, the date of the balance sheet of Figure 10.2, it has been depreciated for five years or by £50 000. The modified balance sheet, which omits the details of Figure 10.2, is shown in Figure 10.9.

The effect of depreciation has been to reduce the value of the fixed assets by £50 000. The company has decided to keep its capital and current liabilities at the existing level, but has increased work in progress and investments. In addition trade debtors have been allowed to increase. Clearly, some of the new investments could go towards a replacement machine. Also, the increase in revenue, caused by increased stock and WIP, could generate greater profits in subsequent years, which could be used to fund a replacement machine.

Although depreciation has not been used directly to fund a replacement asset, the directors of the company know from the balance sheet that at the end of the next year the plant and equipment will be fully depreciated and will probably need to be replaced, so they have to plan for this.

Liabilities		Assets		
	£	*Fixed assets*		£
Capital	150 000	Property		50 000
Long-term liabilities	20 000	Plant and equipment	70 000	
		Depreciation	50 000	20 000
				70 000
		Current assets		
		Stock and WIP		60 000
Current liabilities	60 000	Trade debtors		30 000
		Investments		60 000
		Cash		10 000
				160 000
	£230 000			£230 000

Figure 10.9 The balance sheet of Figure 10.2 with depreciation included.

			£
Sales			700 000
Cost of sales	£		
Labour	300 000		
Materials	250 000		
Overheads	50 000		
Depreciation	10 000		610 000
Gross profit			90 000
Administration expenses	50 000		
Interest on mortgage	2 000		52 000
		Net profit	38 000
Tax			19 000
Profit for distribution			19 000
Dividends			
Preference	3 000		
Ordinary	11 000		14 000
		Balance	5 000
Profit retained from previous years			35 000
Total profit retained			40 000

Figure 10.10 The profit and loss account of Figure 10.2 with depreciation included.

The effect of depreciation on the profit and loss account of Figure 10.7 is shown in Figure 10.10.

Depreciation is added as a cost of sales item and it therefore reduces the gross profit and eventually the profit available for distribution to shareholders. The company decides to retain the same level of profits as before (£5000), so the ordinary dividend is reduced to £11 000 (13.8 per cent), although the preference dividend is unchanged at 10 per cent.

10.2.4 Source and application of funds

The source and application of funds statement is an optional requirement, although most organizations include this in their financial statements. Its aim is to show how the company's operations have been financed and how the financial resources have been used. It indicates changes in net funds.

The source of funds is usually achieved by an increase in creditors and loans, or by a decrease in assets, debtors and cash. Application of funds requires the reverse of these: an increase in assets, cash or debtors, or a decrease in assets and cash.

Figure 10.11 shows a source and application of funds statement. The sources of funds during the period have been profit generated, depreciation and short-term loans. These funds have been applied to increase the fixed assets by £15 000, to increase stock and WIP by £3000, to paying taxes and interest on loans and to paying dividends to shareholders. It is seen that for the period under consideration there is a net flow of funds into the company of £9000, which can be held as cash or be invested.

Source	£
Profit before tax	38 000
Depreciation	10 000
Short term loans	3 000
	51 000
Application	
Increase in capital	15 000
Increase in stock and WIP	3 000
Taxes paid	10 000
Dividends paid to shareholders	12 000
Interest on loans	2 000
	42 000
Net flow of funds in/(out)	9 000

Figure 10.11 Source and application of funds statement.

10.3 Inflation

In the considerations so far the effect of inflation has been ignored. Although many items are affected by inflation, only the problems of determining capital utilized and the cost of sales are described in this section.

In an inflationary period the value of stocks will appreciate so that when the goods are sold they will fetch a larger profit. This will, however, attract tax, and the replacement cost of the stocks will also have increased, so the company may not have enough cash remaining to continue trading unless the cost of sales initially took the inflationary factor into account.

A similar problem arises when depreciating a machine to determine the cost of sales. Historical accounting methods depreciate on the initial costs even though the price of the machine might have increased substantially in subsequent years. This again results in higher profits (and tax) and insufficient funds to buy a replacement machine. Over two decades the average rate of inflation can be about 10 per cent. An asset may have a useful life of between eight and twenty years. After eight years, with an inflation rate of 10 per cent, the original 'cost' of the asset will be understated by over 50 per cent.

During inflation the government is not unduly concerned, since its tax revenue is higher. The management of a company are also relatively unconcerned since their performance is measured in terms of profit on capital employed, and this also increases in an inflationary period. It is only the shareholders who suffer, since they will eventually need to find more capital to replace their depleted assets.

Two methods have been proposed in the past to introduce the effects of inflation into a company's accounts. These are known as the Current Purchasing Power (CPP) and the Current Cost Accounting (CCA) methods. The CPP method reduces all historical data to their current purchasing power using an inflation factor based on a retail price index. Monetary items, such as cash, debtors, creditors and loans, are not affected by inflation in that their numerical value remains unchanged even though their current purchasing power changes. If a company holds cash and debtors it loses. It gains by having loans and creditors, since, when the time comes to settle debts, the purchasing power value of the money will have fallen.

As an example of the use of CPP consider the simple trading accounts shown in Figure 10.12(a). At the start of the year (the end of the previous year) the company had finished stocks worth £2000 and an issued capital of £2000. During the year these stocks are sold for £3000 and the profit and loss account for the year is given by Figure 10.12(b). The company pays out £500 in tax and is therefore left with cash amounting to £2500. Assuming that all the profits after tax are retained, the balance sheet at the end of that year is as shown in Figure 10.12(c). There has been a capital appreciation of £500 when the effect of inflation is ignored.

Suppose that during the year being considered in Figure 10.12 the price index

(a)

Liabilities		Assets	
Share capital	2 000	Stock	2 000

(b)

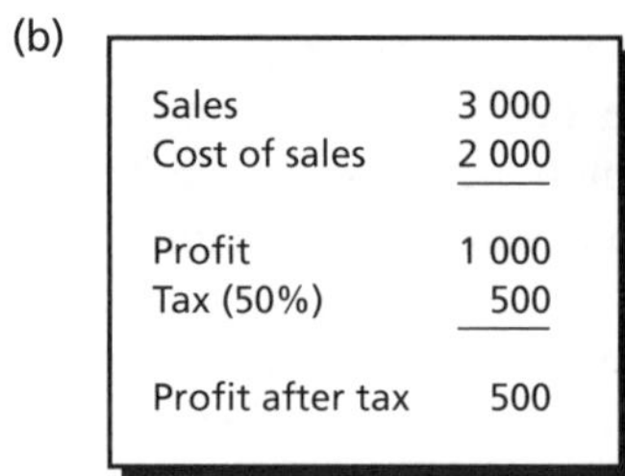

Sales	3 000
Cost of sales	2 000
Profit	1 000
Tax (50%)	500
Profit after tax	500

(c)

Liabilities		Assets	
Share capital	2 000	Cash	2 500
Retained profits	500		
	2 500		

Figure 10.12 Trading accounts ignoring inflation: (a) balance sheet at beginning of year; (b) profit and loss account for the year; (c) balance sheet at end of year.

Time	Price index	Inflation factor for end of year adjustment
Begining of year	100	150/100 = 1.5
Middle of year	125	150/125 = 1.2
End of year	150	150/150 = 1.0

Figure 10.13 Calculation of inflation factors for the example of Figure 10.12.

rose from 100 to 150, as in Figure 10.13. The inflation factors for the beginning, middle and end of the year can be calculated as in the third column, and these are applied to the trading accounts of Figure 10.12, as shown in Figure 10.14.

At the beginning of the year the monetary value of the stocks and share capital was £2000. However, these same items will be worth £3000 at the end of the year, oowing to a 50 per cent inflation rate, and this is shown in Figure 10.14(a).

The sales are assumed to have taken place equally throughout the year, so that an average or middle of the year inflation factor is used to convert their value to that at the end of the year. The cost of sales is again the stock value, which is £3000 after adjustment for its end of year value. Figure 10.14(b) gives the profit

(a)

Liabilities	Assets
Share capital 2 000 × 1.5 = 3 000	Stock 2 000 × 1.5 = 3 000

(b)

Sales 3 000 × 1.2	=	3 600
Cost of sales 2 000 × 1.5	=	3 000
Profit		600
Tax (50%)		300
Profit after tax		300

(c)

Liabilities		Assets	
Share capital 2 000 × 1.5 =	3 000	Cash	2 700
Defficiency of assets	(300)		
	2 700		

Figure 10.14 The example of Figure 10.12 after adjustment for inflation, using the factors of Figure 10.13: (a) balance sheet at the beginning of the year; (b) profit and loss account for the year; (c) balance sheet at end of year.

and loss account, which has been amended to cater for inflation. Note that £300 is paid in tax, so that although the cash received from the sales was £3000, the value remaining is £2700. This is shown in the closing balance sheet for the year, as in Figure 10.14(c). The share capital was £3000, so that in effect the company has suffered a capital depreciation of £300 during the year.

This example illustrates the two-fold effects of inflation. Profit before tax has been reduced (from £1000 to £600) and the capital has also shown a decrease instead of an increase (−£300 instead of +£500).

As a further example of the effects of inflation, suppose that two companies A and B buy assets at the beginning of years 1, 2 and 3 as shown in columns 4 and 6 of Figure 10.15. Using historical data both companies will have employed £7000 worth of assets at the end of year 3. However, if the inflation factors are used then the adjusted columns 5 and 7 are obtained by multiplying the initial costs by the inflation factors, and this shows that Company A is using more assets than Company B, since most of its assets were bought when money had greater purchasing power. It is these adjusted asset figures which should strictly be used in calculating accounting ratios to compare company performance and to calculate depreciation.

Year (1)	Price index (2)	Inflation factor (3)	Cost of assets bought			
			Company A		Company B	
			Initial (4)	Adjusted (5)	Initial (6)	Adjusted (7)
1st beginning	100	1.6	4 000	6 400	1 000	1 600
2nd beginning	120	1.3	2 000	2 600	1 000	1 300
3rd beginning	140	1.1	1 000	1 100	5 000	5 500
3rd end	160	1.0				
			7 000	10 100	7 000	8 400

Figure 10.15 The effect of inflation on capital employed.

The CPP method uses historical cost data, which is then adjusted for inflation by reference to a consumer or retail price index. The disadvantage of this method is that this index is applied uniformly to all companies, although inflation does not affect all industries in the same way and by the same amount. In CCA the replacement cost is used to measure the resources that are used up in carrying on a business. In the historical accounting system a fixed asset was bought and depreciated by a figure based on the initial cost of the asset, even though this may have gone up in the meantime, so that the profit and loss account does not reflect the actual consumption of the asset that has occurred. The CCA system uses 'value to a business' as a measure of assets, and this is the loss which the firm would suffer if it was deprived of the asset at that time. This is usually calculated as the current replacement cost of the asset less any depreciation that has been allowed for.

10.4 Profitability

In order to determine the profitability of a company, several accounting ratios have been devised. In practice these ratios are primarily useful when comparing the performance of the company from one year to the next, or in comparing the company with other similar organizations in the industry.

Almost any ratio can be used so long as it provides management and investors with the information they need, although only a few of the more commonly used ratios are described in this section. Generally these ratios can be divided into management ratios, which are useful for managing and running the company, and investment ratios, which provide shareholders and new investors with a view on how well their investment is likely to perform. Many ratios, such as market share, are useful to both managers and investors, but they are operational rather than financial ratios and are not considered here.

It is important to exercise caution when determining the performance of a company based on its financial ratios. Most ratios are obtained from the company's published reports, and, as seen earlier, these represent a snapshot of the company at any one instant in time. Also, a ratio contains a numerator and a denominator, and if the ratio is low it may be that this is because the numerator is low or the denominator is high; and vice versa.

Because ratios are usually taken from balance sheet figures they represent an 'average' of all the products and activities within the company. If the organization was involved in manufacturing several product lines, or providing several services, or in operating in different markets, it might wish to know how profitable each of these was, and in this case management would need to set up special data collection processes to obtain individual information from which the ratios could be calculated.

In evaluating ratios there is no absolute indicator of good or bad. There is only a comparative measure of how one is performing compared with past results and with others within the industry. Many companies belong to organizations that take information from their members and then produce statistics for them, based on averages and ranges, in which the individual results cannot be identified. This often allows the companies to compare their performance with others within their industry (who may be competitors) as well as comparing their own performance over several periods. In comparing performance over time it is important, however, that the effect of inflation is fully accounted for, so as not to produce misleading information.

10.4.1 Management ratios

Some of the management ratios used are shown in Figure 10.16. This is referred to as the pyramid of ratios, with the 'prime ratio', given by Equation (10.5), at the top.

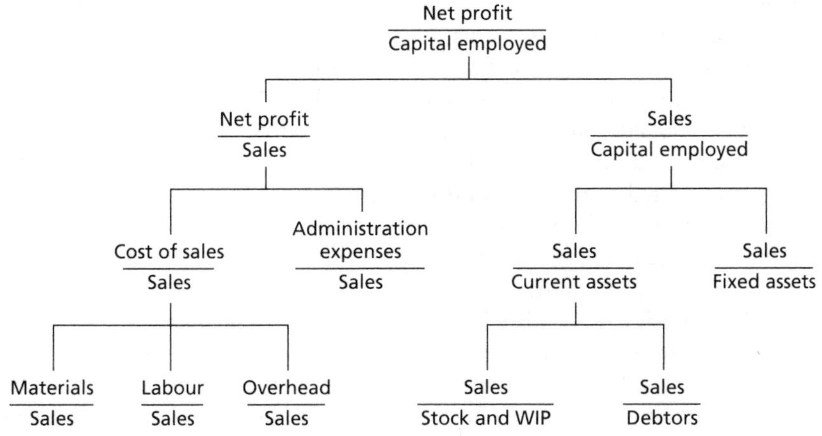

Figure 10.16 A pyramid of ratios.

$$\text{Prime ratio} = \frac{\text{Net profit}}{\text{Capital employed}} \tag{10.5}$$

The prime ratio represents the return on capital, used in the business and is therefore an indicator of how well the company is using its resources. It is also know as the 'return on capital employed' or ROCE ratio. Other similar measures are 'sales (turnover) per employee' and 'profit per employee'.

A high prime ratio compared to others in the industry, or one which increases over several years, indicates that the company is using its resources effectively. However, since the capital employed figure used for this ratio does not include current liabilities, a high ratio could also mean that the business is undercapitalized and is depending heavily on current liabilities for its operations, and this will be shown by other ratios. Furthermore, the capital employed figure taken should be an average of that used throughout the year, although usually the figure in the balance sheet (which is the amount on the last day of the year) is used. It may be preferable to take the mean of the present year (end) and previous year (end) to arrive at an average for the current year.

The prime ratio can be further investigated by working down the pyramid of ratios, as shown in Figure 10.16 and illustrated by Equation (10.6).

$$\frac{\text{Net profit}}{\text{Capital employed}} = \frac{\text{Net profit}}{\text{Sales}} \times \frac{\text{Sales}}{\text{Capital employed}} \tag{10.6}$$

Each item on the right-hand side of Equation (10.6) can then be further divided, as in Figure 10.16.

The ratio of net profit to sales shows the profit margin under which the company has been operating. If this ratio is small or decreasing its cause can be investigated further by determining the ratio of the costs to sales. It may be due to manufacturing costs being too high, too much being paid for bought-in materials, selling price being too low and so on.

A low ratio of net profit to sales may be due to low sales or to high costs. The cost ratios are divided into cost of sales (direct) and administration (indirect) costs. These costs can be broken down into further ratios in order to investigate them. Cost of sales has been shown to be split into materials, labour and overheads, and administration can also be subdivided, for example into clerical; management and supervision; R&D; and selling and marketing. Many indirect expenses are fixed in nature and do not change proportionally with the volume of sales, so that the administration expense to sales ratio should fall if the sales volume increases; if this did not happen the ratio would indicate that further investigation was required.

Further division may also indicate the cause of a high cost of sales to sales ratio. For example, this may be due to a high labour to sales ratio, which could be investigated still further by calculating the sales per employee ratio. If this was also low it might be due to poor productivity and indicate to management that greater investment should be made in automation.

The sales to capital employed ratio shows how often the profit margin, described by the profit to sales ratio, is obtained, since the two ratios together lead to the prime ratio, as in Equation (10.6). This ratio can be increased by several means, such as by working shifts in order to better utilize capital. However, this could push up labour costs, especially if premium overtime rates are being paid, and this would reduce the net profit to sales figure. The sales to capital employed ratio can be broken down by the two elements of capital, fixed and current assets, to work down the pyramid of ratios, as in Figure 10.16.

The ratio of stock and work in progress to cost of sales is usually stated in months. It is an important ratio since it shows the stock position of the company. How long this period should be depends on many factors, such as how effective one's suppliers are in operating just-in-time systems; the state of the economy; and the demand for goods. Figures over several periods may show a trend. Clearly, running out of stock can be expensive, but high levels of stock may also indicate wasteful inventory, and much of the stock may have deteriorated and not be usable.

The debtors to sales ratio (inverse of sales to debtors ratio) is again usually given in months, and it indicates how efficient the company is in collecting its debts. One would expect this ratio to be one month, since customers are usually given 30 days to settle an account. A large debtor ratio indicates poor collection processes within the company, and it may be hiding a high level of bad debts.

Several other ratios are commonly used by management in monitoring the performance of their organizations. The current ratio is given by Equation (10.7).

$$\text{Current ratio} = \frac{\text{Current assets}}{\text{Current liabilities}} \qquad (10.7)$$

This ratio illustrates the company's ability to meet its liabilities and is another indicator of the level of stocks held. It is generally believed that this ratio should be about 2. Too high a value may indicate a large stock level.

Because stock and work in progress take time to convert into cash, the liquidity ratio, which removes these from current assets in Equation (10.7), is a better indicator of the company's ability to finance its day-to-day operations. Generally, a liquidity ratio of just over 1 is considered to be desirable, although it will depend on several other factors, such as how old the debtors and creditors are; the amount of bad debt; and how much of the current assets are held as cash. A very low liquidity ratio may mean that the company is in serious difficulties, since it is depending too heavily on current liabilities to finance its operations and may not be able to repay its debts.

The calculation of management ratios is illustrated in the case study 'A few ratios'.

Case study

A few ratios

Based on the balance sheet of Figure 10.2 and the profit and loss statement of Figure 10.7, Johnny Bacon, Financial Director of the FloRoll Manufacturing Company, decided to produce a set of ratios to brief the other directors regarding the performance of the company.

Johnny calculated some of the ratios, as shown in Table 10.3. He found the prime ratio to be 28.2 per cent, which was an improvement on last year, although his intelligence reports told him that this was still slightly low compared with others within the industry.

Table 10.3 Analysis of management ratios based on Figures 10.2 and 10.7

Description	Calculation	Value
Net profit/capital employed	(48 000 × 100)/170 000	28.2%
Net profit/sales	(48 000 × 100)/700 000	6.9%
Cost of sales/sales	(60 000 × 100)/700 000	85.7%
Materials/sales	(250 000 × 100)/700 000	35.7%
Labour sales	(300 000 × 100)/700 000	42.9%
Overheads/sales	(50 000 × 100)/700 000	7.1%
Administrative expense/sales	(50 000 × 100)/700 000	7.1%
Sales/capital employed	700 000/170 000	4.1
Sales/current asset	700 000/110 000	6.4
Sales/(stock and WIP)	700 000/40 000	17.5
Sales/debtors	700 000/20 000	35.0
Sales/fixed assets	700 000/120 000	5.8
(Stock and WIP)/cost of sales	(40 000 × 12)/600 000	0.8 months
Debtors/sales	(20 000 × 12)/700 000	0.34 months
Current asset/current liability	110 000/60 000	1.8
(Current asset – Stock and WIP)/current liability	(110 000 – 40 000)/60 000	1.2

The net profit to sales figure of 6.9 per cent was considerably below that of other companies, so Johnny decided to investigate it further. The cost of sales to sales figure of 85.7 per cent was very poor, and when Johnny Bacon broke this down into its elements he found the labour to sales ratio to be 42.9 per cent, which was almost twice that of other comparable organizations within the industry. Johnny was certain that when Jake Topper, the Chief Executive, saw this he would want to accelerate the automation programme within the labour-intensive Dinsher Shop, in order to cut labour costs.

Johnny was satisfied with the other ratios. The sales to debtor ratio of 35 was a bit low, but acceptable since they were getting very good service from their suppliers.

The debtors to sales figure of 0.34 months was very good, and indicated that the new debt-collecting process was working. The current ratio of 1.8 was also satisfactory and this was borne out by the liquidity ratio of 1.2, which showed a healthy state, especially as a large amount of the current assets was held as cash and investments which could be quickly converted into cash.

10.4.2 Investment ratios

The ratios discussed in the last section are referred to as management ratios, since their main purpose is to help managers to run the company. They also, to a limited extent, provide a guide to investors in the company. The ratios in the present section are called investment ratios, since their primary purpose is to help investors.

The capitalization ratio represents the proportion of the different types of share which make up the capital of the company, and is given by Equation (10.8). It is often referred to as leverage or gearing.

$$\text{Leverage} = \frac{\text{Ordinary equity}}{\text{All interest-bearing capital}} \tag{10.8}$$

The leverage is important in determining the dividend return on ordinary shares. For example, if a company has its capital made up of ordinary shares of £1000 and 10 per cent preference shares of £50 000, then a profit of £5000 would result in the preference shareholders being paid in full (£5000) while the ordinary equity-holders would receive no dividend. However, if the profit level now increases by 20 per cent to £6000, the preference shareholders would still receive a 10 per cent dividend, while the ordinary shareholders could receive £1000: a dividend rate of 100 per cent. It is because of this large fluctuation in dividends, caused by small changes in profit that companies with low ordinary equity to total capital are said to have high leverage or high gearing.

Low gearing usually indicates a stable company, although the prospect for substantial dividends is correspondingly less.

The interest cover ratio indicates how easily the company can pay its interest charges and whether borrowed money is being put to good use. The ratio is given by Equation (10.9).

$$\text{Interest cover} = \frac{\text{Gross profit} - \text{Indirect expenses}}{\text{Loan interest}} \tag{10.9}$$

The inverse of the interest cover ratio, expressed as a percentage, shows the percentage of profits that is paid out as interest before shareholders get any return.

The dividend cover ratio indicates the number of times over that the company has earned dividends in the year. The income after all interest and tax has been paid is now divided by the dividends distributed, as in Equation (10.10).

$$\text{Dividend cover} = \frac{\text{Net profit} - \text{Tax and preference dividend}}{\text{Dividend distributed}} \tag{10.10}$$

The dividend cover gives an indication how well the dividend is protected for future distributions. In calculating it the effect of all extraordinary items should be removed.

The dividend cover ratio usually refers to ordinary shares. The preference dividend cover can also be calculated, in which income after tax is divided by the preference dividends paid.

The earnings per share ratio indicates how much each ordinary share has earned, which is not the same as the dividend distributed (Blair, 1997; Thomson, 1995b). It is equal to the profit available for distribution, after preference dividends have been paid, divided by the number of ordinary shares, as in Equation (10.11).

$$\text{Earnings per share} = \frac{\text{Profit after tax} - \text{Preference dividends}}{\text{Number of ordinary shares}} \tag{10.11}$$

Several other ratios give the earnings potential of shares compared with their market value at the time. The earnings yield ratio is given by Equation (10.12) and the dividends yield ratio by Equation (10.13).

$$\text{Earnings yield} = \frac{\text{Earnings per share}}{\text{Market price of share}} \times 100 \tag{10.12}$$

$$\text{Dividend yield} = \frac{\text{Dividend per share}}{\text{Market price per share}} \times 100 \tag{10.13}$$

The price to earning ratio (PE ratio) is the inverse of the earnings yield ratio, and it indicates the number of times the share price exceeds the earnings potential. In theory this would be the number of years after which investors would have recouped their outlay, assuming dividend levels are unchanged and the share price has not changed. A high PE ratio usually indicates a growth company. It may also mean that the company has had a temporary sharp fall in its profits, but that investors still expect it to grow in later years.

The last investment indicator to be considered here is the net book value or the net asset value, which is the value at which an item is carried on the company's account books. For example, the net book value of the ordinary share is the amount of money each share would get if the company was liquidated. It is found by taking away all liabilities, such as current and long-term liabilities and preference shares, from the company's assets and then dividing by the number of shares, as in Equation (10.14).

$$\text{Net book value} = \frac{\text{Assets} - \text{All liabilities} - \text{Preference shares}}{\text{Ordinary shares}} \tag{10.14}$$

The net book value is only an indicator of what the shareholders would get if the company went bankrupt. In reality, in an enforced sale assets rarely fetch their book value. Calculation of investment ratios is illustrated in the case study 'A good investment'.

Case study

A good investment

Jake Topper was thinking of proposing a rights issue in order to raise finance to fund the FloRoll Manufacturing Company's proposed new automation programme. He asked Johnny Bacon to carry out an analysis of the investment ratios for the company, to see how attractive it would be to an investor.

Johnny based his analysis on the latest published figures for the company, given by its balance sheet and profit and loss statement, as in Figures 10.2 and 10.7. These are summarized in Table 10.4.

Table 10.4 Analysis of investment ratios based on Figures 10.2 and 10.7

Description	Calculation	Value
Leverage	80 000/(80 000 + 30 000)	0.73
Interest cover	(10 000 – 50 000)/2000	25.00
Dividend cover	(48 000 – 24 000 – 3000)/16 000	1.31
Preference dividend cover	(48 000 – 24 000)/3000	8.00
Earnings per share	(24 000 – 3000)/80 000	£0.26
PE ratio (price = £3.90)	3.9/0.26	15.00
Net book value	(230 000 – 60 000 – 20 000 – 30 000)/80 000	£1.50

From Table 10.4 it is seen that the leverage is relatively high, which pleased Johnny since it indicated a stable company and suited the company's image. Interest cover was also healthy. The dividend cover was typical of other companies in the industry and the preference dividend cover was also good.

Earnings per share were attractive, especially when compared to the issued share price of £1.00. However, the current share price, as quoted on the stock exchange, was £3.90, so the price to earning ratio was lower, at 15.00, which was still good compared with competitors. Finally, the net book value was calculated as £1.50. Therefore, theoretically, investors would receive about 50 per cent more then their initial investment of £1.00, although they would be getting less than 40 per cent of their current investment, if this was made at £3.90.

Overall Johnny Bacon was confident that the company was attractive to new investors and set about putting a price on the rights issue.

10.5 Budgets and controls

A budget is a financial statement of the future plans and intentions of an organization. It also represents its future predictions on financial matters and is a plan of action. Controls are required on budgets to check whether these financial plans are being achieved and to take corrective action, if necessary.

The purposes of a budget are to:

- Define the financial objectives and responsibilities of the organization, in terms of objectives for individuals and functions, relative to the overall company objectives.
- Provide a picture of the overall financial state of the company, on a continuous basis.
- Measure the financial performance of the company, to reveal problem areas and to increase efficiency of the financial processes used.
- Motivate managers by setting clearly defined financial goals, relative to which they can be measured.
- Help in cost control.

Budgets are formed by a 'roll upwards' and a 'work down' process; what is needed and what can be afforded by the organization (Bryan, 1996). This is illustrated in Figure 10.17. The work down process determines the expenditure that is allowable, usually based on the expected level of sales in the period for which the budget is being developed. The roll up process works the other way, and sets out to determine the amount of expenditure required to meet the company's objectives. Where the required expenditure exceeds the allowable expense, as in Figure 10.17, an adjustment is required. This may be in the form of increasing the revenue forecast to increase allowable expense, or reducing the expenditure level to meet the allowable expense.

Because many budgets are cut back from the expenditure level initially requested, most managers tend to add a 10–15 per cent excess factor on to their requests in anticipation of these cuts. This is done at every stage so the final figures for the organization could have a sizeable amount added on to them. Since this is a known practice most budgets are often automatically cut back by this amount to start with, so the net impact is low. Unfortunately, this practice creates an attitude of mind which indicates that the budget is of limited value (Secrett, 1994). Contingencies can be added to budgets, but this should only be done after a careful consideration of all the factors that may arise, and the contingency budget should be kept separate from the main budget. An alternative

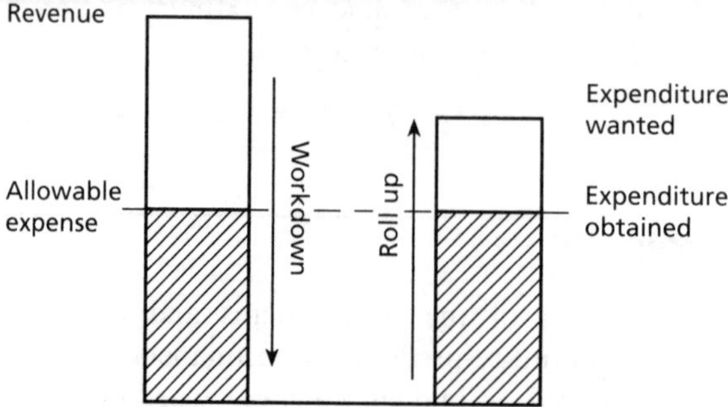

Figure 10.17 The simple roll up, work down budget preparation process.

way of treating this is to have an operational budget and an opportunity incremental target. The operational budget is followed unless circumstances change, when the contingency or opportunity budget is brought into action.

Budgets should normally be flexible, so that they can vary during the budgeting period, as circumstances change. They could also be linked to some other item, such as sales; for example, the greater the level of sales during a period the higher the expenditure budget allowed during the next period. Unfortunately many activities, such as R&D and investment in manufacturing automation, require long-term planning and commitments and cannot be readily turned on and off. Others, such as manufacturing output, should always be linked to items such as sales to avoid large inventories.

Budgets are usually short-term, covering periods from three months to one year. They serve to break down the long-term financial plans into these shorter intervals and then to measure progress against them. There are several types of budget encountered in a typical organization:

- **Cash budget**. This is clearly very important to manage, in order to maintain the overall liquidity of the company.
- **Revenue budget**. Examples are sales and profit.
- **Expense budget**. Examples are travel, small tools, training, materials and subscriptions to professional bodies.
- **Headcount budget**. This measures the number of people and their costs. Loaded labour rates are usually used; that is, salaries of the staff concerned plus compensations such as National Insurance and company pension contributions, plus a portion of overheads.
- **Capital budget**. This can cover a wide spectrum of items, from heavy machinery used in manufacturing to a computer or test equipment used within an R&D department.

In controlling the budget, speed of feedback is much more important than absolute accuracy. If a budget is being grossly overspent then it is important to know this quickly, so that action can be taken. Whether the overspend is 40 per cent or 100 per cent becomes important later. The longer the delay the larger it is likely to be.

The cost of monitoring and controlling should also be balanced against the benefits to be gained. Controls should usually be easy to implement and to understand. The activities which are controlled should also be relevant: those that show the greatest benefit.

It is usually very difficult to determine the budget requirements accurately. This is especially true in R&D, where the level of problems met when developing new products or technologies cannot be determined in advance. Usually these budgets are based on experience of similar work done, and frequent checks are made against plans, as the work progresses, to determine the accuracy of the budgets.

To be successfully implemented a budget requires the following:

- A well-defined organizational structure, with those within the organization having a clear understanding of their responsibilities.
- Sound accountancy procedures in place to set and monitor budgets within the organization.
- Support from all levels of management within the organization, to implement the budget.
- Good control and feedback procedures, with corrective actions in place.
- A willingness to modify the budget as circumstances change. For example, a small overspend on the budget in the year being considered may be required in order to give a greatly improved product and vastly increased sales in subsequent years. Clearly the budget should be modified to allow for this increased spend.

Usually budgets are controlled by regular review (weekly or monthly) at all levels, the figures at the various levels being rolled up to the top to give an overall picture for the whole organization. This is usually the reverse of the budget setting process, as shown in Figure 10.18. From the level of sales required for the organization the sales from each geographical area may be set. This can, in turn, be broken down further into the sales budget for each product within the different geographical areas. For monitoring purposes the sales from each area are rolled up. If the budget figures are missed then it is important that enough detail is available to determine the cause and to take corrective action. In the case of the example of Figure 10.18, a low sales achievement could be tracked to the geographical area and then on to the product within that area having low sales.

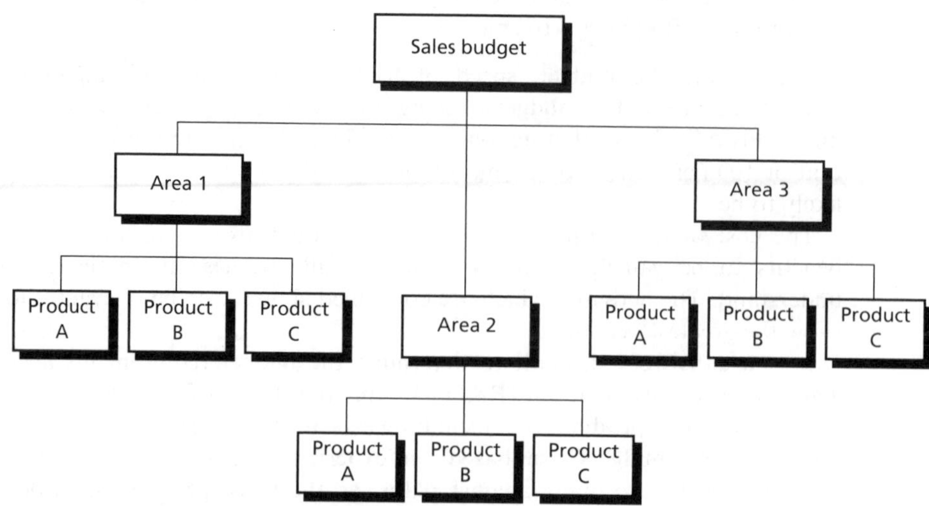

Figure 10.18 Example of a sales budget setting and review procedure.

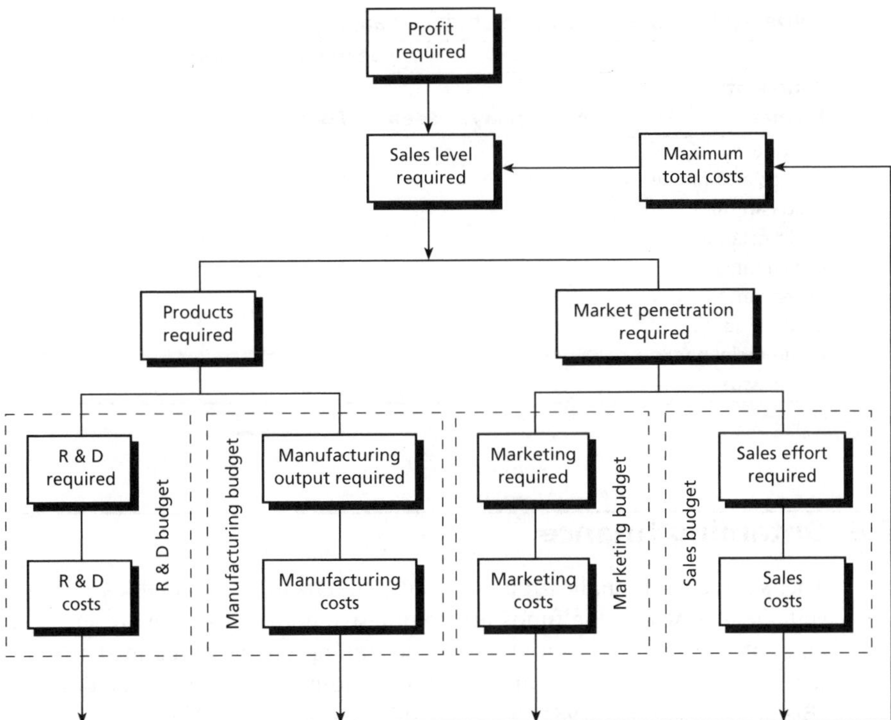

Figure 10.19 A simplified scenario for budget setting.

Figure 10.19 illustrates a typical budget-setting process. Profit required determines the sales level, assuming an associated total maximum cost. This sales level requires activity to support it, such as products and market penetration. Products need to be developed, and these have associated costs which result in the R&D budget. Similarly, the manufacturing output needed to support the assumed level of sales, together with its associated costs, form the manufacturing budget. Market penetration would require a level of marketing and sales activities, and these, along with their costs, give the marketing and sales budgets respectively.

Total costs are then rolled up and subtracted from the expected sales level. If the profit is below the value desired an adjustment is needed and the budgeting process is again worked through. Several iterations may be required before a satisfactory value for profit is obtained, one which meets the company's aims with an affordable level of costs.

Although profit is taken as the target for the budget in Figure 10.19, any other parameter may be used, depending on the company's aims. For example, this could be sales level, cost level, market penetration and so on.

Table 10.5 Finance at various stages of an organization's development

Source of finance	**Stage of development**				
	Pre-company	**Seed**	**Formation**	**Established**	**Sustained growth**
Personal savings	*	*			
Friends and family		*	*		
Seed capital		*	*		
Business angels			*	*	
Non-financial organizations			*	*	
Venture capital				*	
Commercial banks				*	*
Public issue					*

10.6 Obtaining finance

The source and application of funds statement, described earlier, provides a statement of where the funds are coming from and how they are being used. The amount of funding required varies, depending on the stage in the organization's development, and the source of this funding also varies, as described below (Bruno and Tyebjee, 1985; Roberts, 1990; Thomson, 1995a).

A company, especially one whose products are based on new technology, goes through several stages when it first starts up. These can be classified as pre-company stage, seed stage, formation stage, established stage and sustained growth stage. The funding requirements and the most common source of these funds will vary during these stages, as illustrated in Table 10.5 and described in the sections below.

10.6.1 Pre-company stage

During this stage the idea for the new company first begins to germinate (Chang, 1996). Usually this is in the minds of its founders, who may be in full-time employment in another larger organization. The potential founders of the new company will be knowledgeable in the technology they wish to exploit, and they will often be working in an allied field in their present company.

This stage of the future company's development will be completed once the principle for the new product has been roughly defined by the new founders and a decision taken to start a new venture.

Very little funding is required at this stage and any needed is met from the personal savings of the founders.

10.6.2 Seed stage

During this stage the company is formed and the founders start work on their original idea. Prototypes are developed in order to prove the principles and to show to potential backers.

In parallel with the technical work the company carries out an estimate of the market size and develops a business plan.

The finance required at this stage is usually quite low. Some materials and capital will be needed to develop the prototypes and the founders will want a basic wage. Usually they will be earning a fraction of their previous salaries and will be putting in a considerable amount of 'sweat equity' to get the organization off the ground.

The organization is in a high risk position where it has neither proven products nor proven management skills. It can fail at any time. Sources of finance for this stage are not readily available and usually come from personal savings or from family and friends. Any potential backer will expect the founders to have committed most of their own savings to the venture first, as a sign of good faith, before they are likely to invest. It is also a good idea to limit the extent of outside investment at this stage to avoid over-dilution of the founders' holdings in the company. This leaves room for dilution to occur at a later stage, when growth accelerates and greater external investment in the company is required, to fund this growth effectively.

Personal savings are still the most common source of funds in these early stages of a company's life, although the amount that can be raised in this way is limited unless the founders have personal wealth.

Family and friends may be willing to invest in the new company, since they know the founders and have confidence in their ability. However, the founders may be unwilling to use this form of finance, especially if they believe that the risk of failure is high. Another disadvantage of this form of finance is that the investors may interfere unduly in the running of the company, by way of advice. The investment is also almost all short-term.

Some seed funds may be available to the start-up company. Usually this is also relatively small in amount and is available from private investors. There are several seed capital groups who work together to share the risk of investing in small start-up companies. These seed finance groups may also be able to provide other help to the fledgling company, such as advice on business plan preparation. They normally work very closely with the founders.

10.6.3 Formation stage

During this stage the first product development is well advanced and prototypes may even have been delivered to customers. The customer base is usually small, the company working closely with selected customers who have a special interest in the product.

The company has also made considerable progress in defining its business plan and the management team is in place, along with an organization to support it.

Most of the time within the organization is still spent on product development, and the capital investment requirements are still small. Apart from materials and meeting the salaries of its few employees, funding will also be needed for low-cost premises, working capital and plant to do limited manufacturing for sale.

The company is losing money at this stage and this needs to be financed. No money is available for inward investment, so the company must look outside for funding.

The company is still in a very high risk state, so investors will be looking for high rates of return to compensate for this risk. They are being asked to put their trust in the management of the company and in its product, neither of which has been proven. Potential investors also need to be very patient, since they are unlikely to see any return on their investments for several years.

Sources of finance at this stage are shown in Table 10.5 and include investments from wealthy private investors, called business angels (Thackray, 1993; Freear *et al.*, 1994). Since these investors are private individuals, without accountability to anyone else, they are usually willing to take greater risks than financial institutions. The business angels may make their investment decisions based on factors other than financial ones. For example, they may be excited by the new company or the product being developed, and they may also invest out of a sense of social responsibility.

Business angels usually depend on advice from associates and friends regarding the company in which to invest. They are influenced by the quality of the management in the new company. Usually they will invest as equity, expecting to make a gain when the company grows. For the new start-up this has the advantage that it is not burdened with debt repayment when it can least afford it.

Government help is also available for newly formed companies. To encourage investment by individuals in small companies, many governments offer tax incentives. Government grants, including those from the EU, are also available for new businesses, although these are usually on a regional basis, such as the grants for investment in areas affected by run-down of existing industries (such as ship building) and attracting industry to areas of high unemployment.

Small new companies may also receive funding from larger non-financial organizations. Usually these companies are interested in the technology being developed by the smaller company and look to them to supplement their own internal R&D. Sometimes this investment may be made via a venture capitalist, as described in the next section. Since the larger organization is mainly interested in the technology being developed, it is less concerned with the quality of the smaller company's management or in its welfare. It may, however, provide the smaller partner with other benefits, such as marketing and technical help.

The disadvantage of this arrangement, to the smaller company, could be that the larger partner would want a say in its day-to-day operation, with representation on its board. It may also oppose any plans the small company might have for going public, since it may prefer to eventually merge it into its own operation. Although many small start-ups have dreams of eventually going public, most of those that survive end up selling out to larger organizations.

10.6.4 Established stage

During this stage of its development the company has completed development of its first product, has manufactured it and has obtained substantial sales. It now has expectations of rapid growth.

Other considerations, apart from developing the product, now require management attention. These include quality matters; growing the management team to include ancillary functions such as manufacturing, sales and marketing, quality and personnel; and installing an effective management control system.

Competitors begin to enter the market and an effective strategy for dealing with them needs to be developed. The customer base grows, but is still relatively small. Funding is required to develop other products and also for investment in more plant and equipment to enable volume manufacture. Sales and marketing activities also need to be funded. As cash flow is still not sufficient to enable these investments to be funded from internally generated profits, external sources of finance are required.

The risks associated with the start-up company have been considerably reduced by this stage but are still high. Apart from some of the sources of finance mentioned earlier, such as business angels and non-financial organizations, other sources at this stage are venture capital and the commercial banks (Maier and Walker, 1987; Timmons and Bygrave, 1986; Weyer, 1995). Some companies could also go for a public flotation or be taken over by a larger organization.

Venture capitalists may be wealthy families or financial institutions. Wealthy families are usually prepared to take greater risks, so they would invest at an earlier stage in the company's development. Commercial venture companies often raise funds from banks, pension funds, wealthy individuals and non-financial companies.

All venture capitalists employ professional staff who go through a detailed vetting process before investing in any new projects. Only between 1 and 3 per cent of applicants receive any funds, and these are usually those who are relatively advanced in their development stage. In its application the new company will need to provide very detailed information on its business operation and plans. Once venture capital has been obtained, however, other sources of finance may be more readily available, since any new investor will know that the company has been through close financial scrutiny.

Usually venture capitalists take an equity stake in the company and depend on the growth that results from this to make their gain. A high rate of return is expected for the risks taken and to compensate for the many businesses that fail.

Venture capitalists usually want to have a say in the running of the company.

They specialize in different sectors of industry, several investing in high-technology start-up companies.

The big commercial banks are still the most important source of finance for small businesses (Wilsher, 1994). Loans provided may be via an agreed overdraft, where interest is paid on the amount borrowed only, or via a term loan. Usually the term loan is for a fixed period and there may be a penalty for early repayment, although the interest rates are lower.

Loans usually start off by being short to medium term, but as they are frequently renewed they can eventually end up being long-term. The banks cultivate small companies in the hope that when they grow bigger they will continue to deposit their money with them, so providing a cheap source of funds as well as generating income by way of charges. They will work very closely with the new company, providing advice and guidance on many different matters, including finance management.

Many companies prefer finance from the commercial banks rather than from venture capital, primarily because this avoids dilution of their equity. Ashton-Tate, a prominent software house, for example, refused venture capital in its early development stage and went instead for a $6M bank credit (Roberts, 1990).

10.6.5 Sustained growth

The company is now well established and is going through a period of growth. The annual sales revenue is high and several product lines have been developed and are being successfully marketed. The company has a large number of employees and may be operating from multiple sites and even multinationally.

The customer base is large and diverse, covering a wide geographical area, including overseas. The company is occupied in those activities associated with a large organization, such as planning and control, communications, staff morale building, and setting the future strategic direction for the organization.

Substantial profits are being generated, so the company is to a large extent self-financing, although external funds may still be needed for opportunity growth.

Investors in the company now carry low risk and therefore expect a lower return on their investment. Sources of new finance still remain the large commercial banks, and the loans are usually of a long-term nature, since the company now has sufficient assets to serve as collateral.

At this stage the company may agree to be taken over by a larger organization or merge with it, especially if the original founders wish to sell up or retire. Alternatively, it may wish to go public, and such a move would also ensure that the founders' and backers' holdings are liquid, enabling them to withdraw from the company at a later date. Another reason for going public might be the prestige associated with being quoted on the Stock Exchange.

Raising funds by equity may represent the most economic way for a company to obtain long-term finance. Once the investment has occurred the company is under no obligation to repay the investment or to provide interest (dividends), both of which can drain a new business of the cash it requires for growth.

The disadvantage of equity funding is the dilution of ownership and that profits are now shared between all equity-holders. (However, absolute profits are what count. Therefore, for example, if a 50 per cent increase in equity results in a 100 per cent increase in profits, the founders are still receiving more on their investment.) Equity funding can also be an expensive form of finance if the company is growing very rapidly (Costley-White, 1991). For example, suppose the value of a company is £50 000 and another £50 000 is invested in it by way of equity, raising the overall equity to £100 000. If the company now doubles its growth in one year it will be worth £200 000, so the original founders will have to pay £100 000 if they wish to buy back their original company; that is paying £100 000 on an initial 'loan' of £50 000. If a bank loan had been taken out for the £50 000 needed, and the interest rate was 20 per cent, then only (£50 000 + £10 000) or £60 000 would be due at the end of the year.

If the organization is already a public company it can go for a rights issue, where existing shareholders have the option to buy new shares as a proportion of their current holdings. Launching a new issue is a complex and risky business and many professional organizations exist who can help the company go public.

10.7 Valuing a company

When raising funds it is often necessary to place a value on the company. If the company is quoted on the stock market then this provides a guide for its share price. However, many factors affect this price and it is therefore not a completely reliable method (Outram, 1997). For example, the price may fluctuate short-term due to:

- Unexpected changes in the current year's dividend. If the company announces a bad result for the year its share price is likely to fall. This is why companies often leak bad results ahead of the actual results, so that the market discounts the results. When these results are eventually announced the share price may actually rise if the actual results are not as bad as anticipated.
- Perception among investors of the future earning potential of the shares. As explained before, the price to earnings ratio (PE ratio) is one of the main indicators which investors use when buying shares.
- Press recommendations to buy or sell a particular share.
- The activities of speculators. Since there is usually a settlement period between buying and selling shares and actually having to pay for them or handing them over, there are two types of speculator. One anticipates a falling market and sells shares he or she does not possess in the hope of buying them back cheaply when the market falls, and before the shares are to be delivered, so making a profit. These speculators are know as 'bears' and

> the falling market is referred to as a 'bear market'. A rising or 'bull market' gives rise to 'bulls' who buy shares at a low price, with money they do not have, in the hope of selling them again at a higher price, before they have to be paid for.

In the long term the price of a share is primarily affected by the future dividends that it is perceived the company will be able to pay. This is illustrated in Table 10.6, where it is assumed that investors buy and hold shares for five years each before selling them on. Each purchaser buys on the basis of the expected dividend over the five years plus the sales value at the time he or she comes to sell the shares. Therefore Purchaser A buys at the purchase price on year 1 and obtains dividends over years 1 to 5. She sells at the sale price in year 5. Purchaser B then buys at this price, holds the shares for five years, collecting dividends over years 6 to 10, and sells at the price existing in year 10.

This process continues through the line of buyers and sellers over the 40 years illustrated in Table 10.6. It is seen from this table that, since the purchase price in the second column exactly equals the sales price in the fourth column, apart from the purchase price in year 1, these cancel out. Therefore the purchase price in year 1 equates to the third column only: the accumulation of dividends expected over the years 1 to 40.

Several other methods of valuing a company exist, especially when the company is not quoted on the stock market. It must be remembered, however, that the method of valuation depends on the purpose for which it is to be used. If a bank wants this, to determine whether it should extend overdraft facilities or provide a loan, then the main points of interest are that the company's assets are sufficient that they can be sold to repay the loan, if necessary, and that the company's income is adequate to cover the loan charges. A supplier evaluating a company for credit is primarily interested in its liquidity state, whereas an investor is concerned with items such as return on capital and the PE ratio.

An alternative method of valuing a company is known as the net asset method. In this the assets and liabilities of the company are valued prior to the pricing and the net assets per share determined. (See the case study 'Putting a price on

Table 10.6 Illustration of share price based on dividend expectancy

		Return expected	
Purchaser	**Purchase price**	**Dividends**	**Sale price**
A	At year 1	Years 1–5	At year 5
B	At year 5	Years 6–10	At year 10
C	At year 10	Years 11–15	At year 15
D	At year 15	Years 16–20	At year 20
E	At year 20	Years 21–25	At year 25
F	At year 25	Years 26–30	At year 30
G	At year 30	Years 31–35	At year 35
H	At year 35	Years 36–40	At year 40

it'.) This method of share valuation is not often adequate, since it does not value a company as a going concern. Any business that has been going for some time will have acquired valuable assets that are difficult to price, such as customers, goodwill and management expertise. The net asset method is useful if the assets are to be sold and the company forced out of business.

To value a company that has been established for some time, the future earnings method is often used. In this the maintainable earnings of the company have first to be established. Taking one year's profits is not a good guide to maintainable earnings, and usually a weighted average of several years' figures is considered. The return obtained from other companies of a similar size and working in the same industry is next obtained, usually from their stock market quotation, and this is assumed to apply to the company being valued, with slight adjustments if necessary. Based on this ratio and the return from the company being valued, its capital can be determined. (See the case study 'Putting a price on it'.)

An alternative to the future earnings method is the price to earnings ratio technique for valuing a company. Once again it is important to establish the maintainable earnings of the company under consideration, but now instead of the return on capital employed it is necessary to determine the PE ratio of other similar companies quoted on the stock-market, the earnings considered being the profits available for distribution to ordinary shareholders. (See the case study 'Putting a price on it'.)

The various methods of company valuation yield different results. This is primarily due to the fact that it is very difficult to determine accurately many of the parameters needed to make the calculations, such as property prices and asset values, which are open to interpretation. This is one of the reasons why so much controversy is generated when share valuation is required, such as in a takeover action.

Case study

Putting a price on it

Johnny Bacon, Financial Director for the FloRoll Manufacturing Company, decided to value the company. He used the balance sheet of Figure 10.9 as the basis for this valuation. He obtained a revaluation of the property, which was now placed at £80 000. He also valued current assets at £120 000 allowing £40 000 for bad debts and some level of stock deterioration. With plant and equipment standing at £2000 this gave overall assets as £220 000, as in Figure 10.20.

Liabilities were obtained from the balance sheet of Figure 10.2 as the preference shares, long-term loans and current liabilities, and these totalled £110 000, as in Figure 10.20. The net assets are therefore £220 000 – £110 000 or £110 000.

Since there are £80 000 ordinary shares in the company the assets per share were found as £110 000/80 000 or £1.38.

Assets		
Property	80 000	
Plant and equipment	20 000	
Current	120 000	
	220 000	
Liabilities		
Preference shares	30 000	
Long term	20 000	
Current	60 000	
	110 000	
Net assets		110 000

Figure 10.20 Calculation of net assets for valuation of the FloRoll Manufacturing Company (from Figures 10.2 and 10.9).

Johnny Bacon then decided to value the company on the basis of future earnings. He used the past four year's profits as in Table 10.7, the current year being year 4, and the profit for this was obtained from the profit and loss account of Figure 10.10. It is seen that the weighted average profit is £366 000/10 or £36 600 per annum. Johnny Bacon also established that for companies of a similar size to the FloRoll Manufacturing Company, who were quoted on the stock-market, the return on capital employed was generally 25 per cent. Therefore he decided to use this figure and valued the company's assets as £36 600/0.25 or £146 400. This gave a price per share of £146 400/80 000 or £1.83.

Table 10.7 Weighting the profits for the FloRoll Manufacturing Company

Year	Profits (£)	Weight	Weighted profits (£)
1	28 000	1	28 000
2	33 000	2	66 000
3	40 000	3	120 000
4	38 000	4	152 000
Total		10	366 000

Johnny Bacon now used the PE ratio method to value the company. He again weighted the profits over the past four years and obtained a weighted average of £15 500, as in Table 10.8, where year 4 is the current year and the earnings for this were obtained from Figure 10.10 (profit available for distribution less preference share dividends). The PE ratio for similar companies quoted on the stock market was calculated to be 8, so the price of the FloRoll Manufacturing Company was £15 500 × 8 or £124 000. This valued the shares at £124 000/80 000 or £1.55.

Table 10.8 Weighting the earnings for the FloRoll Manufacturing Company

Year	Earnings (£)	Weight	Weighted earnings (£)
1	12 000	1	12 000
2	14 000	2	28 000
3	17 000	3	51 000
4	16 000	4	64 000
Total		10	155 000

The three methods of share valuation gave Johnny Bacon three different results: £1.38, £1.83 and £1.55. Since he only wanted a rough price indication and the three valuations were close enough, he took an overall average to arrive at a price of £1.59 per share.

Summary

Accounting can usually be classified as financial or management accounting. Financial accounts deal with historical data. They produce information on items such as profit and loss, which is of interest to investors. Management accounts are produced primarily to help managers to run the company, and as such they are an aid to decision making.

The financial account should present a fair and true view of the company at any instance in time. Its main elements are the balance sheet and the profit and loss statement. The balance sheet is required by law and the key entries within it are the assets and liabilities of the company. This is known as a double entry system, since entries on one side balance out those on the other.

The profit and loss account is also required by law. It determines the profit made by the company during the year, this being the difference between revenue and expenses. Expenses can be of two types, direct and indirect. Direct expenses are those which can be clearly identified as going into a product, such as labour and raw materials. Indirect expenses usually cannot be directly attributed to the product concerned, such as the cost of advertising that promotes the company's image rather than any one product.

Depreciation of fixed assets, such as plant and machinery, needs to be taken into account in the financial accounts. Its effect is to reduce company profits and tax payable, so enabling increased earnings to be retained, which could, if required, be put towards a replacement machine. Two methods are used to calculate depreciation, the straight line method and the reducing balance method. The straight line method assumes that the value of the fixed asset falls uniformly through its working life, while the reducing balance method depreciates it faster in earlier years. Both methods take as their base the original cost of the fixed asset, which may be lower than its replacement value, so depreciation should not be looked on as the prime method of replacing assets.

Inflation has the effect of inflating the value of stocks and providing a false impression of increased profits. It also causes the price of replacement fixed assets to increase rapidly. Several methods have been used to introduce the effect of inflation into company accounts, the two considered are the Current Purchasing Power (CPP) and the Current Cost Accounting (CCA) methods. CCP reduces all historical financial data to their current purchasing power,using an inflation factor based on the retail price index. CCA uses value to a business as a measure of its assets and calculates this as the current replacement cost of the assets used up less any depreciation that has been allowed for.

Several measures of company profitability are used in order to compare the performance of an organization over several years and with others within the industry. These are usually classified as management ratios, which are used for managing the company, and investment ratios, which are useful to the investors in the company. Management ratios can be considered to form a pyramid, as in Figure 10.16, with the ratio of net profit to capital employed as the prime ratio. This can be broken down into other ratios to investigate the overall performance of the company in various areas. Investment ratios include items such as interest cover, dividend cover and earnings per share, which indicate to investors the returns they can expect from the company, as well as its financial state.

Budgets are set within a company as a process of bottom-up and top-down analysis of what is needed and what can be afforded. Budgets cover a variety of areas, such as cash, revenue, expense, headcount and capital. They are an important method for setting financial objectives for the whole organization and for managing and controlling it.

Several sources of funds exist for a company, depending on its stage of development. For example, when it is first set up finance will usually come from the personal savings of the founders. At the seed stage, when prototypes are being developed, some seed capital may be available from individual backers. When prototypes have been completed, and perhaps delivered to customers, the company enters the formation stage, where funding may be available from business angels and some non-financial organizations. Government help may also be available in the form of grants and tax incentives. At the established stage the company has made sales of its products and the investment risk has been reduced. Money can now be attracted as venture capital and in the form of bank loans. In the final stage the company shows sustained growth and can be self-financing. Bank loans will be the main source of additional funds, although a public flotation to raise funds is also likely.

When raising funds it is often necessary to place a value on the company, and several methods may be used for this, although none of these provides a definitive value. The net asset valuation method determines the net asset per share based on the assets and liabilities of the company. This does not value the company as a going concern. The future earnings method of valuation overcomes this problem by determining the maintainable earnings of the company and then determining the price by comparing this with other companies having a similar figure. Similarly, the price to earnings ratio method determines the maintainable earnings of the company and compares this with other companies quoted on the stock market that have a similar figure.

Case study exercises

The exercises given below are based on an extract from the balance sheet and profit and loss account for the Skeldale Manufacturing Company as given in Figures 10.21 and 10.22.

10.1 What is understood by the terms 'balance sheet' and 'profit and loss account'? What is the difference between preference shareholders and ordinary shareholders? Why does a company retain profits and what effect does this have on the value of its share price? Who owns the assets of the company? Note that in the case of the Skeldale Manufacturing Company the dividends for the year were paid out of retained profits. Do you consider this to be a valid use of these profits?

Liabilities		**Assets**	
Capital		*Fixed assets*	
Ordinary shares of £1	100 000	Property	80 000
Retained profits	20 000	Plant and equipment	90 000
10% preference shares	40 000		
Current liabilities		*Current assets*	
Trade creditors	20 000	Stock and WIP	10 000
Company tax	10 000	Trade debtors	30 000
Dividends	20 000		
	210 000		210 000

Figure 10.21 Balance sheet for the Skeldale Manufacturing Company.

Sales		800 000
Cost of sales		600 000
Gross profit		200 000
Administration expenses		180 000
Net profit		20 000
Tax		10 000
Profit for distribution		10 000
Dividends:		
Preference	4 000	
Ordinary	20 000	
		24 000
Balance		(14 000)
Profit retained from previous year		34 000
Total profit retained		20 000

Figure 10.22 Profit and loss account for the Skeldale Manufacturing Company.

10.2 Why should assets be depreciated and what is the effect on the balance sheet and profit and loss account? Describe two methods by which assets can be depreciated. Assuming that the plant and equipment for the Skeldale Manufacturing Company have a life of eight years, and a residual value of £1000 at the end of this time, calculate the depreciation per year, using the straight line method, and the revised profit for the year.

10.3 What are accounting ratios? Calculate as many ratios as possible using the balance sheet and profit and loss account of the Skeldale Manufacturing Company and prepare a report to the Board on the profitability of the company.

10.4 What do you understand by the term 'leverage'? Why does leverage affect the stock market price of a company and what other factors also affect this price?

10.5 Find the value per share for the Skeldale Manufacturing Company using three different methods and the following additional information. (a) Earnings have remained substantially constant over the past five years. (b) A return on capital employed of 15 per cent is expected. (c) A price to earnings ratio of 15 is usual for similar companies. (d) The current value of the fixed assets is as shown in the balance sheet.

References

Barfield, R. (1994) Think like an investor, *Professional Manager*, March, pp. 17–19.
Blair, A. (1997) Watching the new metrics, *Management Today*, April, pp. 48–50.

Broadbent, M. (1993) *Managing Financial Resources*, Butterworth-Heinemann, Oxford.

Bruno, A.V. and Tyebjee, T.T. (1985) The entrepreneur's search for capital, *J. Bus. Ventur.*, **1**, 61–74.

Bryan, K. (1996) Challenge of managing devolved budgets, *Professional Manager*, September, pp. 20–1.

Chang, M. (1996) Maximize value in your company before you sell, *Laser Focus World*, February, pp. 95–9.

Costley-White, A. (1991) Equity finance for small businesses, *Engineering Management Journal*, February, pp. 37–41.

Freear, J. *et al.* (1994) The private investor market for venture capital, *IEEE Engineering Management Review*, Fall, pp. 91–7.

Holmes, G. and Sugden, A. (1994) *Interpreting Company Reports and Accounts*, Prentice-Hall, Englewood Cliffs NJ

Holmes, G. and Dunham, R. (1994) *Beyond the Balance Sheet*, Prentice-Hall, Englewood Cliffs NJ.

Johnston, D. (1991) Finance – an introduction, *Engineering Management Journal*, February, pp. 33–6.

Maier, J.B. and Walker, D.A. (1987) The role of venture capital in financing small business, *J. Bus. Ventur.*, **2**, 207–14.

Maynard, R. (1994) Profit goes west, *Professional Manager*, March, pp. 24–7.

Oldcorn, R. (1980) *Understanding Company Accounts*, Heinemann Pan, London.

Outram, R. (1997) For what it is worth, *Management Today*, May, pp. 70–2.

Pocock, M.A. and Taylor A.H. (1981) *Handbook of Financial Planning and Control*, Gower, Aldershot.

Roberts, E.B. (1980) New ventures for corporate growth, *Harvard Business Review*, July–August, pp. 134–42.

Roberts, E.B. (1990) Initial capital for the new technological enterprise, *IEEE Trans. Engineering Management*, May, pp. 81–94.

Secrett, M. (1994) Making your budget a success, *Professional Manager*, March, pp. 22–3.

Sizer, J. (1985) *An insight into management accounting*, Penguin, London.

Thackray, J. (1993) Hatching Uncle Sam's entrepreneurs, *Management Today*, November, pp. 54–6.

Thomson, R. (1995a) The funding feast, *Management Today*, March, pp. 84–91.

Thomson, R. (1995b) Who needs earnings? *Management Today*, June, pp. 56–9.

Timmons, J.A. and Bygrave, W.D. (1986) Venture capital's role in financing innovation for economic growth, *J. Bus. Ventur.*, **1**, 161–76.

Weyer, M.V. (1995) The venture capital vacuum, *Management Today*, July, pp. 60–3.

Wilsher, P. (1994) Capital ways for business, *Management Today*, March, pp. 72–8.

11

Control through costing

Introduction

This chapter introduces the fundamentals of costing used within industry, primarily as a method of control. These include total absorption costing, standard costing and marginal costing, and their applications, such as for the valuation of stock.

11.1 Cost accounting

Costing covers many aspects, such as valuation of materials drawn from stock to determine the overall cost of the product being made; costing of other elements of a product, such as labour and overheads; and product marginal costs to compare the contribution made to cost recovery by different products (Balderston, 1984; Sizer, 1985). Traditional techniques of costing include:

- **Total absorption costing**. In this method all the costs, direct and indirect, associated with the product are absorbed into the units.
- **Marginal costing**. The primary purpose of this method is to help in management decision making. The method considers only those items of costs that are incurred exclusively by the product; overheads are not considered.
- **Standard costing**. In this method targets are first set for future costs and then actual costs are measured, as they occur, using historical costing methods. Standard costing techniques allow managers to plan the future cost burden on the company and also provide a target against which management performance can be judged. It is clearly important to set the standard costs carefully, to ensure that they are realistic and provide a challenge and incentive to management. This technique, however, can only be used where repetitive processes are involved and standardized products are being made.

The purpose of cost accounting is to provide management with information for decision making on items such as pricing, product selection and process monitoring. Costs are, however, only one factor on which management decisions are based, others being items such as market conditions, customer value and competitor activity (Davies, 1996).

For example, if a company is submitting a bid for a contract it is important to know the costs that are likely to be incurred in order to help in pricing the contract and make a profit. Furthermore, once the contract is being worked, costs must be measured to know whether they are in line with the profits predicted. Investment decisions also require costing, for example to determine whether the investment will make sufficient cost savings to justify the initial outlay. There are, however, other factors which often determine the outcome of management decision making, such as gaining market share or getting faster to market with a new product.

Cost is the value of a resource used up when carrying out the task that is being costed. For example, it is the amount of material and labour that goes into the manufacture of an item. It is not always easy to determine these parameters. For example, if 1 kg of material is bought at £100 and its market price rises to £150 by the time the work is finished, the actual cost or 'value' of the material used is open to debate.

Cost is made up of two main ingredients, amount and price. Costs apply to products and services, although it is also possible to speak of implicit costs, for example the cost of lost opportunity incurred by not investing the money in an alternative project. Costs may also be related to different 'cost centres', which can be items of production, people, departments or whole factories. The key questions to be answered when collecting costs are 'why is it required?' and 'to what purpose is it to be put?'

There are several principles associated with costing:

- Costs must be associated as closely as possible with the cause or item. This ensures that the costs can be investigated later, to determine their cause and allow management to take action if cost reduction is required. For example, if a machine on the factory floor is used to make items A, B and C only, then its costs should not be spread out as a general overhead and allocated to all other products made in the factory.
- Costs should only be charged when they are incurred, even if they are expected to occur in the future. Therefore if it is expected that some stock will have to be written off against the product being made, because of future deterioration, this write-off should only occur when the stock is actually scrapped.
- If the event causing the cost has passed and the costs were not recovered at that time, then they should not be moved forward and charged to another item that was not responsible for the cost. To do so may provide a convenient way of balancing the accounts, but it would give a false picture regarding the profitability of the new product being made.
- Costs allocated must be as accurate as possible. It was stated in the last chapter that in the company's financial statement the costs used should be conservative, so as not to overstate the profitability of the company. However, for actual product costing purposes this must not be done, since it would give management a false picture regarding the profitability of a specific product line and might result in the product being unjustifiably discontinued.
- Costs which are atypical should not be included or they will give false management information. For example, if a fire occurs in the factory and damages the machine that is being used to make products A, B and C, then the cost of repairing the machine should not be charged to these products.

Costs are a key management consideration. In times of high inflation or high demand costs may not get management priority, since they can be passed on to the customer in the form of higher prices. However, when demand falls the company that does not have its costs under control will find that it is not competitive.

There are several key cost considerations for any business (Ames and Hlavacek, 1990):

- The company must ensure that costs, adjusted for inflation, are continually decreasing. It must strive to be a low-cost supplier.
- The company must know the true costs and profit associated with each product, cost centre, market or customer, so that it can take effective management decisions.
- The company must manage its costs so as to manage cash flow, which is crucial to any business. High costs can lead to a vicious cycle, where

increasing costs lead to reducing profits, so that less money can be spent on increasing sales (such as by investing in R&D, sales and marketing) and on cost reduction (such as by investing in new plants and machinery). This will mean that profits are reduced further and even less cash is available for investment. This vicious cycle is illustrated in the cost/profit cycle of Figure 11.1.

Products are usually costed on the basis of job costing or process costing. Job costing applies to a product that is built on a batch basis, usually to a specific order, so that each job has a clearly defined cost associated with it. It is important that all costs in this method can be identified with the batch being made, and this includes a proportion of overhead costs.

Process costing applies when products are being built for stock. The manufacturing process is geared to building a large amount of very similar products, which are made on a continuous or mass production basis. In these circumstances it is not possible to identify costs with individual batches. Costs are accumulated against the production line, and at the end of an accounting period they are averaged out over the number of units built to determine the cost of each item.

There are several classifications used to define costs. Generally they can be summarized as direct and indirect costs. Direct costs relate only to those items that are associated closely with the product being made or costed and are not shared with any other product.

Examples of direct costs are materials used to make a product, the labour which goes into the manufacture of the product, and any expense directly associated with the product, such as packing and shipping. These are called the prime costs. Usually the direct costs are variable: the greater the number of products made the higher the direct costs. This variable may be a direct proportionality or may vary in a non-linear fashion.

Indirect costs are those items which are not dependent only on the products

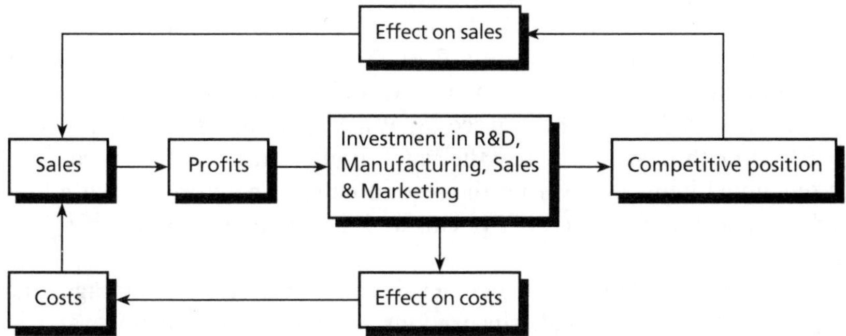

Figure 11.1 The cost/profit cycle: increasing costs lead to a cycle of reducing profits and further cost increases.

being made or costed. Therefore there is some sharing of costs across several products. Usually these costs do not vary in proportion to the amount of product being made; instead they often vary as a function of time. Examples are the rents and rates associated with a factory, which will be determined by the size of the factory and the period for which the factory is occupied, rather than the quantity of goods made in it. These costs are also often referred to as fixed costs or overheads.

Fixed costs are not strictly fixed, and will also vary with the quantity of products being made, but not in a proportionate manner. For example, a machine may have the capacity to make 10 000 items per hour, and its cost, whether it is making 1 item or 10 000, is fixed. However, if the production capacity is to be increased to 10 001 items an hour then another machine will be needed, and there will be a step increase in costs.

11.2 Valuation of stock

With increasing automation the labour costs per item manufactured are continually falling, so that material costs make up an increasing amount of the overall cost of the product. This clearly calls for effective procurement policies, but it also requires a reliable method of valuing material used on products.

During production stock is withdrawn from stores. Generally this stock is built up from purchases made over a period of time and at different prices, so it is not always easy to determine the actual cost of the material being used. Many techniques have been developed for determining this cost, a few of these being discussed here.

11.2.1 FIFO

In the FIFO (first in, first out) method of stock valuation the earliest purchase price is used to determine the value of the material being withdrawn from stores, until all the stock at this level is used up, and then the next purchase price is used. This is illustrated in Table 11.1. At time T_1 quantity A of material is bought at a price of £x. At time T_2 quantity P of this is withdrawn, and using the FIFO method of valuation it is costed as £x. At another time T_3 a further quantity B of the same material is bought at price £y. When quantity Q is withdrawn at time T_4 the FIFO method requires that the batch bought at £x is first used up and then the remaining items taken from the second batch. Therefore this withdrawal from stock is valued as £x for $(A - P)$ items and £y for the remainder of $(Q - (A - P))$ items.

The FIFO method of stock evaluation has the advantage that it is straightforward and logical, since goods are valued at the price at which they were bought. However, when prices are rising rapidly the method can give a false picture of profitability by generating 'paper profits' due to the increased selling

Table 11.1 Methods of stock evaluation

Time	Items into stores: Number	Items into stores: Unit price (£)	Items issued from stores	FIFO unit price (£)	LIFO unit price (£)	Average: Store content	Average: Store value (£)	Average: Unit price (£)
T_1	A	x				A	Ax	x
T_2			P	x	x	$(A-P)$	$(A-P)x$	x
T_3	B	y				$(A+B-P)$	$(A-P)x+By=v_1$	$\frac{(A-P)x+By}{A+B-P}=u_1$
T_4			Q	$(A-P)@x$ $(Q-(A-P))@y$	$B@y$ $(Q-B)@x$	$(A+B-P-Q)$	$v_1-Qu_1=v_2$	$\frac{v_1-Qu_1}{A+B-P-Q}=u_2$

price resulting from the materials bought going up in price compared with the original costs. For example, if material is bought at £10 and its price rises by 20 per cent due to inflation to £12, then if it is sold a profit of £2 will be realized even though no value has been added. However, this is not a true profit, since the higher price of £12 will now need to be paid to buy replacement material.

11.2.2 LIFO

The LIFO (last in, first out) method of stock valuation uses the latest purchase price paid for the stock until that is used up, and then the earlier price is used.

For the example of Table 11.1, at time T_2 the withdrawal from stores is priced by the LIFO method as £x, since this is the latest price paid for the materials in stock. At time T_4 the latest purchase price is £y, so the initial withdrawal is made at this price (B at £y), and when this has been used up the remainder ($Q - B$) is then made at the older purchase price of £x.

The advantage of the LIFO method of stock valuation is that the cost allocated to goods withdrawn from stores is kept close to their market price, so that paper profits are avoided. However, this method is not logical since it puts costs on items that are the reverse of those actually incurred when buying them for stores.

Using a LIFO method of stock valuation does not mean that goods have actually to be withdrawn from stores in that order. In the case of perishable goods it is desirable to adopt a FIFO method of stores issue, irrespective of the system used to value such issues.

11.2.3 Weighted average

In the weighted average pricing system of store valuation, the average price of the store contents at the time of issue is used to cost the job. For example in Table 11.1 at time T_1 the value of the store content is Ax, and since there are A items in stock the unit price is £x. Therefore when P units are withdrawn at time T_2 the issue price is £x. This leaves a store value of $(A - P)x$, and the addition of B items at time T_3 at a price of £y per item results in a store value given by Equation (11.1) and a unit cost given by Equation (11.2).

$$v_1 = (A - P) \times x + B \times y \tag{11.1}$$

$$u_1 = \frac{(A - P) \times x + B \times y}{A + B - P} \tag{11.2}$$

When Q items are now withdrawn from stores in time T_4 they are priced at u_1. This leaves the stores with the store value given by Equation (11.3) and a unit price as in Equation (11.4), which is the price for any subsequent withdrawals, unless further purchases are made before then.

$$v_2 = v_1 - Q \times u_1 \tag{11.3}$$

$$u_2 = \frac{v_1 - Q \times u_1}{A + B - P - Q} \tag{11.4}$$

The advantage of the weighted average method is that it is logical, since the same price is applied to all identical items in store at any time. It also smoothes out purchase price fluctuations and does not result in a paper profit or loss situation. Its disadvantage is that the price of stock issued is not related to the current market value, although it may be close to it. In fact, the issue price does not relate to the actual purchase price at any time except for the very first purchase.

The three methods of stock valuation give different material values, as illustrated in the case study 'Pricing the stock'. The most appropriate method for use will depend on the overall aims of management.

11.2.4 Other methods

Two other methods are used to value stock: the replacement price method and the standard price method.

In the replacement price method the withdrawals from stores are valued at the actual market price that exists at that time. Clearly this is the best method to use at times of violently fluctuating prices, and it is also simple to calculate. The problem arises in implementing the system, since the market price for each commodity in stores must be continually obtained, probably on a daily basis, and used to cost withdrawals. Also, this method will result in paper profit or loss depending on whether the market price of the materials in store is rising or falling.

The standard price method uses the initial standard price calculated for the material as the price for all subsequent withdrawals. The standard price may be periodically reviewed and changed, but this would be done relatively infrequently. This technique is easy to apply and has the benefit of not changing excessively over time. It also provides a check on the efficiency of the buying function and provides an incentive to management to buy more effectively. Considerable effort, however, must go into setting an accurate standard price, and since this price is unlikely to match the actual value at any time, a paper profit or loss situation will arise. The standard price method has the further disadvantage of having no connection with market price trends.

Case study

Pricing the stock

In order to compare the effect of the different methods of stock issue pricing on the overall material costs of products being made, Ted Court, Stock Control Supervisor at the FloRoll Manufacturing Company, decided to issue goods out of stores using the current method of costing, which was weighted average, but at the same time to keep a record of what these items would have cost if FIFO or LIFO had been implemented.

Table 11.2 shows these records for six periods of time. The calculations can be followed with reference to Table 11.1. It is seen from this that at every issue FIFO gives the lowest cost and LIFO the highest. Weighted average costs are in-between these two. The total issue costs for the 800 items varies by 7 per cent between the LIFO and FIFO methods and by between 3 per cent and 4 per cent between these and weighted average. Clearly the difference will increase as more time periods are considered.

Table 11.2 Stock valuation for the case study 'Pricing the stock'

Time period	Items into store: Unit (1)	Items into store: Unit price (£) (2)	Total number issued (3)	FIFO unit price (4)	LIFO unit price (5)	Weighted average: Store content (6)	Weighted average: Store value (£) (7)	Weighted average: Unit price (£) (8)
1	500	1.00				500	500	1.00
2			300	£1.00	£1.00	200	200	1.00
3	400	1.10				600	640	1.07
4			300	200 @ £1.00 100 @ £1.10	£1.10	300	320	1.07
5	100	1.50				400	470	1.18
6			200	£1.10	100 @ £1.50 100 @ £1.10	200	235	1.18
Total issue			800	£830	£890		857	

Ted Court decided to run the trial over a longer period and then to recommend that the company review the method it uses for valuing stock, to ensure that it is the most appropriate for its needs.

11.3 Allocation of overheads

Overheads are generally taken to include all costs that cannot be identified exclusively with a specific product or service, and therefore cannot be easily allocated to individual units. Overheads are 'recovered' by charging out a proportion of the overhead to the various items that shared in causing it to occur.

In a factory this is usually done on the basis of the amount of production resources used by each item. It is important that overheads are only allocated to items of production when they occur. For example, selling and distribution overheads cannot be added to items that are still being made, since clearly they could not as yet have contributed to these overheads.

Overheads tend to be an emotive cost, since many managers associate them with inefficiencies within their organizations. The first reaction of these managers to rising overheads is to cut costs. Unfortunately, this usually results in loss of capacity and loss of market share. The right way to tackle rising overheads is to improve the processes used in the organization (Blaxill and Hout, 1991) since this would eventually lead to a reduction in costs and greater competitiveness.

Allocation of overheads poses several problems. For example, should all the overheads associated with building the first batch of a new product be charged to that batch? Some of the time may have been used in training operators, and also time is likely to have been wasted if design faults were found and needed correction.

Overheads can be apportioned in several different ways, depending on the item being considered. For example: overheads associated with the rent and rates of a factory can be apportioned according to the area occupied by different product lines; overheads covering the personnel department and the canteen costs can be apportioned by the number of employees in departments or used to make different products; and the overheads associated with materials procurement and storage can be apportioned according to the materials usage on products.

Generally overheads are allocated to items of production according to six methods: by the number of units made; by the cost of the material used; by the labour costs associated with making the units; by the labour hours used in production; by the machine hours; and by the prime cost of the units. The calculations associated with these methods are illustrated by reference to Table 11.3.

In the units method of overhead allocation, if O is the overall overhead and the number of units of two product lines A and B are U_a and U_b then the overhead allocation per unit for products A and B are given by the last two columns of Table 11.3. This method of overhead allocation is very good but it can only be applied if the units being made are identical and therefore result in identical overhead costs.

The materials method of overhead apportionment, illustrated in Table 11.3, is based on the cost of the materials M_a and M_b used by the two products A and B. The allocation of overhead on this basis is shown in the last two columns. This method is generally unsuitable, since it has very little connection with factory overheads. For example, one product may be using expensive materials but it may be on the factory floor for a very short time. However, because of its high material costs, it would be allocated a large proportion of the overall overheads.

Table 11.3 Methods for overhead allocation

Allocation method	Total overhead	Item	Total per product		Total for factory	Overhead allocation per product		Overhead allocation per unit	
			A	**B**		**A**	**B**	**A**	**B**
1	0	Units	U_a	U_b	$U_a + U_b$	$\frac{O\,U_a}{U_a + U_b}$	$\frac{O\,U_b}{U_a + U_b}$	$\frac{O}{U_a + U_b}$	$\frac{O}{U_a + U_b}$
2	0	Material	M_a	M_b	$M_a + M_b$	$\frac{O\,M_a}{M_a + M_b}$	$\frac{O\,M_b}{M_a + M_b}$	$\frac{O\,M_a}{(M_a + M_b)\,U_a}$	$\frac{O\,M_b}{(M_a + M_b)\,U_b}$
3	0	Labour cost	L_a	L_b	$L_a + L_b$	$\frac{O\,L_a}{L_a + L_b}$	$\frac{O\,L_b}{L_a + L_b}$	$\frac{O\,L_a}{(L_a + L_b)\,U_a}$	$\frac{O\,L_b}{(L_a + L_b)\,U_b}$
4	0	Labour hours	H_a	H_b	$H_a + H_b$	$\frac{O\,H_a}{H_a + H_b}$	$\frac{O\,H_b}{H_a + H_b}$	$\frac{O\,H_a}{(H_a + H_b)\,U_a}$	$\frac{O\,H_b}{(H_a + H_b)\,U_b}$
5	0	Machine hours	P_a	P_b	$P_a + P_b$	$\frac{O\,P_a}{P_a + P_b}$	$\frac{O\,P_b}{P_a + P_b}$	$\frac{O\,P_a}{(P_a + P_b)\,U_a}$	$\frac{O\,P_b}{(P_a + P_b)\,U_b}$
6	0	Prime cost	$L_a + M_a$	$L_b + M_b$	$L_a + L_b + M_a + M_b$	$\frac{O\,(L_a + M_a)}{L_a + L_b + M_a + M_b}$	$\frac{O\,(L_b + M_b)}{L_a + L_b + M_a + M_b}$	$\frac{O\,(L_a + M_a)}{(L_a + L_b + M_a + M_b)U_a}$	$\frac{O\,(L_b + M_b)}{(L_a + L_b + M_a + M_b)U_b}$

Alternatively a second product, which used inexpensive but bulky material, and which was on the factory floor for a very long time, would be allocated only a small proportion of overheads.

The third method of overhead allocation shown in Table 11.3 relates to labour costs, L_a and L_b. This method of overhead allocation is not as good as labour hours, since one product may only take a short time to make, but may require skilled staff with high labour rates, while another product may require a long time to manufacture but only use low-paid labour. Overheads are associated with the number of people employed and not their absolute costs.

Labour hours is usually the best all-round method for overhead allocation, since it relates directly to the number of people used in the manufacture of products and it is these people who are the prime cause of overheads. If, however, the factory is totally automated then overhead allocation based on machine hours may be a better method. However, if there are any products which are made manually then, in this method, they will not carry their share of the overheads.

The final method of overhead allocation shown in Table 11.3 is that based on prime costs of the products. Because prime costs are made up of the material and labour costs, this is not a representative method of overhead allocation.

The different methods of overhead allocation are illustrated by the case study 'Sharing the overheads'.

Overheads are usually related to time, such as the rent and rates associated with a factory. When allocating overheads, the actual overheads incurred during a period can be used. However, this means that it is not easy to cost a job, as it is being done, since it is necessary to wait until the end of the period before the overhead costs can be collected and apportioned. Another disadvantage of this method is that if there are wide fluctuations in production, such as to meet seasonal demand, then when very few items are being made, the overhead costs allocated to these items will be disproportionately high, since many overheads, such as rents and rates, are likely to be independent of the volume of products.

Overheads are therefore usually allocated on the basis of an estimate of future costs and future production volumes over a relatively long period. This is known as the predetermined overhead rate (P_r) and is given by Equation (11.5).

$$P_r = \frac{\text{Budgeted overhead for next period}}{\text{Budgeted units for next period}} \tag{11.5}$$

When the actual overheads and units manufactured in the period have been determined, it is likely that they will be different from the budgeted values. This will result in an over-absorption or under-absorption of overheads, which can be covered by an adjustment in the profit and loss account for the period. (See the case study 'Over the top'.)

Case study

Sharing the overheads

Joe Plant, Manager of the Assembly Shop in the FloRoll Manufacturing Company, wanted to know how overhead allocation would affect the cost of two new products which were to be built there, Floroll A and Floroll B. Joe decided to allocate the expected factory overheads of £10 000 for the next period on six different bases, as in Table 11.3.

Joe Plant obtained the various parameters associated with the two new products, as in Table 11.4.

Table 11.4 Information on Floroll A and Floroll B for the case study 'Sharing the overheads'

Item	A	B
Units to be made in the year	10 000	15 000
Materials cost per unit	£0.20	£0.40
Labour hour per unit	3.5 hours	1.0 hour
Labour cost per unit	£2.50	£1.00
Machine hours per unit	1.0 hours	2.0 hours

Based on these parameters Joe calculated the overheads to be allocated per unit of the two new products as in the last two columns of Table 11.5. From this it is seen that there is a wide spread in overheads, depending on the method chosen. Joe dismissed the method based on number of units, since the two florolls were very different in structure. The material method was also not selected because the material for Floroll B was much more expensive than for Floroll A. Machine hours were not considered to be reliable, since both products had a mix of manual and automated processes.

Table 11.5 Method of overhead allocation for Floroll A and Floroll B from Table 11.4

Method	A	B	Total	Over-head	Total allocation		Per unit allocation	
					A	B	A	B
Units	10 000	15 000	25 000	10 000	4000	6000	£0.40	£0.40
Material	2 000	6 000	8 000	10 000	2500	7500	£0.25	£0.50
Labour cost	25 000	15 000	40 000	10 000	6250	3750	£0.63	£0.25
Labour hours	35 000	15 000	50 000	10 000	7000	3000	£0.70	£0.20
Machine hours	10 000	30 000	40 000	10 000	2500	7500	£0.25	£0.50
Prime costs	27 000	21 000	48 000	10 000	5625	4375	£0.56	£0.29

The labour cost method was considered to be fairly reliable, since labour rates on both products were similar. The best method was the labour hours method and, because of similar labour rates, this gave overhead allocations which were very close to labour costs. This was the method Joe Plant decided to select.

Case study

Over the top

John McCaully, Manager of the Dinsher Shop in the FloRoll Manufacturing Company, was working out the overheads for his Shop over the next period. At the start of the period the labour hours spent on products, and the overheads in the Dinsher Shop, had been budgeted as in Table 11.6. This had given a predetermined overhead rate of P_r, calculated using Equation (11.5), as in Equation (11.6).

$$P_r = \frac{40\,000}{20\,000} = £2.00 \text{ per hour} \tag{11.6}$$

At the end of the period the actual labour hours and overheads were found to be as in the last column of Table 11.6 and John McCaully decided to calculate the absorption variation, as requested by Johnny Bacon, the Financial Director.

Table 11.6 Budgeted and actual overheads for the case study 'Over the top'

Item	Budgeted	Actual
Labour hours	20 000	19 000
Overhead (£)	40 000	41 000

The overheads actually charged to the Dinsher Shop during the period, based on the actual labour hours of 19 000 and a predetermined labour rate of £2.00 per hour is equal to 19 000 × 2 = £38 000. Since the actual overheads were found to be £41 000, John calculated the under-absorption of overheads to be equal to £41 000 − £38 000 = £3000.

11.4 Standard costing

There are two basic methods for costing a product or service:

1. Determining the cost after the activity has been completed. This involves collecting costs from the various tasks undertaken and either allocating them totally to the product or service or allocating a proportion of the costs. This method of costing is known as historical costing, since it determines costs after the event.
2. Estimating the costs that are likely to occur before the activity is started. Usually these costs are projected for a period into the future. This method of costing is known as standard costing, since it sets 'standards' for the costs involved. Historical costing is then used, after the activity has been completed, to determine the actual costs and to compare these with the standard costs.

Since standard costing compares predicted with actual costs it is a cost plan, similar to budgeting, as described in the last chapter. It differs from budgeting in that it is primarily concerned with costs applied to products and services, whereas budgets are normally associated with departments, companies and even countries.

Setting up a standard costing system involves the following:

1. Establishing cost centres. These may be products, persons, locations and so on, for which costs can be determined and controlled.
2. Clearly establishing the responsible person for each standard cost.
3. Preparing a classification system for cost collection. Usually this will be on the basis of direct materials, direct labour and so on used on the product or service.
4. Determining the type of standard to be set up. There are three types: ideal, expected and basic. Ideal standards are those which assume a perfect organization, without inefficiencies, wastage of materials or time and so on. It gives a measure of what to expect if everything went perfectly. Expected standards set results which would be expected from a highly efficient organization, but allowances are made for unavoidable efficiencies and process defects. It sets high standards, but ensures that targets can be attained. Basic standards are those which were set a relatively long time ago. They rarely change and provide a measure of past history.
5. Determining standard costs for each activity. This involves a considerable amount of work, and key personnel must be involved in this, especially if they are to be measured against these standards.

Standard costs are usually entered on standard cost cards, per product or service, which also contains the historical cost data, as it is collected, for comparison.

Standard costing has several advantages over historical costing:

- It provides a yardstick or standard against which cost performance can be measured.
- It clearly defines responsibility for each standard and provides a target, which can act as an incentive, for improvement.
- The activity involved in setting standards requires a detailed analysis of the processes being used within the organization. This usually identifies areas for improvement and gives management a better understanding of the operation of their own company. However, setting standards, especially if done properly, can take up a considerable amount of management time.
- Standard costs help to identify problem areas, which can then be investigated using variance analysis, as described in the next section, to determine their cause.

Usually standard costs, once set, should not be changed over the period covered, normally one year. This is to ensure stability in the targets aimed for and to reduce the workload, since it takes a considerable amount of time to determine a standard.

If a key item changes permanently then it is permissible to revise the standard. For example, if a product is being bought competitively from four suppliers and three of them go out of business, then the remaining supplier may substantially increase the material price, requiring a change in the material standard cost being used. Even in these circumstances it can be argued that standard cost should not change since it provides a measure of absolute efficiency.

The standard costs that are usually set are for direct materials, direct labour, variable and fixed overheads, selling and distribution costs, and sales. These are then subdivided into other costs. The standard material cost is the material used during the period to make the number of items specified. This can be further subdivided into quantity costs, which gives the standard quantity of each type of material needed for the product after taking into account any process losses, and price cost, which is the standard price for each material.

To set the standard material costs requires a knowledge of several items such as:

- The standard material specification, such as the mix of materials used on the product and their quality.
- The standard material price. The actual price which is paid is likely to fluctuate over the period being considered, so the standard price is the forecast for the average price to be paid over this period. It should take into account all price-related matters, such as quantity discounts expected.
- The standard material usage on the product or service being considered. This should include expected wastage in material handling and account for any other losses such as breakages or loss in storage.

The standard labour costs are those which are used in the manufacture of the product. These are subdivided into rate cost, which is the standard rate of pay for each class of labour employed, and time cost, which is the standard time for each grade of labour needed to make the product when the job is carried out in the most efficient way. Some of the information required to set the standard labour costs is:

- The standard skills needed to make the product or provide the service. The availability of the right skills and whether training can be provided to reach the level required are some of the prime considerations.
- The standard labour rates. The rates per skill required over the period need to be determined. The impact of skill shortages on pushing up rates and on overtime must be considered.
- The standard labour hours needed to produce the product or service must be forecast. Time lost due to uncontrollable events, such as sickness and strikes, are factors to be considered.

Overheads are normally divided into variable and fixed. For both these the overhead absorption rate needs to be determined. Since labour hours are normally taken as the basis for allocation of overheads, this is usually calculated as the ratio of the budgeted overhead (variable or fixed) to the budgeted direct labour hours. In addition, the standard overhead usage is to be established, which is usually on the basis of standard labour hours, established earlier.

In setting standards for selling and distribution costs several factors need to be taken into account, such as:

- The cost of material used for packing and transport of the products.
- The cost of transport, such as hire of trucks and freight.
- The selling costs, including a share of any common costs, such as administration and advertising.
- All labour costs in distribution and selling. Sales commissions, overtime and so on should be included.

The standard selling price is normally the budgeted selling price. It is again an average of that expected over the period being considered. It should include any free promotional material or discounts, and also take into account price erosion caused by competitor activity.

11.5 Variances

Having established standard costs, as described in the last section, the differences between these and the actual costs are computed as the job progresses. These differences are know as variances and they are used to determine the areas which have altered since the cost plan was set and to identify the person responsible, from whom an explanation can be obtained.

There are two basic types of variance: price and volume. A price variance relates to such items as the price of materials, labour rates, overhead expenditure and selling price. Volume variances relate to the amount of raw material used, hours worked, plant utilization and quantity sold.

Figure 11.2 illustrates some of the most commonly used variances, which are discussed further in this section. Further subdivisions are possible to investigate problems. For example, the usage variance under direct materials can be divided into mix variance, which indicates how the different materials have been intermixed in making the product, and the yield variance. At the head of the pyramid is the profit variance, since this is the item which usually most concerns managers, and this can then be subdivided to find the cause of an adverse or favourable profit variance.

It is not always easy to make comparisons between the output of different products, especially if they are very different. It is therefore usual to convert all the output to that of output in standard hours (OSH) as illustrated in Table 11.7. This is obtained by multiplying the number of units produced for each product and the standard hours per unit.

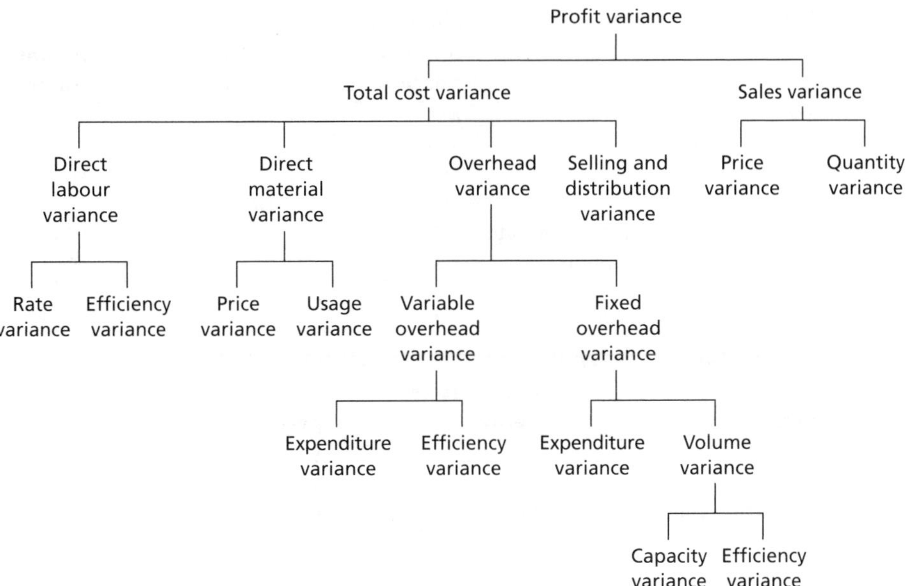

Figure 11.2 The variance pyramid.

Table 11.7 Output in standard hours (OSH)

Product	Units produced	Standard hours per unit	Output in standard hours (OSH)
A	100	3	300
B	200	4	800
C	300	5	1500
Total			2600

11.5.1 Direct labour variance

The direct labour variance is caused by the actual labour costs, calculated once the activity has been completed, differing from the standard or planned labour cost, set at the start of the activity. This is given mathematically by Equation (11.7).

$$\text{Labour variance} = \text{Actual labour cost} - \text{Standard labour cost} \tag{11.7}$$

The mathematical symbols used in this section are given in Tables 11.8 and 11.9 and using these the direct labour variance can be restated, as in Equation (11.8).

$$\text{Labour variance} = LH_{\text{ta}} \times LR_{\text{a}} - PV_{\text{a}} \times LR_{\text{s}} \times LH_{\text{s}} \tag{11.8}$$

Table 11.8 Symbols used in calculating variances

Item	Standard cost per unit	Actual cost per unit	Actual total cost for period
Material price	MP_s	MP_a	
Material usage	MU_s	MU_a	
Labour hours	LH_s	LH_a	LH_{ta}
Labour rate	LR_s	LR_a	
Variable overhead absorption rate	VO_s		VO_{ta}
Fixed overhead absorption rate	FO_s		FO_{ta}

Table 11.9 Further symbols used in calculating variances

Item	Budgeted for period	Actual for period
Sales price	SP_s	SP_a
Sales volume	SV_s	SV_a
Production volume	PV_s	PV_a

For example, suppose that to produce one unit of product a standard of 20 hours at a labour rate of £10 per hour has been defined. When the work is actually done it is found that it takes 22 hours to make one unit and the labour cost is £9 per hour. Then, in Equation (11.8), $LH_{ta} = 22$; $LR_a = 9$; $PV_a = 1$; $LR_s = 10$; and $LH_s = 22$. The direct labour variance is given as in Equation (11.9).

$$\text{Labour variance} = 22 \times 9 - 20 \times 10 = 2\,(F) \tag{11.9}$$

The (F) means that the variance is favourable: below the standard value. When it is above the standard value it is known as an adverse variance and is indicated by (A).

As shown in Figure 11.2, the direct labour variance can be broken down further into the direct labour rate variance and the direct labour efficiency variance, in order to determine its cause. The direct labour rate variance is due to the actual labour rate being different from the standard labour rate. It can be calculated from Equation (11.10).

$$\begin{aligned}\text{Rate variance} &= (\text{Actual rate} - \text{Standard rate}) \times \text{Actual hours} \\ &= (LR_a - LR_s)\,LH_{ta}\end{aligned} \tag{11.10}$$

Therefore in the example considered earlier the direct labour rate variance is given as in Equation (11.11).

$$\text{Rate variance} = (9 - 10) \times 22 = 22\,(F) \tag{11.11}$$

The labour efficiency variance is caused by the actual labour efficiency being different from the planned efficiency. It is a usage variance, and is given by Equation (11.12), where OSH is the output in standard hours as defined earlier.

$$\text{Efficiency variance} = (\text{Actual hours} - \text{OSH}) \times \text{Standard rate}$$
$$= (LH_{ta} - PV_a LH_s)\, LR_s \qquad (11.12)$$

Therefore the direct labour efficiency variance for the example given earlier is as in Equation (11.13).

$$\text{Labour efficiency} = (22 - 20) \times 10 = 22\,(A) \qquad (11.13)$$

The sum of the direct labour rate variance and the direct labour efficiency variance should equal the direct labour variance. This is the case, as can be confirmed by adding Equations (11.10) and (11.12), which gives Equation (11.8); or by the example where the direct labour variance of 2(F) is obtained from the direct labour rate variance of 22(F) and the direct labour efficiency variance of 20(A). In this example the favourable direct labour variance is explained by the fact that labour of a lower skill than originally planned for was used, so that although it took longer the wages paid to the operators were lower, leading to an overall favourable result. (See also the case study 'Making it faster'.)

Case study

Making it faster

The Coolit Distribution Company, who used to be a supplier to the FloRoll Manufacturing Company when it first started up, employed 150 people in its factory making coolits. They worked a standard five-day week with eight hours per day. The standard wage rate was £2 per hour, and the standard output was 20 coolits per day. During week 26 the factory produced 115 coolits and the actual labour rate was £2.25 per hour. Based on these figures the factory manager, Tom Wise, wished to calculate the direct labour variance.

Table 11.10 shows the basic parameters calculated from the information given above. From this the direct labour variance can be found from Equation (11.8) as in Equation (11.14).

Table 11.10 Parameters for the case study 'Making it faster'

Description	Symbol	Value
Standard rate	LR_s	£2 per hour
Standard output per week	PV_s	$20 \times 5 = 100$ coolits
Standard hours	LH_s	$(150 \times 8)/20 = 60$ hours
Actual output per week	PV_a	115 coolits
Actual rate	LR_a	£2.25 per hour
Actual total labour hours	LH_{ta}	$150 \times 8 \times 5 = 6000$ hours
Output in standard hours	$OSH = PV_a LH_s$	$115 \times 60 = 6900$ hours

$$\textbf{Labour variance} = 6000 \times 2.25 - 115 \times 2 \times 60 = 300(F) \qquad \textbf{(11.14)}$$

In order to determine the cause of this favourable variance Tom Wise decided to explore further and calculate the direct labour rate variance and the direct labour efficiency variance. The direct labour rate variance is given by Equation (11.10), as in Equation (11.15), and the direct labour efficiency variance by Equation (11.12), as in Equation (11.16).

$$\textbf{Rate variance} = (2.25 - 2.00) \times 6000 = 1500\,(\text{A}) \qquad \textbf{(11.15)}$$

$$\textbf{Efficiency variance} = (6000 - 115 \times 60) \times 2 = 1800\,(\text{F}) \qquad \textbf{(11.16)}$$

Tom Wise was very pleased with these results. After a few weeks of production he had gambled on the fact that if he used higher skilled operators it would put up the labour rate, but he could then introduce more advanced production techniques which would increase production efficiency and give an overall favourable labour cost variance.

11.5.2 Direct material variance

The direct material variance is due to the actual cost of materials used in production differing from those predicted in the standard costs. It is given by Equation (11.17).

$$\begin{aligned}\text{Material cost variance} &= \text{Actual material cost} - \text{standard material cost}\\ &= (MP_a MU_a - MP_s MU_s)PV_a \qquad (11.17)\end{aligned}$$

Therefore if, for example, a factory predicted that it would use 10 kg of materials at £20 per kg to make one unit of product (standard material costs) and it actually used 12 kg of material at a cost of £21 per kg (actual material costs) then the direct material cost variance is given by Equation (11.18).

$$\text{Cost variance} = 12 \times 21 - 10 \times 20 = 52(\text{A}) \qquad (11.18)$$

The direct material cost variance can be broken down further, as in Figure 11.2, into the direct material price variance and the direct material usage variance. The direct material price variance determines how the price paid for the materials differ from the standard price and is given by Equation (11.19).

$$\begin{aligned}\text{Price variance} &= (\text{Actual price} - \text{Standard price}) \times \text{Actual usage}\\ &= (MP_a - MP_s)MU_a PV_a \qquad (11.19)\end{aligned}$$

Therefore in the earlier example the material price variance will be given as in Equation (11.20).

$$\text{Price variance} = (21 - 20) \times 12 = 12(\text{A}) \qquad (11.20)$$

The direct material usage variance measures the actual quantity of materials used to the standard value, both items being converted into monetary terms by multiplying by the standard price, as in Equation (11.21).

$$\begin{aligned}\text{Usage variance} &= (\text{Actual usage} - \text{Standard usage}) \times \text{Standard price} \\ &= (MU_a PV_a - MU_s PV_a)MP_s \qquad (11.21)\end{aligned}$$

Therefore in the example used earlier the direct material usage variance can be calculated as in Equation (11.22).

$$\text{Usage variance} = (12 - 10) \times 20 = 40(\text{A}) \qquad (11.22)$$

The direct material usage variance and the direct material price variance sum to the total direct material variance, as expected. The adverse material variance, in the example, can be explained by the fact that not only did the material cost more than planned, but a greater amount of material was also used, probably due to excessive wastage. The actual cause for both can now be investigated and corrected. (See also the case study 'Paying more for less'.)

Case study

Paying more for less

In the Coolit Distribution Company it has been estimated that the factory can make 100 coolits per week and that each coolit will use 2 kg of material at a cost of £10 per kg. During week 26 it is calculated that the factory makes 115 coolits and uses 207 kg of material at £11 per kg. Tom Wise, the factory manager, decides to calculate the direct material variances and summarizes the parameters, as in Table 11.11. From this he calculates the direct material cost variance from Equation (11.17), as in Equation (11.23).

Table 11.11 Parameters for the case study 'Paying more for less'

Description	**Symbol**	**Value**
Standard material used	MU_s	2 kg per coolit
Standard material price	MP_s	£10 per kg
Actual material used	MU_a	207/115 = 1.8 kg per coolit
Actual material price	MP_a	£11 per kg
Standard output	PV_s	100 coolits
Actual output	PV_a	115 coolits

$$\textbf{Cost variance} = (11 \times 1.8 - 10 \times 2) \times 115 = 23(\text{F}) \qquad \textbf{(11.23)}$$

Tom Wise then explores these variances further by calculating the direct material price variance and the direct material usage variance, from Equations (11.19) and (11.21) respectively, as in Equations (11.24) and (11.25).

Price variance = (11 − 10) × 115 × 1.8 = 207(A) **(11.24)**

Usage variance = (1.8 × 115 − 2 × 115) × 10 = 230(F) **(11.25)**

Tom is pleased with the overall favourable direct material cost variance. He is also pleased that his drive to reduce wastage in the factory, and to reduce rejects during inspection, is paying off, as evidenced by a very favourable direct material usage variance. However, the price paid for the materials is much higher than originally planned. Tom tackles the procurement manager on this and is told that the cause is a contract buyer (who had since left) who was employed to cover for staff sickness in the department. Corrective action is now in place to reduce the price and to ensure that it is tightly controlled in future.

11.5.3 Overhead variance

Overhead variance occurs when the overhead absorbed is less than that originally planned. Overhead absorption is usually determined by using the output in standard hours, as in Equation (11.26). The overhead variance is then given by Equation (11.27).

$$\text{Overhead absorbed} = OSH \times \text{Overhead absorption rate} \quad (11.26)$$

$$\begin{aligned}\text{Overhead variance} &= \text{Actual overheads} - OSH\,(\text{Fixed overhead absorption rate} \\ &\quad + \text{Variable overhead absorption rate}) \\ &= (VO_{ta} + FO_{ta}) - OSH(VO_s + FO_s) \quad (11.27)\end{aligned}$$

For example, suppose that the fixed and variable overheads in a factory are as in Table 11.12, which also gives the labour hours for the product made in the factory and the output during the period in standard hours.

The fixed overhead absorption rate (FOAR) and the variable overhead absorption rate (VOAR) are calculated in this table. The value of the overhead variance is then given by Equation (11.27) as in Equation (11.28).

$$\text{Overhead variance} = (2500 + 1200) - 500 \times (2 + 4) = 700(\text{A}) \quad (11.28)$$

Table 11.12 Information used to illustrate overhead variances

Parameter	**Standard for period**	**Actuals for period**
Fixed overhead	£2000	£2500
Variable overhead	£1000	£1200
Labour hours	500 hours	450 hours
Output in standard hours (OSH)	–	500
Fixed overhead absorption rate (FOAR)	2000/500 = £4 per hour	
Variable overhead absorption rate (VOAR)	1000/500 = £2 per hour	

The overhead variance can be divided further, as in Figure 11.2, into the variable and fixed overhead variances. The variable overhead variance is the difference between the actual and standard variable overheads and is given by Equation (11.29).

$$\text{Variable overhead variance} = \text{Actual variable overhead} - OSH \times VOAR$$
$$= VO_{\text{ta}} - OSH\ VO_{\text{s}} \quad (11.29)$$

Therefore in the example of Table 11.12 the variable overhead variance is given by Equation (11.30).

$$\text{Variable overhead variance} = 12\,000 - 500 \times 2 = 200(\text{A}) \quad (11.30)$$

The fixed overhead variance is given by a similar expression to Equation (11.29), except that the variable overhead absorption rate is replaced by the fixed overhead absorption rate. This is shown in Equation (11.31).

$$\text{Fixed overhead variance} = \text{Actual fixed overhead} - OSH \times FOAR$$
$$= FO_{\text{ta}} - OSH\ FO_{\text{s}} \quad (11.31)$$

Therefore in the example of Table 11.12 the fixed overhead variance is given by Equation (11.32).

$$\text{Fixed overhead variance} = 2500 - 500 \times 4 = 500(\text{A}) \quad (11.32)$$

As expected the fixed and variable overhead variances sum to the total overhead variance.

The variable overhead can be further divided into the expenditure variable overhead variance and the efficiency variable overhead variance. The expenditure variance arises due to the actual variable overhead differing from the planned variable overhead. Since overhead is usually on the basis of hours worked, the expenditure variable overhead variance is given by Equation (11.33).

$$\text{Expenditure variance} = \text{Actual variable overhead} - (\text{Hours worked} \times VOAR)$$
$$= VO_{\text{ta}} - (LH_{\text{ta}} VO_{\text{s}}) \quad (11.33)$$

Therefore for the example of Table 11.12 the expenditure variable overhead variance is given by Equation (11.34).

$$\text{Expenditure variance} = 12\,000 - (450 \times 2) = £300(\text{A}) \quad (11.34)$$

The variable overhead efficiency variance is due to the difference between standard hours and actual hours worked. It is given by Equation (11.35).

$$\text{Efficiency variance} = (\text{Actual hours worked} - OSH) \times VOAR$$
$$= (LH_{\text{ta}} - OSH) VO_{\text{s}} \quad (11.35)$$

Therefore for the example of Table 11.12 the variable overhead efficiency variance is given by Equation (11.36).

$$\text{Efficiency variance} = (450 - 500) \times 2 = 100(\text{F}) \quad (11.36)$$

As expected, the expenditure and efficiency variances sum to the overall variable overhead variance.

Figure 11.2 shows the fixed overhead variance broken down into two further levels. The fixed overhead expenditure variance is due to the difference between the actual and budgeted overheads, as in Equation (11.37).

$$\text{Expenditure variance} = \text{Actual fixed overhead} - \text{Budgeted fixed overhead} \tag{11.37}$$

Therefore in the example of Table 11.12 the fixed overhead expenditure variance is given as in Equation (11.38).

$$\text{Expenditure variance} = 2500 - 2000 = 500(\text{A}) \tag{11.38}$$

The fixed overhead volume variance is due to the actual volume of production differing from the budgeted volume, where production volume is measured in standard hours. The fixed overhead volume variance is given by Equation (11.39).

$$\text{Volume variance} = (OSH - \text{Budgeted hours}) \times FOAR \tag{11.39}$$

Therefore for the example of Table 11.12 the fixed overhead volume variance is given by Equation (11.40).

$$\text{Volume variance} = (500 - 500) \times 4 = 0 \tag{11.40}$$

As expected the fixed overhead expenditure and volume variances add to the total fixed overhead variance.

The fixed overhead volume variance can be divided further into the fixed overhead volume capacity variance and the fixed overhead volume efficiency variance. The fixed overhead volume capacity variance is due to the actual capacity varying from the planned capacity, both measured in hours, and is given by Equation (11.41).

$$\text{Capacity variance} = (\text{Actual hours} - \text{Budgeted hours}) \times FOAR \tag{11.41}$$

Therefore in the example of Table 10.12 the fixed overhead volume capacity variance is given by Equation (11.42).

$$\text{Capacity variance} = (450 - 500) \times 4 = 200(\text{A}) \tag{11.42}$$

The fixed overhead volume efficiency variance is due to the actual hours worked differing from the standard hours of production and is given by Equation (11.43).

$$\text{Efficiency variance} = (\text{Actual hours} - OSH) \times FOAR \tag{11.43}$$

Therefore for the example of Table 11.12 the fixed overhead volume efficiency variance is given by Equation (11.44).

$$\text{Efficiency variance} = (450 - 500) \times 4 = 200(\text{F}) \tag{11.44}$$

The fixed overhead volume capacity variance and the fixed overhead volume efficiency variance sum to the fixed overhead volume variance, as expected.

An example in the derivation of overhead variances is provided in the case study 'Analyzing the overheads'.

Case study

Analyzing the overheads

Tom Wise, factory manager for the Coolit Distribution Company, decided to analyze the factory overheads. The budgeted and actual values of the fixed and variable overheads for the period (week 26) are shown in Table 11.13 and, based on the fact that there are 150 production personnel and a five day, eight hours per day week is worked, he calculated the total standard hours and the VOAR and FOAR as shown in this table.

Table 11.13 Parameters for the case study 'Analyzing the overheads'

Parameter	Budgeted (standard)	Actual
Variable overhead	£1200	£1400
Fixed overhead	£2400	£3000
Total standard hours	150 × 8 × 5 = 6000	
Variable overhead absorption rate (VOAR)	1200/6000 = £0.20 per hour	
Fixed overhead absorption rate (FOAR)	2400/6000 = £0.40 per hour	

Tom then calculated the pyramid of overhead variances, as shown in Table 11.14. From this he deduced that the adverse overhead variance was primarily due to the fixed overhead variance. Although the fixed overhead volume variance was favourable, the fixed overhead expenditure variance was considerably adverse, and Tom decided to explore this further.

Table 11.14 Calculation for the case study 'Analyzing the overheads'

Variance	Equation number	Calculation	Value
Overhead variance	11.27	(1400 + 3000) − 6900 × (0.2 + 0.4)	260(A)
Variable overhead variance	11.29	1400 − 6900 × 0.2	20(A)
Fixed overhead variance	11.31	3000 − 6900 × 0.4	240(A)
Variable overhead expenditure variance	11.33	140 − (6000 × 0.2)	200(A)
Variable overhead efficiency variance	11.35	(6000 − 6900) × 0.2	180(F)
Fixed overhead expenditure variance	11.37	3000 − 2400	600(A)
Fixed overhead volume variance	11.39	(6900 − 6000) × 0.4	360(F)
Fixed overhead volume capacity variance	11.41	(6000 − 6000) × 0.4	0
Fixed overhead volume efficiency variance	11.43	(6000 − 6900) × 0.4	360(F)

11.6 Selling and distribution variances

The selling and distribution variances arise when the selling and distribution costs differ from those planned. They are based on the volume of sales, not production, and are given by Equation (11.45).

$$\text{Variance} = \text{Actual costs} - (\text{Units sold} \times \text{Standard cost per unit}) \quad (11.45)$$

11.7 Sales variance

The sales margin variance is the difference between the actual margin from sales and the budgeted margin (profit). It is given by Equation (11.46).

$$\text{Sales margin variance} = (\text{Actual sales} - \text{Cost of sales}) - \text{Budgeted margin} \quad (11.46)$$

For example, suppose that a company has planned and actual sales and costs as in Table 11.15. From these figures the sales margin variance can be found as in Equation (11.47).

$$\text{Sales margin variance} = (8000 \times 11 - 8000 \times 8.50) - 10\,000 \times 2 = 0 \quad (11.47)$$

The sales margin variance can be analyzed into the sales margin price variance and the sales margin quantity variance, as in Equations (11.48) and (11.49).

$$\text{Price variance} = (\text{Actual margin} - \text{Standard margin}) \times \text{Units sold} \quad (11.48)$$

$$\text{Quantity variance} = (\text{Units sold} - \text{Budgeted margin}) \times \text{Standard margin} \quad (11.49)$$

For the example of Table 11.15 the sales margin price variance and the sales margin quantity variance are given by Equations (11.50) and (11.51).

$$\text{Price variance} = (2.50 - 2.00) \times 8000 = 4000(\text{F}) \quad (11.50)$$

$$\text{Quantity variance} = (8000 - 10\,000) \times 2 = 4000(\text{A}) \quad (11.51)$$

Table 11.15 Information used to illustrate sales margin variances

Parameter	Budgeted (standard)	Actual
Sales (units)	10000	8000
Price per item	£10	£11
Total costs per item	£8.00	£8.50
Sales margin	10 – 8 = £2.00	11 – 8.50 = £2.50

11.8 Marginal costing

Historical or absorption costing allows verification of what the actual costs were. Standard costing provides management with an analysis of performance against budget. It gives actual costs and shows how good or bad these are compared with standard values. Standard costing is difficult to implement in a jobbing environment and it does not provide management with information to help in future decision making.

Marginal costing differentiates between fixed and variable costs and its prime use is in decision making. Fixed costs are assumed to be those that remain constant.

11.8.1 Costs and contributions

Costs can be broadly divided into two categories:

1. Those related to the level of activity, such as production or sales.
2. Those that are fixed; that is, not related to the level of activity. Generally even fixed costs vary, although this normally occurs over a relatively long period of time and would be due to some other factor, unrelated to the activity being considered.

Figure 11.3 graphically illustrates these two costs. It is assumed that over the time frame being considered the fixed costs are truly fixed and the variable costs vary linearly with the level of activity.

Fixed costs are ignored in marginal costing exercises, but an important consideration is the 'contribution' that a product makes towards meeting the fixed cost element. The contribution is numerically equal to the difference between the sales value and the variable costs.

Figure 11.4 graphically illustrates the concept of contribution. The greater the level of activity associated with the product or service the higher its overall contribution. The ratio of contribution to sales is an important measure of a product's profitability and is know as the Profit/Volume or P/V ratio. It is given by Equation (11.52).

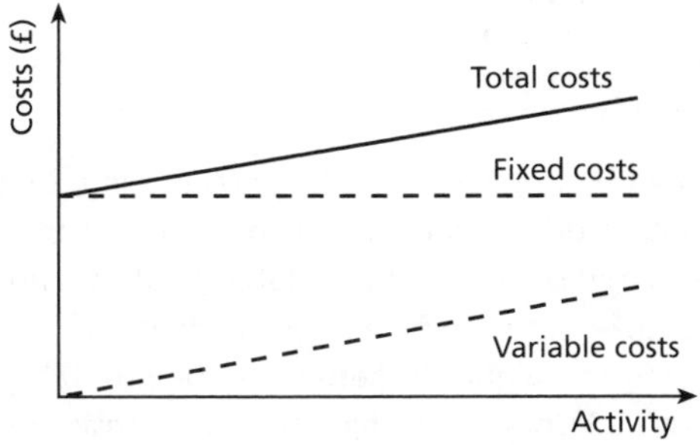

Figure 11.3 Breakdown of total costs as fixed costs and variable costs.

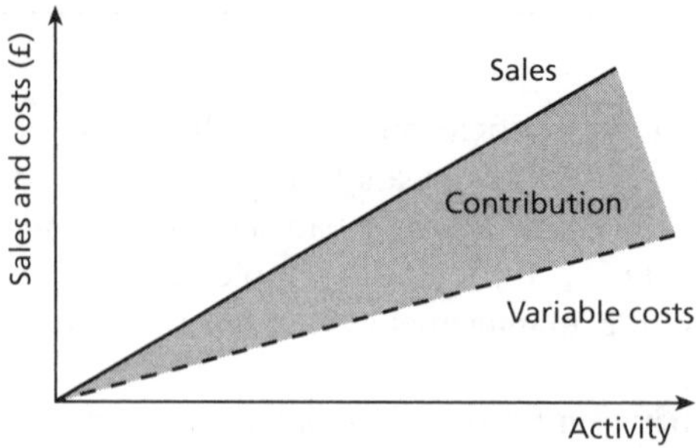

Figure 11.4 The concept of 'contribution'.

$$\text{P/V ratio} = \frac{\text{Contribution}}{\text{Sales}} \tag{11.52}$$

The P/V ratio is seen from Figure 11.4 to be constant, irrespective of the level of activity, since the contribution is proportional to the increase in sales. Any money left over after fixed costs are subtracted from this contribution is the profit.

The use of contribution in management decision making is illustrated in the case study 'One product has to go'.

Case study

One product has to go

Two new experimental products were introduced by the FloRoll Manufacturing Company, called simply 'P' and 'Q'. The sales forecasts for the next year and the direct material and labour costs are given in Table 11.16. The fixed overhead costs for the company are estimated as £100 000 over this period.

Table 11.16 Sales and costs for the case study 'One product has to go'

Item	P (£)	Q (£)
Sales	80 000	120 000
Direct materials cost	40 000	10 000
Direct labour cost	10 000	40 000

Unfortunately it has been determined by Jack Fox, Director of Marketing, that the market will not be able to support both products, and one needs to be discontinued. It was agreed by the Board that this should be the least profitable product, and Johnny Bacon, Financial Director, was asked to prepare a recommendation.

Johnny first used historical accounting techniques to determine the profitability of both products. This is shown in Figure 11.5. Johnny assumed that fixed costs are allocated to each product in proportion to their direct labour content.

		P		Q
Sales (£)		80 000		120 000
Costs (£)				
Material	40 000		10 000	
Labour	10 000		40 000	
Overheads	20 000		80 000	
		70 000		130 000
Profit/(loss)		10 000		(10 000)

Figure 11.5 A calculation of profitability (for the case study 'One product must go') based on historical costing.

From Figure 11.5 Johnny Bacon deduced that that product P makes a profit and product Q a loss, so that one would assume that product Q should be discontinued. However, Johnny then decided to calculate the marginal costs associated with each product, as in Figure 11.6, where the allocation of fixed overheads has been ignored to start with. Contributions are the amount the product makes towards meeting the factory overheads and the contribution to sales ratio should be as large as possible. Product Q has the larger ratio of the two, which means that it is more profitable, and Johnny Bacon decided that this was the product the company should make and he would recommend to the Board that the other product, P, should be discontinued.

		P		Q
Sales (£)		80 000		120 000
Variable costs (£)				
Material	40 000		10 000	
Labour	10 000		40 000	
		50 000		50 000
Contribution		30 000		70 000
Contribution/sales		0.38		0.58

Figure 11.6 A calculation of profitability (for the case study 'One product must go') based on marginal costing.

11.8.2 Break-even chart

The break-even point for any product is the volume beyond which profit is realized and below which profit turns into loss. The break-even point is illustrated graphically by the break-even chart, as in Figure 11.7. The fixed and variable costs result in total costs, as in Figure 11.3, and on this is plotted the sales line, which is again assumed to vary linearly with the level of activity. (In this case 'activity'

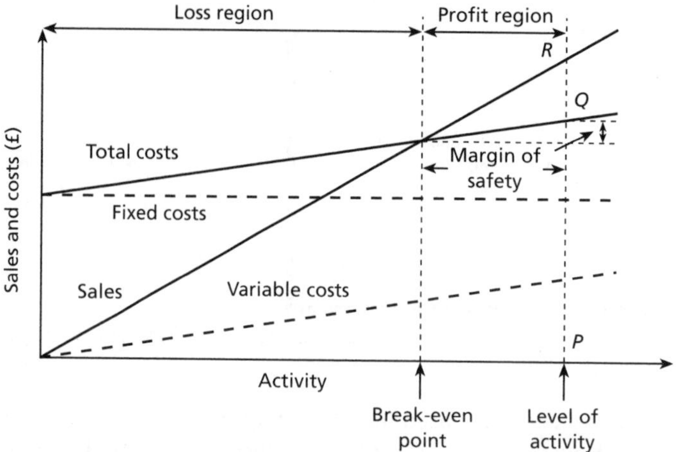

Figure 11.7 Break-even chart.

could be the volume of sales and variable costs would relate directly to the amount of items made to support this sales.)

The break-even point is shown on this graph where the sales and total costs curves intersect. Below this point costs exceed sales and there is net loss. Above it sales exceed costs resulting in profit. The cost at any point, along with sales, can be read off from this graph. Therefore at level of activity P, the costs equals Q and sales equal R, resulting in a profit of $(R - Q)$. The margin of safety can be measured in terms of both the activity and money. It gives the amount that the activity (or money) needs to fall before loss occurs.

The break-even chart is also only accurate for relatively small levels of variation in activity. The fixed costs will vary over time, usually in a step function, for example as extra levels of activity result in the requirement for larger premises or more plant. Variable costs will also not vary linearly with activity, since higher levels of activity can result in a greater increase, for example due to overtime having to be worked. The sales curve also may not be linear; for example, at higher levels of activity (sales) greater quantity discounts may have to be given.

An alternative form of the break-even chart can be obtained by plotting the P/V line, as in Figure 11.8(a). At zero sales all the costs are fixed costs and amount to losses. At the break-even point loss turns to profit. The profit and margin of safety can be read off for any level of sales volume.

Figure 11.8(a) also shows the effect of a decrease in fixed costs on the break-even point and on profit. The slope of the P/V line is unchanged, since the sales price and variable costs are assumed to be the same. If variable costs change then the slope will change, as in Figure 11.8(b). In Figure 11.8(c) all the costs are assumed to be unchanged, but the volume of sales decreases so that the margin of safety and the profit are both reduced.

The use of the break-even chart is illustrated in the case study 'Breaking even'.

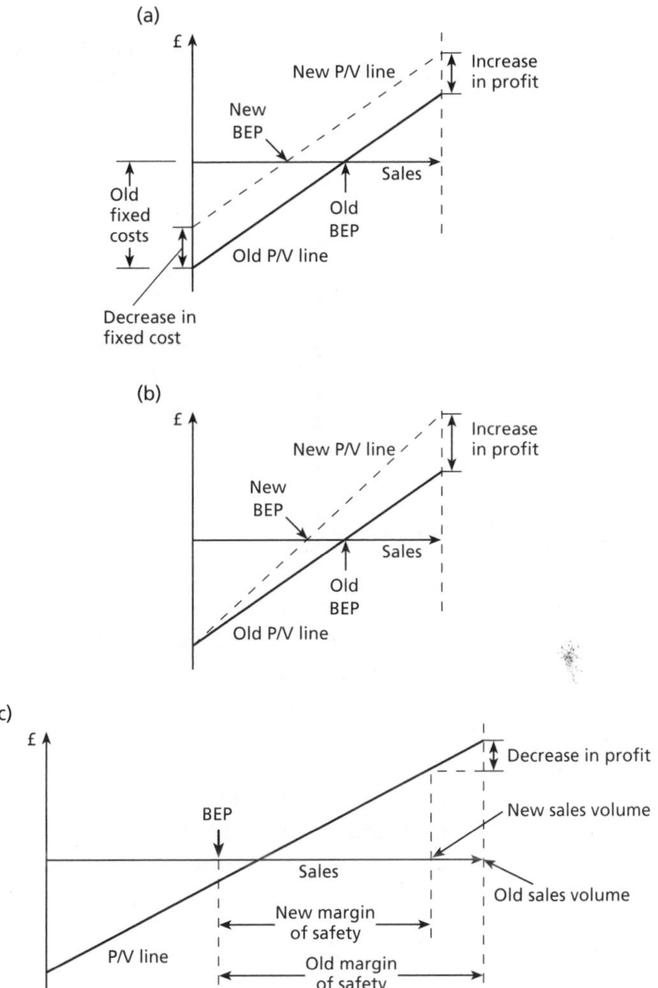

Figure 11.8 Analysis using the P/V chart: (a) decrease in fixed cost; (b) decrease in variable cost; (c) decrease in sales volume.

Case study

Breaking even

Johnny Bacon, in his analysis of products P and Q (see the case study 'One product has to go') now decided to investigate the profitability of product Q further. He determined that the selling price for each unit of Q was £1.20 and therefore restated the analysis of Figure 11.6 for product Q as in Table 11.17. The variable cost per unit is equal to total variable costs divided by the number of units, as in Equation (11.53).

Table 11.17 Parameters for the case study 'Breaking even'

Parameter	Value
Sales (£)	120 000
Sales (Units)	100 000
Sales price per unit (£)	1.20
Variable cost per unit (£)	0.50
Contribution per unit (£)	0.70
P/V ratio	0.58
Fixed cost (£)	100 000

$$\textbf{Cost per unit} = \frac{\textbf{10 000 + 40 000}}{\textbf{100 000}} = \textbf{0.50} \qquad \textbf{(11.53)}$$

Contribution per unit is the total contribution (that is, 70 000 divided by the number of units (100 000)), and the contribution to sales ratio, also referred to as the P/V ratio, or Profit/Volume ratio, is given by Equation (11.54).

$$\textbf{P/V ratio} = \frac{\textbf{0.7}}{\textbf{1.2}} = \textbf{0.58} \qquad \textbf{(11.54)}$$

The break-even chart for this table is shown in Figure 11.9. From this the break-even point can be read off (about £172 000 or 140 000 units). The margin of safety for any level of activity can also be obtained from this chart, along with the profit.

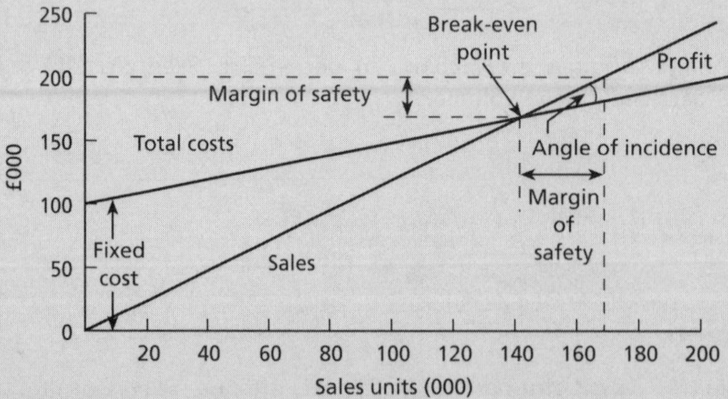

Figure 11.9 Break-even chart for the case study 'Breaking even'.

11.8.3 Mathematical considerations

The graphical analysis given earlier is useful for obtaining a 'feel' for the problem. Analysis can also be done mathematically, and this is important for more accurate results and computer-based calculations.

Variable costs are proportional to the level of activity and can be stated mathematically as in Equation (11.55), where k_1 is a constant and represents a measure of how fast the variable costs increase relative to level of activity.

$$\text{Variable costs} = k_1 \times \text{Level of activity} \tag{11.55}$$

Sales are also related to the level of activity and are given by Equation (11.56), where k_2 represents another constant and measures how fast sales increase relative to level of activity.

$$\text{Sales} = k_2 \times \text{Level of activity} \tag{11.56}$$

At the break-even point total costs (fixed plus variable) equal sales, as in Equation (11.57), from which Equation (11.58) can be obtained.

$$\text{Fixed costs} + k_1 \times \text{Level of activity} = k_2 \times \text{Level of activity} \tag{11.57}$$

$$\text{Level of activity} = \frac{\text{Fixed costs}}{(k_2 - k_1)} \tag{11.58}$$

The level of activity to get to the break-even point is therefore proportional to the fixed costs and inversely proportional to how fast sales increase relative to the variable costs associated with these sales. This is evident from Figure 11.7.

The contribution is equal to the difference between the sales and variable costs and is given by Equation (11.59), so that the P/V ratio is given by Equation (11.60).

$$\text{Contribution} = (k_2 - k_1) \times \text{Level of activity} \tag{11.59}$$

$$\text{P/V ratio} = \frac{k_2 - k_1}{k_2} \tag{11.60}$$

The break-even point can be redefined by Equations (11.61).

$$\text{Break-even point} = \frac{\text{Fixed cost}}{\text{P/V ratio}} = \frac{\text{Fixed costs} \times \text{Sales}}{\text{Contribution}} \tag{11.61}$$

Therefore in the case study 'Breaking even', since the P/V ratio for product Q is 0.58 and fixed costs are £100 000, the break-even point is equal to 100 000/0.58 or £172 414, as found graphically.

The case study 'Determining the contribution' provides a further example of calculation of the break-even point and contribution.

Case study

Determining the contribution

Ajax Ltd, who provide low-cost fixings to the FloRoll Manufacturing Company, sold 1000 units of its product during the past period at £5 each. It had fixed costs of £1500 over this period and the variable costs associated with the product were calculated as £3 per unit.

The contribution per unit is therefore £5 – £3 = £2.

Since 1000 units are sold the total contribution is £2000.

Total sales (number of units times selling price) equals £5000, so the P/V ratio is 2000/5000 or 0.4. This could have also been calculated on a unit basis: £2/£5 or 0.4.

Therefore the break-even point is equal to fixed cost divided by the P/V ratio, or 1500/0.4, or £3750. The contribution in terms of number of units is then £3750/5 or 750 units.

The profit at 1000 units is the sales less the fixed and variable costs: £5000 – £3000 – £1500 = £500.

11.8.4 Key factors

In the case studies so far it has been assumed that as much labour and material required is available, and that the limiting, or key factor, is the amount of goods which can be sold. This is usually the case (that is, sales is the key factor), and the contribution to sales is widely used in the analysis of marginal costs. The contribution per key factor, whatever that may be, should be considered in the calculations.

Application of key factors other than sales is illustrated in the case studies ‘Labour is the key’, ‘Materials are the key’ and ‘The best mix’.

Case study

Labour is the key

When Johnny Bacon (see the case study ‘One product has to go’) reported back to the Board on his recommendations for discontinuing manufacture of product P, he was informed by Adrian Elton, Manufacturing Director, that there was a critical shortage of skilled labour and that this determined the amount of product P or product Q which could be built.

Johnny realized that labour hours were the key factor rather than sales, as he had originally assumed. After the Board meeting he decided to recalculate the contribution made by products P and Q. He determined that the labour rate was £10 per hour for both P and Q. Therefore the number of labour hours used in product P is 100 and for product Q it is 400. The contribution per labour hour for P is therefore 30 000/100 = £300 and for Q it is 70 000/400 = £175. Therefore on this basis it is product Q which should be discontinued and product P which should be manufactured.

Johnny Bacon decided to make this recommendation at the next Board meeting.

Case study

Materials are the key

Ajax Ltd makes two other types of fixing, which are designed to withstand greater strain, but cost proportionately more. Fixing A sells at £20 per unit and the direct labour and material costs per unit are equal to £5 and £10 respectively. FIxing B is priced at £15 and the labour and materials costs are £10 and £3 respectively.

The contribution and contribution to sales figures for these two products are shown in Table 11.18. This indicates that Fixing A is the more profitable product. However, materials used in the manufacture of these fixing are in short supply so materials become the key factor. The contribution to materials ratio is higher for Fixing B so that, in the present circumstances, this is is really the more profitable product.

Table 11.18 Illustration of the case study 'Materials are the key'

Parameter	**Fixing A**	**Fixing B**
Selling price (£)	20	15
Labour cost (£)	5	10
Material cost (£)	10	3
Contribution (£)	5	2
Contribution/Sales (£)	0.25	0.13
Contribution/Materials	5/10 = £0.5	2/3 = £0.67

Case study

The best mix

Ajax Ltd decided to introduce a new range of low-cost fixings, LC, VLC and XLC. The proposed selling price and maximum sales potential over the next year, along with estimated costs, for each product is given in Table 11.19. The fixed factory costs are estimated at £5000. The production manager has been asked to determine the best product mix assuming that total labour hours are restricted to 10 000.

Table 11.19 Illustration of the case study 'The best mix'

Item	LC	VLC	XLC
Selling price per unit (£)	9	7	8
Labour cost per unit (£)	3	2	1
Material cost per unit (£)	4	3	5
Labour rate per unit (£)	0.3	0.4	0.8
Maximum sales forecast (units)	4000	3000	6000

Table 11.21 shows the analysis based on Table 11.20. XLC has the highest contribution per labour hour and it is therefore the first one that is considered for manufacture. 6000 units of this fixing are made (the maximum that can be sold) and this absorbs 7500 hours of labour and gives a total contribution of £12 000.

Table 11.20 Data for the case study 'The best mix'

Item	LC	VLC	XLC
Selling price per unit (£)	9	7	8
Variable costs per unit (£)	7	5	6
Contribution per unit (£)	2	2	2
Labour hours per unit (£)	10	5	1.25
Contribution per labour hour (£)	0.2	0.4	1.6

Table 11.21 Analysis of the case study 'The best mix'

Product	Hours needed	Contribution per labour hour (£)	Total contribution (£)
XLC	6000 × 1.25 = 7500	1.6	12 000
VLC	500 × 5 = 2500	0.4	1000
	10 000		13 000

There are 2500 labour hours left and since each unit of VLC needs 5 hours, 500 units can be made only and no units of LC can be made. The total contribution is the product of the contribution per labour hour to the labour hours used on each product. The overall contribution is £13 000, and taking the fixed costs of £5000 from this gives a profit of £8000. Therefore to maximize profits during the year the company should make 6000 units of XLC, 500 units of VLC and none of LC, and this will give a total profit of £8000.

11.8.5 Advantages of marginal costing

There are several advantages to marginal costing techniques, compared with standard costing:

- Fixed costs, which are time-related rather than product-related, are kept separate from production costs. This prevents misleading results being obtained regarding profitability of products.
- Marginal costing avoids the false results that can arise if fixed costs are absorbed into products which later cannot be sold, for example they are spoilt.
- Marginal costing avoids the situation of over- or under-absorption of fixed costs, which can arise with standard costing.
- Contribution is a much better indicator of profitability of individual product lines than profit obtained from total absorption costing. It is therefore a more effective decision-making method.

The main disadvantage of marginal costing is that if fixed costs are totally ignored, and too much sales is taken at marginal costs plus a margin of profit, it will result in lower overall profit or even a loss.

Summary

Costs are a key management consideration in determining the competitiveness of a company, and as such they are often used as a management control tool. Products are generally costed on the basis of job costing, which applies to those built on a batch basis, usually to a specific order, or process costing, where the product is built in larger volumes using mass production techniques, usually for stock.

Costs can be classified as direct costs, which are closely related to the product being made, and indirect costs, which are not so related. There are three main methods of costing: total absorption costing (or historical costing), standard costing and marginal costing. In total absorption costing, all the costs, direct and indirect, associated with the product are absorbed into the units produced.

An application of costing is valuation of stock. This can be done as FIFO (first in, first out), where the price of the items first bought into store is taken to cost the items first withdrawn; LIFO (last in, last out), where the price of the items last into store is taken to cost the items first withdrawn; and weighted average, which takes an average of the price of items in store to cost the items withdrawn at any time. Other methods of costing stock include replacement price, where the current market price is used to cost the item withdrawn, and standard price, which is the initial price calculated for the item and which may not relate to the actual price paid or its current price.

In most operations overheads occur. These can be variable in nature (varying in relation to the amount of activity) or fixed (relatively insensitive to the level of activity). In absorption costing overheads can be allocated to products in several ways, the common methods being in proportion to the number of units made, the amount of materials used, the labour costs of the products, the labour hours spent, the machine hours, or the prime cost.

In standard costing the targets are first set for future costs (known as standard costs) and then the actual costs are measured, as they occur, using historical costing methods. Standard costs are therefore used as a planning and a measurement tool. Generally standards are set for various elements of cost, such as direct materials, direct labour, variable and fixed overheads, selling and distribution costs, and sales. These can then be subdivided further into other costs. The difference between the standard cost and the actual cost, measured when the activity has been completed, is known as a variance. There are many different variances, illustrated in Figure 11.2, starting with the profit variance. This can be broken into the cost variance and the sales variance, and so on in order to understand any variance which occurs and so take management action to correct the problem.

Standard costing provides management with an analysis of performance against budget by giving actual costs and showing how good or bad these are compared with budgeted or standard values. By contrast, marginal costing is primarily used in decision making and it largely ignores fixed costs in the first analysis, determining instead the contribution that each product makes towards absorbing these fixed costs. Contribution in this context is calculated as the difference between the sales and the variable costs associated with each product. The product that makes the greatest contribution is assumed to be the most profitable. This contribution, however, should be measured against the key factor, which is usually the item associated with the limiting resource. Often this limiting factor is sales (the ability to sell all of the product that can be made), but it could be other items such as the labour or machine time, or the material availability. An important concept in marginal costing is break-even analysis, which involves determining the break-even point at which the product sales exceed costs, resulting in a profit. This can be calculated mathematically or graphically, by means of a break-even chart, as shown in Figure 11.7.

Case study exercises

11.1 A company makes purchases of Unit X, as in Table 11.22. It then issues 500 units from stores. Place a cost on these issues using the FIFO, LIFO and weighted average methods of stock valuation.

Table 11.22 Purchases of Unit X

Quantity	Price per unit (£)
100	5.00
200	6.00
400	7.00

11.2 A company has £90 000 fixed overheads and has three production departments employing 50, 100 and 150 operators, and utilizing capital equipment of £100 000, £500 000 and £300 000 respectively. Allocate the fixed overheads to the three departments based on staff levels and on capital equipment.

11.3 A company has fixed overheads of £30 000 and makes two products, FX and FY. The sales price and costs of each are as given in Table 11.23. The company can make a total sales of £300 000 of either product. Which one should it sell and what would be the break-even point and profit?

Table 11.23 Sales and costs for FX and FY

Item	FX	FY
Unit sales price (£)	3	5
Material cost per unit (£)	1	2
Labour cost per unit (£)	1	2

11.4 A company has fixed costs of £30 000 per month. It makes a single product which has a direct cost of £4 per unit. Variable overheads, which are related to the production machine, are calculated as £1 per unit. The selling price is £10 per unit and sales equal 10 000 per month. Calculate sales at the break-even point.

11.5 The standard and actual costs for one period, related to a product, are shown in Table 11.24. The budgeted and actual sales and production figures are given in Table 11.25. In addition the total fixed overheads are found to be equal to £6200 at the end of the period, and the total actual variable overheads are found to be £3500.

Calculate variances and use the variance pyramid to explain these.

Table 11.24 Actual and standard costs per unit

Item	Standard cost	Actual cost
Materials (£)	3.50	3.20
Labour hours	1.00	1.20
Labour rate (£/hour)	8.00	8.00
Variable overhead absorption rate (£/hour)	0.50	
Fixed overhead absorption rate (£/hour)	1.00	
Selling and distribution	1.50	1.20

Table 11.25 Budget and actual sales and production

Item	Budget	Actual
Sales price (£)	21	19.50
Sales volume	5000	5500
Production volume	6000	5500

11.6 Movements into and out of a store are shown in Table 11.26 for eight periods. For each withdrawal calculate the stock price using the LIFO, FIFO and weighted average methods.

Table 11.26 Valuation of stock

	Into stock		Qut of stock
Period	**Quantity**	**Price (£)**	**(Quantity)**
1	150	1.10	
2	250	1.20	
3			180
4			100
5	100	1.30	
6			200
7	100	1.40	
8			80

11.7 Table 11.27 shows key parameters relating to three products, A, B and C, made in a factory. The fixed factory overheads are £6000 per month, which can be broken down into £1000 for material handling, material waste and procurement; £3000 for general overheads such as rent; and £2000 for personnel related costs such as safety and canteen costs.

Allocate these overheads using as many different methods as possible.

Table 11.27 Allocation of overheads

Item	A	B	C
Units made per month	250	400	900
Total material cost (£)	500	800	400
Labour hours per unit	4.0	3.5	1.5
Labour rate per unit (£/hour)	5.0	4.0	7.0
Machine hours per unit	1.0	1.0	3.0

References

Ames, B.C. and Hlavacek, J.D. (1990) Vital truths about managing your costs, *Harvard Business Review*, January–February, pp. 140–7.

Balderston, J. *et al.* (1984) *Modern Management Techniques in Engineering and R&D*, van Nostrand Reinhold Co. Inc., NY.

Blaxill, M.F. and Hout, T.M. (1991) The fallacy of the overhead quick fix, *Harvard Business Review*, July–August, pp. 93–101.

Davies, T.A. (1996) Focus on customer service at all costs, *Professional Manager*, July, pp. 12–13.

Sizer, J. (1985) *An Insight into Management Accounting*, Penguin, Harmondsworth.

12

Investment decisions

Introduction

This chapter introduces the types of financial investment made by managers and the evaluation process used to determine their viability. This includes techniques such as ranking, payback period, average rate of return, net present value and internal rate of return.

12.1 Defining the investment

Making investment decisions is a key management task. It is a risky business, since relatively large sums of money will be committed over several years by these decisions, and the benefits expected may not materialize at the end of that time. These investment decisions affect everyone within and outside the organization: shareholders, employees and customers. Too many companies squander scarce resources on projects which fail to produce long lasting profits (Hall *et al.*, 1993).

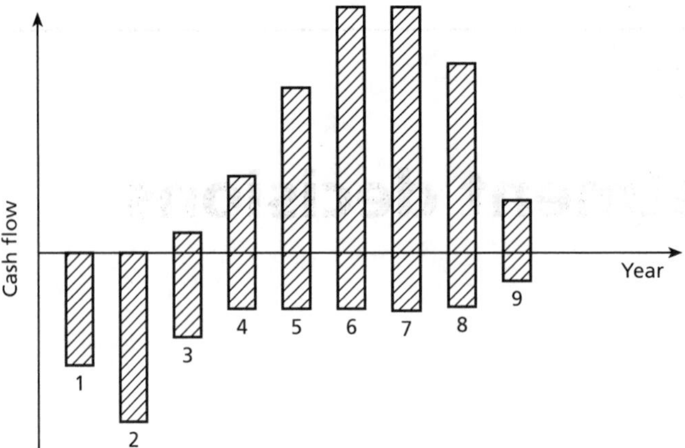

Figure 12.1 Possible cash flow due to an investment decision.

Figure 12.1 shows a possible cash flow scenario from an investment decision. During years one and two large amounts of money are being invested without any return. It is only in year three that there is some payback. The amount of revenue continues to grow from year three onwards, and eventually the cumulative sum of the cash received exceeds that spent in previous years. There is positive cash flow. The key, however, is to know whether the initial investment has been successful. This can only be determined once the measurement criteria for a successful project have been defined (Dixit and Pindyck, 1995). Too much concentration on financial and other evaluation aspects, however, can detract from the main aim of any investment selection process: finding worthwhile projects in which to invest.

12.1.1 Types of investment

There are many different types of product and project investment decision that are required to be made by managers within an organization. Examples of these are:

- Investment in new plant or machinery, or upgrades to existing equipment. This is usually needed to increase production capacity or to make the manufacturing process more efficient, so reducing costs. Plant and equipment have to be depreciated and usually have an end value. The machines also have an ongoing maintenance cost. However, revenue will be generated almost immediately by their use.
- Investment in R&D, usually to develop a new product to meet a market need or to hold market share against competitive pressure. All R&D carries considerable risk, and risk reduction is important (Alston, 1986; Baker and Freeland, 1975; Benzin *et al.*, 1991; Cook and Scifford, 1982; Foster, 1982, 1985; Krawice, 1984). Generally R&D spending is not directly related to

sales, but is influenced by other factors, such as the market requirements (DeYoung, 1992). Many methods have been proposed for selecting R&D projects, for example by determining the contribution that R&D makes towards overall profits (Kuwahara and Tekeda, 1990). Very few of these, however, are based on financial analysis (Hall and Nauda, 1990).

R&D costs can be relatively high and stretch over several years. Often no revenue is generated until the end of the process, when the concept has been proved and the product is built and sold.

- Product- and service-related activities, such as marketing, advertising and promotion. These activities require very large amounts of investment if they are to be successful, and in a competitive environment they are essential for success. The costs associated with these activities can usually be curtailed at relatively short notice, although for maximum impact they should be looked on as long-term investments.
- Strategic activities, such as buying-in a product rather than developing one (Venkatesan, 1992). Even if no development costs are involved, the investment needed to buy-in a product can be high, covering the training of staff, marketing, badging of products, customer support activities and so on.

In all these investment considerations it is important to remember that the investment itself is not the key factor; it is the profit that is generated at the end that matters.

12.1.2 The evaluation process

The process used in an investment decision is illustrated in Figure 12.2.

The first, and most important, activity is to identify suitable investment opportunities. This is the most critical phase, since if only bad projects have been identified for investment then, even though some of these may meet the minimum investment criteria, the end effect will be to get mediocre results. The project investment selection techniques described here cannot produce good projects; they can only show how good or bad the selected projects are.

Having identified the investment opportunities the alternative methods for achieving these must be determined and clearly defined. It is these alternatives that will be evaluated in further detail in the next stage.

The parameters associated with each alternative investment opportunity must now be determined. Quantifiable and non-quantifiable parameters must be considered. This is not an easy task, especially if the parameters are likely to occur some time into the future. For example, when building a product out of relatively standard materials it may be possible to predict its cost a few years away, but it will not be easy to predict its selling price, since this will be dependent on the economic climate at the time and competitor activity, both of which cannot be easily predicted. Alternatively, when building a product, such as a ship, to a contract, then the price and quantity are fixed and it is the cost that is likely to vary and need predicting.

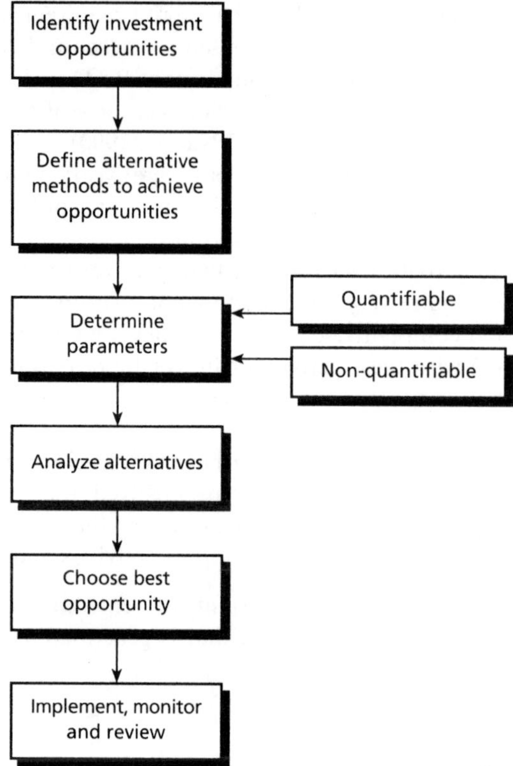

Figure 12.2 The investment process.

Many projects also have to consider other parameters, apart from financial ones. For example, a project may be necessary to meet health and safety requirements, or employee welfare projects may be required to contribute to the well-being and morale of staff.

The parameters of the alternative project must now be analyzed. Several methods can be used for this, as described in later sections of this chapter. These may be subjective or objective and may be based on ranking methods or meeting a threshold of acceptance. For example, a single objective criterion, such as financial return on investment, can be used. The problem is that if several projects pass this criterion then some other method may be needed to select between them. Objective methods such as this also cannot take into account intangible benefits, such as improvement in customer relationships or staff morale.

An important consideration when evaluating alternatives is to remember to value the cost of lost opportunities if a project is selected, since the resources used on this cannot be used on any other project. It is often said that many senior managers lack an appreciation of the opportunity cost of resources (Venkatesan, 1992).

The next stage of the investment process is to determine the best opportunities for investment. Before this can be done it is important to keep clearly in mind:

- Why the investment is to be made; for example, productivity increase; new product opportunity; or lower cost product.
- How the investment is to be funded, since this could influence the projects chosen. For example, funding could come directly from the customer, such as for a customer specific project on a cost-plus basis. Alternatively, the funds may be taken from company overheads, being spread over a range of other products and services, or from the capital of the company, in which case it would come from revenue reserves, which would not go out as dividends to shareholders.
- The real cost of the investments over time, in terms of cash and lost opportunities.
- The real benefits expected over time, both in financial terms and as intangibles.
- Environmental factors over time, such as interest rates, taxes, inflation and the state of the economy.
- Details of the alternatives in which investments are needed.

The selection process is often associated with a high degree of risk, and the willingness to take risk would often affect the investment decision. For example, Figure 12.3 illustrates the risk associated with two projects.

Project A has a modest payback, illustrated as G_1, but the spread in uncertainty around this mean is low and the probability of success (P_1) is relatively high. Project B, on the other hand, has a much higher expected gain (G_2), but the spread in gain is wide, even going into a possible loss, and the probability of success is lower. Usually the higher the risk the higher the expected gain from the investment, and management must decide on the level of risk it is willing to take.

In the example of Figure 12.3, if a high target for gain is set then Project A would not be considered, while a low target would select both projects, which would then go forward for selection on some other criterion.

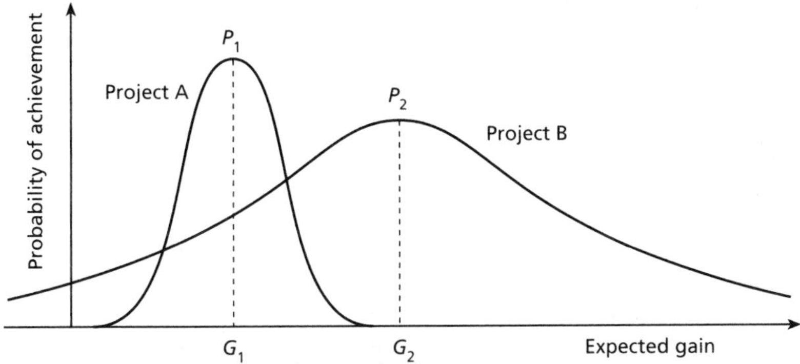

Figure 12.3 Risks associated with two projects.

Luck plays an important part in selecting a successful project or product investment. Therefore some companies often invest in several projects in the hope that some of these will strike it lucky and succeed, so paying for the many which fail. However, in a competitive environment, projects can be very expensive to implement and there are no guarantees of success.

12.2 The ranking process

The aim of project selection procedures is to introduce an element of objectivity so that personal prejudice and bias are avoided. However, subjectivity is an important characteristic of all selection processes, and often the manager has nothing better to go on but experience and 'gut feel'.

Selection of a project can be based on a single or multiple criterion ranking process. Some of the parameters used for ranking are shown in Table 12.1. Each item is considered in turn and the various projects are given a rank, which could

Table 12.1 Some parameters used in project ranking

Parameter	**Products**		
	A	**B**	**C**
Marketing sales and support considerations			
Market size	3	1	2
Competition	1	3	2
Market trend	1	2	3
Impact on existing products	2	3	1
Distribution and sales channels	3	2	1
Support infrastructure	3	1	2
R&D considerations			
R&D skills required	1	3	2
New technology involved	3	1	2
Development risk	3	2	1
IPR situation	1	2	3
Manufacturing considerations			
Plant requirements	1	3	2
New processes needed	3	1	2
Labour skills	2	1	3
Storage facilities	1	3	2
Material suppliers	2	1	3
Financial considerations			
Total costs	2	1	3
Sales volume	1	2	3
Payback period	1	3	2
Cash flow	2	1	3
Rate of return (DCF)	2	1	3
	38	37	45

be a number out of 10 or a relative position of each project against the other (as illustrated in the table; see the case study 'One out of three'). The project(s) that attain a rank above the minimum can then be selected.

Marketing, sales and support parameters often include:

- The potential size of the market that the project or product will address, and the market trend.
- The competition within this market.
- The impact that the new product will have on existing products sold by the company. For example, the new product could compete and take market share from an existing product.
- The distribution and sales channels required for the new product and whether the company is able to implement these.
- The infrastructure needed to support the new product and the costs involved in setting up and maintaining this.

Some of the engineering or R&D considerations are as follows:

- The engineering skills needed to develop the new product.
- Whether any new technology is involved in developing the product and whether the company has access to it.
- The estimated development risks. Usually the longer the development time the greater the risk of overspend and delays, resulting in loss of market share.
- Intellectual property rights issues, and whether any patents are involved in the development of the product.

Manufacturing considerations include:

- Investment needed in new plant and equipment.
- New manufacturing processes, which may need to be developed or introduced to make the new product.
- Labour required to make the product and whether the right skills are available within the company.
- Storage facilities required, in addition to manufacturing space.
- Suppliers and whether any new ones are involved, who will have to be contacted and approved.

Financial considerations are often the most commonly used selection criteria for projects, but they should not be used exclusively. Some parameters considered are:

- The total costs involved in the project, which will include R&D, sales, marketing, manufacturing and support. This is the money that the company needs to find before it can successfully launch the project.
- The total sales volume expected and the timing.
- The payback period, which represents the time when the company will get back its original investment.
- Cash flow, which is an important liquidity consideration.

- Rate of return (discounted cash flow). This is considered in greater detail later in this chapter.

Financial ranking has the problem that actual magnitudes are not considered in the figures, just relative size. Therefore the difference between two projects could be £10 or £10 000, but the result will be the same. Generally there are three key financial ranking methods:

1. Payback period.
2. Average rate of return.
3. Discounted cash flow, which includes net present value and internal rate of return (yield).

These three methods are described in greater detail in the following sections.

Ranking projects is important if for no other reason than the fact that it sets up a dialogue between functions. The specialists assigned within the functions should rank their own parameters, not the managers making the choice between projects. The ranking process can highlight weaknesses that may need to be fixed, especially if the project is eventually selected.

Ranking can be done as a multi-step process. For example, the low-risk projects may first be selected, and from these those that need low capital investment can be short-listed, and finally those that best fit within the company's product portfolio are chosen.

Case study

One out of three

The Board of the FloRoll Manufacturing Company wish to develop a product to complement their existing range of florolls. Three potential new products have been identified and the Board have to choose between these, since investment is available for only one product.

In order to aid in this selection process a list of parameters has been drawn up, and specialists within each function have been asked to rank the three projects against each parameter, with 1 being the best and 3 the worst. The parameters and their ranking are given in Table 12.1.

Product B has the largest potential market size, although it is also the one with the highest number of competitors. The market trend favours Product A, which has the fastest market growth rate of the three. Product B will have the greatest impact on the company's existing products, since it will compete with some of them in certain market sectors. Product C is the best match with the company's existing distribution and sales channels, while Product B best fits its support structure.

Different aspects of R&D favoured different projects. Product A requires very few new R&D skills, although it involves development of the largest amount of new technology and therefore represented the highest risk. In spite of this Product A is also the best regarding IPR, in that the company owns most of the intellectual property rights associated with the new technologies involved.

Product A requires very little new manufacturing plant, although new processes are involved. Labour skills availability favours Product B along with availability of suppliers of raw materials. Product A, however, needs the least storage space for raw materials and finished goods.

In financial considerations Product B needs the lowest overall investment (lowest cost), whereas Product A shows the highest expected sales volume and the shortest payback period. In spite of this the cash flow position is slightly better with Product B, as also is the discounted rate of return.

Based on this assessment the Board decide that Products A and B should be investigated further since they have scores that are close together and are better than Product C.

12.3 Payback period

The payback period is a measure of the time taken for the investment to generate enough return to pay back the original investment. It is the time, measured from the original investment, within which the profits will equal the capital sum spent and will therefore repay the original outlay on the project. Depreciation is not normally taken into account in these calculations; that is, it is not subtracted from the return, although any tax due is deducted.

Figure 12.4 illustrates the calculation for payback period. It is assumed that the investment is in the form of a large initial sum followed by a steady incremental

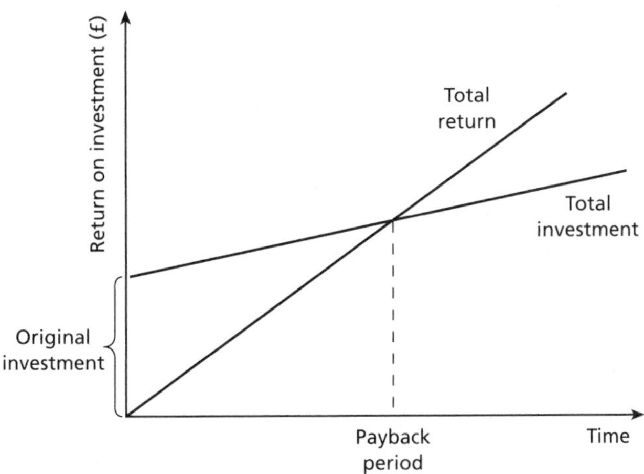

Figure 12.4 Payback period.

amount, as shown by the investment line. Return is assumed to start at zero at the beginning of the project and then increase, also linearly, over time. The payback period is reached when the total return exceeds the total investment made to date. (See the case study 'Payback or return'.)

The advantages of the payback method are:

- It is easy to understand and operate.
- It is good for applications where rapid changes are likely to occur, such as new technology making earlier investments obsolete within a short time. Under these circumstances the payback method gives preference to projects with the shortest payback periods.
- It is a good method to use if the company needs to ensure high liquidity, since it gives preference to projects that bring in cash earlier.

The disadvantages of the payback period are:

- It does not consider time factors in investments and returns: no discounting is involved. Usually the value of money will vary over time. This can be overcome by discounting all investments and returns in the payback method, as described later.
- It does not take into account the picture after payback is reached. For example, a project may have a very steep increase in return after the payback point and another may have no further increase, but these factors will be disregarded.

The payback method is widely used because of its simplicity. Generally it is a useful method for eliminating projects that have an unduly long payback period, say over five years, and then using other methods to choose between the short-listed projects.

12.4 Average rate of return

The average rate of return (ARR) is calculated as the average return over the life of the project as a percentage of the total investment over this period. It is given by Equation (12.1), where items such as depreciation are subtracted to arrive at the return.

$$\text{ARR} = \frac{\text{Average return during period}}{\text{Total investment during period}} \times 100 \tag{12.1}$$

The advantages of this method of selecting an investment are:

- It is easy to understand and derive.
- It takes into account all investments and returns over the life of the project.

The disadvantages of this technique are:

- There is no consideration of the time effects on money received or spent; that is, no discounting.

- The technique considers percentage and not absolute values. Therefore a project may have very low earnings (say £100 per year) but also low total investment (say £100) and give a high ARR (100 per cent), although in reality the project may not be worthwhile due to its very low absolute return.

The use of the payback and average rate of return methods for investment appraisal are illustrated in the case study 'Payback or return'.

Case study

Payback or return

Johnny Bacon, Financial Director of the FloRoll Manufacturing Company, has been asked to prepare a financial appraisal for two projects between which the Board has to choose.

Both projects have an estimated life of five years and Johnny Bacon obtains the investment and expected return for each project as in Table 12.2. From this table Johnny plots the cumulative investment and spend, as in Figure 12.5, and obtains the payback period for Project 1 as a little over 3.5 years and for Project 2 as well over 4.5 years. Therefore, on the face of it, Project 1 looks like the better investment.

Table 12.2 Investent and return on two projects (for the case study 'Payback or return')

Year	Investment (£000)		Return (£000)	
	Project 1	**Project 2**	**Project 1**	**Project 2**
1	8 000	15 000	1 000	–
2	2 000	5 000	2 000	–
3	500	1 000	5 000	–
4	500	–	5 000	14 000
5	–	–	2 000	23 000
	11 000	21 000	15 000	37 000

Johnny Bacon then calculates the average rate of return for each project. The average return for Project 1, over the five years, is given by 15 000/5 = £3000, so the ARR is given by Equation (12.2).

$$\textbf{ARR (Project 1)} = \frac{3000}{11\,000} \times 100 = 27.3\% \qquad \textbf{(12.2)}$$

For Project 2 the average yearly return is 37 000/5 = £7400 and the ARR is given by Equation (12.3).

$$\textbf{ARR (Project 2)} = \frac{7400}{21\,000} \times 100 = 34.2\% \qquad \textbf{(12.3)}$$

Therefore, on the basis of the ARR, Johnny Bacon concludes that Project 2 is the better investment.

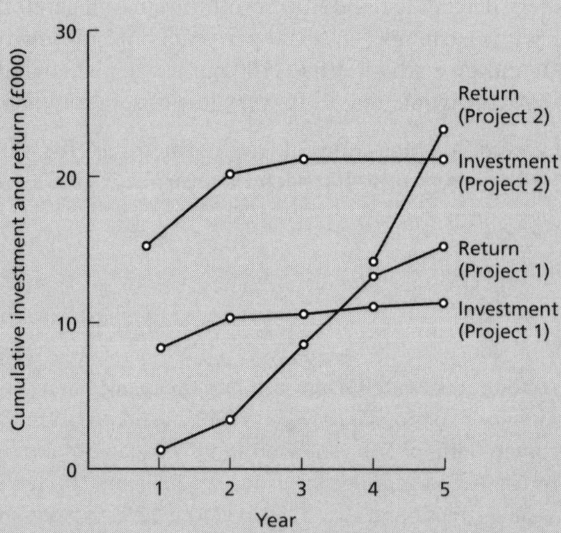

Figure 12.5 Payback method (from Table 12.2).

Since the company has a healthy reserve of funds, Johnny Bacon does not consider that the earlier payback of Project 1 is significant enough to compensate for the greater ARR of Project 2. He decides to put these figures before the Board so that they can be considered along with other factors in the selection process.

12.5 Discounted cash flow

12.5.1 The principle of discounting

If money is spent now on an investment and returns are obtained some time in the future, then it is important to have some means for comparing the two sums, taking 'time' into account. Money 'now' is worth more than 'in the future' for several reasons:

- Money can be invested now to yield interest in the future. This is known as the opportunity cost of money.
- Inflation will reduce the value of money received in the future. This is known as the purchasing power of money.
- Obtaining money in the future involves a measure of risk, whereas money now is already available. This is known as the risk factor of money.

If money is invested then its value can be found at some future date by compounding. This is given by Equation (12.4), where A_0 is the value of money

Table 12.3 Value of £1000 compounded at a rate of 10 per cent

Year	Value (£)
0	1000.00
1	1100.00
2	1210.00
3	1331.00
4	1464.10
5	1610.51

invested at Year 0 (start date); r is the rate of interest stated as a decimal; n is the number of years after the investment is made that the value of the money is to be calculated; and A_n is the value of the original sum at Year n.

$$A_n = A_0(1 + r)^n \tag{12.4}$$

Therefore, as an example, if £1000 is invested in Year 0 at a compound interest rate of 10 per cent, then its value over the next five years can be found from Equation (12.4), where $A_0 = £1000$; $r = 0.1$; and n varies between 1 and 5. This is given in Table 12.3.

This process is known as compounding, and it calculates the future value of money. The process can work in reverse; that is, finding the present value of money received in the future. This is known as discounting and is illustrated in Figure 12.6.

The equation for discounting is the reverse of that given by Equation (12.4), and is shown in Equation (12.5).

$$A_0 = \frac{A_n}{(1 + r)^n} \tag{12.5}$$

Therefore, in the example illustrated in Table 12.3, if £1610.51 is received in Year 5, then this can be discounted back to Year 0 as in Equation (12.6), to give the expected result of £1000.

$$A_0 = \frac{1610.51}{(1 + 0.1)^5} = £1000 \tag{12.6}$$

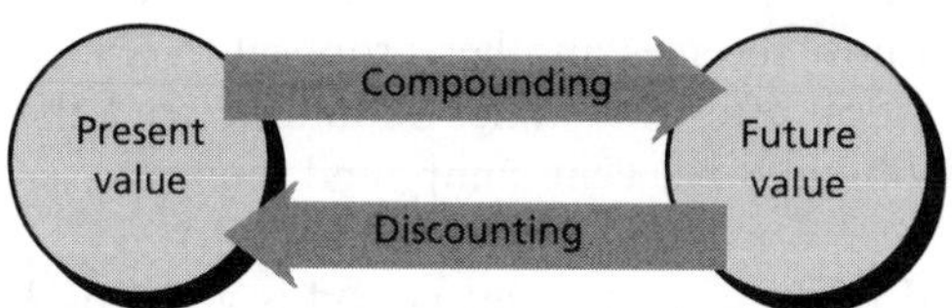

Figure 12.6 Compounding and discounting.

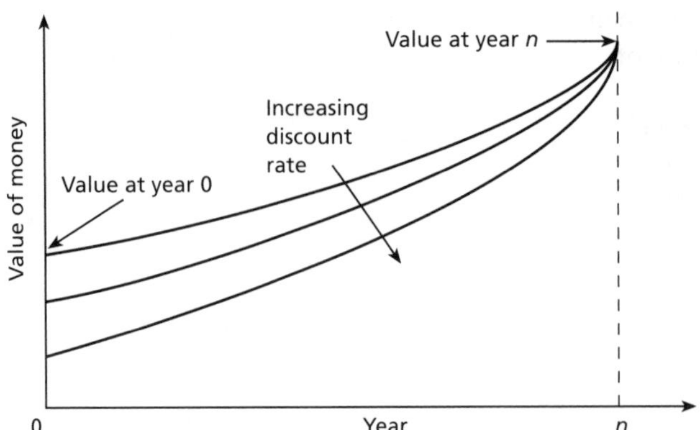

Figure 12.7 The effect of the discounting rate on the present value of money.

Factor r is now known as the discounting rate. The present value of money received in the future will depend on the discounting rate adopted, as shown in Figure 12.7.

Two techniques are used in the Discounted Cash Flow (DCF) method of investment appraisal: the Net Present Value (NPV) and the Internal Rate of Return (IRR), also called the DCF yield method. These are described in the following sections.

12.5.2 Net Present Value

The NPV method of DCF appraisal relates all investments and returns to the present time: Year 0. For example, suppose a person is asked to lend £1000 and is promised that, in return, he will receive £1100 after one year. The current interest available, with comparable risk, is 8 per cent and the person needs to know if he should lend the money.

The present value of £1100 which will be received in one year's time is first discounted back to the present year at an 8 per cent discount rate, as in Equation (12.7).

$$A_0 = \frac{1100}{(1 + 0.08)} = £1019 \tag{12.7}$$

Therefore the person lends £1000 and in effect receives back £1019 at his required interest rate. The NPV of the investment is £1019 – £1000 = +£19, which is positive so the person should lend the money.

Suppose that the same person now receives another proposal to lend £1000 for two years and to receive back £1150 after that time. Should he lend the money, assuming that he still wished to obtain 8 per cent compound interest over the two years?

The value of £1150 discounted back to the present period is given by Equation (12.8).

Table 12.4 Net Present Value factors for discount rates of 1 per cent to 14 per cent

Years	1%	2%	3%	4%	5%	6%	7%	8%	9%	10%	11%	12%	13%	14%
1	.9901	.9804	.9709	.9615	.9524	.9434	.9346	.9259	.9174	.9091	.9009	.8929	.8850	.8772
2	.9803	.9612	.9426	.9246	.9070	.8900	.8734	.8573	.8417	.8264	.8116	.7972	.7831	.7695
3	.9706	.9423	.9151	.8890	.8638	.8396	.8163	.7938	.7722	.7153	.7312	.7118	.6931	.6750
4	.9610	.9238	.8885	.8548	.8227	.7921	.7629	.7350	.7084	.6830	.6587	.6355	.6133	.5921
5	.9515	.9057	.8626	.8219	.7835	.7473	.7130	.6806	.6499	.6209	.5935	.5674	.5428	.5194
6	.9420	.8880	.8375	.7903	.7462	.7050	.6663	.6302	.5963	.5645	.5346	.5066	.4803	.4556
7	.9327	.8706	.8131	.7599	.7107	.6651	.6227	.5835	.5470	.5132	.4817	.4523	.4251	.3996
8	.9235	.8535	.7894	.7307	.6768	.6274	.5820	.5403	.5019	.4665	.4339	.4039	.3762	.3506
9	.9143	.8368	.7664	.7026	.6446	.5919	.5439	.5002	.4604	.4241	.3909	.3606	.3329	.3075
10	.9053	.8203	.7441	.6756	.6139	.5584	.5083	.4632	.4224	.3855	.3522	.3220	.2946	.2679
11	.8963	.8043	.7224	.6496	.5847	.5268	.4751	.4289	.3875	.3505	.3173	.2875	.2607	.2366
12	.8874	.7885	.7014	.6246	.5568	.4970	.4440	.3971	.3555	.3186	.2855	.2567	.2307	.2076
13	.8787	.7730	.6810	.6006	.5303	.4688	.4150	.3677	.3262	.2897	.2575	.2292	.2042	.1821
14	.8700	.7579	.6611	.5775	.5051	.4423	.3878	.3405	.2992	.2633	.2320	.2046	.1807	.1597
15	.8613	.7430	.6419	.5553	.4810	.4173	.3624	.3152	.2745	.2394	.2090	.1827	.1599	.1401
16	.8528	.7284	.6232	.5339	.4581	.3936	.3387	.2919	.2519	.2176	.1883	.1631	.1415	.1229
17	.8444	.7142	.6050	.5134	.4363	.3714	.3166	.2703	.2311	.1978	.1696	.1456	.1252	.1078
18	.8360	.7002	.5874	.4936	.4155	.3503	.2959	.2502	.2120	.1799	.1528	.1300	.1108	.0946
19	.8277	.6864	.5703	.4746	.3957	.3305	.2765	.2317	.1945	.1635	.1377	.1161	.0981	.0829
20	.8195	.6730	.5537	.4564	.3769	.3118	.2584	.2145	.1784	.1486	.1240	.1037	.0868	.0728

Continued on page 356

Table 12.4 (continued) Net Present Value factors for discount rates of 15 per cent to 50 per cent

Years	15%	16%	17%	18%	19%	20%	25%	30%	35%	40%	45%	50%
1	.8696	.8621	.8547	.8475	.8403	.8333	.8000	.7692	.7407	.7143	.6987	.6667
2	.7561	.7432	.7305	.7182	.7062	.6944	.6400	.5917	.5487	.5102	.4756	.4444
3	.6575	.6407	.6244	.6086	.5934	.5787	.5120	.4552	.4064	.3644	.3280	.2963
4	.5718	.5523	.5337	.5158	.4987	.4823	.4096	.3501	.3011	.2603	.2262	.1975
5	.4972	.4761	.4561	.4371	.4190	.4019	.3277	.2693	.2230	.1859	.1560	.1317
6	.4323	.4104	.3898	.3704	.3521	.3349	.2621	.2072	.1652	.1328	.1076	.0878
7	.3759	.3538	.3332	.3139	.2959	.2791	.2097	.1594	.1224	.0949	.0742	.0585
8	.3269	.3050	.2848	.2660	.2487	.2326	.1678	.1226	.0906	.0678	.0512	.0390
9	.2843	.2630	.2434	.2255	.2090	.1938	.1342	.0943	.0671	.0484	.0353	.0260
10	.2472	.2267	.2080	.1911	.1756	.1615	.1074	.0725	.0497	.0346	.0243	.0173
11	.2149	.1954	.1778	.1619	.1476	.1346	.0859	.0558	.0368	.0247	.0168	.0116
12	.1869	.1685	.1520	.1372	.1240	.1122	.0687	.0429	.0273	.0176	.0116	.0077
13	.1625	.1452	.1299	.1163	.1042	.0935	.0550	.0330	.0202	.0126	.0080	.0051
14	.1413	.1252	.1110	.0985	.0876	.0779	.0440	.0254	.0150	.0090	.0055	.0034
15	.1229	.1079	.0949	.0835	.0736	.0649	.0352	.0195	.0111	.0064	.0038	.0023
16	.1069	.0930	.0811	.0708	.0618	.0541	.0281	.0150	.0082	.0046	.0026	.0015
17	.0929	.0802	.0693	.0600	.0520	.0451	.0225	.0116	.0061	.0033	.0018	.0010
18	.0808	.0691	.0592	.0508	.0437	.0376	.0180	.0089	.0045	.0023	.0012	.0007
19	.0703	.0596	.0506	.0431	.0367	.0313	.0144	.0068	.0033	.0017	.0009	.0005
20	.0611	.0514	.0433	.0365	.0308	.0261	.0115	.0053	.0025	.0012	.0006	.0003

$$A_0 = \frac{1150}{(1 + 0.08)^2} = £986 \qquad (12.8)$$

The NPV of the investment is now equal to £986 – £1000 = –£14. Therefore the NPV is negative (the lender gets less, in real terms, than he lends) so the loan should not be made.

The present value of money can be obtained by multiplying the amount received by the factor $1/(1+r)^n$, as in Equation (12.5). This is often shown calculated in tables for convenience, as in Table 12.4.

For the example given earlier, the net present value factor for an 8 per cent discount rate and two years is obtained from Table 12.4 as 0.8573, so that the present value of £1150 is £1150 × 0.8573 = £986, as found by Equation (12.8).

As illustrated by Figure 12.7, a very important consideration of the NPV method is the rate of return for the investment appraisal. Generally this will be fixed by a variety of methods, such as:

- The rate at which money can be borrowed to finance the project, plus any margin for profit.
- The rate that the company would receive if it invested the money in projects outside the company.
- The rate that ensures that investment projects selected are those that have a high profit margin.
- The cost to the company of servicing its invested capital. For example, suppose that a company has a capital structure as in Table 12.5. To service this capital the company wishes to obtain a return of 20 per cent on all capital and to plough back half this amount, after debenture interests have been paid. Therefore the amount needed is found as in Figure 12.8. This gives an interest rate of 52.9 per cent before tax, calculated as in Equation (12.9).

$$r = \frac{3700 \times 100}{7000} = 52.9\% \qquad (12.9)$$

The following case studies illustrate the use of the NPV method.

Table 12.5 Capital structure of a company

	Amount (£)
Ordinary shares	5000
Retained profits	1000
10 per cent debentures	1000
Total	7000

20% of £6000	1 200
Retained (50% of 1200)	600
	1 800
Tax (50%)	1 800
	3 600
Debenture interest	100
	3 700

Figure 12.8 Calculation of the amount needed to service the capital of Table 12.5 for a 20 per cent return.

Case study

A sound investment

Mike Smith, Mechanical Design Engineer with The FloRoll Manufacturing Company, read about a 'sure fire' investment, where for every £1000 invested the return over five years would be as in Table 12.6. He decided to consult his manager, Jill Tait, who had recently been on a finance course.

Table 12.6 Return received for £1000 invested (case study 'A sound investment')

Year	Income (£)
1	50
2	100
3	100
4	200
5	1000
Total	1450

'It looks pretty good to me', said Mike. 'I put in £1000 and I get back £1450.'

'It all depends', replied Jill Tait, pulling out her course notes. 'We did something on the course on investment appraisal, which is after all what this is. Now, let me see. What sort of interest rate do you think you would get if you invested your £1000 in something else?'

'I suppose about 10 per cent', said Mike.

'Yes, I think so as well. Now let's calculate the present value of the money you will get over the five years.'

Jill got out her calculator and soon had the results, as in Table 12.7. (Present value factors were found from Table 12.4 and the present value of income in Table 12.7 was calculated by multiplying the present value factor by the income.)

Table 12.7 Calculation of NPV from Table 12.6 for a 10 per cent discounting factor

Year	**Present value factor (10%)**	**Income**	**Present value of income**
1	0.9091	50	45.5
2	0.8264	100	82.6
3	0.7513	100	75.1
4	0.6830	200	136.6
5	0.6209	1000	620.9
Total		1450	960.7

'You see', said Jill, triumphantly. 'If you discount all the money back to the present time, which is when you will be putting in your £1000, you will only be getting £960.7 in return, which is an NPV of £960.7 – £1000 = –£39.3. This is negative, so you should not make this investment.'

Mike was impressed with Jill's advice and decided to go on the finance course as well.

Case study

Less service

The Link assembly machines in the Dinsher Shop of the FloRoll Manufacturing Company cost about £6000 each per year to service. John McCaully, manager of the Dinsher Shop, has received a proposal from his contractors to upgrade the machines at a cost of £18 000 each, which would then reduce the service charges to £2000 each per year (a saving of £4000 per year). John estimates that the machines have a life of five years, and he wants to know whether the upgrade should be made.

The rate of return used by the company for estimating investments of this type is 8 per cent. John therefore calculated the present value of the savings of £4000 per year over the next five years, as in Table 12.8. From this he found that the NPV of the proposal is equal to £15 970.4 – £18 000 = –£2029.6, which is negative. Therefore John decided not to accept the proposal.

Table 12.8 Calculation of Net Present Value for the case study 'Less service'

Year	Present value factor (8%)	Saving (£)	Present value of saving (£)
1	0.9259	4 000	3 703.60
2	0.8573	4 000	3 429.20
3	0.7938	4 000	3 175.20
4	0.7350	4 000	2 940.00
5	0.6806	4 000	2 722.40
Total	3.9926	20 000	15 970.40

It should be noted that where fixed amounts are involved every year the present value factors can be added together, as in Table 12.8. The total present value can be calculated by multiplying this by the amount each year: 3.9926 × £4000 = £15 970.4, as before.

Case study

Project selection

The FloRoll Manufacturing Company decided to invest £10 000 in one of two projects, each of which gave a different level of profits over the next five years, as in the third and fifth columns of Table 12.9.

Table 12.9 Illustration of the case study 'Project selection'

Year	PV factor	Project 1		Project 2	
	(10%)	Profit	PV of profit	Profit	PV of profit
1	0.9091	2 000	1 818.2	4 000	3 636.4
2	0.8264	3 000	2 479.2	5 000	4 132.0
3	0.7513	4 000	3 005.2	2 000	1 502.6
4	0.6830	4 000	2 732.0	1 000	683.0
5	0.6209	2 000	1 241.8	1 000	620.9
Total		15 000	11 276.4	13 000	10 574.9

In order to select one of the projects the present value of the profits was calculated, using the company's return requirements of 10 per cent. This is also shown in Table 12.9. From this it is seen that Project 1 has an NPV of £11 276.4 – £10 000 = £1276.4, and Project 2 has an NPV of £10 574.9 – £10 000 = £574.9. Both projects have a positive NPV and so either can be selected. However, Project 1 has a higher NPV, so if no other criterion is used for project selection then this is the one that should be selected.

11.5.3 Internal Rate of Return

In the Net Present Value method of DCF all the outgoings and receipts are converted back to the original year and compared. The investment proposal is considered to be acceptable if the present value of the income exceeds the present value of outgoings (that is, if the NPV is positive) at the rate of return or yield required by the company from its investments.

In the Internal Rate of Return (or DCF Yield) method trial and error is used to find a discount rate that, when applied to the cash flows in the project, produces a zero NPV. This discount rate is the IRR of the project and is used as the figure for comparison between projects, or to determine whether an investment proposal is acceptable. (See the case study 'Determining the rate'.)

Managers usually prefer the use of IRR over NPV, since it gives a single number for comparison between projects. However, when comparing two projects one may think that the project with the higher yield is better, but this may not always be the case. If money can be borrowed at a lower rate, then the project with the lower IRR may have a higher NPV and be preferable.

Case study

Determining the rate

The Board of the FloRoll Manufacturing Company was to consider a proposal to invest £1 000 000 in a new project which is expected to yield £300 000, £400 000 and £500 000 over the next three years.

The minimum yield expected from projects is 10 per cent and Johnny Bacon, Financial Director, has been asked to determine the IRR for this project.

In order to find the yield at which the profits discount to the original investment of £1 000 000 (i.e. NPV equals zero), Johnny takes two discount rates, 5 per cent and 15 per cent and calculates the NPV, as in Table 12.10. From this Johnny sees that for a 5 per cent yield the NPV is positive and for 15 per cent yield it is negative. Johnny therefore knows that the IRR lies between 5 per cent and 15 per cent.

Table 12.10 Calculation of IRR for the case study 'Determining the rate'

		5% rate		15% rate	
Year	**Profit (10%)**	**PV factor**	**PV of profit (£000)**	**PV factor**	**PV of profit (£000)**
1	300	0.9524	286	0.8696	261
2	400	0.9070	363	0.7561	302
3	500	0.8638	432	0.6575	329
Total			1081		892

He can plot the two points (1081, 5) and (892, 15) on a graph to find when they pass through 1000, or he can calculate it from the equation for a straight line passing through these points. Johnny finds the equation of the straight line as in Equation (12.10) where *Yield* is in per cent and *PV* in thousands.

$$PV = -18.9\,Yield + 1175 \qquad \textbf{(12.10)}$$

Therefore for a present value of £1 000 000 the value of *Yield* is found as 9.3 per cent.

Johnny Bacon presented this to the Board with the recommendation that the proposal be rejected, since it did not meet the company's minimum investment criteria.

Summary

Making financial investments is a risky, but necessary, activity for managers. Large sums of money are committed over several years without any guarantee of benefits. These investments can cover different items, such as new plant and machinery; research and development into a new product or technology; a marketing programme to sell a new product or to enhance the company's image; or strategic investments, such as buying in products or even taking over another company.

Usually several different projects within a company compete for the limited financial investments available. Different techniques exist for determining the attractiveness of the various investments and for selecting between them. They all start with the key step of identifying suitable investment opportunities, since the selection process cannot in itself come up with new investment ideas. All the techniques also make an assumption about the level of benefits, or return, that the organization requires from its investments. Generally the greater the risk associated with the investment the greater the return that is required.

The ranking method of project selection uses a systematic approach in ranking, based on what are considered to be important parameters for the projects being evaluated. The selection criteria, and the allocation of ranks, are done by specialists in the various areas to ensure that the selection is made objectively. Although this technique of investment appraisal is useful, in that it sets up a dialogue between the different functions affected by the investment, it does not provide any information on the actual magnitude of the financial benefits obtained from the projects that are competing for the investment.

The payback period of investment appraisal measures the time from the original investment until the profits are equal to the capital sum spent and so repay the original outlay on the project. It can be illustrated and calculated graphically, as in Figure 12.4. This method of investment assessment is good for instances where rapid changes are likely to occur, such as new technology making earlier investments obsolete, since it gives preferences to projects which show early payback. However, the method does not consider the time factor, which affects the value of money, nor the picture after the payback point has been reached.

The average rate of return technique of investment assessment arrives at a single number to denote the attractiveness of the investment. It is calculated as the ratio of the average return over the life of the product to the total investments made during the period. It has the advantage that it takes into account the whole of the product life, but still suffers from the fact that time elements are not considered.

The Net Present Value (NPV) method of investment appraisal discounts all the investments and returns back to the present time, so it ensures that the effect of time on the value of money (in terms of its opportunity cost, its purchasing power and its risk factor) is fully considered. A key consideration is fixing the rate of return required for the investment, and this can be determined by taking several factors into account, such as the rate at which money can be borrowed to finance the projects; the rate which the company would receive if it invested the money in projects outside the company; the rate which ensures that investment projects are those with a high profit margin; and the cost to the company of servicing its invested capital. The Internal Rate of Return (IRR) method of investment appraisal works on a similar principle to NPV in that it uses discounted cash flow techniques. In the NPV method an investment is considered to be acceptable if the present value of the income exceeds the present value of outgoings, based on the selected discounting rate. For IRR the discounting rate at which the NVP is equal to zero, for the project, is actually calculated and this is then compared between competing projects.

Case study exercises

12.1 The FloRoll Manufacturing Company wishes to invest £100 000. There are two projects, which have returns on investment as shown in Table 12.11. Which project should the company choose, based on the payback method? Use the average rate of return and discounted cash flow methods and compare the results, assuming that the company requires a 15 per cent return on its investments.

Table 12.11 Return on investment for case study exercise 1

Year	Return (£)	
	Project 1	**Project 2**
1	20 000	30 000
2	20 000	40 000
3	30 000	50 000
4	50 000	20 000
5	80 000	10 000

12.2 Joe Plant, Manager of the Assembly Shop in the FloRoll Manufacturing Company, is considering buying a new machine for £20 000. The machine has an effective life of three years and its use will reduce scrap per year by £10 000. Should Joe make the investment, considering that the company requires a 20 per cent return on all its investments this year?

12.3 Micro Dowling sat at his desk looking out of the window. It was a glorious day and Micro was thinking of what he would do when he retired. 'It would be nice to have a little nest-egg', sighed Micro, 'not too much, about £100 000. I wonder how much I would have to invest today in order to have £100 000 in ten years time, assuming a compound rate of interest of 8 per cent.' Can you help Micro by calculating the present value of his retirement nest-egg, assuming: (a) Micro makes a single payment; (b) Micro invests equal amounts yearly?

12.4 Jack Fox, Director of Marketing at the FloRoll Manufacturing Company, wants to recommend to the Board that they enter the Latin American market with their new product. Considerable investment will be needed for this. The market is also expected to be open for five years and to yield a revenue of £10 million per annum. The company considers that a yield of 25 per cent is necessary for projects such as this. How much can Jack Fox spend on entering the market?

12.5 Adrian Elton has put two proposals before the Board for producing the new floroll product. The first consists of manufacturing the product in-house and it would cost £500 000 and yield profits as in the first column of Table 12.12. The second proposal consists of subcontracting manufacture to a third party. This would also cost £500 000, but the profits would be lower in earlier years, as shown in the second column of Table 12.12, due to the mark-up from the third party. However, the profits in later years increase considerably because of a reduction of overheads and support needed. Which of the two proposals should the Board adopt, assuming that it is looking for a 10 per cent yield on its investment? (Use as many different financial methods as possible and also consider non-financial issues.)

Table 12.12 Return from the make or buy proposal of case study exercise 5

Year	Profits (£000)	
	Make in house	**Buy in**
1	150	50
2	200	150
3	250	200
4	150	300
5	100	200

12.6 The Board of the FloRoll Manufacturing Company is considering a new R&D proposal from Micro Dowling, Engineering Director, for a technology project. This would require an investment of £5 million this year and would yield profits of £1 million per year for ten years. Johnny Bacon has been asked to calculate the IRR for the project and the annual returns that would be needed if the expected annual yield were 25 per cent.

12.7 Adrian Elton, Manufacturing Director, has extra warehousing capacity and has suggested that this be rented to a reliable company close by for £20 000 per year. Assuming that the company expects a yield of 15 per cent on such investments, what is the present value of the rent for 10 years? If the tenancy agreement is for ten years and the rent is increased by 50 per cent after five years, what is the present value of the rent?

12.8 Mike Smith, Mechanical Designer at the FloRoll Manufacturing Company, went to his manager Jill Tait with a personal problem. He wants to buy a new hi-fi which costs £800. He has been given the option of either paying for this in full or paying in four yearly instalments of £250, starting from this year. Assuming that Mike believes that he can get a return of 10 per cent on his money, which option should Jill advise him to take?

12.9 If the FloRoll Manufacturing Company has a requirement for a minimum rate of return of 20 per cent on its investments, what is the maximum amount they should invest in a machine if the expected yield is £50 000 per year for five years? Assume first that the machine has no end of life value and then that the machine has a scrap value of £1000.

12.10 James Silver, Materials Manager at the FloRoll Manufacturing Company, has been asked by Adrian Elton, the Manufacturing Director, to secure the best terms he can for the purchase of some production equipment. James gets two quotations. The cost of the machine from both companies is the same, equal to £160 000, but the payment terms vary. One quotation requires an up front payment of £80 000, with the balance of £80 000 paid in yearly instalments of £20 000 over four years. The second quotation spreads the total payment equally over the next four years, with £40 000 paid at the end of each year. Which quotation should James Silver accept, assuming that the average cost of capital is 10 per cent?

References

Alston, F.M. (1986) Recovering R&D costs, *NCMA J.*, Summer, pp. 55–9.

Baker, N. and Freeland, J. (1975) Recent advances in R&D benefit measurement and project selection methods, *Manag. Sci.*, June, pp. 1164–75.

Benzin, R.W. *et al.* (1991) Evaluating production readiness: improving returns on research and development investment, *IEEE Proceedings Annual Reliability and Maintainability Symposium*, pp. 4–9.

Cook, W.D and Scifford, L.M. (1982) R&D project selection in a multidimensional environment: A practical approach, *J. Oper. Res. Soc.*, **33**, 397–405.

De Young, H.G. (1992) Making R&D pay off better and quicker, *Electronic Business*, December, pp. 61–4.

Dixit, A.K. and Pindyck, R.S. (1995) The options approach to capital investment, *Harvard Business Review*, May–June, pp. 105–15.

Foster, R.N. (1982) Boosting the payoff from R&D, *Res. Manag.*, January, pp. 22–7.

Foster, R.N. (1985) Improving the return on R&D, *Res. Manag.*, January–February and September–October.

Hall, D.L. and Nauda, A. (1990) An interactive approach for selecting R&D projects, *IEEE Transactions on Engineering Management*, May, pp. 126–33.

Hall, G. *et al.* (1993) How to make reengineering really work, *Harvard Business Review*, November–December, pp. 119–31.

Krawice, F. (1984) Evaluating and selecting research projects by scoring, *Res. Manag.*, March, pp. 21–5.

Kuwahara, Y. and Tekeda, Y (1990) A managerial approach to research and development cost effectiveness evaluation, *IEEE Transaction on Engineering Management*, May, pp. 134–8.

Venkatesan, R. (1992) Strategic sourcing: to make or not to make, *Harvard Business Review*, November–December, pp. 98–107.

Part V

Project and operations management

13

Project planning and control

Introduction

This chapter introduces the concepts of project management within industry and the techniques, such as activity networks and Gantt charts, that are used to control medium- and large-sized projects.

13.1 Projects and management

Projects can vary in size and complexity. For example, a project could involve the development of a new product with costs exceeding several million pounds, or it could be a project for setting up a stall in a local exhibition (Archibald, 1995; Shtub, 1994). The common factor in all projects is a clearly defined aim and a start and a finish.

Most people have been involved in projects, either formal or informal, and either as the manager or organizer of the project or as a team member. It is the responsibility of the manager of the project to ensure that the project is completed within the planned time, to a given cost and meeting the project specification and quality requirements (Cooper, 1991).

Figure 13.1 shows the typical steps in the formulation of a project (Marion and Riddleberger, 1991; Siell, 1991; Squires, 1994). The project is first defined. This includes the overall aims of the project, and could include a detailed specification that the product has to meet, such as the technical and commercial specification for a new product development. In addition, an initial study into the feasibility of the project will be undertaken, to ensure that it can be completed.

The tasks that the project needs to go through in order to meet its aims are next defined. These may be major steps and could constitute subprojects, all linking into the main project. All dependencies and risks (Mustafa, 1991) need to be clearly defined and understood, and will include those involving outside

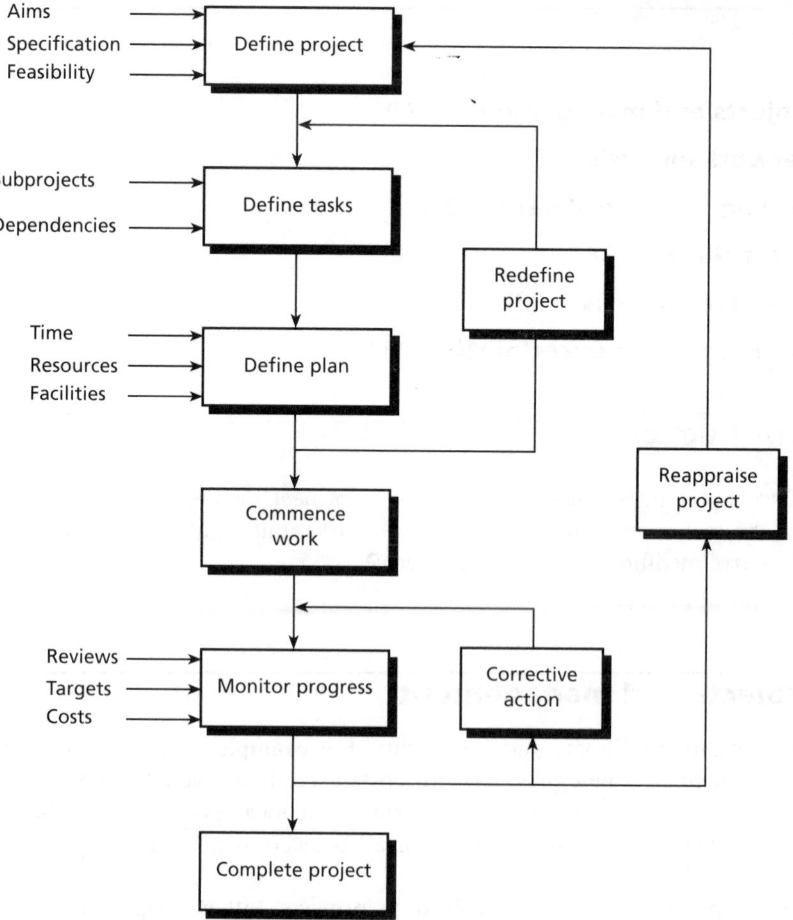

Figure 13.1 The steps in a typical project.

suppliers and even the customer. For example, the customer may have asked to be in the approvals loop for the technical specification, and therefore one of the dependencies is timely approval of this specification by the customer.

Once the tasks have been defined plans need to be put into place to achieve these tasks. The plan, when completed, will show the time needed to complete the project and the resources required. Resources include staff, skills, materials and capital, such as space and tools. Although the project manager will be the custodian of the project plan it must be done in conjunction with all the functions involved in the project if it is to gain credibility and buy-in from project members. (See the case study 'The lonely planner'.)

The plans may not meet the original project definition, for example regarding costs or time-scales. The project may therefore need to be redefined, as shown in Figure 13.1, and the tasks again defined to meet the modified project goals.

Having determined the plans the work can commence, with many tasks carried out concurrently (Evans, 1991; Markowitz, 1991; Noore and Lawson, 1992; Stewart, 1993; Turino, 1992). The job of the project manager is now one of monitoring progress. Reviews can be organized with functions and targets, and costs monitored. Results must be fed back to functional units. The project manager ensures that communication flows freely within the project team. Most of the major decisions will be taken within the functions, since they have the best knowledge and are closest to the work, but the project manager must ensure that decisions are being taken at the right time and within the correct framework.

In monitoring project costs and achievements it is important to obtain the correct information from the data. For example, Figure 13.2(a) shows that the project costs are actually below budget at time t_1 when the project is monitored. This is clearly good if target achievement is on or above schedule. However, Figure 13.2(b) shows that target achievement is behind plan and, as seen from Figure 13.2(c), the cost per target is higher than originally planned, so the project is running above its cost plan.

The project manager will aim to be proactive, anticipating problems and taking corrective action to prevent problems from occurring. However, in some instances the problems encountered may be major, such as the technical feasibility of certain tasks or the availability of key resources, which may require a major reappraisal of the project. Therefore the project can go round in a loop, as in Figure 13.1. It is important that the project manager anticipates these rework cycles in the plans (Cooper, 1993).

It is planned that the project will be successfully completed. However, projects fail for a variety of reasons (Pinto and Mantel, 1990) and in these instances the project manager must ensure an orderly close down. This may be an immediate cut, to move resources on to other more urgent projects, although it normally takes the form of a wind-down, where the results of the project are archived for future use. Staff morale must be kept high in these instances and the obvious incentive is a move to more interesting work.

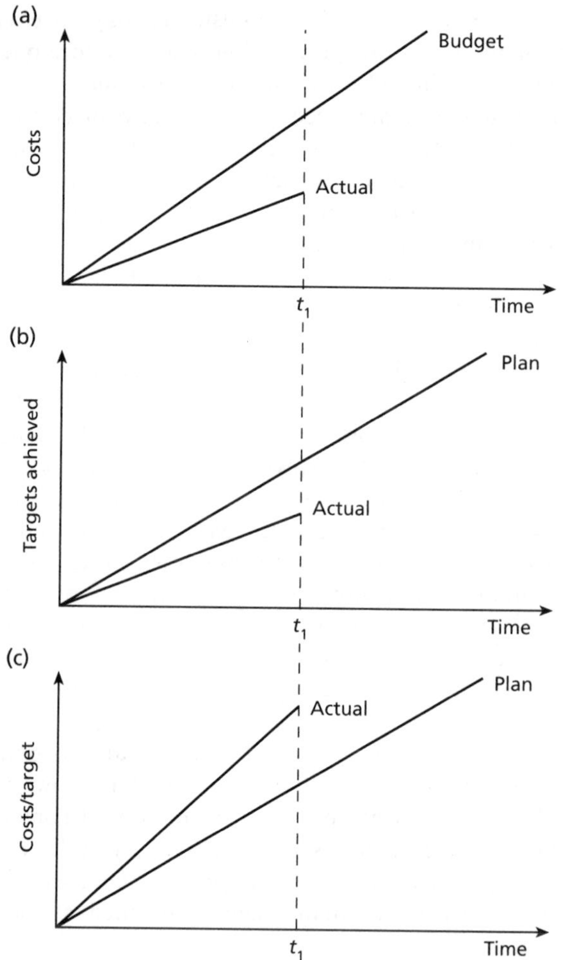

Figure 13.2 Measures of (a) costs and (b) targets achieved; (c) shows the costs per target.

Figure 13.3 illustrates a typical project organization within a company. The project office interfaces with the customer's project office and provides the necessary interface for functions within the company.

The main task of a project manager is to coordinate the various functions working on the project to meet a common goal. The project manager is a conductor; the functions working on the project make up the orchestra. The project manager must show considerable skill in influencing and directing the resources, which are not directly controlled.

The main skills required from a project manager are:

- Organizational and planning skills, to set the plan for the project and to control its development.

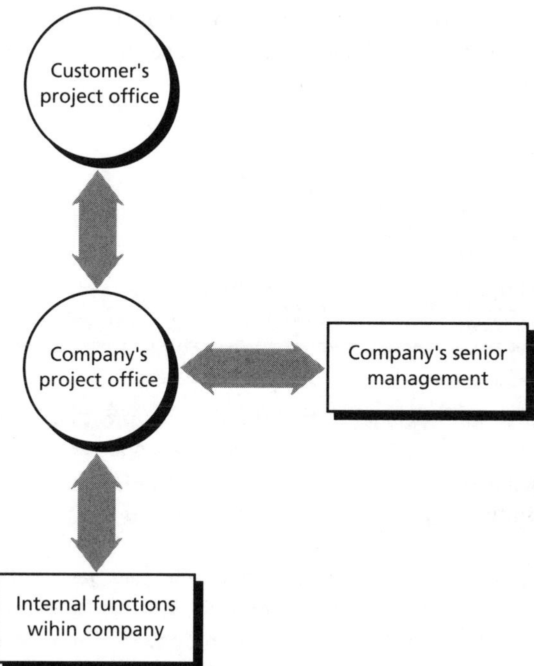

Figure 13.3 Typical project organization

- People management skills, to be able to influence the people working on the project. Usually the project manager will be the team leader, who leads by strength of personality rather than organization position.
- Ability to follow problems down to a level of detail, while at the same time standing apart so as to see the whole picture. Often a project manager will be able to guide the functions, who are working close to the problem, because of this remote view. The project manager must also show considerable imagination and be able to think laterally around problems.
- Communication skills to ensure that everyone on the project is briefed and is communicating with others. The project manager must be able to communicate at several levels, with project staff, senior directors within the company, and with customers.
- Drive and energy, with an optimistic outlook, to push the project along in spite of the many problems that are likely to arise.
- Goal orientation and customer focus. The project manager must have a very strong will to succeed and must continually keep the overall project aims in view.
- Good technical knowledge covering many disciplines. Although project managers do not need to be specialists it is important that they are able to communicate with the functions involved on their projects in their own language.

- Change orientation. Every project introduces change and the project manger must be an advocate of change and be able to introduce this successfully into the organization (Barnes, 1993).

Case study

The lonely planner

In its early days the FloRoll Manufacturing Company did not have a project organization. This was before David Champion joined the company as its first Programme Director and set up a highly successful project organization within the company.

Peter Kean, who was Engineering Manager at that time, came up with a modified design for florolls which would reduce their costs by using lower priced materials. Because of time pressures he decided to introduced this immediately into Manufacturing without consulting anyone. He told his staff to modify all the drawings accordingly.

'I don't think we have done enough tests on the new material', said David Fuller, his mechanical design manager. 'The bend radius looks a bit tight for the new plating.'

'We have tried all that out', said Peter Kean, 'and we don't have any more time to waste. Put the drawings out to Manufacturing and let them get on with it.'

David Fuller did as he was told, but privately confided to Jimmy Excel, the Manufacturing Engineer. 'We haven't done a lot of testing on this design, but the Old Man seems determined to become a hero. Let's hope he is right.'

After a few days of experimenting Jimmy Excel was convinced that the new design was faulty. 'I don't think it can be made in our present machines without a fairly major modification to the whole structure', he told David Fuller.

'Well, ours is not to question why', replied David Fuller. 'It's your problem now Jimmy old boy. Best of luck!'

The project failed. David Fuller and Jimmy Excel had expected this and were not surprised. They were not aware of the fact that they were the prime cause of its failure; their skills were needed to overcome the obvious problems which the project had. But it was not their project; it belonged to Peter Kean.

13.2 Network analysis

A small project can usually be planned and controlled with very few aids. Larger projects need more elaborate techniques to make optimum use of the resources available and to ensure that project objectives are met (Wheelwright and Clark, 1992). However, project management methodologies and computer-based programmes are no substitute for good project management skills (Farish, 1994).

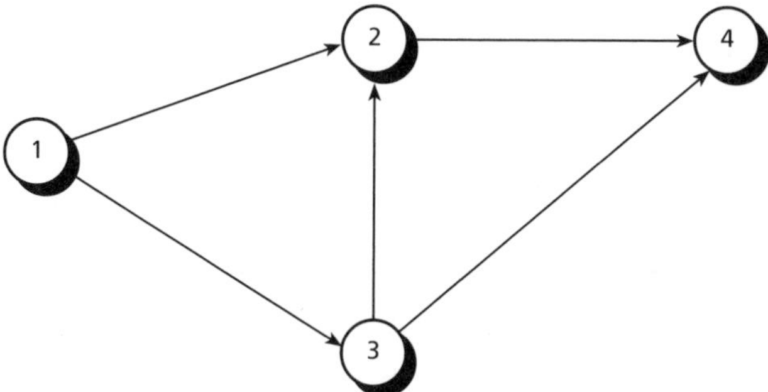

Figure 13.4 A simple activity network

The control technique described in this section is often known as network analysis, the two main methods within this being Programme Evaluation and Review Technique (PERT) and Critical Path Analysis (CPA). Although these techniques had minor differences when they were first developed, they now provide the same facilities, and the terms are often used interchangeably.

There are many different types of network, but they all have points, or nodes, which are connected together by lines called routes, arcs or edges. A simple network is shown in Figure 13.4, in which the nodes have been numbered and the arc can be referenced by the nodes at its two ends. An activity network is not drawn to scale, and the lengths of the lines or sizes of the circles are not significant.

The nodes within an activity network are often referred to as events and they are the instances where something happens, such as an order being placed, a design review completed or a product shipped to a customer. The arcs are referred to as activities, and they are the elements on a network that take time and use resources, for example the process of raising paperwork for an order, doing the design for a product, or producing product ready for shipment to a customer. An activity network shows the sequential and parallel work, with their relationships, of all the tasks that have to be done from start to finish of a project; it is a project map.

The major steps in creating an activity network for a project are:

1. The objectives of the project must first be clearly defined and understood by everyone working on the project or affected by the outcome of the project.
2. The project is broken down into major work areas or groups. For a large project it is important to ensure that each sub-project is of manageable size. The relationships between these work areas must be clearly defined.
3. The tasks, or activities, needed to complete each sub-project are defined. The linkages between each task are clearly specified, including all dependencies. The order in which the tasks are to be undertaken is also defined, for example whether tasks must be done sequentially or whether they can be done in parallel.

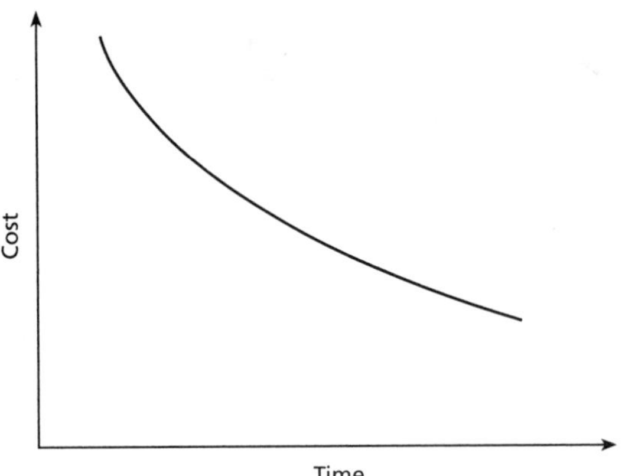

Figure 13.5 The exponential relationship between project cost and time.

4. The time needed to complete each task is estimated, along with the resources needed. There is often a trade-off, such that the time can be shortened if the resources on the task are increased. However, the relationship is not linear. Beyond a certain point the resources needed to reduce project time rise exponentially, resulting in an exponential increase in costs, as shown in Figure 13.5. This also shows that if the resources are below a critical mass the time factor may also increase unduly.
5. The activity network can be drawn, knowing the tasks and the relationship between tasks. The time and resource information is used to analyze the activity network, as described later.

The event at the start of an activity is called the tail event and the event at the end of the activity is the head event. Therefore in the network of Figure 13.4, event 1 is the tail event and event 2 the head event for the activity linking events 1 and 2. The tail and head events are often referred to as the i and j events respectively, and the activity between them is labelled A_{ij}.

Some important laws need to be observed when drawing and analyzing an activity network:

- The network must not contain loops, as shown between events 3, 4 and 5 in Figure 13.6(a). Loops are a repetition of activity sequences and they will be repeated endlessly, never leading to the final event.
- The network should not have any 'dangles'. These are events that go nowhere, as for event 3 in Figure 13.6(b). Dangles can be avoided by the use of dummy activities, as shown by the broken line.
- In an activity network an event cannot be completed until all the activities leading to it have been completed.
- In an activity network an activity cannot start until its tail event has been completed.

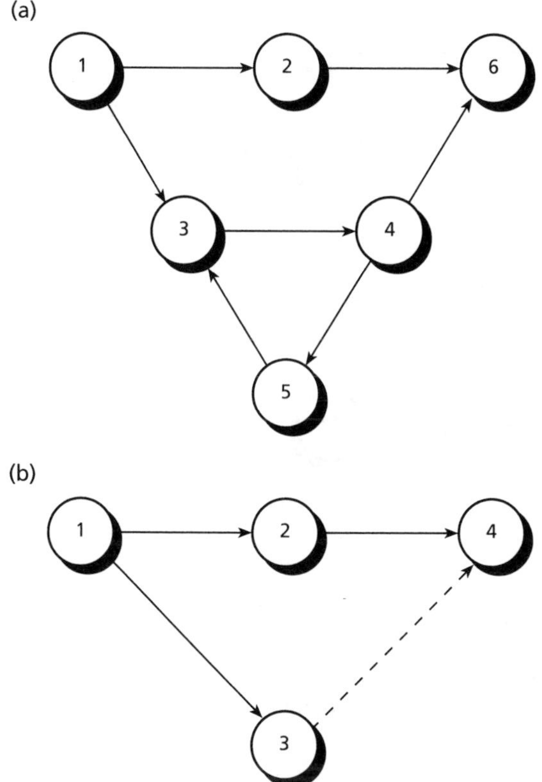

Figure 13.6 Examples of faulty network structures: (a) a loop; (b) a dangle.

Therefore for the network of Figure 13.4, event 2 cannot be completed until activities A_{12} and A_{32} have been completed and this prevents activity A_{24} from starting. Note that in any network the events do not need to be numbered sequentially and it is also not required that all the numbers are used.

All the activities described so far represent an actual action that contributes to the final event. It is often necessary to use activities that do not represent actual work done and do not use up any resources or take any time to complete. These activities are called dummy activities and they are primarily used for three reasons:

1. To avoid parallel activities with the same beginning and end events. This is illustrated in Figure 13.7(a), where activities A and B go between the same events but have different work contents. Figure 13.7(b) shows the use of a dummy activity, indicated by the broken line, going from a new event 14, to overcome this problem.
2. To show one-way dependencies. For example, in the network of Figure 13.8(a) suppose that activity C cannot start until activity A has been completed, but the start of activity D depends on both activities A and B

(a)

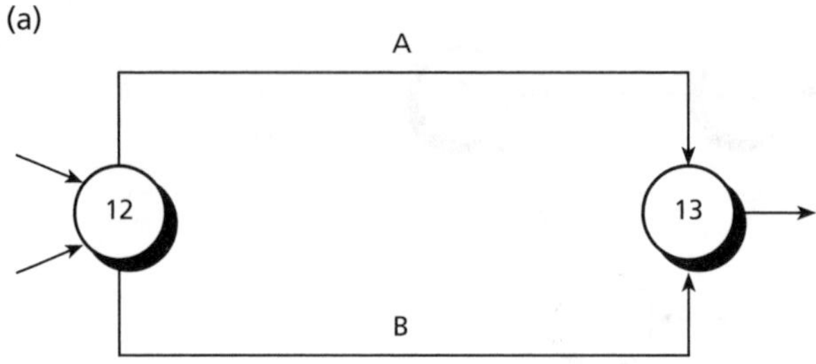

(b)

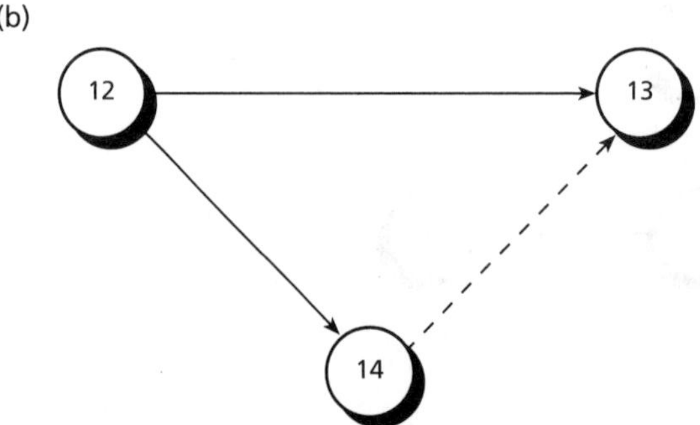

Figure 13.7 Using dummy activities to avoid parallel activities: (a) parallel activity; (b) dummy activity.

having being completed. The representation in Figure 13.8(a) is clearly not correct, since it indicates that both activities C and D are constrained by activities A and B. Figure 13.8(b) shows how this problem can be overcome by using a dummy activity and a new event 21. Now activity C can start as soon as activity A has been completed and event 20 reached, although activity D must wait until both activities A and B have been completed.

3. To introduce a constraint, as shown in Figure 13.9, where activity from node 6 has been constrained by the dummy activity A_{76}, shown dotted, and it cannot start until this dummy activity has been completed along with activity A_{56}.

Many computer-based project planning programmes exist, which enable large activity networks to be drawn and analyzed. The use of these networks has several advantages:

- They force management to think logically about the objectives and tasks of a project. The tedious work of completing the network is then done by a computer.
- They show critical activities and enable management to focus on these.
- The process of creating an activity network requires a multi-departmental input and this builds up a better understanding of interdepartmental tasks and encourages teamwork.

(a)

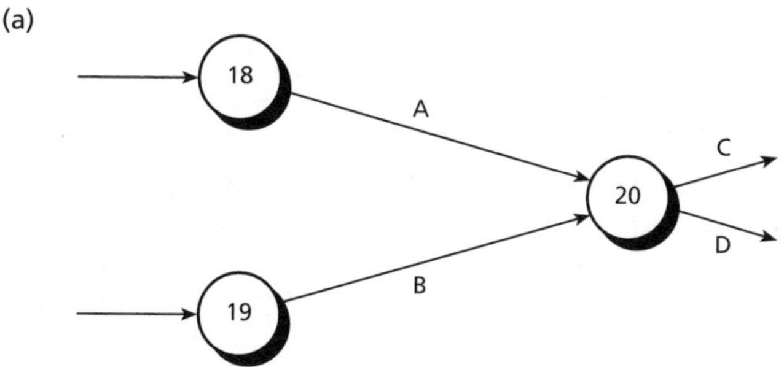

(b)

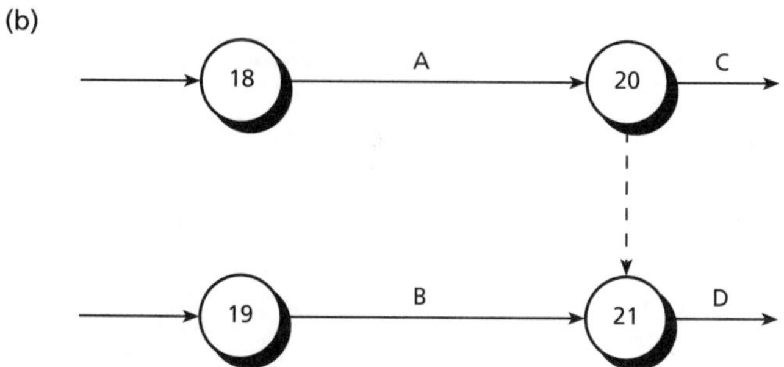

Figure 13.8 Using a dummy activity to show one-way dependencies: (a) problem; (b) dummy activity representation.

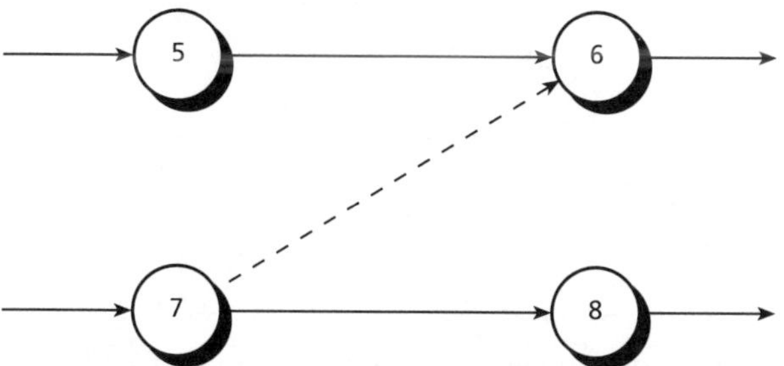

Figure 13.9 Use of a dummy activity to introduce a constraint.

- The activity network enables 'what-if' analysis to be done, in order to study the impact of changes.
- It provides an ongoing project record. It gives a method for progressing tasks and determining the effects of slips as they occur and of plan changes.

The case study 'The floroll network', shows how a simple activity network can be used to plan a project.

Case study

The floroll network

The FloRoll Manufacturing Company is developing a new finishing paint for florolls and the Project Manager, Patricia (Pat) Doer, has prepared a plan for the project in conjunction with the various functions within the organization, as shown in Figure 13.10. The activities include buying special test equipment to do laboratory experiments on the new finish, hiring and training staff to use the equipment, and modifying existing production plant ready to manufacture the new finish. The numbers on the activities indicate the times needed to complete the tasks and they are in months.

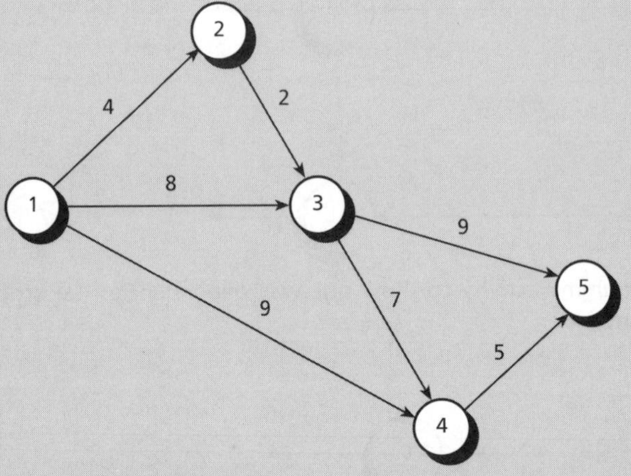

A_{12} Hire staff
A_{23} Train new staff
A_{13} Order and receive test equipment
A_{14} Development new finish
A_{34} Install test equipment
A_{35} Modify production equipment
A_{45} Test new finish and bring to production stage

Figure 13.10 Network for the development of a new finish for florolls (see the case study 'The floroll network').

Having arrived at the network logic the first analysis that Pat does is to establish the earliest and latest times for each event. It is assumed that event 1 starts at time 0. The earliest event 2 can start is therefore 4 months later. Event 3 can start after activities A_{12}, A_{23} and A_{13} have been completed and the earliest that this will occur is 8 months after event 1. In a similar manner the earliest event times of events 4 and 5 are found. The earliest time of event 5 is also the total project time.

To obtain the latest project time Pat works backwards through the network, starting with event 5. The latest event time of event 5 is 20 weeks and this time can be obtained so long as event 4 is reached five months earlier, which is the duration time of activity A_{45}. Now the latest event time of event 3 is (20 – 9) if activity A_{35} is considered and it is (15 – 7) if activity A_{34} is considered. If the larger figure of 11 is taken then it will mean that event 4 will not be reached until month 18, which means that event 5 is reached in month 23, giving a slip in the project time of 3 months. Therefore the smaller time of 8 is taken as the latest event time for event 3. This gives latest times of 6 and 0 for events 2 and 1 respectively. Patricia Doer summarizes the earliest and latest times as in Table 13.1, so that it can be used later for further analysis. She also redraws her network of Figure 13.10 to show the earliest and latest event times, as in Figure 13.11.

Table 13.1 Earliest and latest times for the case study 'The floroll network'

Event	**Event time**	
	Earliest	**Latest**
1	0	0
2	4	6
3	8	8
4	15	15
5	20	20

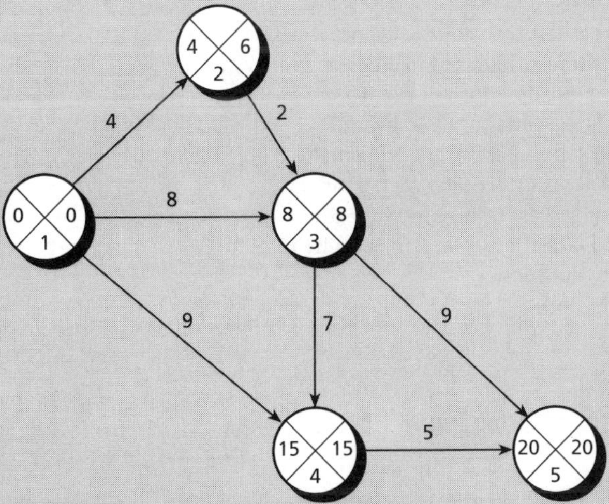

Figure 13.11 The activity network of Figure 13.10 redrawn with earliest and latest event times added.

13.3 Finding the critical path

A project manager needs to determine the critical activities within any project. These are the tasks which, if they slip, will affect the overall project time-scale. This is usually done by determining the slack within project activities.

Table 13.2 defines some of the parameters used to describe activities, and Table 13.3 gives the formulae used to calculate some of the network parameters. The difference between the earliest and latest times represents the freedom which exists in delaying or spreading out the programme without affecting the overall project time. This is known as slack, and the slack in the start and finish times are the same. The critical path is that which has zero slack, since a delay on this activity will mean a delay to the overall project.

It is important to note the difference between an event time and an activity time. The earliest start time of an activity is the same as the earliest time of its tail event, and the latest finish time of an activity is the same as the latest time of its head event. But it is only on the critical path that the latest start time of an activity is equal to the latest time of its tail event and the earliest finish time of the activity is equal to the earliest time of its head event.

The case study 'What's critical?' illustrates the process used to calculate the critical path.

Table 13.2 Parameters used to describe activities

Parameter	**Description**
EE_i	Event earliest time for tail event
EE_j	Event earliest time for head event
EL_i	Event latest time for tail event
EL_j	Event latest time for head event
AES_{ij}	Activity earliest start time for activity ij
ALS_{ij}	Activity latest start time for activity ij
AEF_{ij}	Activity earliest finish time for activity ij
ALF_{ij}	Activity latest finish time for activity ij
AD	Activity duration
AS	Activity slack

Table 13.3 Formulae for determining activity parameters

Parameter	**Formula**
AES_{ij}	EE_i
AEF_{ij}	$AES_{ij} + AD$
ALF_{ij}	EL_j
ALS_{ij}	$ALF_{ij} - AD$
AS	$ALS_{ij} - AES_{ij}$ or $ALF_{ij} - AEF_{ij}$

Case study

What's critical?

Having drawn the activity network, as in Figure 13.11, Patricia Doer, Project Manager within the FloRoll Manufacturing Company, decided to determine the activities that were on the critical path, so that these could be closely monitored. Table 13.4 shows the parameters obtained by using the formulae given in Table 13.3.

Table 13.4 Calculation of activity slack for the network of Figure 13.11

Activity	Duration	Start time		Finish time		Slack
		Earliest	Latest	Earliest	Latest	
(1)	(2)	(3)	(4)	(5)	(6)	(7)
A_{12}	4	0	2	4	6	2
A_{23}	2	4	6	6	8	2
A_{13}	8	0	0	8	8	0
A_{14}	9	0	6	9	15	6
A_{34}	7	8	8	15	15	0
A_{35}	9	8	11	17	20	3
A_{45}	5	15	15	20	20	0

The first column of Table 13.4 lists the activities in no special order and the second column gives their duration.

Patricia then calculated the third column, which is the earliest possible time that an activity can start, and this is given by the earliest time of the tail event. For example, the earliest time that activity A_{34} can start is the earliest time that event 3 is reached, which is 8 months after event 1.

The earliest finish times were then calculated, as in column five. This is the earliest possible time that an activity can finish and is equal to the earliest start time of the activity plus the duration of the activity. So for event A_{23} it is $4 + 2 = 6$. The latest finish time, given in the sixth column, is found as the latest time of the head event of that activity. Therefore for activity A_{12} it is the latest time of event 2, that is 6 months after event 1.

The latest start times, shown in the fourth column, were then calculated. These are the latest possible times that activities can start and are equal to the latest finish times minus the activity durations. Therefore for activity A_{35} it is equal to $20 - 9$ or 11 months after event 1.

The difference between the earliest and latest times represents the slack, and this is shown in Table 13.4. The critical path is that which links the activities of zero slack and Patricia saw from the table that this was activities 1–3– 4–5.

13.3.1 Project float

It can be seen from Table 13.4 that the earliest time that activity A_{23} can start is 4 months after event 1 and the latest time that it can finish is 8 months after event 1. Therefore the maximum time available to complete A_{23} is 4 months. However, the duration of A_{23} is only 2 months so that it is possible to move the activity or expand it by two months, without affecting the critical path. This time of two months is called the total float of activity A_{23}.

Activity A_{12} has a total available time of 6 months and a duration of 4 months so that it also has a total float of 2 months. However, if this activity takes up all its float of 2 months, for example by reducing the resources on A_{12} so that its duration increases to 6 months, then the earliest time of event 2 will become 6 and this will give zero float in activity A_{23}. Float that affects subsequent activities, if taken up, is called interference float. Therefore activity A_{12} has a total float of 2 months and all of this is interference float. If a float does not affect subsequent activities it is called free float.

In some cases, where the activity duration is changed so as to take up some of its float, it can affect the float of earlier stages, although it does not affect subsequent stages. If absorption of the float does not affect earlier or later activities, the float is referred to as independent float. For network analysis it is usual to consider total float (TF), free float (FF), which does not affect subsequent activities, and independent float (IF), which does not affect previous or subsequent activities.

The slack in an event is the difference between its earliest and latest times. The free and independent floats can be calculated from the slack using the relationships of Equations (13.1) and (13.2).

$$\text{Free float} = \text{Total float} - \text{head slack} \tag{13.1}$$

$$\text{Independent float} = \text{Free float} - \text{tail slack} \tag{13.2}$$

When assessing a network it should be remembered that although float and slack often represent a case of underutilized resources, this is not always a bad thing, since if all the resources were used to the maximum then all the paths of the network would become critical and project time-scales could be at risk.

13.4 Gantt charts

Gantt charts, known more commonly as bar charts, were developed by Henry Gantt during the First World War. In this method of planning, the activities, which are based on the work breakdown structure for the project, are drawn as bars across a time map. There are many variations of the Gantt chart, an example being shown in Figure 13.12 for five time periods. It is seen that Activity 1 is planned to start in period 1 and extend for two periods, to time period 3. Activity 2 also starts at period 1 and runs parallel to Activity 1 for one time

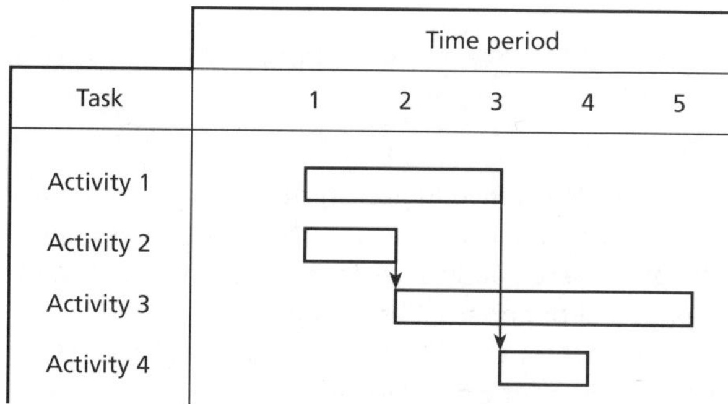

Figure 13.12 A Gantt chart.

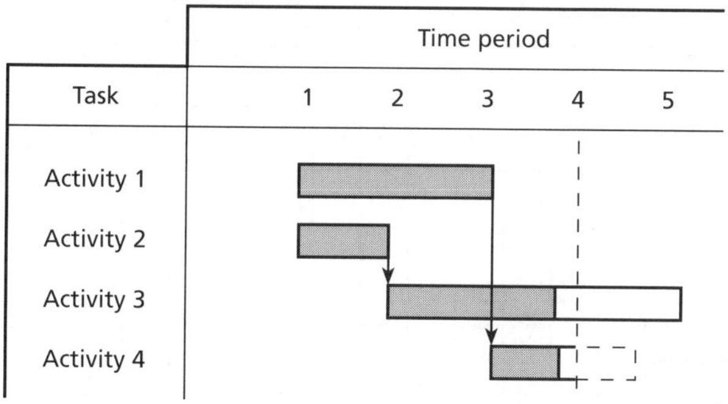

Figure 13.13 Progressing the Gantt chart of Figure 13.12.

period. Activity 3 starts when Activity 2 completes and runs for three time periods. Activity 4 starts when Activity 1 completes and runs for one time period.

Figure 13.13 shows one method for progressing the tasks of Figure 13.12 in time period 4. The shaded areas indicate task completion, while a dashed bar indicates the new completion plan. Therefore Activity 1 completed on time in period 3. Activity 2 completed on time and so allowed Activity 3 to start on time. However, Activity 3 has not reached its level of completion in period 4, although it is expected to catch up during periods 4 and 5, so no slip in the overall activity is expected. Activity 4 started on time, after Activity 1 completed, but it has also not reached its level of completion at time period 4. A slip in this activity is expected as shown by the dashed extension to the activity bar.

Gantt charts can be used to determine the critical path, as illustrated in the case study 'Charting it'. They are also easy to understand and provide a visual presentation of the length of tasks and their interrelationships, and of project slips. Gantt charts are very popularly used, since minimal training is required to draw a bar chart. However, they are not suitable for large projects, since the charts can then become very cumbersome.

Case study

Charting it

Patricia Doer was preparing for a project meeting at which she intended to present her plans to the project team. It was important that she obtained buy-in from all team members and that the plans were clearly understood. Since it was a relatively straightforward project she decided to present it as a Gantt chart.

Figure 13.14 shows the Gantt chart for the activity network of Figure 13.11. To draw the Gantt chart Patricia first listed the activities in increasing order of head numbers, and where the activities have the same head number they were listed in increasing order of tail numbers. The bar corresponding to each activity was then drawn, starting with event 1.

Time / Activity	Months																			
	1	2	3	4	5	6	7	8	9	10	11	12	13	14	15	16	17	18	19	20
A_{12}	1			2																
A_{13}	1							3												
A_{23}					2	3														
A_{14}	1								4											
A_{34}									3						4					
A_{35}									3								5			
A_{45}																4				5

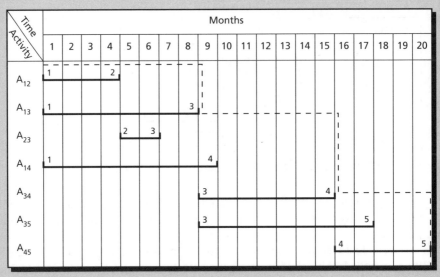

Figure 13.14 The Gantt chart for the case study 'Charting it'.

The tail number of each activity, apart from those starting from event 1, was matched to the farthest right point of the same number. For example, for activity A_{23} the tail (2) starts at the same month as the head of activity A_{12}. Any dummy activities are shown as single uprights, since they take up zero time.

Having completed the bar chart Patricia found the critical path by starting with the farthest right-hand point and then drawing a vertical line to meet the activity whose head number is the same as the tail number of the first activity. Therefore a line was drawn from A_{45} to A_{34} and in this manner the dashed line shown in Figure 13.14 was obtained.

The critical path was given by the activities that touch this dashed line, and it lies along A_{13}, A_{34} and A_{45}.

Patricia determined the float of activities from the Gantt chart, again starting at the bottom of the chart. Activity A_{45} cannot move since it is bounded by the critical path and by A_{34}, which is itself on the critical path. A_{35} can move three months, and in doing so it does not affect preceding or succeeding activities, so that it has an independent float of 3.

Activity A_{14} has a float of 6 months, and again this is independent float. A_{23} has a total float of 2 up to the start of A_{34}, since any more float will delay the critical activity. In floating by two months A_{23} does not affect subsequent activities, but it does affect A_{12} so that A_{23} has a free float of 2 months. Activity A_{12} has a float of 2 months, but then A_{23} is also delayed by 2 months, and this later activity loses its float. Therefore A_{12} has a total float of 2 but no free float. Patricia summarized her findings, as in Table 13.5.

Table 13.5 Calculation of float from Figure 13.14

Activity	Duration	Start		Finish		Float		
		E	L	E	L	T	F	I
A_{12}	4	0	2	4	6	2	0	0
A_{23}	2	4	6	6	8	2	2	0
A_{13}	8	0	0	8	8	0	0	0
A_{14}	9	0	6	9	15	6	6	6
A_{34}	7	8	8	15	15	0	0	0
A_{35}	9	8	11	17	20	3	3	3
A_{45}	5	15	15	20	20	0	0	0

13.5 Resource analysis

In the discussions so far, time has been considered to be the most important parameter. Projects are also restrained by the limitations of other types of resource, such as people (often referred to as men or heads) and machines. It is important to be able to analyze a network to obtain information on the best method for allocating these resources to activities. This is known as resource allocation or loading.

It is usual to specify the work content of any activity in terms of the resources required and not the work to be done. For example, if an activity involved designing a widget it would be meaningless to specify the resources as one widget-design. If people are the resource the activity will be specified as needing 6 person months, that is one person working 6 months or 2 people working 3 months each, and so on. Clearly this cannot be taken to its limits, and sometimes activities are bounded by time. For example, it is unlikely that 24 people could design the widget in 1 week.

In optimizing resources the problem often arises of deciding which resource should be optimized, for example people, machines or raw materials. Often there are also a very large number of alternative ways of doing something, so that even with a computer it is not always possible to try them all and so arrive at the optimum solution. Once an acceptable solution is reached it is adopted.

Project times can be reduced by increasing the resources for it or, alternatively, the costs expended on a project can be reduced by allowing the time to increase. Generally a balancing act is needed where money is saved on non-critical activities, and this is used to reduce the time of activities on the critical path.

The indirect costs associated with a project generally decrease as the activity duration decreases, but the direct costs increase. There is therefore a minimum cost point, and this is the normal cost of the activity. The activity duration can be reduced by increasing the resources, but the costs increase exponentially as the project time becomes small and one is working in a crash time period. This is because activity duration is reduced by decreasing the time spent on the critical activities, and as the crash situation is reached the project obtains more and more parallel critical paths. The cost slope of an activity is given by Equation (13.3). This is a measure of the cost involved in reducing the activity time.

$$\text{Cost slope} = \frac{\text{Crash cost} - \text{Normal cost}}{\text{Normal time} - \text{Crash time}} \tag{13.3}$$

A histogram represents a very effective method of representing the loading on a project. For example, suppose that there are four people within a department and during the next six months the expected loading is as given in Table 13.6.

Figure 13.15 shows the loading histogram from Table 13.6 and this shows that the workload exceeds the capacity during two months and that resources are underutilized for three months. The project manager can now try to smooth out the loading to match the available resources.

The best way to draw a loading histogram from a network is first to draw a Gantt chart and then run down each time slot and add up the resources used.

Table 13.6 Example of monthly loading in a department

Month	**Load (person months)**
1	2
2	6
3	3
4	4
5	6
6	3

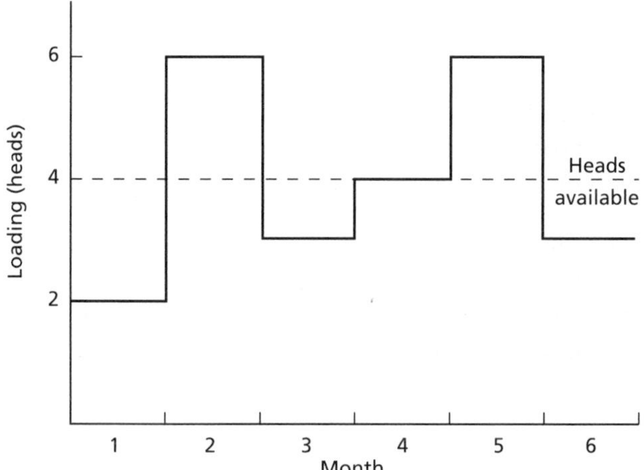

Figure 13.15 Histogram for the loading of Table 13.6.

13.6 Planning under uncertainty

When preparing plans it is often necessary to work on estimates caused by uncertainties in the estimation process. For example, the technology may be new and unpredictable; the tools that are needed for the project may be new or still under development; the members of the project team may have little previous experience of the environment to be used; the success of any recruitment drive may be uncertain and there may also be a risk of a high number of leavers during the project life.

Uncertainties mean that estimates given for time and resources for an activity can be in error, and therefore the planned project time-scale and costs are at risk. It is important that the project manager know the risks involved. Usually, therefore, functions are asked to provide three figures for each activity: the most likely estimate (T_m), the shortest or most optimistic estimate (T_s) and the longest or most pessimistic estimate (T_l).

If the activity estimates are assumed to be a random variable with a type of distribution known as a beta distribution (the most usual assumption) then the mean (M) and the standard deviation (S) are given by Equations (13.4) and (13.5).

$$M = \frac{T_s + 4 \times T_m + T_l}{6} \tag{13.4}$$

$$S = \frac{T_l - T_s}{6} \tag{13.5}$$

If M_1, M_2, M_3 are the mean times and S_1, S_2, S_3, are the standard deviations of the activities A_{13}, A_{23}, A_{34}, then for the path 1–2–3–4 the overall mean time and standard deviation are given by Equations (13.6) and (13.7).

$$M = M_1 + M_2 + M_3 \tag{13.6}$$

$$S = \left(S_1^2 + S_2^2 + S_3^2\right)^{1/2} \tag{13.7}$$

If a path contains many activities, as would be the case in a large project, then it can be assumed that the parameters for the path are as for a normally distributed curve, and the area under this curve, which gives probability factors, can be found from tables, as in Table 13.7.

When activity times are uncertain it is important to remember that it is no longer sufficient to monitor activities along the critical path alone, since other paths may have a lower probability of being completed within the critical time. (See the case study 'An uncertain future'.)

Table 13.7 The area under a normal curve

ω	**0.00**	**0.02**	**0.04**	**0.06**	**0.08**
0.0	0.500	0.508	0.516	0.524	0.532
0.1	0.540	0.548	0.556	0.564	0.571
0.2	0.579	0.587	0.595	0.603	0.610
0.3	0.618	0.626	0.633	0.640	0.648
0.4	0.655	0.663	0.670	0.677	0.684
0.5	0.692	0.700	0.705	0.712	0.719
0.6	0.726	0.732	0.739	0.745	0.752
0.7	0.758	0.764	0.770	0.776	0.782
0.8	0.788	0.794	0.800	0.805	0.811
0.9	0.816	0.821	0.826	0.832	0.837
1.0	0.841	0.846	0.851	0.855	0.860
1.1	0.864	0.869	0.873	0.877	0.881
1.2	0.885	0.889	0.893	0.896	0.900
1.3	0.903	0.907	0.910	0.913	0.916
1.4	0.919	0.922	0.925	0.928	0.931
1.5	0.933	0.936	0.938	0.941	0.943
1.6	0.945	0.947	0.950	0.952	0.954
1.7	0.955	0.957	0.959	0.961	0.963
1.8	0.964	0.966	0.967	0.969	0.970
1.9	0.971	0.973	0.974	0.975	0.976
2.0	0.977	0.978	0.979	0.980	0.981
2.1	0.982	0.983	0.984	0.985	0.985
2.2	0.986	0.987	0.988	0.988	0.989
2.3	0.989	0.990	0.990	0.991	0.991
2.4	0.992	0.992	0.993	0.993	0.993
2.5	0.994	0.994	0.995	0.995	0.995
2.6	0.995	0.996	0.996	0.996	0.996
2.7	0.997	0.997	0.997	0.997	0.997
2.8	0.997	0.998	0.998	0.998	0.998
2.9	0.998	0.998	0.998	0.998	0.999

Case study

An uncertain future

Patricia Doer, Project Manager for the FloRoll Manufacturing Company, wanted to determine the probability of a given overrun on her project, as calculated from the network drawing of Figure 13.11. She asked the functions to give her three estimates for each activity: most likely time, most optimistic time and most pessimistic time. From these she calculated the mean and standard deviation for each activity. The times on Figure 13.11 are the mean durations. The earliest and latest event times are found as before from the mean (expected) durations as in Table 13.4, but these are now the expected times.

From Table 13.4 Patricia found the path with zero slack as A_{13}, A_{34}, A_{45}, as before. She had calculated the standard deviation for each path as equal to half the activity duration values shown in Figure 13.11. Therefore for the critical path Patricia calculated the mean or expected completion time (M_c) and the standard deviation (S_c) using Equations (13.6) and (13.7) as in Equations (13.8) and (13.9).

$$M_c = 8 + 7 + 5 = 20 \qquad (13.8)$$

$$S_c = (4^2 + 3.5^2 + 2.5^2)^{1/2} = 5.87 \qquad (13.9)$$

Although the critical path only contained two activities Patricia made the assumption that these were normally distributed. She wished to know the probability that the project would be completed in 27 months. Patricia calculated the value of ω as in Equation (13.10).

$$\omega = \frac{27 - 20}{5.87} = 1.2 \qquad (13.10)$$

Therefore from Table 13.7 Patricia found that the probability of completing the project in 27 months is 0.89 or 89 per cent.

Patricia suspected that the standard deviations on some of the activities had been incorrectly estimated by the functions. For example, she though that for activity A_{35} this could be 8, so that for A_{13} and A_{35} the mean and standard deviation are as in Equations (13.11) and (13.12).

$$M = 8 + 9 = 17 \qquad (13.11)$$

$$S = (4^2 + 8^2)^{1/2} = 8.9 \qquad (13.12)$$

Therefore the value of ω is given as in Equation (13.13).

$$\omega = \frac{27 - 17}{8.9} = 1.1 \qquad (13.13)$$

The probability that this path is completed within 27 months is therefore (from Table 13.7) about 86 per cent. This is less than the probability of completing the critical path so that if the project time exceeds 27 months, the delay is more likely in path 1–3–5 than the critical path 1–3–4–5. Therefore Patricia decided to monitor all the paths on the project.

Summary

The project office represents the focal point for all medium- and large-sized projects, and it acts as the interface between the customers and the company's internal functions. The project manager's prime task is to coordinate the many activities that occur on a project. This includes monitoring progress and acting proactively to identify and resolve problems before they can have a major impact.

The most frequently used tool for managing large projects is network analysis, such as PERT (programme evaluation and review technique) or CPA (critical path analysis). In this technique the project is drawn as a series of actions, made up of nodes, representing end events, and lines joining these nodes, representing the activities to go from one event to the other. The resource, usually specified as time, needed for each activity is indicated on the drawing. Several rules exist for drawing up network diagrams, such as the prevention of loops and dangles.

The activity network can be analyzed to determine parameters such as the resources needed at each stage and the critical path: the activities which, if they slip, will affect the whole project. Float can exist on other activities not on the critical path, indicating that limited slip on these can be tolerated. Techniques also exist for planning under uncertainty, when the exact resources needed for each activity are not known or might change. Computers are used for analysis of large project activities, and software for this is available from several suppliers.

Gantt charts are also used for project planning. They are easy to use and give a visual picture of the activities, although they are not suitable for use with large or complex projects. In a Gantt chart each activity is represented by a bar or line, and these can be interlinked to indicate that the start of one activity is dependent on others finishing. Resources can also be indicated on a Gantt chart, and these can be totalled to give a histogram of the total resources needed at various phases of a programme.

Case study exercises

13.1 Time was clearly of vital importance if the company was to get the new contract. Marketing had made this abundantly clear, and Patricia Doer, Project Manager at the FloRoll Manufacturing Company, was wondering how to meet Marketing's requirements. The estimates for completion time, provided by the various functions, would result in the product being delivered six months later than the customer's requirements, and these estimates were based on the most optimistic dates. Patricia knew that if she did not meet the customer's dates the contract would be given to a competitor. Discuss what Patricia should do.

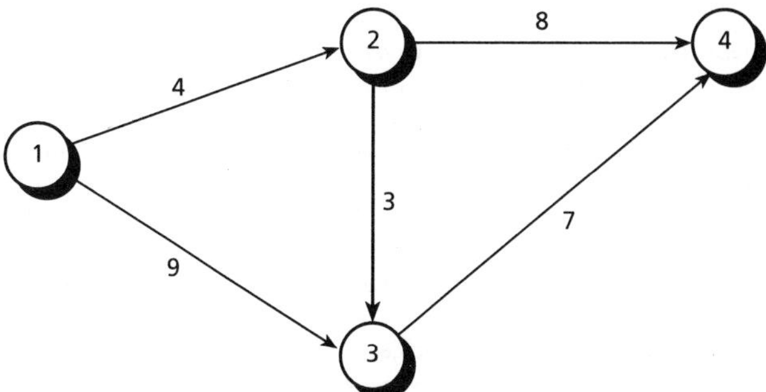

Figure 13.16 Part of an activity network for the case study of Exercise 13.2.

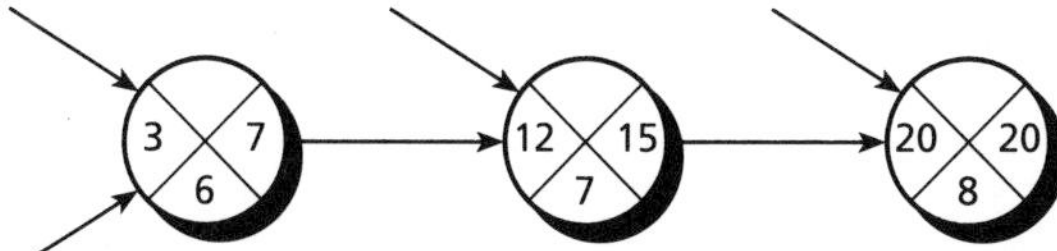

Figure 13.17 Part of an activity network for the case study of Exercise 13.3.

13.2 Figure 13.16 shows part of an activity network. Redraw the network with the earliest and latest times for each event clearly indicated. Draw up an activity table and from this determine the critical path.

13.3 In the network shown in Figure 13.17 find the total, free and interference floats for activity A_{67}.

13.4 Draw a Gantt chart for the network of Figure 13.16 and from this find the critical path.

References

Archibald, D. (1995) Projecting success, *Project Management*, January, pp. S-3–S-4.

Barnes, M. (1993) Project management: the way to get ahead, *Professional Manager*, March, pp. 16–18.

Cooper, K.G. (1993) The rework cycle: why projects are mismanaged, *EMR*, Fall, pp. 4–12.

Cooper, W.P. (1991) The project manager's tool kit, *Engineering Management Journal*, February, pp. 27–32.

Evans, S. (1991) Changing your way to a better business, *Engineering Computers*, May, pp. 12–17.

Farish, M. (1994) Avoid creating the illusion of control, *Engineering Computers*, May, pp. 12–14.

Marion, E.D. and Riddleberger, E.J. (1991) Modular project management, *AT&T Technical Journal*, March–April, pp. 49–62.

Markowitz, M.C. (1991) Concurrent engineering journey starts with the first step, *EDN*, July 18, pp. 110–14.

Mustafa, M. (1991) Project risk assessment using the analytic hierarchy process, *IEEE Transactions on Engineering Management*, February, pp. 46–52.

Noore, A. and Lawson, M. (1992) Electronic product design and manufacture in a concurrent engineering environment, *IEEE Transactions on Consumer Electronics,* August, pp. 666–70.

Pinto, J.K. and Mantel, S.J. (1990) The causes of project failure, *IEEE Transactions on Engineering Management*, November, pp. 269–76.

Shtub, A. *et al.* (1994) *Project Management – Engineering, Technology and Implementation*, Prentice-Hall, Englewood Cliffs NJ.

Sieli, E.M. (1991) Managing a project as a process, *AT&T Technical Journal*, March–April, pp. 33–9.

Squires, A. (1994) Milestones and tollgates on project management road, *Communications Networks*, July, pp. 38.

Stewart, D. (1993) Concurrent development puts projects on target, *Electronic Design*, May 13, pp. 61–6.

Turino, J. (1992) Concurrent engineering speeds development time, lowers costs, *EDN*, pp. 191–5.

Wheelwright, S.C. and Clark, K.B. (1992) Creating project plans to focus product development, *Harvard Business Review*, March–April, pp. 70–82.

14

Manufacturing operations

Introduction

This chapter introduces the concepts of Manufacturing Operations, such as the aims of manufacturing, the different manufacturing methods used, the experience curve, global operations, logistics, design for manufacture, quality within manufacturing operations, and information technology (IT) systems.

14.1 The manufacturing environment

Manufacturing, or production as it is often called, strives to operate in a controlled, uniform environment, to ensure consistent volume output (Adam and

Ebert, 1986; Gaither, 1987; Render and Heizer, 1994; Schmenner, 1987; Schroeder, 1989). Change is inevitable but it is usually controlled, to minimize its impact on the manufacturing process.

There are several important components in a manufacturing operation:

- The manufacturing processes, such as procurement, production scheduling and stock control. These are often computer-based systems, so Information Technology (IT) plays an important role in any modern manufacturing operations (Bird, 1995a; Bartlett and Ghoshal, 1995). It is also important to remember that manufacturing starts at the design stage, especially since use of computer-aided design and computer-aided manufacture (CAD/CAM) often means that products are transferred from design into manufacture as a computer database. Design to enable ease and quality of manufacture is a key requirement (Morley, 1996).
- The manufacturing systems employed to produce the goods or services. These systems must be able to meet the volume output as well as producing the mix of product required. Emphasis was originally placed on obtaining large manufacturing runs to reduce unit costs. This has now shifted to flexible manufacturing systems (FMS), where product lines can be changed relatively quickly and easily to meet changing market needs (Upton, 1995).
- Materials management, which covers the process of materials into and out of the factory. Activities under this include the ordering of raw materials, determining their quantity and timing, and the management of work in progress and finished goods. The larger the inventory the higher the cost, although it reduces risk since raw materials can be drawn from store to meet production requirements, and customer orders can be met from the finished stock.
- People management. A wide variety of different skills are employed within a production unit, and the management of these different groups requires considerable skill. In addition, trade unions are normally active within a factory, and this also requires attention.
- Quality management. As discussed in earlier chapters, quality is an essential requirement for all functions, and this also applies within manufacturing. Quality is everyone's responsibility, not only that of the inspector at the end of the production line. Quality cannot be inspected into a product; it must be built into it at every stage, from the drawing board to when it reaches the customer.

Figure 14.1 shows a typical organization chart for Manufacturing Operations. The Manufacturing Director has different titles in different organizations, such as the Plant Manager or VP of Manufacturing. The main function of the post is to determine the manufacturing strategy for the organization.

The Facilities Manager plans and controls the layout of the production facilities and designs and controls the manufacturing process. Several vital decisions need to be made, such as the manufacturing technology to be used, the type of plant and equipment and their location. There is close cooperation

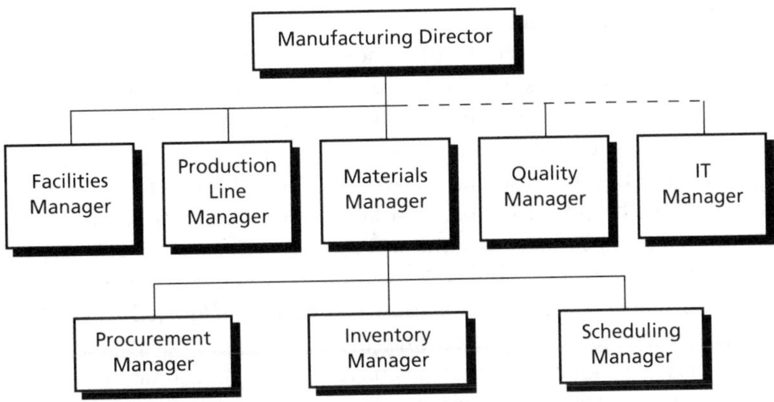

Figure 14.1 A typical Manufacturing Operations organization chart.

between the Facility and Production Line Managers, since the Line Manager is responsible for managing the workforce that uses the facilities. The Line Manager must be a skilled people manager, determining the best organization for the workforce and improving their performance on the job.

The Materials Manager has overall responsibility for materials, from when they enter the factory as raw materials to when they leave as finished product. Reporting to the position are the Procurement Manager, Inventory Manager and Scheduling Manager. The Procurement Manager, or Purchasing Manager as the position is sometimes called, is responsible for placing the orders on external suppliers. Prices are negotiated and vendors selected and their performance monitored. Often the Procurement function will work in close partnership with the company's suppliers, so that the performance of both organizations is improved and both obtain financial benefits.

The Inventory Manager's prime responsibility is to monitor the inventory of goods and raw materials and to ensure that they are kept as low as possible, consistent with meeting production demands. The Inventory Manager is responsible for determining what to order and when the orders are placed.

The Scheduling Manager is responsible for controlling the loading on production, so as to balance the workload and make the most efficient use of available production plant and people, consistent with meeting customer orders.

The Quality Manager and IT Manager usually report to the Manufacturing Director on a dotted line basis, having a solid reporting line into their own functional directors. The Operations IT Manager is responsible for developing the many IT systems which improve the efficiency of the manufacturing process and help in planning, control, budgeting and decision making. The Quality Manager has overall responsibility for advising on quality matters relating to product quality, and to process quality standards and systems. In addition, the Quality Manager would measure the overall performance of the Manufacturing Operations and provide feedback and help in problems analysis.

Although Manufacturing Operations usually relates to the production of goods, it can also apply to service provision. Companies often do both, such as a manufacturer who provides after-sales service as part of the sales package. There are several differences between the operations required for services compared with those for goods. For example, the process used is different, since the end products differ; goods are tangible, whereas services are often intangible.

Services cannot be stored for use at a later date, such as is possible with goods. There is no inventory for services and if a service is not used it is wasted. There is also very little freedom regarding the location at which services can be produced. Goods can be made at a remote location and then transported to consumer sites, but services cannot be transported and must be generated at the point of use. Therefore service production must be moved to the consumer or the consumer moved to where the service is being produced. This means that service sites, such as shops, banks and repair centres, are often more widely scattered than goods production units, being located close to consumer locations.

Manufacturing Operations go through various phases within an organization. The two main phases are shown in Table 14.1. During Phase 1, Manufacturing Operations is tolerated by senior management from other functions as a necessary evil. The aim is to minimize its impact on overall profit margins by reducing its costs. Investment made in it is kept to a minimum and the Operations Manager is expected to meet goals set by other functions. Generally operations management is preoccupied, at this phase, with day-to-day issues associated with the physical process, such as plant and machinery. The processes used are generally those that are standard within the industry, the manufacturing managers being content to follow the leaders.

Table 14.1 The two stages of Manufacturing Operations development

Phase 1	Phase 2
Operations tolerated as necessary evil	Operations treated as giving strategic advantage
Minimize impact of operations, for example reduce its costs	Use operations as a key competitive weapon
Operations to meet short-term goals set by others	Operations to set long-term strategic goals
Operations management preoccupied with structural issues – equipment, facilities	Operations management concerned with strategic issues – meeting customer needs and company mission
Operations management preoccupied with day-to-day fire fighting	Operations management concerned with long-term issues and strategies
No new process or technologies developed – buy in and follow industry	Industry leader – new processes and practices developed
Workforce policies follow industry norms: wages, training, skills	Innovative workforce policies: bonus, quality partnership
Low investment in plant and people – minimum to meet short-term aims	High investment to meet long-term objectives – continuous improvement

During Phase 2 the organization realizes the competitive advantage it can gain from its Manufacturing Operations, and management takes part in long-term strategic planning. Operations managers are no longer content to follow others within the industry. They develop processes and practices which lead the industry and provide advantage over competitors (Pisano and Wheelwright, 1995). The level of investment made in Operations is high and it results in continuous improvement.

Manufacturing Operations is often internally focused, and concentrates on activities such as cost cutting. However, like other functions within the organization, it must be customer facing and be prepared to meet customer requirements, such as the need for change. Change can occur due to several factors:

- Change in customer requirements. This may result in a change in the amount of product required as well as the type of product. To some extent customer requirements can be controlled by Marketing, such as by advertising and promotions.
- Change in technology. This can result in a change in the product as well as in the manufacturing process.
- Change in the workforce. Changes will occur over time, such as skills shortage, increase in wages due to higher living standards and changes in social attitudes.
- Changes in competition, affecting prices (and costs), the level of demand and the requirements for new products.
- Changes in raw materials, such as in price, quality and the availability of new substitute materials.
- Changes in the environment, such as legal and political considerations. For example, the greater awareness of environmental issues would result in the company using environmental friendly materials during the manufacturing process.

All the plans formulated by Manufacturing Operations must be related back to the organization's mission, as illustrated in Figure 14.2. For example, the organization's mission may be simply stated as meeting customer's needs for new products at competitive prices. Manufacturing Operations' mission could then be translated into introducing products quickly in order to meet customer needs. This mission statement would lead to a definition of the key competences which need to be developed in order to meet this aim.

Operations' objectives are essentially its mission translated into meaningful and measurable tasks. For example, these could be objectives for reducing costs, such as manufacturing costs to be a certain percentage of sales; objectives for improving quality, such as reducing warranty costs by a certain amount; or objectives for reducing time to market, perhaps measured in the number of months needed to increase manufacturing capacity by a certain amount, or the time needed to introduce a new product.

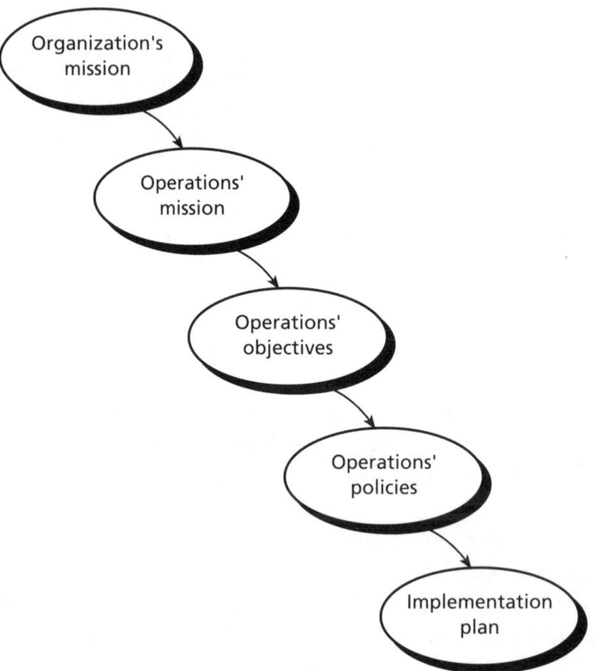

Figure 14.2 Steps in the formulation of Manufacturing Operations' plans.

Operations' policies define how its objectives are to be met. For the policies given earlier, for example, manufacturing costs could be reduced by the use of greater automation; quality could be improved by using more highly skilled operators; and manufacturing time to accommodate changes could be reduced by introducing flexible manufacturing systems (FMS).

Finally, the implementation plan defines in detail the actions needed to realize the policies. They would define, for example, how much automation is to be introduced; when it should be introduced and how it would be funded; and the steps needed to recruit skilled people and the training courses to improve the skills of existing operators.

Manufacturing Operations must focus on areas it can do well and gain competitive advantage. There are several areas for focus, such as on product type; manufacturing technology; manufacturing process (batch, continuous flow and so on); differentiation by new products or mature products; volume of output; or according to whether products are made for stock or to customer orders. Each Operations activity must be treated as a separate strategic line with its own requirements. Economies of scale do not always work, especially when they are not optimized for each area or need more support. (See the case study 'The wall'.) Large operations can also become complex and difficult to manage, resulting in inefficiencies.

Case study

The wall

Although florolls were the main product of the FloRoll Manufacturing Company, the organization also manufactured the special chains, called holders, that were used to restrain the florolls before they were operated. Holders were relatively low cost, but because they were disposable the volume used, compared with florolls, was high.

Although the company had been making holders for over five years it had consistently made a loss on them and the Board was considering whether the product should be discontinued.

'We can buy these holders for 10 per cent less than we can make them', said Jack Fox, Marketing Director, 'and they are not a strategic line, like florolls. I think we should stop making them and come to an arrangement with a supplier to have them delivered to our customers' sites every time we deliver a floroll.'

Adrian Elton, Manufacturing Director, was reluctant to give up production of the product, since he was convinced that it was covering a useful part of his factory overheads. However, he received an action from the Board to look at alternatives to making the holders in house. After the meeting he called Joe Plant, Manager of the Floroll Assembly Shop, where the holders were made, and John McCaully, Manager of the Dinsher Shop, together for a meeting.

'Looks like we have a problem here', began Adrian Elton. 'Unless we can show a profit we don't have any option but to stop manufacture and buy in the holders. Any suggestions?'

Joe Plant had nothing to say; he had tried for five years to make the holder line show a profit and had failed. John McCaully had a suggestion.

'I think the problem is mainly the difference in the characteristics of florolls and their holders. Whereas a floroll is a high-cost, high-technology product, a holder is fairly simple and low cost. The way we have it, both these are made in the same plant, using the same processes and quality systems and using the same people. Is it right to treat them in the same way?'

'It helps with my overheads to have a high volume of products going through the same line', said Joe Plant. 'This way both florolls and holders benefit.'

'I think John may be right,' said Adrian Elton, anxious to make a success of holder manufacture and so prove the Board wrong. 'There is certainly a big difference between florolls and holders. In fact the process used to make holders is much closer to that for dinshers than that for florolls. I think we should look to make the holders in the Dinsher Shop.'

'That may not be necessary,' said John McCaully. 'What is needed is not a physical move, which can be quite expensive, but to treat holders as a strategic product in their own right, with their own special requirements. Why not build a physical partition around the holder production area? I will be happy to manage this and will apply the same level of skill as I do for dinshers.'

Adrian Elton and Joe Plant agreed to give this a try. A wall was built and John McCaully took over management of the holder line. It was treated as a separate strategic product with its own special Operations requirements. Within four months the product was breaking even and after a year it was making a healthy profit. Even the floroll line benefited, since Joe Plant could now concentrate on this strategic product.

14.2 The experience curve

Learning is important in any manufacturing operation, since it reduces the time required to carry out the operation and also improves its quality. The importance of repetition has been recognized in manufacturing operations and the aim is to produce volume products, so as to drive the accumulated volume down along the experience curve, as shown in Figure 14.3.

The experience curve describes how costs vary with volume. The lower down the curve the production can operate the lower will be the cost per unit and the greater the profit for the same selling price. Usually, however, price is determined by customer perception of value and by competitor activity, and these pressures cause the price to decrease with time. Therefore reduction in costs, by moving down the learning curve, is a necessary condition if the company is to continue to operate profitably.

There are several reasons for the cost reduction with volume of output:

- Manufacturing tasks are learned by operators, so they take less time to complete.

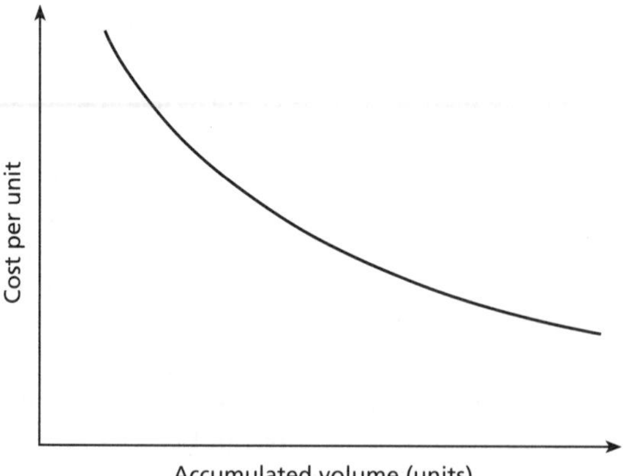

Figure 14.3 A typical experience curve.

- Larger volumes enable specialization among operators: there is a reduction in the amount of the total product which is made by each individual. This too aids in learning and it gives a reduction in operating time. However, too much specialization can result in monotony and a subsequent slowing down of operations.
- Increasing the volume of production usually means that machines are being used to full capacity, so improving their utilization and reducing the cost allocated per unit. However, as production volume increases a point may be reached where the production capacity of existing machines is exceeded and a new machine needs to be added, giving a step increase in costs.
- Increasing volumes enable greater investments to be made in plant and machines. Usually these are improvements on existing machines, enabling greater efficiencies to be made in production.
- Materials costs decrease as the volume of production increases and bulk buying becomes possible. However, the cost reduction eventually reaches a plateau where further cost reductions can only be obtained by the use of new, cheaper materials.
- Traditional manufacturing overheads, such as procurement, supervision, and factory heating and lighting, are usually recovered over the units made in the factory, and the cost per unit will be lower as the volume of product increases.

Experience can be transferred between products provided there are not too many changes between successive products. This is illustrated in Figure 14.4, where the cost per unit increases when going from Product A to Product B, although there is some transfer of learning, so the cost per unit for Product B starts at a lower value than that for Product A. Therefore many companies try to ensure that product changes happen in increments, although if this is done too slowly there is a danger that the product will fall behind those of competitors and not incorporate necessary innovative changes.

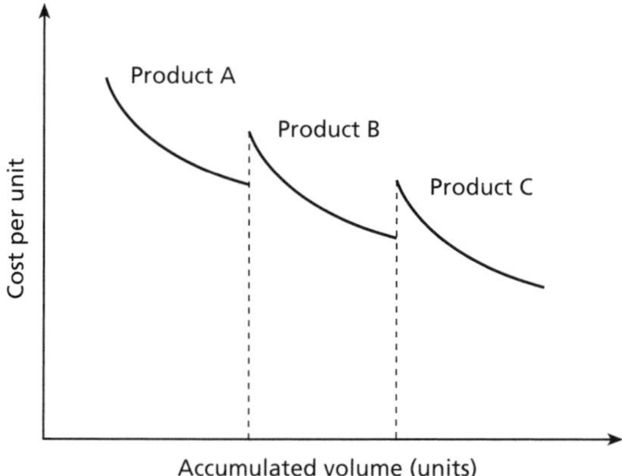

Figure 14.4 Transfer of experience between similar products.

Table 14.2 Illustration of the 90 per cent experience curve

Units produced	**Cost per unit**
100	£1000
200	1000 × 0.9 = £900
400	900 × 0.9 = £810
800	810 × 0·9 = £729

Experience curves are often specified in terms of the experience rate. This gives the reduction in cost per unit for each doubling in the production volume. For example, if a product has a 90 per cent experience rate it means that the unit cost will decrease to 90 per cent each time output doubles in volume. This is illustrated in Table 14.2, where the cost when producing 100 units is £1000 per unit, and when the number of units produced increases to 800 units the cost per unit falls to £729.

In practice experience rates vary between 60 per cent and 90 per cent. The lower rates usually apply to those processes that are easy to learn and to improve: relatively easy manufacturing tasks. Figure 14.5 shows the effect of various experience rates. The lower rates give steeper experience curves and reach a lower eventual cost.

Experience curves are usually more important during early product stages, when costs are decreasing steeply. They provide an indication of cost changes and should be used when formulating pricing strategies. However, other factors must also be taken into account, such as the problems that can arise when production volumes become very large, for example due to unwieldly organizations, long lines of communication or operator dissatisfaction.

As explained earlier, one of the reasons for unit cost reduction as volumes increase is the decrease in the overhead allocation per unit. Overheads have

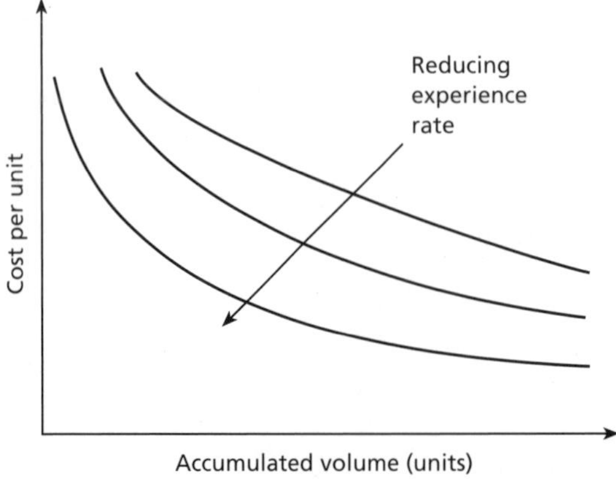

Figure 14.5 The effect of experience rate on the experience curve.

traditionally been allocated on the basis of labour hours, but as the amount of direct labour within modern products decreases it is important that overheads are not based on it. Other factors become more important, such as batch size, frequency of machine set-up and problems of materials procurement (Hepworth and Stevenson, 1993). The factory manager must identify the activities that generate the overheads concerned and ensure that products that carry out these activities, and use the corresponding resources, are allocated their share of the overheads. (See the case study 'The inexperience curve'.)

Case study

The inexperience curve

It was clear to Joe Plant, Manager of the Floroll Assembly Shop, that florolls, because of their high technology content and relative low volume, would take a long time to get down the experience curve. He was therefore very pleased when it was suggested to him by Adrian Elton, the Manufacturing Director, that he take on the manufacture of holders within the Floroll Assembly Shop.

'A high-volume product, at last', thought Joe Plant. 'That should help me to get down the experience curve quickly.'

Although florolls and holders were very different products, Joe Plant treated them in the same way and measured the movement on the experience curve by their accumulated volume. Progress was erratic and not as Joe expected. Although volumes increased rapidly, cost per unit fell only slightly. Unfortunately this also resulted in the cost of holders being very high (see the case study 'The wall'). The problem was evident to John McCaully when he looked into the process being used. Florolls were a complex, expensive product, while holders were simple and low cost, and yet they were being treated in the same way.

For example, machine overheads were allocated to florolls and holders by labour hours, although holders, being a very simple product, required very little machine set-up time, most of which was taken up by the far more complex floroll. Similarly, florolls and holders were treated equally in the way the cost of the procurement department was recovered: on the basis of volume of product. However, florolls used expensive materials, which required considerable negotiations with several suppliers in order to arrive at an economic price, whereas holders used standard low-cost materials, which were routinely bought as repeat orders from two suppliers.

By separating the floroll and holder products into two separate lines, the cost of holders fell rapidly as the product moved down its learning curve, whereas the cost of florolls fell much more modestly. Joe Plant had failed to realize that holders had a low experience rate while florolls had a much higher experience rate.

14.3 Manufacturing technology

A company would like, ideally, to produce one standard product and lots of it, in order to prove the manufacturing line and to progress rapidly down the learning curve. Unfortunately, customers often require variable volumes and product mix, so that compromises are necessary.

It is possible to decouple production volumes from market demand by holding stocks of finished goods or work in progress (WIP), but now this results in other costs, and trade-offs are required.

Figure 14.6 shows the trade-off that can be made between WIP and manufacturing plant. A high investment in plant would mean that there is very little queuing of goods, and this would give a low investment in WIP.

Alternatively, the investment in plant can be reduced, and this would increase the WIP to meet production schedules. The total investment curve shows that there is usually a minimum cost, at point A, where there is an optimum balance between plant and WIP.

There are several types of production system:

- **Job-lot or one-off production**. In this system each contract is unique and differs in some aspect from previous ones. An example is the construction of a railway station, which will be different from other railway stations, even though there will be many similarities. Job-lot production systems require careful project planning, with limited opportunity to learn between projects. Many different skills are required for job-lot manufacture, which can be subcontracted, especially for manufacture of sections of the work.
- **Batch production**. In this method a batch of a given product is built, after which the manufacturing equipment is set up for a run on a second product, and so on. An earlier product may be made again when demand exceeds stock. The size of the batch is determined by equipment set-up costs

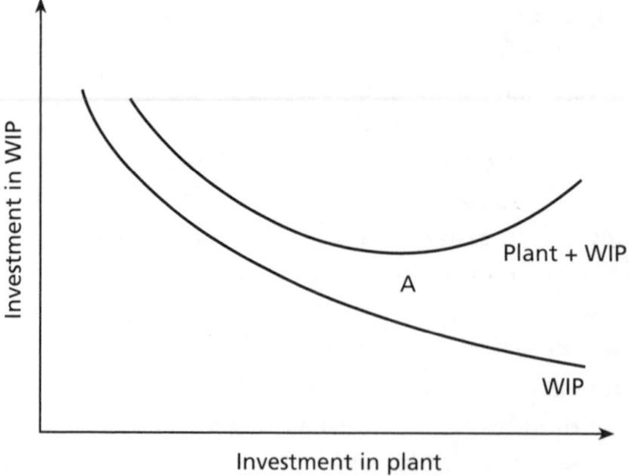

Figure 14.6 The trade-off between plant and work in progress (WIP).

compared with inventory costs. Some learning can be transferred between products and standard processes can be developed more easily than in the case of job-lot manufacture. Resources required, measured in terms of people and machines, are also more constant during the life of the project. However, changing market demand has a greater impact on batch production than on job-lot production, where the product is, in effect, being built to a customer's order.

- **Mass production**. In this production technique the same product is made continuously on a dedicated production line. The problems associated with this system relate mainly to keeping equipment going, covering for service periods and breakdowns, and getting maximum output from production operators.
- **Continuous or process production**. This method is mainly associated with the chemical and oil industries, where the plant runs continuously, 24 hours a day, 365 days a year, with the same inputs and output. Very little operator intervention is needed. Management has little control once the process has started (dealing only with breakdowns and shortages in raw materials), which must be planned to meet the forecast demand.

Figure 14.7 summarizes the major differences between the manufacturing systems.

Trends in manufacturing are towards shorter delivery times and smaller lot sizes (Chatha, 1994). Manufacturing is often driven to waiting for orders before committing materials, capacity and people. The move is from make-to-forecast to make-to-order. Production is driven from sales order systems. This is shown in Figure 14.8. Customer orders and sales forecasts provide the input into

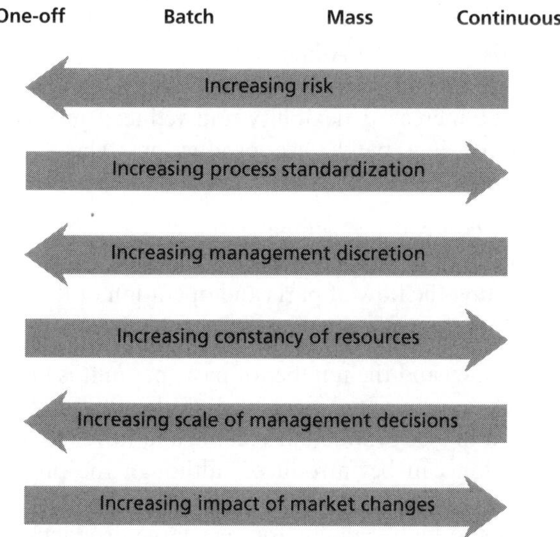

Figure 14.7 The major differences between manufacturing systems.

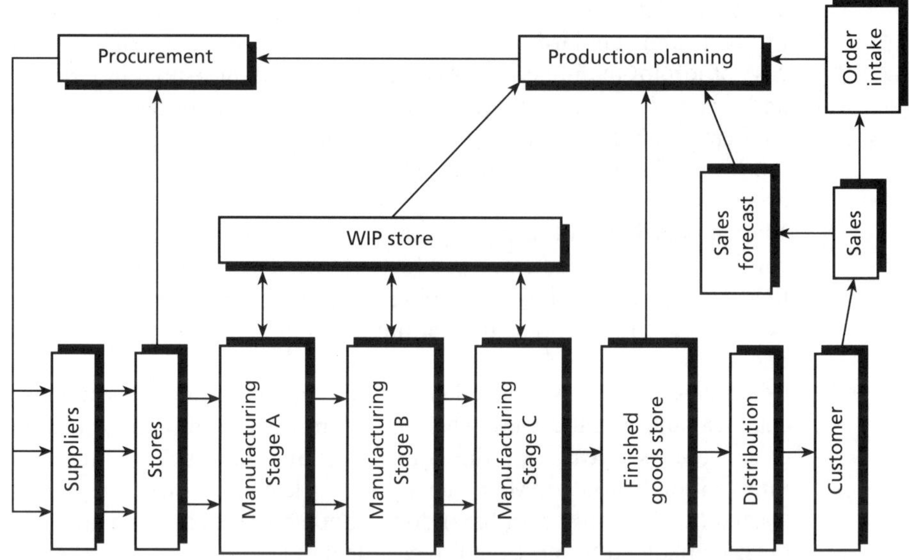

Figure 14.8 An overall manufacturing control process.

production planning. Based on the knowledge of finished goods and work in progress, the level of production output is controlled and information is fed to procurement to obtain replacement raw materials from suppliers.

Flexible Manufacturing Systems (FMS) provide several advantages in an environment where customer requirements are changing. The financial and strategic benefits produced from use of FMS must, however, be evaluated on a long-term basis, since it rarely shows sufficient payback on traditional financial investment evaluation methods (Chen and Adam, 1991).

FMS investment leads to a reduction in labour costs, increased output, decreased manufacturing costs, increased flexibility and reduced production lead times. Generally it is used for low batch size production. There are several components of an FMS system:

- Computer- or numerically controlled machines.
- Conveyance network for moving parts between machines.
- A control system coordinating the flow of parts and operation of the machines.

Figure 14.9 shows how FMS is positioned relative to other manufacturing methods. When production volumes are low, and the number of parts per unit is high, manual (job-shop) methods are most appropriate, since very little automation can be economically introduced. For a higher production volume and lower part count per unit, some degree of automation can be introduced, although the process is still predominantly manual. FMS is best used when the production volumes are relatively large and number of parts not too high. Finally, for very large production volumes and a low number of parts per unit, dedicated automated lines prove economic.

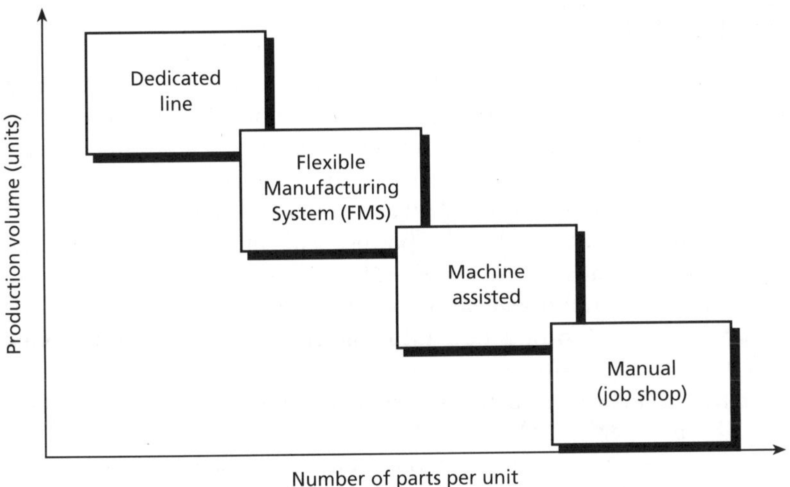

Figure 14.9 Positioning of FMS relative to other manufacturing methods.

It is important to remember that automation cannot substitute for good management, and automation and other advanced manufacturing technologies benefit those manufacturers most who are already efficient. It is also essential to link manufacturing strategy to an organization's business strategy (Hill, 1993). Three elements of this strategy can be considered (Grant *et al.*, 1992):

- **The organization's strategic goals**. For example, the goal may be to meet differing customer requirements in order to target niche markets. Under these circumstances the company would need flexibility in manufacture, such as by using FMS systems or by a flexible layout of its plant. Alternatively, the company may want to reduce manufacturing costs and be a cost leader within its industry. High volumes would now be needed, with a few dedicated product lines, enabling a high level of automation.
- **Resources available to the company**. This should include both outside resources and those within the organization itself. For example, the company may have access to large numbers of operators at low wages, in which circumstance it may wish to operate a largely manual production line. Alternatively, it may have access to high engineering skills and have spare capital to invest, and it may then decide to go for automation, linking design and manufacturing (CIM or Computer Integrated Manufacture).
- **Product and market characteristics**. For example, if the market demands that products are changed frequently it may be necessary to build in small batch sizes using FMS techniques. Alternatively, competitive pressures may force a low-cost operation, with concentration on a few product lines and mass production in highly automated plants.

14.4 Global operations

The trend during the 1980s was towards multinational operations. Large organizations set up subsidiaries in different countries, these subsidiaries usually operating as autonomous units, decisions being taken locally based on local conditions. The different units sold local products and operated their own processes. Central control was relatively haphazard, being managed by an army of staff personnel with little real power. The main requirement was for the separate business units to meet their sales and profit targets, and to contribute positively to the overall organization bottom line.

Rationalization of production into bigger and bigger plants took place in the 1980s. However, with increasing competition in multinational markets, the 1990s have begun to see a move towards greater globalization of operations (McGrath and Hoole, 1992). The separate units around the globe are starting to operate as one large homogeneous unit, controlled by common procedures and production control systems. This should enable local manufacture to meet specific local conditions, while still achieving the scale of production needed to get down the learning curve and obtain low cost with increased quality.

Many products are global in nature, such as photocopiers, telephone handsets and cars, and the proportion of such products is increasing every year. Global operations can be carried out in many ways (see the case study 'Going global'). Examples are:

- Setting up international coordination teams for materials management, such as procuring products on a global basis to use the best suppliers and obtaining volume discounts.
- Carrying out critical activities in one area only, such as scheduling all products centrally.
- Making use of the strengths of any one area, such as making labour-intensive parts in areas where labour is cheap.

There are many advantages in this approach to globalization:

- Costs can be reduced, such as by buying materials in bulk, building products where costs are lowest, or making use of capital grants and incentives available from some governments.
- Local skills in an area can be exploited, especially if this is not readily available in other areas where production is being done.
- Overheads can be reduced for some activities, such as procurement, by doing this centrally for all production units.
- Products can be engineered for global markets, so avoiding the re-engineering costs required for individual markets.
- Central control of stock, spare parts and so on can result in faster deliveries to customers, since loads can be balanced between individual plants.

Case study

Going global

RGU International have three subsidiary companies operating in the USA, UK and Japan. Between them these three organizations cover most of the world, with development and manufacturing plants in several countries. Each subsidiary is given considerable autonomy in the way it operates. This is partly historic, since the subsidiaries have been acquired over time (as had the UK subsidiary, the FloRoll Manufacturing Company).

It is becoming clear that, in order to succeed in global markets, the organization needs to change the way in which it operates. 'Big' Al, President of RGU International, has set up a Global Study Group to report on the changes that are necessary to make the organization more competitive globally.

Members have been nominated to this study group from all three subsidiary companies, and from multiple disciplines: engineering, finance, marketing, sales and manufacturing. The group has worked on the project for three months, talking to many personnel within the company and visiting other organizations who have begun to operate globally. Their findings are presented in a report, which is summarized in Table 14.3.

The study group has considered four main areas for initial implementation: product development, procurement, production and process management. Their key recommendation for product development is that a lead house is nominated for every component of a product. For example, the drive mechanism of a floroll could be developed in a laboratory in the UK, the propulsion system in the USA and the bodywork in Japan. Each lead house is to take its input from the other markets and engineer the product to meet the most stringent requirements. This will not only ensure that re-engineering to meet local conditions is minimized, but also that the product exceeds the requirements in many markets, so providing competitive advantage. The extra costs involved in meeting the 'worst case' specification should be minimal and can be offset by the fact that common components are now used, leading to greater volumes and lower costs. To aid in this process the number of different parts used should also be limited, so that procurement from a few global suppliers is possible, again giving a cost reduction. The study group estimated an overall cost saving of £400 million per year by implementing these recommendations.

The study group recommended that purchase negotiations for most materials be done centrally. The only exception would be very low-cost, low-volume materials, which would normally be relatively bulky, and these could be negotiated on a local basis. Central negotiations are expected to lead to lower costs since the volumes from all plants could be pooled, giving greatly increased quantity discounts. Once the central contracts had been negotiated each plant could call off materials against these contracts to meet local inventory requirements. Central procurement of expensive materials was expected to show a saving of over £500 million in a full year.

Table 14.3 Recommendations from the Global Study Group (see the case study 'Going global')

Task	Recommendations	Benefits/savings
Product development	Lead house design for global markets	Minimize re-engineering
	Limit number of different parts	Procurement from fewer global suppliers
	Engineer to meet most stringent specification	Improved performance
	Pooling of market knowledge	Essential for global markets
Procurement	Central selection/review of suppliers	Improved quality of products
	Central negotiation of contracts	Bulk buy giving lower costs
	Local procurement against contracts	Tighter management of local inventories
Production	Lead house for component manufacture	Reduce cost and improve quality
	Product assembly close to final customer	Meet local needs and higher visibility
	Common procedures	Reduce transfer costs – taxes
		Flexibility of product transfer
Process management	Central order intake and production planning	Lower costs due to economies of scale
		Better load balancing
		Faster to customer

Central contract negotiations should be linked to central selection and appraisal of vendors. This would mean that the best suppliers could be chosen globally and their performance continually monitored, leading to increased quality of supply and implementation of systems such as just in time delivery.

The Global Study Group also recommended that a lead house is nominated for the manufacture of the various components of a floroll. They linked the design and manufacture of these components together, so that the UK was to make all the drive mechanism for global markets, the USA the propulsion system, and the bodywork was to be made in Japan. This would ensure that each plant would quickly move down the learning curve for new products by having increased volumes. It would also be possible to manufacture some of the high labour content components in low labour cost areas, and to use skilled labour for some other components.

The study group recommended, however, that assembly of the finished product should take place close to individual markets. This would ensure that final test and control could be applied to meet local conditions, and it would also ensure that the stringent requirements of the defence markets in the different countries are met. Local assembly would classify the floroll as a local product, giving higher visibility to local customers and reducing the amount of tax imposed by certain countries.

In order to meet the strategy of transferring products between plants the study group felt that it was essential that common procedures were used on all sites, so as to give consistency of quality and to make transfers easier. Transfers of two types were felt to be necessary: vertical transfers, from component to assembled product; and horizontal transfers, where transfers occurred between plants making similar components in order to balance the load on the various plants.

Centrally administered process management was also considered to be a cost saver and covered items such as the use of market and sales forecasts to set worldwide quotas for various plants and for production planning and scheduling. Local information was to be electronically fed into global systems, and consolidated results fed back to the local plants. Customer orders could be linked to the factory most likely to meet the customer's time-scale, so resulting in better load balancing and faster delivery.

All the recommendations of the Global Study Group were accepted by the RGU organization and implementation teams were set up to carry through the individual recommendations.

14.5 Logistics

Logistics is the total concept encompassing the flow of goods from the supplier, through the manufacturing plant, and out to the customer (Cooper, 1994; Quayle, 1993; Ward, 1994). It covers many areas within a manufacturing environment, such as procurement, goods receiving, work in progress, stock control, finished goods stores and distribution to the customer. Several techniques are included in this, such as JIT (just in time) and MRP (materials requirement planning).

There are many considerations within procurement. These include the specification against which the materials or components are to be bought; the price of the goods with quantity breaks; the delivery date and delivery method; and the method of payment and any penalty clauses. The procurement function needs to be involved in supplier management, as described later, and to be aware of the technical specification of components. If similar parts are bought under different part numbers then the company may be losing out on quantity discounts, as well as incurring unnecessary overheads in buying and stocking different parts and in inventory holding costs.

The components received from the supplier are checked in goods receiving to ensure that they meet the requirements of the order regarding type, quantity and quality. Many companies are now developing relationships with their suppliers so that the goods incoming function can be dispensed with, the components going straight from the supplier's plant to the company's manufacturing floor.

The goods receiving function is also involved in checking parts that have been returned by the customer as being defective. Quick action is now needed to confirm the defects and to ensure that the customer receives replacements, if appropriate, in order to maintain good customer relationships.

The stores or stock control function is responsible for storing materials, work in progress and finished goods. Costs must be minimized and considerations such as health and safety, fire risks, deterioration of goods during storage and security (theft) taken into account. In addition, it is important to ensure that goods are available at the geographical location where they are needed, and at a time when they are required.

The goods outgoing function must ensure that the product leaving the factory meets the delivery requirements regarding type of goods, quantity, reliability and timing of delivery. It is part of the overall distribution function, delivering products to the customer.

14.5.1 Materials control

Two systems are usually used to control materials flow within a factory: JIT and MRP. Both of these are available in the form of computer programs from several suppliers, although they involve more than just IT systems; they require a fundamental change in the way that a company conducts its business if they are to succeed.

Just in time (JIT) is a technique in which stock held by the company is measured in terms of hours of production rather than in days or months (Karmarkar, 1989; Ward, 1994). A close working relationship is developed and maintained with the company's suppliers, so that materials and components are delivered to the company when they are needed, and go straight from the suppliers' lines to the company's manufacturing floor. Costs involved in product handling and storage are therefore minimized, although the risk of late delivery from a supplier affecting the company's manufacturing line is increased.

Materials resource planning (MRP) is a method for ordering and scheduling materials needed in production. This has now been extended to other areas of the shop floor, apart from materials, and is often referred to as manufacturing resource planning or MRP II (Bywater, 1994; Holder, 1991). It is an IT-based tool for managing a business, encompassing planning, setting goals, measuring performance and so on. It links day-to-day tasks of all the functions within an organization to overall goals and plans, as in Figure 14.10, so giving a view of the integrated business.

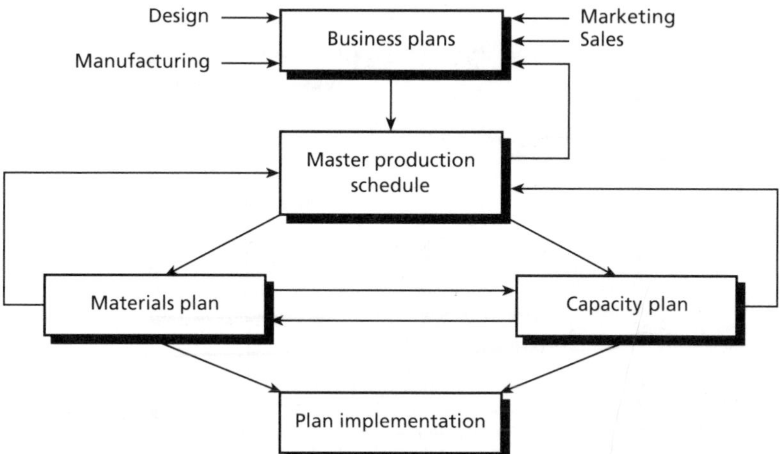

Figure 14.10 Example of an MRP II structure.

MRP II programs available commercially help in predicting inventories, cash flow, wages and salaries, sales order processing, purchase order processing, stock and WIP, materials planning, capacity planning and so on. They allow progress to be measured against plan and for corrective actions to be incorporated. 'What-if' studies can be undertaken to determine the impact of certain actions.

One of the key functions of materials control is to minimize the amount of inventory held by the company in terms of materials, work in progress and finished goods (Tersine, 1994). Surplus inventory ties up capital which can be usefully employed elsewhere within the business. It is no longer sufficient to be able to record the amount of inventory held within a manufacturing facility; systems must be used that are able to forecast the level of sales and to manage inventory so as to keep stock levels to a minimum value (Knutton, 1993).

Figure 14.11 shows the production flow within an organization (compare this with Figure 14.8) and the flow of information, which enables inventory to be controlled. Customer demand, which is the overall driver for inventory levels, must be measured from the sales order processing system, which will show the sales received by the company as well as the sales that have been lost, since this measures the overall level of customer demand.

14.5.2 Supplier management

The cost of materials and bought in items often exceeds 80 per cent of the total cost of a product in most modern technology-based industries, and the trend is upwards, as illustrated in Figure 14.12. As the percentage of bought-in items increases, the in-house manufactured content falls and so also do the manufacturing overheads associated with these activities. This is one of the

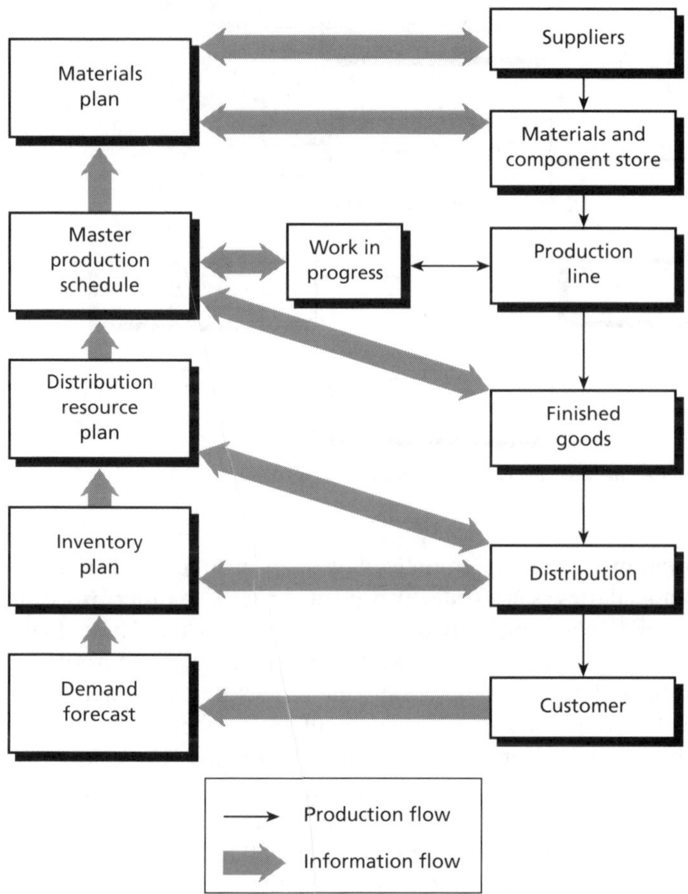

Figure 14.11 Production and information flow within a plant.

reasons why so many companies are turning to external sourcing for an increasing amount of their materials and components.

With this trend in supplier dependence, supplier selection and management is the key to the success of a modern strategic manufacturing plant (Burt, 1989). Traditionally the procurement department has been concerned with multiple sourcing of a product, driving hard price bargains to ensure that products are bought at the lowest price possible.

However, concentration on low initial prices has often meant that products received have been of sub-standard quality, requiring the company to invest in large inspection facilities and resulting in cost increases due to rework, scrap, warranty returns, line down time and so on. The modern aim is to manage suppliers to get the lowest overall costs, not only a low initial procurement price.

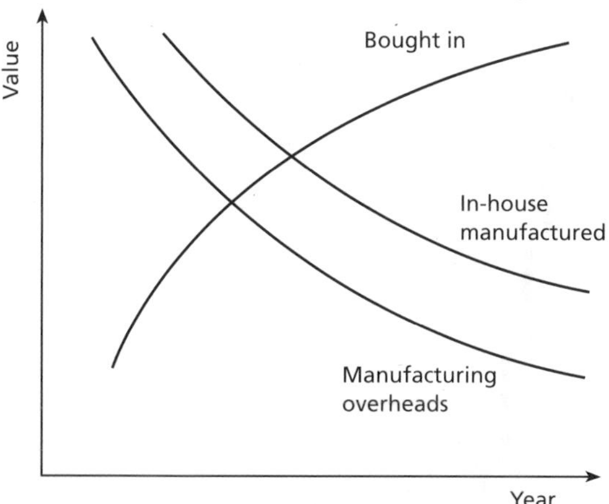

Figure 14.12 Trends in the content of a manufactured product.

Techniques such as JIT require a company to have a few smart suppliers, those that it can work with closely and who are able to change their schedules and technologies quickly to meet the company's changing needs. The relationship between the company and its suppliers is one of partnership. The supplier must work with the company's internal design teams to develop new technology products.

Large manufacturers are often nervous about buying critical components from a single small supplier. However, this is usually the best method for ensuring continuity of supply. The manufacturer must work closely with the supplier, obtaining commitment from the supplier to operate as an extension of the company's own organization. The manufacturer can work with the supplier to improve overall quality and to ensure that its supply needs are met. In this way the manufacturer can eliminate many of its own internal inspection and control functions, where these duplicate those of the supplier. Cost savings made in this way can be shared with the supplier so that both benefit from the closer working relationship.

Everyone within an organization is involved in developing and managing supplier relationships, as illustrated in Figure 14.13. Design must ensure that the suppliers are involved in the selection of materials and components. The temptation is for the company's design function to give the supplier detailed design information regarding the component required, rather than a specification that has been developed jointly with the suppliers, so that the suppliers can carry out their own optimum design.

Purchasing is involved in the selection of suppliers and in managing them regarding vendor rating, delivery and so on. The company's quality function will work with its suppliers to ensure that their quality meets the company's requirements and to improve this if required.

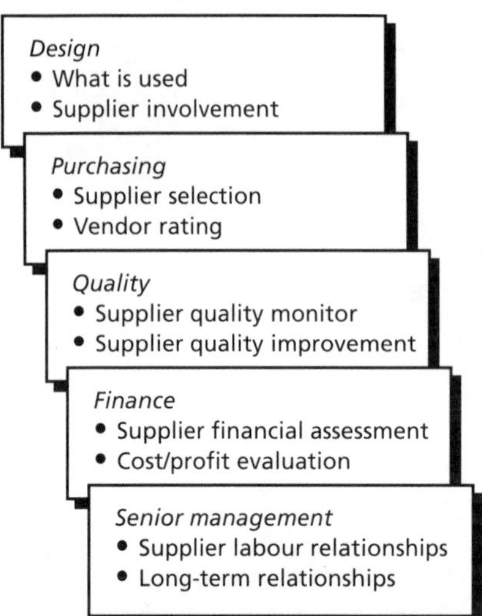

Figure 14.13 Organizational involvement in supplier management.

Finance is responsible for assessing the financial strength of suppliers, including an assessment of their costs and profits. By this method the company can ensure that its suppliers make a fair profit and remain financially viable. Senior management must also develop relationships with the supplier's management, assisting with factors such as labour problems, and building long-term relationships.

14.6 Design for manufacture

In a typical company design changes move back and forth between engineering and manufacturing as the product struggles through the manufacturing plant. This has a major impact on costs and on time to market, which can affect market share and the chances of product success.

Designers believe that manufacturing considerations can constrain their creativity. This is probably true, since it would limit their choice of parts, packaging options, process technology and so on. However, it is important that the product can be made, using the manufacturing technology available, rather than being confined to the drawing board and engineering prototype laboratory.

Manufacturing starts on the engineering drawing board, and manufacturing engineers need to be part of the design team, inputting their requirements from the very start of a product (Leonard-Barton *et al.*, 1994; Morley, 1996). This is known as DFM or design for manufacture (Knutton, 1992a; Maliniak, 1992). The design should take advantage of the facilities available within manufacture, such as automation, in order to improve process yields, reduce parts cost, reduce the amount of new tooling needed, and reduce WIP and handling costs.

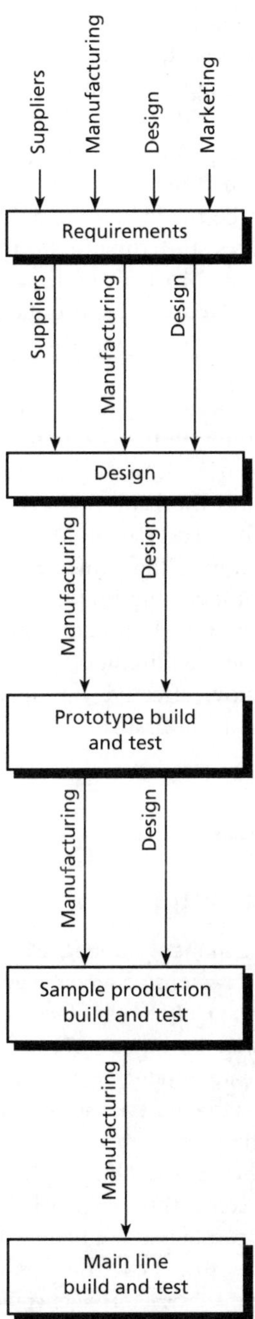

Figure 14.14 Functional involvement in the design and manufacturing stages.

The aim is to move manufacturing engineering from the shop floor into the design laboratories. IT-based design tools, which incorporate manufacturing rules, help the design engineer, but for maximum effect projects should be organized such that there is direct involvement by manufacturing. It also helps if designers understand the manufacturing process and its constraints. Designers should be judged not on how quickly they complete their designs but on how quickly the product can be made in volume, off the production line.

Figure 14.14 illustrates the functional involvement in a typical design for manufacture environment. Most of the functions, such as Marketing, Design and Manufacturing, are involved at the product requirement stage. It is important that the suppliers are also involved at this and the subsequent design stage, since, as stated earlier, the cost of the final product is likely to be determined very largely by the bought-in materials and components, and it is important that these are chosen in conjunction with the company's prime suppliers.

Design and Manufacturing lead during the prototype build and test stages, and then Manufacturing takes the lead at the sample production stage and the main build stage. Many of the processes used can be shared between engineering and manufacturing. For example, the same test programmes used during design test can be reused after manufacturing build, so saving on development time and cost.

At no stage of the DFM process must the customer requirements be compromised. If customer requirements are not met then the product will not be successful, no matter how effectively it has been manufactured. (See the case study 'Putting Manufacturing first'.)

Case study

Putting Manufacturing first

A classical case study, used in many business schools, concerns the unsuccessful launch of Floroll Vee. For obvious reasons no actual names are mentioned in this study.

Manufacturing within the FloRoll Manufacturing Company went through the traditional phased stages, as illustrated in Table 14.1. When the company was in Phase 1 it introduced an early product, Floroll Flyer. This was a high-technology machine, which was by far the most advanced on the market. The designers spent a considerable amount of time ensuring that its technical superiority exceeded that of its rivals. Marketing was extremely pleased and wanted to rush the product out into the market-place, before competitors could catch up.

Unfortunately Manufacturing had not been involved in the design or prototype stage. When the design got tossed over the metaphorical design wall, into Manufacturing, they found that the product could not be made on their conventional machines without major changes. After these were completed it was found that the components used were so delicate that the automatic assembly process would cause a high level of failures and give very low manufacturing yield.

What was worse was the unreliability introduced in the field, so that even the few products shipped were returned as defective.

The product eventually ended up being a year late into the market, even though the design stage was on time, and it gained a poor reputation, being quickly overtaken by competitors.

When a replacement product, Floroll Vee, was being formulated it was decided to avoid the problems experienced with Floroll Flyer. A multi-disciplinary team was set up, consisting of Manufacturing, Engineering and Marketing, with strong input from suppliers. The aim was to ensure that Manufacturing requirements were considered from the very start.

The representative from Manufacturing had very conservative views and believed that very little change should be introduced. She vetoed any new ideas put forward by Design Engineering.

'But that's the only way we can meet Marketing requirements', the designers would moan.

'Then let's change the requirements', insisted Manufacturing.

Marketing was steamrollered into changing the product requirements to suit Manufacturing, even though it did not meet what the customers wanted. The aim was to get to market as quickly as possible with a safe, reliable product.

Floroll Vee rolled off the manufacturing lines on time and in the volumes planned. Unfortunately it no longer met customer requirements, and the product went straight from the production line into warehouse storage. It soon became evident that the product was not going to sell. It was discontinued and the stock from the warehouses sold for scrap. This episode very nearly caused the premature end of the FloRoll Manufacturing Company.

14.7 Quality

Various definitions have been applied to the term 'quality'. Generally this term is used for products or services if they meet the expectations of the customer. The customer must feel that value for money is being obtained. This is different from the oft-quoted definition of quality, as a product or service which meets its specification, since the customer's perception of the specification may be very different from that of the designer or manufacturer.

The concept of quality started within manufacturing, and it was the Japanese who first showed the benefits of applying quality principles to production (Garvin, 1983). Manufacturing quality was initially assumed to be the responsibility of the inspector at the end of the production line, who operated as policeman on behalf of the customer, rejecting any goods that were outside of specification (Lesser, 1983). To ensure that the inspector's role was independent of the production line, inspectors reported to an independent quality directorate.

With time this arrangement changed. It was realized that the responsibility for quality lay not with the line inspector but with each operator on the line. The organization was changed so that the inspector reported to the line manager and took on the function of a coach rather than policeman. The customer was the final judge of quality, not the inspector, and the whole line worked to meet customer requirements. Quality circles or teams were set up to improve quality and suggestion schemes introduced to encourage innovation and involvement in change.

A formal quality department was still maintained, but it was much smaller and performed a staff function, reporting in at a high level within the organization, so that the quality director had sufficient influence to ensure that uniform quality procedures were being followed within the organization.

Although Manufacturing Operations makes the most visible contribution to product quality, it is not the only or most important area. Product quality starts at the concept stage and quality is built into the product at every stage from specification, through design and manufacture to delivery to the customer and after-sales service. Quality management must therefore be applied throughout the organization, in every function and every level, starting from the Chief Executive and going down to the operators. This concept is known as total quality management (TQM), as described in Chapter 3. The organization must provide a quality shield around the customer, as illustrated in Figure 14.15. This shield is only as strong as its weakest element; therefore all employees must be trained and supported to reach a high level of quality in their work.

The quality of purchased materials is as important as that for items built in house (Burt, 1989). Supplier quality is usually monitored and controlled by audits and allocation of vendor ratings, which influence the amount of business that the suppliers receive from the company. In the past, emphasis was placed on buying on the basis of the lowest price and using in-house inspection and sample testing to overcome quality problems. This was found to be very expensive, and inspection has largely been replaced by obtaining a certificate of conformance

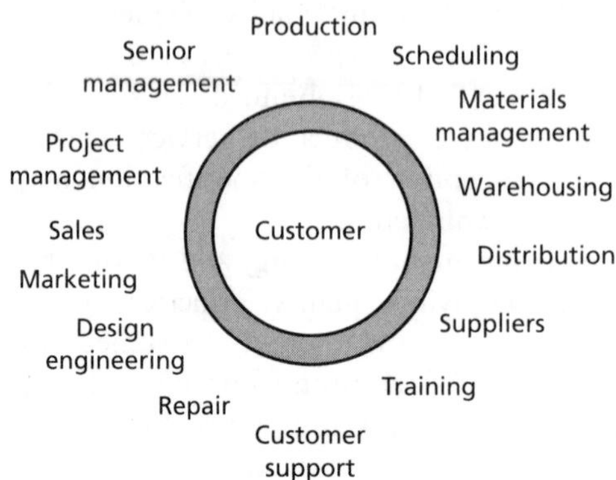

Figure 14.15 The quality shield.

from vendors, so that the requirement for bought-in quality has been shifted from the manufacturer's plant to that of its suppliers.

Organizations have discovered that quality is one of the most effective and potent of competitive weapons, both regarding product and service quality (Lockyer, 1993). An organization cannot have several quality standards, for example a lower standard for 'cheaper' work that it takes on, compared with that for more expensive work. Quality is a state of mind, so that once a lower standard is accepted it soon becomes the norm throughout the organization.

Although implementing a quality process can be expensive, in the long-term it can pay for itself, which gives rise to the phrase, 'quality is free'. Figure 14.16 illustrates the quality–cost balance. Several costs occur when implementing a quality system:

- Prevention costs, such as quality training for staff; quality audits, both internal and of suppliers; and costs associated with developing quality procedures and processes.
- Quality costs incurred in maintaining quality by appraisal methods, such as inspection and testing (Stott, 1996), both on the shop floor and for incoming goods.

The costs of not operating a quality process (know as the costs of unquality) outweigh the costs of implementing quality, as shown in Figure 14.16. These include costs of rework of goods and of scrap; the costs of warranty repair and of providing replacement goods; and the very serious costs associated with losing customers and of replacing them.

A good quality management system should work on targets for improvement as an incentive and a measure of progress. For example, these could be targets for the reduction of the amount of scrap within the plant, the reduction of warranty

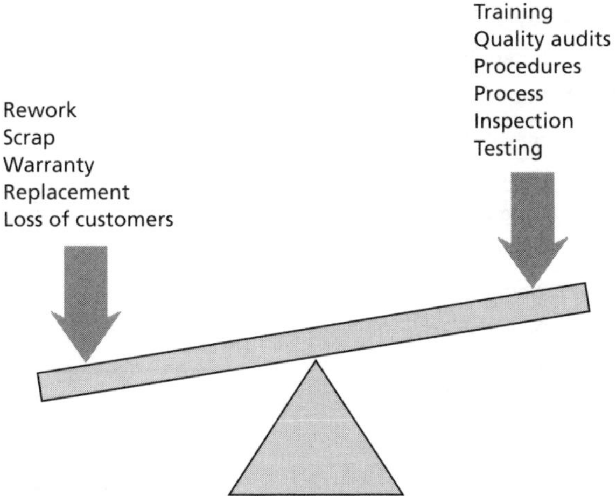

Figure 14.16 The quality–cost balance.

returns and so on. Information systems need to be set up to gather and analyze these data and to track trends.

Testing alone cannot guarantee product quality, especially in high-technology products (Williamson, 1993) and traditional methods of inspection and testing of every part can prove to be expensive and ineffective. For best results quality assurance plans must be implemented throughout the organization, including all suppliers. It is important that the organization works to recognized processes (McDonnell, 1993) such as published within ISO9000, BSI5750, ANSI/ASQC Q90 and so on.

Quality improvement must be ongoing. Once a level of quality is reached there is a need to monitor it and to ensure that it is being maintained or improved. Customers can send out their own inspectors to monitor their suppliers or they can rely on one of the professional bodies, such as the BSI. Once a company obtains BSI accreditation, for example, it can expect regular audits from the BSI to ensure that its quality is being maintained. Usually this is in the form of a visit to the plant, although modern IT techniques are reducing this need. Approved companies often carry out their own certification and store the results on computers. This information can usually be accessed remotely by BSI inspectors over telecommunication links, and a visit to the company's plant is only carried out at infrequent intervals.

A company's quality operation is probably most evident to a customer during the order management stage. An order should be considered to be a surrogate for the customer and be treated with the same respect and urgency (Shapiro *et al.*, 1992). Figure 14.17 shows typical customer–company interaction stages during the order management process and, as seen, it can involve many 'wait' stages during which the customer is cut off and not being served.

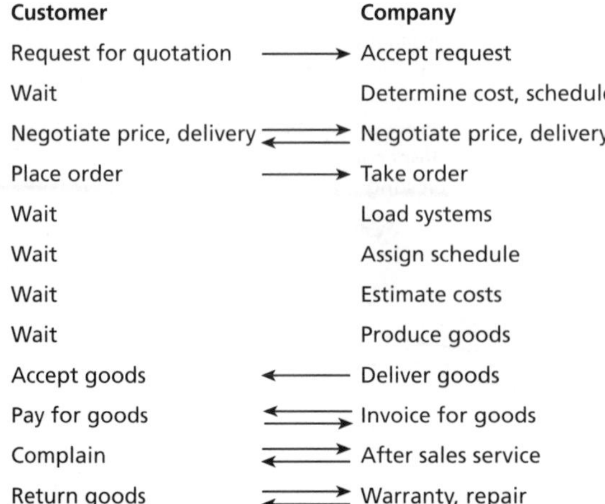

Figure 14.17 Customer–company interaction during a typical order management process.

The process starts with a request for a quotation from the customer. The company can take a considerable amount of time responding to this request, as the different functions enter into internal squabbles on the costs of doing the work and on time-scales. This keeps the customer waiting and can end up providing unreliable dates and prices.

Following this a price and delivery will be agreed between the company and the customer. This is where the company may fall at the first quality hurdle; if the price is not right the product may end up being sold at a loss, and if the delivery date is incorrect the product may be delivered late, resulting in a discontented customer.

Orders will eventually be placed and taken. This information must be distributed throughout the organization, so that everyone knows the commitment entered into. Usually, however, the order is logged by a junior clerk into a computer database, and the key people, including senior management, are ignorant of the commitment.

Orders must be scheduled on the production floor. Important customers, for example those who are large spenders and have been with the company a long time, should be given priority if necessary. However, in practice priorities are often set on the basis of which salesperson shouts the loudest, or even on how quickly the company can bill the customer and receive payment, usually to meet year-end financial targets.

Several factors need to be considered when delivering the goods to the customer, such as quality and quantity, location and time, packaging and method of transport. The delivery should be considered from the customer's perspective, not that of the company (see the case study 'Part deliveries'). After-sales service must investigate customer complaints and act promptly, especially on warranty matters, to prevent dissatisfied customers transferring to a competitor. Usually, however, customers find it difficult identifying a responsible person within the company to complain to, ending up being pushed around the organization, so that complaints are often lost or given a low priority.

Case study

Part deliveries

One of the measures of the manufacturing plant in the FloRoll Manufacturing Company was the extent to which it could make and deliver customer orders on time. Adrian Elton, Manufacturing Director, monitored this measure personally, since it received high visibility during the reviews held by Chief Executive, Jake Topper.

'We are always above 95 per cent on this', Adrian Elton used to boast, and Fred Steel, Quality Director, would agree, consulting his computer printout on the organization's targets for the month.

However, an incident leading to a telephone call from an irate customer to the Chief Executive soon revealed the fallacy of this measure. The company measured its deliveries by the amount of equipment actually shipped out of the plant. Manufacturing policy was to note the number of parts involved in the total order and then to make and ship those parts that could be completed quickly, from either finished goods or work in progress, so that the amount shipped equalled or exceeded 95 per cent of the total order. The remaining parts were scheduled into the production process, often receiving low priority over those orders that had not reached the 95 per cent mark.

The customer in this instance, Kelvin Engineering from Glasgow, Scotland, had placed an order comprising 100 different parts. Materials control discovered that 96 of these parts could be made quickly and these were completed and shipped to the customer as a part delivery. The remaining four parts were scheduled into the production plant as a low priority, to be completed two months after the promised delivery date.

The customer could not install the product, since all 100 parts were required for this to happen and, in spite of continual complaints to a variety of contacts within the company, no satisfactory delivery date was received. When the customer received an invoice for the 96 parts that had been delivered, with a request to pay within the current financial year, the company telephoned the Chief Executive and made its views known in no uncertain terms.

The quality measure on Manufacturing was changed the next day to cover completed orders only, not part deliveries. The company had learned to consider quality as seen by the customer rather than from its own viewpoint.

14.8 Information technology

Organizations of all sizes have been making major investments in information technology (IT) systems this decade, in all areas from Engineering and Manufacturing to Sales and Marketing (Bird, 1995b, 1997; Simson, 1990; Wheatley, 1997; Wilson, 1993). The costs of IT have been falling by a factor of ten every five years or, to put it another way, the power of IT-based systems has increased by a factor of ten over five years for the same cost (Gretton, 1994). Organizations have become broader and flatter, with many of their internal divisions running their own IT-based systems, often located on distributed PCs. This has resulted in the need for improved communications and increased data sharing (Knutton, 1991).

Managers are familiar with the basic IT-based systems, such as word processing, presentation software and spreadsheets. However, many IT projects fail to meet expectations, especially those dealing with executive (or enterprise) information systems (EIS) (Knutton, 1992b). This is partly due to the fact that the power of IT systems has not been fully understood and therefore the system has not been effectively designed (Bird, 1996). Computers can do standard, rational tasks quicker and more cheaply than humans (Aldrich, 1994). However, they lack the intuitive reasoning, which is so important in a business environment. Managers also have a tendency to ask for IT systems to be developed in a self-centred way, to enable them to control their business (Bird, 1994).

An IT system should be designed around how the business should operate, not as to how it has historically worked. IT presents an opportunity for change. One should never set out to computerize what is done now; it is important to find out what should be done and to computerize that. The business process should be analyzed and understood fully, then re-thought and re-engineered.

Effective IT management must begin by thinking about how people use information, not with how people use machines (Davenport, 1994). Changing an IT system will not change a company's information culture; changing the technology only reinforces the behaviours that already exist. Changing the company's information culture is the best way to implement successful IT, but it is also the hardest to carry out. Tools are just tools; new technologies alone will not change anyone's behaviour.

Figure 14.18 illustrates the major steps in implementing a typical IT project (Brunwin, 1994). The organization needs first to define its business objectives, to determine where the business is going and where the added value is coming from. The strengths and weakness of the present IT systems to meet these needs must also be evaluated.

An application review is then required. Such a review should also be done on a regular basis, approximately annually, to ensure that the systems used are current. Generally it is better to use tools for rapid application development rather than using traditional software packages.

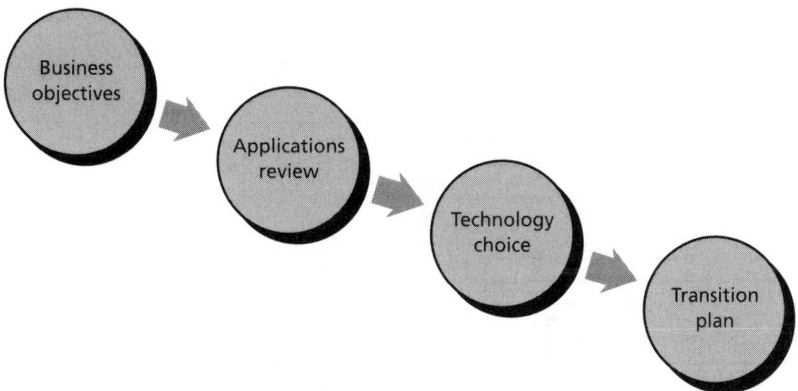

Figure 14.18 The major steps in an IT project.

The technology base chosen for the IT project must be flexible, in order to change with future needs. Many considerations need to be taken into account, such as the use of decentralized systems, based on PCs, compared with a centralized mainframe based system; the use of networks to carry multimedia data; and the use of transmission and storage media.

Finally, a transition plan must be developed to move from what is currently being used to what is needed. This plan is very important, since it will involve many elements, such as investment needs, organizational change and training of staff. It is important that a working IT system is available for the organization to use at all times during this transition phase, since any down time could cause a major exposure to the company.

Managers need relevant and timely information in order to make decisions. The availability of effective, easy to use, IT-based information systems has resulted in responsibility for decision making being pushed further down the organization hierarchy. IT has also been instrumental in linking the different operations within a company, especially in a global company, into one unit. Figure 14.19 illustrates some of the benefits an effective IT system can give to a company, and how these vary with time. In the initial stages the quality of information available to executives is improved and more effective financial and operations controls result. Communication improves, leading to effective teamwork. Decision making becomes more effective and is taken at the appropriate levels within the organization. These developments result in a fundamental change in the organizational structure of the company, making it flatter, leaner and more effective.

IT-based application areas widely used within manufacturing operations include electronic data interchange (EDI) and sales-driven manufacturing systems

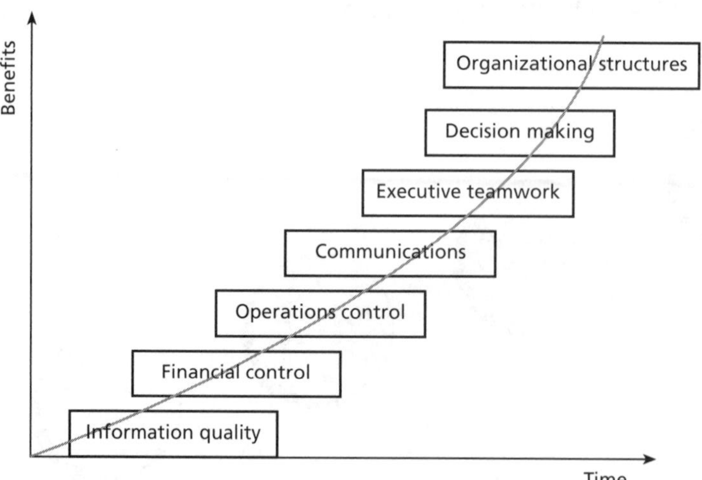

Figure 14.19 The benefits of effective IT systems.

(Sarson, 1995). It is estimated that 70–90 per cent of information put into a typical computer based system has come from another computer. It is therefore better to link the two computer systems together, so that these computers effectively 'talk' to each other. This is the basis of EDI, and its use provides an effective trading medium, although it does require a level of trust between the trading partners.

EDI provides the means for transferring business documents from one computer to another. Internationally agreed standards exist for the message format, in order to enable this exchange to occur. Typical examples of EDI documents are delivery schedules; orders; bar-coded transport labels; schedule amendments; invoices; despatch information; and remittance advice. (Fellows, 1992).

The widespread use of the Internet has provided a boost to paperless communication within and between organizations. Replacing paper with electronic transmission has several advantages, such as:

- Faster transmission of data and its integration into the IT systems, so faster action can be undertaken.
- Lower costs, since less paper is being handled in putting data in and out of computers.
- Fewer errors, since the need to enter the same information into different computers is being avoided.
- A flatter organization, since the clerical levels needed to handle data are no longer needed.

EDI provides competitive advantage, enabling the company to effectively operate systems such as just in time (JIT). With EDI orders can go directly from the

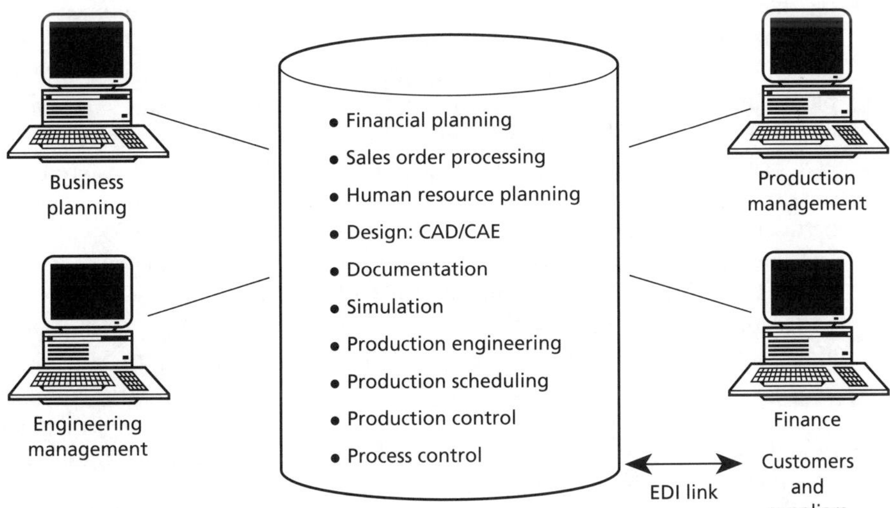

Figure 14.20 An integrated IT system.

customer into the supplier's manufacturing system (Bird, 1995b): stock levels are checked, subcontractor orders placed, product schedules started, credit limits checked and invoices prepared.

Many companies are using sales order-driven manufacturing systems, where the lot sizes are smaller and delivery times shorter. Companies often wait until the order is received before committing materials, capacity and resources (Chatha, 1994). Make-to-order is replacing make-to-forecast. This ensures tight control over inventories, but it needs an effective production management system to be able to implement rapid changes in production schedules. An integrated IT-based system is called for, as in Figure 14.20, where the various functions within the organization share a common database of information, and effective links exist with customers and suppliers.

Summary

There are several components to Manufacturing Operations, such as the manufacturing process, covering procurement, production scheduling, stock control and so on; the manufacturing systems used to produce the goods or services, such as flexible manufacturing systems (FMS); the management of materials, such as work in progress and inventory; management of people involved in the manufacturing operations; and the management of quality. All of these have to be effectively planned for and controlled.

The larger the quantity of goods built or services supplied the lower the unit cost is likely to fall, a process known as moving down the experience curve. This occurs for several reasons, such as the manufacturing tasks being learned by the operators, who become more proficient; increasing volumes ensuring that machines are kept working at full capacity, so reducing idle time; greater sales mean that more money is available for investment in new and better plant and machinery; materials costs being reduced due to increased opportunity for bulk buying; and manufacturing overheads spread over a greater volume of product, reducing the overhead per unit. Cost reductions, obtained by moving down the experience curve, can usually be transferred between similar products.

There are primarily four different types of production system: job-lot systems, used when one-off items are to be built to a customer's specification; batch production, in which different products are built in batches, the manufacturing equipment being changed between batches to suit the new product being built; mass production, which is used to build large numbers of items on a dedicated line; and continuous or process production, which is used to make items such as chemicals, in which the process operates 24 hours a day, 365 days a year, with very little operator intervention.

The trend in manufacturing is moving towards shorter delivery times and smaller lot sizes: building to order rather than to forecast. Flexible manufacturing systems assist in this process, the basic elements of such a system being computer- or numerically controlled machines; a conveyance network for moving parts between machines; and a control system for coordinating the flow of parts and the operation of the machines. FMS is used for applications where medium-sized production volumes are to be built and the number of parts per unit are also in the medium category. When production volumes are high and number of parts per unit is low a dedicated line is preferred, and for low volumes and a large number of parts a manual (job shop) operation is the most effective.

Many products are global in nature and multinational companies are moving towards global operations. This has many advantages, such as cost reductions, since raw materials can be bought in bulk; advantage can be taken of lower labour costs or capital grants in some countries; local skills can be exploited; designing for global markets can reduce re-engineering costs; and loads can be balanced by central control of stocks production.

Logistics within manufacturing is the total concept covering the flow of goods, from the supplier, through the manufacturing plant, and out to the customer. As such it involves many areas, such as materials procurement, work in progress, stock control, finished goods and distribution to the customer. Materials used within a product constitute the largest part of its overall costs and this trend is likely to increase. Materials management is therefore an important function and partnership, should exist between the company and its selected suppliers to implement programmes such as just in time (JIT) delivery. Several tools, such as materials resource planning (MRP), can also be used for ordering and scheduling of materials, although these systems have now been extended to cover many other aspects of operations planning and control.

The key to shortening product development times is to ensure that manufacturing starts at the product design stage, with manufacturability considerations taken into account when the product specification is first formulated. This will ensure that manufacturing processes and practices are taken into account when the product is being designed. However, customer requirements must not be sacrificed in this process. Quality must also be built into the product early on, during the design and manufacturing stages; it cannot be tested into the product after it has been built. There is a cost associated with producing quality products, such as the need to train operators, implementing quality processes, and inspection and testing of products. However, the cost of not meeting quality is much higher, caused by items such as rework and scrap, warranty and replacement costs, and loss of customers.

Information technology is widely used within Manufacturing Operations, as it is within all functions of an organization. Apart from control of production machines and work flow, it is also used for executive information systems

(EIS) and for linking the company with its customers and suppliers using electronic data interchange (EDI) and electronic mail. The widespread use of the Internet has also given a boost to the use of IT and paperless communications, both within and between organizations.

Case study exercises

14.1 The Bearing Standard Company makes three types of bearing used on florolls, called A, B and C. The labour hours, machine hours, material usage per unit and the number of each type made per period are shown in Table 14.4. The manufacturing manager is considering whether to make the three bearings in the same plant using the same processes or to segregate one or more of them. Can you advise him?

Table 14.4 Parameters for three bearings (case study exercise 1)

	A	B	C
Labour hours per unit	50	5	20
Machine hours per unit	10	40	25
Materials cost per unit (£)	10 000	150	600
Number made per period	200	5000	100

14.2 The wing drive for florolls has to be assembled by skilled labour. Some extent of automation is possible, but the manufacturing runs are relatively short, because customers require frequent modifications. What are the alternative production techniques that can be used to make the wing drive and what are the advantages and limitations of each?

14.3 When the Engineering department of the FloRoll Manufacturing Company first saw the Marketing specification of the boosters for the new floroll, they immediately set about improving the specification so that the product more closely met their available technology, and so could be manufactured at low cost. Manufacturing made further changes, in order to make the product fit their automation process and enable fast delivery times. Unfortunately, the product did not sell in sufficient quantities to make it a success. Discuss the possible reasons.

14.4 Adrian Elton, Manufacturing Director at the FloRoll Manufacturing Company, was determined that his manufacturing quality would be second to none within the industry. Over a period of a year he transformed the plant so that scrap and rework was reduced by 50 per cent, and the quality of product leaving the company was acknowledged to have vastly improved. Yet the return rate from the field continued to rise due to product failures after a relatively short time of operation. This problem was traced to a design fault and to a component supplied by one of the company's main suppliers. What lesson in quality had Adrian forgotten?

References

Adam, E.E. and Ebert, R.J. (1986) *Production and Operations Management: Concepts, Models and Behaviour*, Prentice-Hall, Englewood Cliffs NJ.

Aldrich, M. (1994) The information technology challenge, *Professional Manager*, July, pp. 8–9.

Bartlett, G.A. and Ghoshal, S. (1995) Changing the role of top management: beyond systems to people, *Harvard Business Review*, May–June, pp. 132–42.

Bird, J. (1994) Senior service, *Management Today*, May, pp. 85–8.

Bird, J. (1995a) Limited intelligence on the shop floor, *Management Today*, July, pp. 68–70.

Bird, J. (1995b) Connect IT up, *Management Today*, June, pp. 78–80.

Bird, J. (1996) It's all IT to me, *Management Today*, June, pp. 78–82.

Bird, J. (1997) Strategy in a box, *Management Today*, May, pp. 77–83.

Brunwin, V. (1994) Making IT fit your business, *Professional Manager*, July, pp. 10–11.

Burt, D.N. (1989) Managing suppliers up to speed, *Harvard Business Review*, July–August, pp. 127–135.

Bywater, P. (1994) MRP II: planning for success, *Software for Engineers Supplement*, IEE, London, March, pp. S-11–S-13.

Chatha, A. (1994) Production management systems: the missing link between business planning and DCS, *Control Engineering*, August, pp. 55–8.

Chen, F.F. and Adam, E.E. (1991) The impact of flexible manufacturing systems on productivity and quality, *IEEE Transactions on Engineering Management,* February, pp. 33–45.

Chen, F.F. and Everett, E.A. (1991) The impact of Flexible Manufacturing systems on productivity and quality, *IEEE Transaction on Engineering Management,* February, pp 33–45.

Cooper, J. (1994) *Logistic and Distribution Planning: Strategies for Management*, Kogan Page, London.

Davenport, T.H. (1994) Saving IT's soul: human-centred information management, *Harvard Business Review*, March–April, pp. 119–31.

Fellows, D. (1992) Trading information pays in business, *Engineering Computers*, May, pp. 23–5.

Gaither, N. (1987) *Production and Operations Management*, Dryden Press, Chicago.

Garvin, D.A. (1983) Quality on the line, *Harvard Business Review*, September–October, pp. 65–75.

Grant, R.M. *et al.* (1992) Appropriate manufacturing technology: a strategic approach, *ERM*, Summer, pp. 13–21.

Gretton, I. (1994) The promise of IT, *Professional Manager*, July, pp. 20–1.

Hepworth, A. and Stevenson, D. (1993) Discover the real cost of manufacturing, *Engineering Computers*, February, pp. 33–5.

Hill, T. (1993) Manufacturing strategy, *Professional Manager*, July, pp. 20–2.

Holder, R. (1991)) MRP II – Misused or a flawed philosophy? *Engineering Computers*, May, pp. 43–9.

Karmarkar, U. (1989) Getting control of Just-in-Time, *Harvard Business Review*, September–October, pp. 122–31.

Knutton, P. (1991) Plan your way to freedom of choice, *Engineering Computers*, May, pp. 55–59.

Knutton, P. (1992a) Winners build teams not empires, *Engineering Computers*, May, pp. 10–13.

Knutton, P. (1992b) Coming your way, wherever you sit, *Engineering Computers*, May, pp. 10–13.

Knutton, P. (1993) Abandon stock and deliver the goods, *Engineering Computers*, February, pp. 22–7.

Leonard-Barton, D. *et al.* (1994) How to integrate work and deepen expertise, *Harvard Business Review*, September–October, pp. 121–30.

Lesser, D. (1983) Quality control: a primer for the manager, *Machine Design*, September, pp. 63–8.

Lockyer, K. (1993) Quality: the cutting edge that starts with a state of mind, *Management in Industry*, October, pp. 9–10.

Maliniak, L. (1992) Engineers must make manufacturing a priority, *Electronic Design*, 15 October, pp. 39–48.

McDonnell, R. (1993) The basics of BS5750, *TEST*, March, p. 9.

McGrath, M.E. and Hoole, R.W. (1992) Manufacturing's new economies of scale, *Harvard Business Review*, May–June, pp. 94–102.

Morley, M. (1996) Building manufacturability into product development processes, *Professional Manager*, March, pp. 12–13.

Pisano, G.P. and Wheelwright, S.C. (1995) The new logic of high-tech R&D, *Harvard Business Review*, September–October, pp. 93–105.

Quayle, M. (1993) *Logistics – an integrated approach*, Tudor, Eastham.

Render, B. and Heizer, J, (1994) *Principles of Operations Management*, Allyn & Bacon.

Sarson, R. (1995) Get in touch with EDI, *DEC Computing*, January, pp. 29–30.

Schmenner, R.W. (1987) *Production/Operations Management: Concepts and Situations*, SRA, Chicago.

Schroeder, R.G. (1989) *Operations Management*, McGraw-Hill.

Shapiro, B.P. *et al.* (1992) Staple yourself to an order, *Harvard Business Review*, July–August, pp. 113–22,

Simson, E.M. (1990) The 'centrally decentralized' IS organization, *Harvard Business Review*, July–August, pp. 158–62.

Stott, D. (1996) Questions of quality, *The VAR*, April, pp. 35–8.

Tersine, R.H. (1994) *Principles of Inventory and Materials Management*, Prentice Hall, Englewood Cliffs NJ.

Upton, D.M. (1995) What really makes factories flexible? *Harvard Business Review*, July–August, pp. 74–84.

Venkatesan, R. (1992) Strategic sourcing: to make or not to make, *Harvard Business Review*, November–December, pp. 98–107.

Ward, P. (1994) Logistics: a simple guide, *Professional Manager*, May, pp. 10–11.

Wheatley, W. (1997) Hidden costs of the humble PC, *Management Today*, January, pp. 52–4.

Williamson, A. (1993) Remote auditing of QA, *TEST*, March, pp. 11–13.

Wilson, D.A. (1993) *Managing Information*, Butterworth-Heinemann, Oxford.

Part VI Marketing and sales management

15

Markets and marketing

Introduction

This chapter looks at the fundamentals of marketing and its role within an organization. It describes the basic concepts associated with marketing, such as the marketing mix, the importance of marketing information, the techniques and need for market segmentation, and the differences between consumer and industrial markets.

15.1 The market

15.1.1 What is marketing?

The key to marketing is developing a product or service that a customer wants, getting it to the customer at a price the customer expects to pay, and making a profit for the company out of the transaction.

It is important to realize that customers do not buy products – they buy perceived benefits. Therefore they do not buy drills, they buy holes (Sowter, 1993).

Marketing is not primarily about selling or advertising; it is a way of thinking about doing business (Houston, 1986). Marketing is also not solely the province of the marketing department, but of the whole management team. Anyone who has any form of customer contact is part of the marketing process (Mitchell, 1994).

In a competitive environment the product or service is not the main factor influencing the buying decision. Other aspects, such as after-sales service or proven record for delivery, are often more important.

Supplier–customer relationships are important, as for example in a just in time operation (Webster, 1993). However, in the new economic order clear distinctions between companies and markets, and between the company and its external environment, are rapidly disappearing.

Marketing is an expensive process and its effectiveness should be constantly monitored and measured (Mitchell, 1995a). Customers have to be taken from product awareness through to loyalty to the company and its products (Boyd and Walker, 1990; Doyle, 1994; Frain, 1983; Hutt and Speh, 1989; McKenna, 1991; O'Brien and Jones, 1995).[1] Costs are high and incurred over many years, even when specific markets are targeted. It is important to target groups who have no current strong loyalty to an established supplier, since it is very difficult to break the bond between a satisfied customer and its supplier. High spenders should generally be targeted, as well as those with long-term growth prospects. Reliable customers who are dissatisfied with their present suppliers are prime targets, not those who have no fixed loyalty, always going for the lowest price.

Targeting is important, even if one is first into the market. It is important to build up customer loyalty quickly in order to prevent customers being wooed away by competitors (Mitchell, 1995a; Grant and Schlesinger, 1995) and one cannot build this loyalty with every potential customer.

Long-term investment is needed to develop loyal customers and future profits, so marketing must be considered to be a strategic investment (Slywotzky and Shapiro, 1993). Unfortunately, most companies base their marketing budget on their yearly sales forecast, which is wrong. Instead, marketing expenses should be treated like any other strategic capital investment which is expected to show a return over a number of years. It takes time to establish a new product, to win customers, to enter new markets and so on.

Figure 15.1 shows the lag that exists between marketing investment and sales revenue. The revenue in any year is affected by the cumulative marketing investment over previous years. Furthermore, the amount of investment needed in subsequent years needs to be increased to achieve the same level of sales and keep market share. The curves in Figure 15.1 can be affected by unexpected events, such as environmental factors or the entry of a strong competitor.

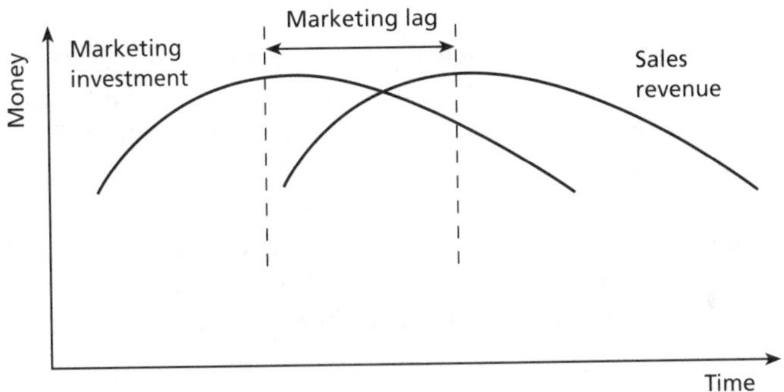

Figure 15.1 The lag between marketing investment and sales revenue.

Trading is essential in modern society – not everyone can make everything needed competitively. Marketing defines what to make, how to ensure customers know about it and buy the product, and how to get it to customers. The aim of marketing is to set up a win–win situation where the company and its suppliers and customers benefit.

Considerable flexibility is called for in marketing, in order to deal with changed situations, as illustrated in the case study 'Thinking laterally'.

Case study

Thinking laterally

It was clear to the Board of the FloRoll Manufacturing Company that one of their star products, Floroll XZ, was nearing the end of its sales life. It had been sold into the defence market for many years and had generated healthy profits, but sales had fallen steadily over the past few years.

'The way things are going', said Peter Sell, Sales Director, 'I would not be surprised to find that the numbers tailed off to zero over the next three years.'

'It is becoming very uneconomical to build in such small batch sizes,' complained Adrian Elton, Manufacturing Director. 'I think we should do the rounds of our customers to get their anticipated demand and ask them to do an all-time buy. We can then make the whole lot in one go and deliver to customers as they need them.'

'That certainly sounds like a good strategy', agreed Jake Topper, Chief Executive. 'What do you think Jack?'

Jack Fox, Director of Marketing, had sat very quietly while the exchanges were going on. He now cleared his throat and, walking over to the whiteboard, began scribbling.

'I think we have several options', said Jack Fox. 'Adrian has outlined one of them, and I must say that I would consider that the least favoured option. I think that instead we should consider the following.

'First we should increase our production volumes over the coming year. Our market intelligence reveals that, because of the decline in the market, two of our biggest competitors have recently dropped out. Several of their customers will be looking for new suppliers and I think we should be sending in our best salespeople to call on them.

'The second thing we need to do is to look for other applications for Floroll XZ. The oil industry is an obvious user, and although Floroll XZ does not fully meet their requirements I believe that we should be able to capture quite a lot of this market on price.

'The third action we need to take is to modify Floroll XZ so that it not only meets the oil industry requirements much more closely, but also that of the other utilities. One of my product managers has prepared a short specification of what we believe is needed, and Engineering have given an estimate of the time and money required to make the modifications. This is entirely reasonable and gives a very attractive business case.'

After some further discussions Jack Fox's proposals were accepted by the Board. What at first sight looked like a loss of business had been turned into a substantial marketing opportunity by some lateral thinking.

15.1.2 Marketing's role

Marketing operates at three strategic levels (Webster, 1993):

- **The corporate level**. Here Marketing leads in the definition of the business the organization is in and the formulation of its mission, structure and shape. What goods should the organization make, which ones should it buy, and which ones should it sell? Customer needs are determined and the customer is championed internally. Emphasis is on the organization's strengths and the whole organization is focused on the customer.
- **The business level**. The key at this level is determining how the organization should compete in its chosen area. Several factors need to be taken into account, such as market segmentation; targeting markets; product positioning; and creating joint ventures and strategic alliances.
- **The operational level**. At this level decisions are made regarding the marketing mix (as explained later) and strategy is implemented. The customer must be managed along with the organization's resellers.

Figure 15.2 illustrates marketing's role in more detail. The first task of marketing is determining the customer's needs or requirements, both immediate and in the future (Jackson, 1985). This is not always easy, since very often customers do not know precisely what they want. It is also easy, in a technical environment, to let technical factors dominate and hide other important customer needs.

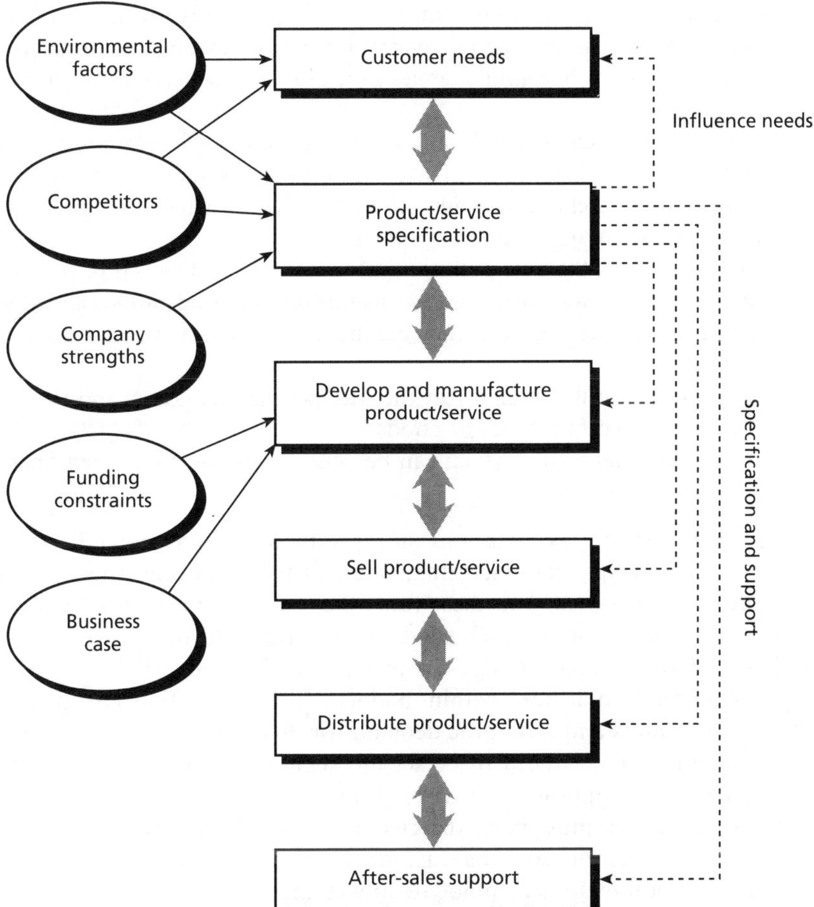

Figure 15.2 Marketing's role.

Customer requirements are always changing. They are influenced by the environment, by competitors and even by the marketing function (Lele, 1994). Continuous market research or intelligence is therefore essential.

The next step it to determine how to meet customer needs. In doing marketing, managers must also take into account environmental factors and exploit competitor weakness. In addition, they need to consider the organization's strengths and how these can be made to match customer requirements.

There are many environmental factors, such as:

- Demographic, for example changes in population towards older people.
- Economic, such as the effect of inflation on the buying power of retired people.

- Legal and ethical. There are many laws governing advertising standards, the labelling of goods, and the misrepresentation of customer information.
- Technical, caused by rapid changes in technology which have occurred in recent years.
- Infrastructure, such as road, rail, air transport and telecommunications. Many companies are in the infrastructure business and are directly affected by infrastructure changes, while others use the infrastructure, and this has a secondary, but major, impact on them.
- Ecology, such as the anti-pollution movements, the disposal of packaging material and the very real need to ensure that all goods are designed so that, when they become obsolete, they can be disposed of in an environmentally friendly way.
- Social and cultural, which affect the way people buy goods and their preference for certain types of goods.
- Political considerations, which can become very important when operating in diverse overseas markets.

Product and service specifications are produced to cover all aspects of the business: engineering, manufacturing, sales, distribution and after-sales support. Assistance is also provided to these functions, such as by the production of sales literature, definition of recommended selling price and discount structures, and the design of suitable packaging. The product must be advertised in the right way and to the right audience, within budget. It is marketing's responsibility to influence customers and create the demand for its products.

Distribution channels need to be set up, such as negotiations with wholesalers and retailers and adoption of overseas agents.

After-sales service must meet the customer's needs and must be monitored so that it continues to do so. The customer's original requirements must also be tracked and modified to accommodate any changes.

One of the most important tasks of marketing is managing the overall business case for the product or service. Adequate funding must be available to meet internal requirements, such as developing the product, setting up a manufacturing line, and training sales and support staff. In addition, the sales over several years must be estimated and the expected profit determined. This must meet the investment requirements of the company.

15.1.3 Marketing organization

Figure 15.3 shows the main groups in a marketing organization. Product Line Management (PLM) is responsible for managing the product business case, for meeting profit targets and for budget control. It identifies customers and provides the top-level customer contact. Progress on any project is monitored via the programme or project management organization, as described in Chapter 13. The product line manager is often referred to as a brand manager in consumer organizations.

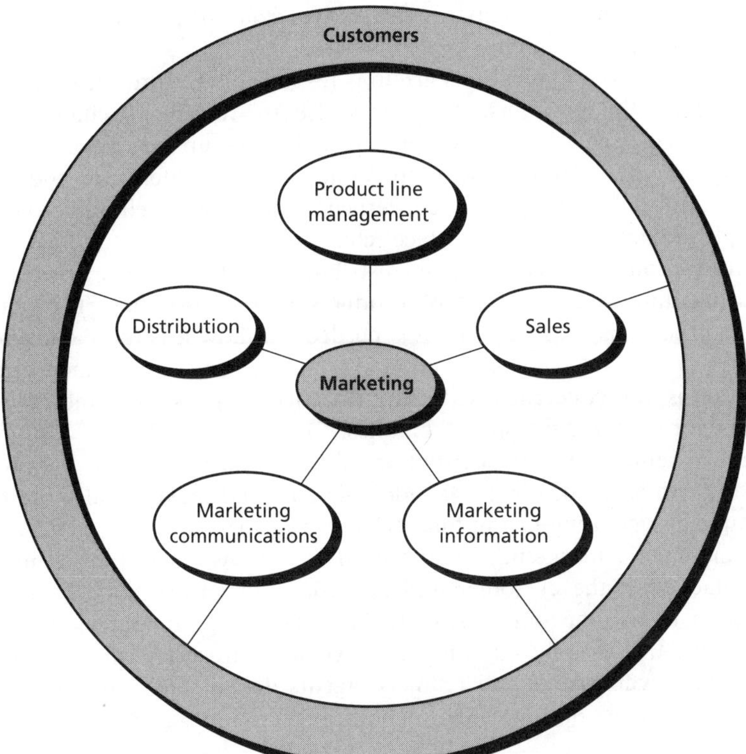

Figure 15.3 Elements in a marketing organization.

Sales is responsible for selling the product and for closing the deal, usually in face-to-face contact with the customer. The sales organization provides the customer with information regarding the product, which is usually a crucial role when it is a complex technical product. In addition the salesperson is best placed to pick up customer requirements and competitor information, and to feed this back to the rest of the marketing organization.

Sales or account managers often set up customer visits, for specialists to visit the customer to discuss technical problems or for the customer to visit the company's plant. In addition, they can determine the need for after-sales support and help their customers to sell the product on to the end user, if there is one.

The marketing information group helps the PLM to identify users for new and existing products and to determine their needs and why they buy specific products. They advise on the literature needed to service this market and sell the product or service. The group can also research the total market size and value, the company's share of this market and the trends in the market. An important role of marketing information is to obtain competitor intelligence data. It is important, however, that competitor information is not taken to extremes. There is a real danger that a company can move from being customer-focused to becoming competitor-focused.

The marketing communications group is responsible for producing the material used in selling the products. They advertise the product, selling it and the company's image, negotiate space in magazines, set up exhibitions and so on. Marketing communications also prepares all the sales literature and often the product handbooks. They are concerned with 'mass selling', whereas the salespeople are involved in face-to-face selling.

The fifth group in marketing is distribution, who are responsible for the storage, handling and transport of products to the customer. They determine where the product is needed, when it is needed and how it is to be delivered, such as by air, sea or land, and using internal transport or the customer's transport. Distribution is also concerned with the most effective packaging and how to pack the goods, for example the size of bundles, for ease of use by the customer. It is important to remember that distribution adds significant value to the product or service and can save customers considerable cost and make them more efficient, for example by operating just in time delivery systems.

Surrounding the marketing organization is the customer. Marketing has the most contact with the customer and its task is to ensure that the rest of the organization is also customer-focused. Pure marketing companies do not make any products; they simply buy and sell. Examples are wholesalers, retailers and agents. Unlike wholesalers and retailers, agents do not take title to goods; they merely get a commission for selling them.

15.1.4 The marketing mix

The marketing mix comprises the four Ps: product, price, promotion and place. It represents a simplified definition of the parameters available to the marketing manager when formulating strategy.

Product represents the physical parameters of the product or service that is being sold. For example, it could refer to the technical features of the product; to its quality and reliability; the handbooks and instructions provided with the product; the warranty and after-sales service; and so on.

Price is clearly the price that is put on the product or service. It includes all aspects of pricing, such as discounts, credit terms, loans, contribution to profits and customer perception of the price.

Promotion is the advertising associated with selling the product. It includes sales promotions, such as competitions and coupons. It also covers sales staff, who carry out face-to-face promotion. Included in promotion is the effort the company puts into selling itself and its image, ensuring that everyone who is likely to come into contact with a customer is well trained. There are many methods of promotion, such as advertising in magazines or on the company's vans and buildings.

Place primarily refers to distribution of the product. It includes the physical delivery of the goods and the various outlets used, such as retailers, wholesalers, agents and catalogue sales. Inventory and storage management also come into the equation, along with the problems of distribution to overseas markets.

There are several considerations when using the marketing mix:

- The combination used must be consistent with the company's image and the product strategy.
- It must be used so as to give competitive advantage in selected areas.
- It must be consistent with the competitor strategy chosen, for example, taking on a competitor head on or avoiding it by operating in a niche market.
- It must be implementable using the resources available to the organization, both in relation to money, skills and people.
- It must meet the needs and expectations of the company's target market.

The marketing mix is usually varied to suit the product or market in which the company wishes to operate. (See the case study 'Mixing it'.)

Case study

Mixing it

Following Jake Topper's Executive Staff Meeting at which the problems with Floroll XZ had been discussed (see the case study 'Thinking laterally'), Jack Fox, Marketing Director, and Peter Sell, Sales Director, sat in Jack's office and considered how the strategy outlined earlier by Jack should be implemented.

'Let's consider the various markets and how we can vary the marketing mix in each area', suggested Jack.

'Okay', agreed Peter. 'Now your suggestion was that we continue to sell Floroll XZ in the present defence market, since many of our competitors have pulled out, and at the same time we target the oil industry with the existing product while developing an enhanced version of Floroll XZ for the wider utilities market.'

'Correct', said Jack. 'Now let's take the first area, selling to the present market. We know the market and our customers know us. The products that our competitors were selling were also very similar to ours. If we want to capture a greater share of this market, price is probably the most effective of the marketing mix ingredients.'

'I agree', said Peter Sell. 'If we could offer an increased quantity discount I think we will not only win a much larger share of the market but also achieve Adrian Elton's desire to have customers place orders in larger batch sizes.'

'Excellent! So price is a very effective marketing mix weapon to use in the first scenario. Now for the second; selling into the oil market with our existing product.'

'I don't think this market is price-sensitive; those boys spend millions getting at the black stuff. No, I think they are much more sensitive to the reliability of the product; it wouldn't do for a failure at the critical instance when life is dependent on it.'

'Good', said Jack Fox. 'In that case we should emphasize the reliability and quality of Floroll XZ. After all, it has been proved in the military field. We will repackage the product, emphasizing its reliability and sell it with a five-year warranty. In exchange I think we should increase the price by 20 per cent. Do you think you will be able to sell that?'

'I think so', said Peter. 'Repackaging is a good idea and the extra price will not be that critical.'

'For that price we should also be able to throw in a 'deliver anywhere' offer for a minimum order batch size', continued Jack Fox. 'We should be able to charter a helicopter relatively cheaply if delivery to oil rigs is needed.'

'That should help me to clinch the deal', said Peter. 'I'll keep it up my sleeve and use it if I think the customer is hesitating.'

'So we have used price (increase), product (reliability and quality) and place (free delivery) for our second market', said Jack Fox. 'That's three from the marketing mix. Let's see if we can use all four for the last scenario.'

'That's modifying the product and selling into the wider utilities market', said Peter Sell. 'The price will certainly need to come down; not all the utilities are as well off as oil seems to be.'

'Yes, I agree', said Jack. 'However, I think we should continue to include free delivery in the price, since this is clearly what is going to be required in this market sector.'

'The market is going to be difficult to reach and is going to need education. We are also relative newcomers so I think quite a lot of promotion is called for to back up my sales team.'

'I'll ask John Low, my Marketing Communications Manager, to have a go at advertising. He is very good at it. I suspect a few full page advertisements in the trade journals, selling the company image, is called for, as well as a stand at the annual utilities exhibition.'

'Good', said Peter, 'that's three out of the four marketing mix items: price, place and promotion.'

'The fourth, product, is easy', said Jack. 'Reliability and quality will still be important selling points, but I think we should also now emphasize the fact that we have specially designed the product to meet the needs of this target market. We will get Micro Dowling to ask one of his boys to put together a technical presentation on the merits of Floroll XZ Mark 2, and you can wheel him up to the customers and let them see for themselves how good it is.'

15.1.5 Markets and products

A market is where sellers come to sell their products and where buyers come to buy in order to meet their requirements. In a simple market consumers and producers meet and do business face to face, such as in a bazaar. In a complex market several layers and levels of complexity are involved.

The strategy to be followed for a product in any market is illustrated in Table 15.1. If selling an existing product into an existing market, then the product and the company are known and the key is to penetrate the market further in order to increase market share. Competitive strategy is important and the aim of all participants must be to grow the market as much as possible.

Table 15.1 Product and market strategies

Product	**Market**	
	Existing	**New**
Existing	Penetrate market further Displace competition Increase market share Expand market	Develop new market Create demand 'Tweak' product Sell company image
New	Build on existing brands Build on reputation	High risk Diversification Needed for growth

Selling existing products into new markets is more difficult, since the company is not known and its image must be enhanced. It is also important to create a demand for the product, and the product may need to be 'tweaked' in order to meet the requirements of this target market more fully.

Selling a new product into an existing market is usually easier, since the company is known and it can build on the reputation of its other products (build on the brand image). However, the highest risk strategy is selling a new product into a new market. This is known as diversification, and it may be the only route that the company can follow if it intends to grow longer term.

Generally products and markets are divided into consumer and industrial. Industrial markets are unique in several ways:

- Industrial buyers are different from consumers, since they are not buying for their own consumption; they buy on behalf of their employers.
- Industrial sellers usually have very few customers compared with consumer producers. This can be dangerous, especially when a supplier becomes dependent on a handful of customers, since a shift in the buying pattern of one of them can have a significant impact on the suppliers' business.
- Industrial products are normally tailor-made to meet the requirements of a few customers. The products also tend to be very technical and the customers are sophisticated and buy on a number of factors: technical, quality, price, delivery and so on.

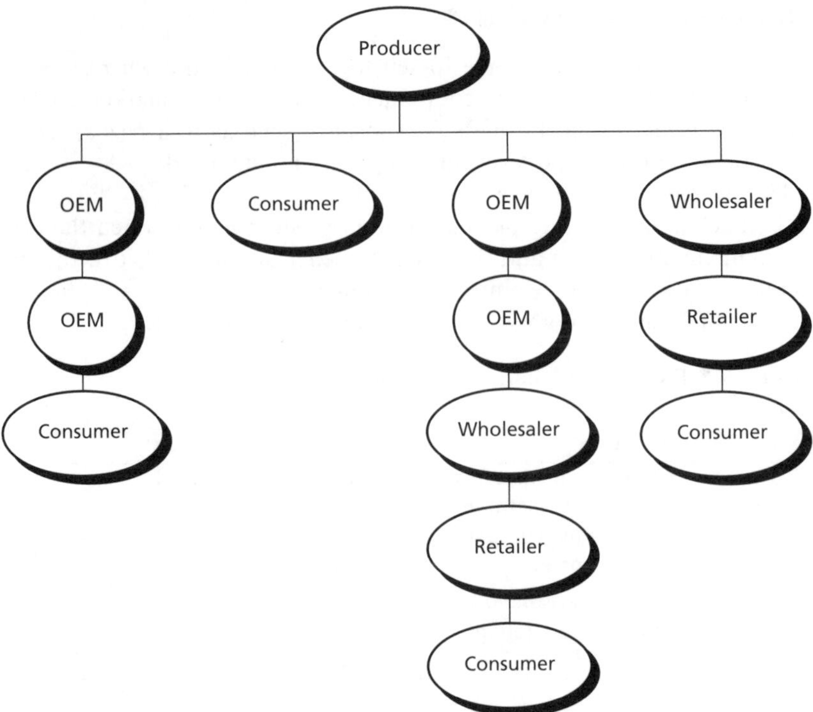

Figure 15.4 Alternative producer–consumer chains.

- Industrial customers usually require a package of services in addition to the product (Cespedes, 1994). It is becoming increasingly common to sell complete solutions (systems) rather than piece parts.
- Industrial products are frequently incorporated into other products by original equipment manufacturers (OEMs) and sold on. Therefore the chain between the producer and end consumer can be quite long. This is illustrated in Figure 15.4. Determining the customer requirements, market size and so on now becomes difficult, since the requirements of all the intermediaries in the chain must be considered, and there is very little contact between the producer and the end consumer.
- Since the chain between producer and consumer can be long there is a time lag between demand and supply. This effect can be seen from Figure 15.5, where it is assumed that a component manufacturer has ten customers, each making equipment; each equipment manufacturer has ten wholesalers who in turn each have ten retailers, and each retailer has ten end customers. Each end customer consumes ten components (priced at £1 each) so the initial demand on the component manufacturer is £1 000 000. If now the end customers reduce their requirements by 20 per cent then the potential inventory up the chain is as seen: from £200 for the retailer to £200 000 for

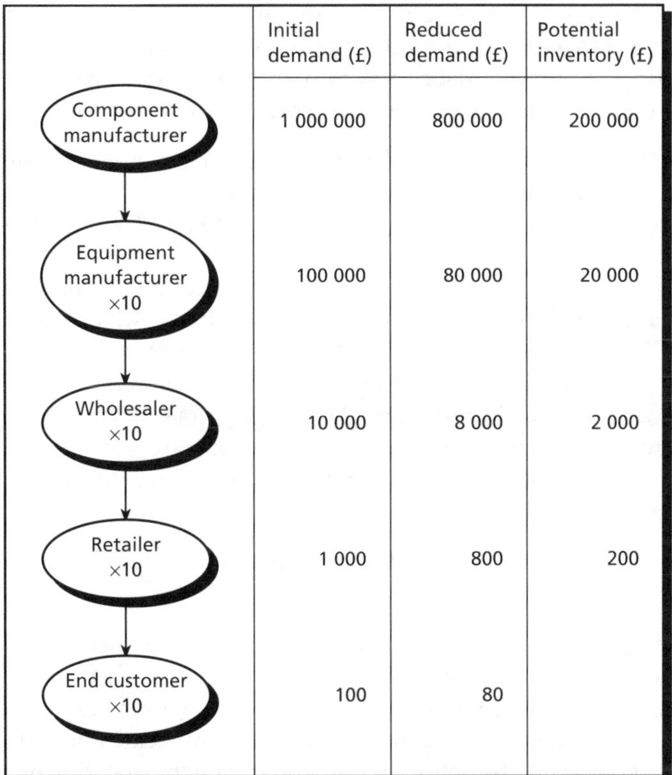

	Initial demand (£)	Reduced demand (£)	Potential inventory (£)
Component manufacturer	1 000 000	800 000	200 000
Equipment manufacturer ×10	100 000	80 000	20 000
Wholesaler ×10	10 000	8 000	2 000
Retailer ×10	1 000	800	200
End customer ×10	100	80	

Figure 15.5 The impact of end-customer demand change on suppliers further up the supply chain.

the component manufacturer. To avoid this magnification effect the component manufacturer must study the end customer market and react quickly to any changes, well before it has filtered up the chain.

Markets are never one homogeneous mass. Usually they consist of many different segments, each with different needs, values and so on. Market segmentation strategy, discussed later in this chapter, considers the best way in which to segment the market and the market mix needed in order to be successful in each segment. For example, one may wish to divide the restaurant market into transport cafés and gourmet restaurants. The marketing mix for cafés is simple wholesome food (the product) at low prices (the price). For gourmet restaurants it is sophisticated food at relatively high prices. Once market segmentation has been carried out it is possible to aim to meet several segments at once with a common marketing mix.

Organizations usually find it more difficult to operate in international markets, compared with their home market. It is certainly true that a company that is successful in its home market is more likely to succeed in the international market. International markets are usually entered for a variety of reasons, such as increasing market size and therefore profits; serving international customers who want to deal with international suppliers; and attacking potential competitors on their home territory, to prevent their entry into one's own home market.

Operating in international markets presents several problems:

- The local culture may be very different from the home culture, making it difficult to understand customer requirements fully.
- Political and ethical considerations may be very different from the home market and be more restrictive.
- The product may need to be modified in order to meet the requirements of the local market. This is often true of technical products, which need to meet a variety of local standards.
- The company will not be known in the international market, so considerable advertising will be needed to promote the company image. Many companies, however, prefer to adopt a new image when they operate overseas, one which is more in tune with that of the host country, so as to give the impression of being a local operator.

There are several ways in which an organization can enter an overseas market. These are:

- By exporting products from their home base. Local wholesalers or agents can be used who deal with the customers.
- By licensing others to make and sell the product in the overseas market being considered. This is in effect a technology transfer arrangement and the supplier would receive a fee, plus a royalty on each item sold. Some countries require technology transfer as a condition for allowing the foreign company to enter its markets.
- By forming joint ventures or alliances with local partners in the overseas country. Again this may be a legal requirement of the country concerned. Joint ventures usually result in shared ownership, with the joint venture or alliance having its own local facilities.
- By setting up a local subsidiary overseas which is wholly owned by the parent company. The local subsidiary could manufacture the product locally or import it from the parent country, but it would almost certainly carry out local sales and support. Local subsidiaries require considerable investment but they do provide the best means for gaining familiarity with an overseas market.

15.2 Marketing information

15.2.1 The need for marketing information

Most companies spend large amounts of money on gathering marketing information, both from internal and external sources. Good information is essential if good decisions are to be made (Crask, 1994; Keller and Staellin, 1987).

Companies usually keep information regarding their internal performance. However, this on its own can be misleading, as illustrated in the case study 'Success or failure?'.

Data are expensive to collect, and one can become swamped by a mass of data, so missing the critical factors. One must therefore plan well and be selective in what is collected. The key questions to be asked are: are these data needed by the business, and how can one ensure that they are accurate?

A company needs to adjust its internal policies relative to external factors, such as environmental changes, competitors and customer needs. These factors continuously change and must be monitored, and items such as the marketing mix adjusted if the company is to survive. Marketing information gathering is a continuous process, which enables the company to react swiftly to changes and to identify opportunities (Figure 15.6).

Marketing information affects the whole organization, although it is the Marketing function, as the keeper of the company's business and product direction, which usually takes the lead in monitoring changes and feeding them back to the organization.

The need for good marketing information has become much more urgent in recent years, because:

- The risks in the market-place have increased and the penalties for failure by not meeting customer requirements are much higher.
- Fierce competition has meant that a company must maximize use of its own strengths and exploit its competitors' weaknesses in order to succeed.
- Product lifecycles have decreased, due primarily to rapid technology changes (Baatz, 1991). Therefore successful new product and service opportunities must be identified on a regular basis if the company is to stay in business.
- Changes in the market structure have meant that the producer of the product or service is often remote from the end user or customer. Feedback is not received on what is required, nor on the performance of the company as

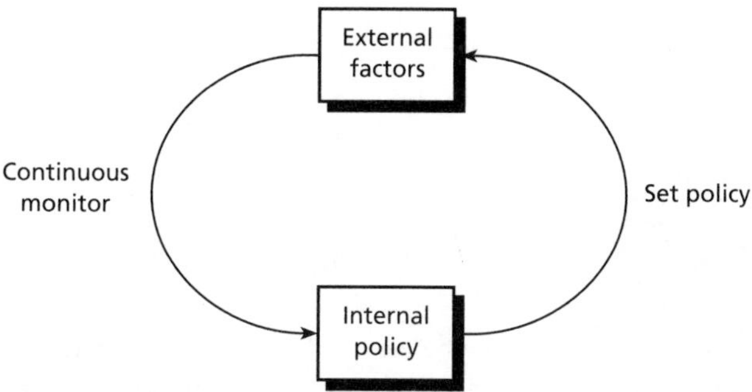

Figure 15.6 The need for continuous marketing information.

perceived by the customer. Marketing information is needed to fill this gap.

- Because of long product development and market penetration times, decisions often have to be made well in advance, so the decision-making process becomes more difficult and must be supported by good information.

Case study

Success or failure

It was budgeting time at the FloRoll Manufacturing Company, and each department was fiercely defending its allocation for the coming year.

'If we cut Manufacturing's allocation any further we will not be able to reduce the manufacturing cost of the new florolls, as I have been tasked to do', said Adrian Elton, Manufacturing Director.

'Well, we have to make cuts somewhere, we all agree on that', said Jake Topper, Chief Executive.

'I know', said Adrian, 'but let's make the cuts somewhere where it doesn't show. Marketing, for example.'

Jack Fox, Marketing Director, raised an eyebrow, to show his contempt, but did not reply.

'After all', continued Adrian Elton, warming to the topic, 'what does Marketing really buy us? Take the amount we spent on Marketing Information last year. What did we get for it?'

'Information on how badly your manufacturing shop was performing', replied Jack Fox.

'Rubbish!', exploded Adrian Elton. 'Our products are second to none. See for yourself. Our prospectus shows how our sales have increased dramatically over the past few years.'

Adrian Elton held up the company prospectus, which showed the sales curve, as in Figure 15.7.

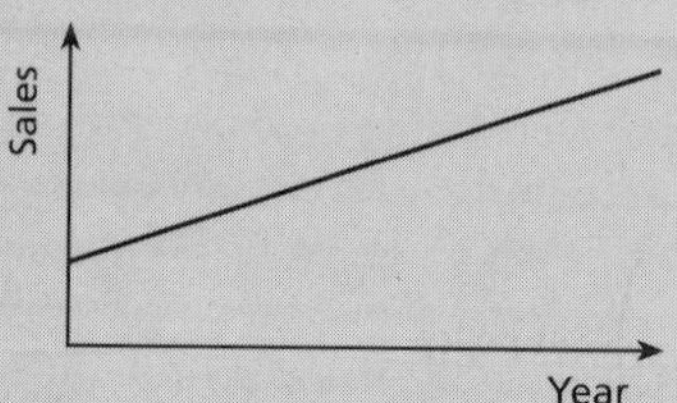

Figure 15.7 Sales curve for the FloRoll Manufacturing Company.

'You have illustrated the precise reason why we need good marketing intelligence', said Jack Fox, rising to his feet. 'In the prospectus we showed the overall sales of the company. What we did not show was how we are performing relative to our competitors. This information has been obtained by the marketing group. I was going to present it when we reached that section in the agenda of the meeting, but you can see it now.' Jack placed a chart onto the overhead projector, as in Figure 15.8.

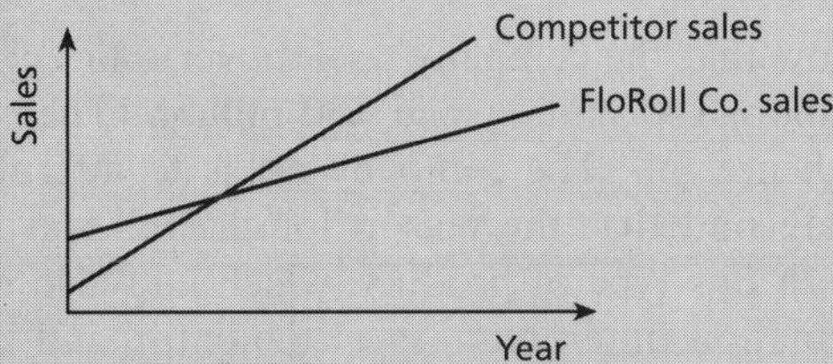

Figure 15.8 Sales of the FloRoll Manufacturing Company compared to competitor sales.

'From this chart you can see that our sales have been increasing much more slowly than our competitors', continued Jack Fox. 'This means that we are steadily losing marketing share, as in this next slide.' Jack put Figure 15.9 onto the screen.

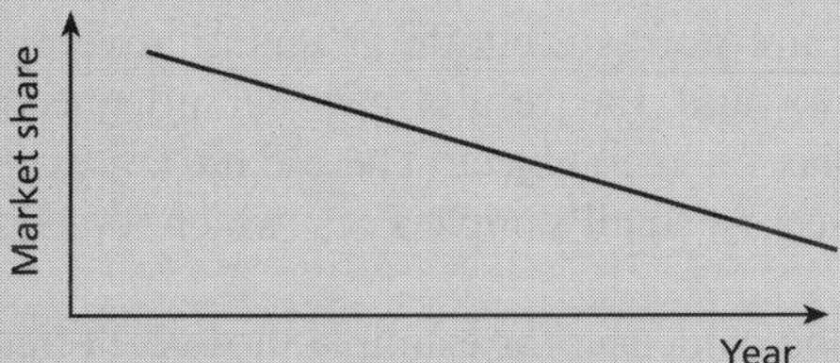

Figure 15.9 Market share of the Floroll Manufacturing Company.

'Where is the problem?', asked Jake Topper.

'The reason is quite simple', said Jack. 'Further analysis of all our competitors has shown that our products are priced quite a lot higher, although our quality is no better. Peter Sell (Sales Director) touched on this when he said at the last meeting that his sales force seemed to be hitting their head against a brick wall trying to get new customers. We are managing to keep our existing customers, who have been with us a long time, but we are not making much impression on the rapidly increasing market of new customers.'

'But we can't reduce our price without cutting our margins severely', protested Johnny Bacon, Financial Director.

'Exactly!', agreed Jack Fox. 'And there lies our problem; further marketing information has shown that our manufacturing costs are some 30 per cent higher than our competitors. At this rate we are going to cost ourselves out of the market.'

15.2.2 Data and information

It is important to define two terms: data and information. Data are the facts and figures that are collected from a number of sources. Information is the interpretation of this data in order to determine what they mean.

Data on their own are worthless; a company can only act on the information it derives from the data.

For example the data may show that the company's sales over four consecutive years were £10 million, £20 million, £40 million and £30 million. These are the raw data. It shows a rise and then a fall. The company needs to interpret these data, and after investigation may find that the reason for the fall was that the market was getting saturated with this type of product, which had also attracted many competitors. This is information (interpretation of the data), and based on this information the company may decide that the time has come to pull out of this market.

Data may be quantitative or qualitative, hard data or soft data. Quantitative data are often expressed in numbers. They can usually be collected relatively easily, and have a good degree of accuracy. Examples are the size of a market or the prices of products.

Qualitative data are based on opinion and have a close link to behavioural science. They represent the feelings and attitudes of people and cannot be expressed as numbers. These data are not as accurate as quantitative data. They are usually collected by interviews and the results can be influenced by the interviewer's style and personal bias on the subject. These data are expensive to collect and are therefore often based on small sample sizes, which also introduces sampling errors.

Hard data are data based on good evidence, for example those collected as part of a well-organized survey or obtained from statistics published by a reliable body. Soft data are based on hearsay, rumour, data obtained at meetings, data picked up by salesmen during their travels and so on. These data have a very high value in providing a lead which can be followed up, but they can be biased and inaccurate and therefore should not, on their own, form the basis of any substantial company action.

Information derived from the data generally falls into four broad groups:

- Market-related information. This includes market size, trends in the market, market structure and composition, and market share.
- Customer-related information. This includes information on customer requirements, changes in buying pattern, customer perception of the company and its products and so on.
- Environmental factors and changes. Examples are the economy, technology development, government policy, competitor activity relating to products, sales, customers and so on.
- Product-related information. This will include competitor product analysis

and customer requirements, but also factors such as the effectiveness of specific advertising campaigns and promotions; the analysis of different promotional media, such as newspapers, journals, exhibitions and their effectiveness in promoting certain products; and the effectiveness of product launches.

15.2.3 Sources of data

The elements of a marketing information system are illustrated in Figure 15.10. Sources of data feed into a central storage area and from there the data can be accessed in a systematic way for producing reports or other documents, to guide the company in its marketing and business activities.

Information can be classified as primary or secondary. Primary information is that which is specifically generated to obtain data on the research being considered, for example a survey of men to determine the type of razor blades they use. Secondary information is that which is generated for some other purpose, but which can be used for the research being considered. For example, the government census of the population can be used to determine the number of males above a given age, in order to assess the potential market for razor blades.

There are a variety of sources of marketing data, including:

- Data generated internally within the company. Examples include financial data on the company's yearly performance; analysis of bids made and won or lost; and feedback from salesmen or service engineers, who make frequent customer visits and can pick up good data.

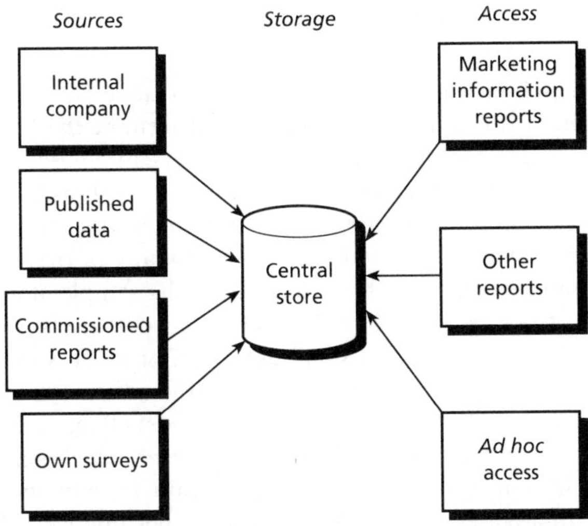

Figure 15.10 Sources, storage and access of data.

- Government statistics, such as the retail price index, the housing construction data, census of population and production, and import and export data.
- Data published by trade associations, professional bodies, banks and so on.
- Data gathered by search in libraries. Apart from public libraries, most professional bodies, trade associations, government bodies and universities have libraries which are open to members and often to the public.
- Published data available in the press and trade journals.
- Data collected from customers, suppliers, wholesalers, retailers and agents. Much of this will be soft data and will need to be verified.
- Intelligence obtained from the company's own marketing research programmes. Some of this may be formal and structured, while others may be informal, such as visiting a trade exhibition to gather data on competitor products.
- Data obtained from external marketing research companies. These may be in the form of multi-client reports, which cover a specific topic of interest, or specially commissioned reports in a particular area. A careful brief must be prepared for commissioned reports, to ensure that the company gets what it wants out of the research programme.

15.2.4 Programme steps

The steps that a marketing information programme can go through are as follows (Figure 15.11):

1. Definition of the objectives of the research. The problem to be solved must be accurately and precisely defined. For example, the problem may be to determine the exact market share between several competitors for a particular type of product.
2. Definition of the research procedure to be used. A clear understanding of the data that must be obtained is important, as this will determine the best method for collecting them. Several research procedures can be used, such as point of sale sampling, consumer panels, *ad hoc* samples, censuses, experiments and observations. These are described in the next section.
3. Selection of the samples. This will include determining many of the sample procedures, such as sample size, representation within the sample, and whether the sample is to be random. In industrial products there is usually a significant difference between large and small users of a product, so in selecting the sample it can be weighted towards the large users.
4. Collecting the data. Often at the start of a survey the most effective questions to ask may not be obvious, especially if the product is highly technical. In these instances the questions can be refined after several trial samples, when the interviewers obtain a better understanding of the issues involved. There is also the problem in a survey of knowing who to contact within an

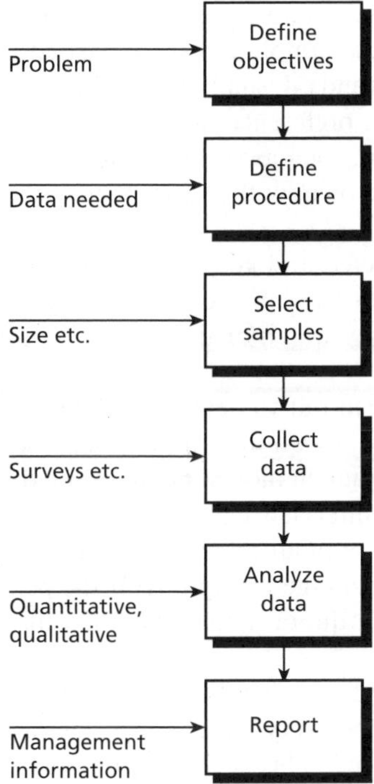

Figure 15.11 Steps in a marketing information programme.

organization in order to get good information. It may be necessary to visit the company on several occasions and see different people to ensure that the data collected are good.

5. Analyzing the data. The technique used in this will vary depending on the type of data obtained (quantitative or qualitative). It is this analysis which turns the data into information on which management can take action.
6. Preparing the report. This should clearly define the management information and indicate the data on which it is based, so that the accuracy of the information is evident. The report must be able to answer the questions which the marketing information research programme set out to solve.

15.2.5 Marketing information techniques

Several techniques may be used when collecting data in a marketing information research programme. These are:

- Point of sale research, also known as shop audit research. This consists of using statistical sampling techniques to cover a large number of retail outlets around the country (and sometimes in several different countries), to determine the usage of products, for example. The sample may be weighted towards large users, and can be conducted on an outlet basis or individually, such as individual users of mobile phones. Sampling is often carried out on a continuous basis in order to determine trends. Usually this technique consists of asking questions about product usage, preferences and so on, although it may take the form of passive observation, such as from which level of a supermarket shelf customers select goods or how they approach certain displays of goods.

- A consumer panel, where a group is selected and keeps a careful diary, on a continuous basis, of certain events in their lives. For example, they may keep a record of the television programmes they watch or the type of food they eat at each meal.
- An *ad hoc* survey, where a survey is carried out to determine data on a specific topic. The survey may be random, or quota sampling may be used. In random sampling every person has an equal chance of being selected. If the selected person is not available when an interview call is made then a call back is necessary. In quota sampling the population is divided into segments: for example, percentage of males, or percentage by regions. If the person from a selected group is not available in a quota sample, then another person can be chosen from the same group.
- A census, in which every person or every factor is taken into account. For example, if selling deodorants for men then every male above a certain age is a potential user and needs to be interviewed. This is clearly 100 per cent sampling and can be very expensive unless the target market is small.
- Test marketing, where a small area is chosen which has similar characteristics to the main market, and marketing research is carried out there. The results found from this sample population are assumed to apply to the main market.
- Experiments, which are set up in a controlled manner to measure certain parameters. For example people could be asked to try two different products and state their preference. Experiments carried out in an artificial environment cannot be assumed to translate into the real world. Field experiments can also be used, such as changing the position of goods in different parts of a supermarket in order to access the effect on sales. However, field experiments can often be affected by other uncontrollable factors, which disguise the main experiment.

When carrying out sampling, further techniques may be used, such as postal questionnaires, personal interviews and telephone interviews. Postal questionnaires have the benefit of being very cost-effective: the cost per answer received is low in spite of the fact that only a fraction of the questionnaires sent out are usually returned. Another advantage of postal questionnaires is that questions can be included which may require the respondent to look up facts before replying. Shyness in the interviewee is absent and full anonymity is maintained. Interviewer bias is also avoided.

There are several disadvantages in postal samples. The questionnaire may be passed down the organization to junior levels, and there is no indication of who has answered it. Questions may also be misunderstood, so complex questions cannot be asked. Follow-up questions are also difficult.

For effective postal surveys the names of the relevant persons in the organization should be determined and the form sent to them. Good presentation of the form is essential, both in the questionnaire and the covering letter. Readers must be able to see very quickly the benefits of replying to the questionnaire; for example, they may be sent a summary of its results, or be able to see the benefits in the form of cheaper and better products. The questions should be easy to answer; ticking yes/no boxes is usually easier than providing lengthy text. The questions asked should also be tested against several respondents, to ensure that they mean the same thing to all readers.

Personal interviews are the best method to use for gauging customer satisfaction (Rice, 1991). They have several advantages. Many questions can be asked, several of them complex. Samples can also be shown as part of the questions and instant feedback is received on the effectiveness of questions, and these can be improved as the survey proceeds. The status of the person being interviewed can also be checked and 'body language' observed for feedback.

A disadvantage of personal interviews is the high cost, both in terms of time and travel to the interview sites. Replies may also be biased, depending on the tone in which the questions are asked and the bias of the interviewer. Because the replies are provided almost instantaneously, respondents may not have enough time to consider them and might change their answers on reflection.

To conduct an effective personal survey the objectives of the survey must first be explained to the respondents. The interviewer must be knowledgeable about the subject concerned in order to gain the respect of the interviewee. The views expressed should be checked to determine whether they are personal views or official company views. Permission should also be asked before taking notes and they should be read through before the close of the interview to check on their accuracy.

Telephone interviews are relatively fast to conduct and have a low cost. During the interview the interviewer can check if the interviewee is the right person relative to the position desired. Complex questions can be asked and a check kept to ensure that they are fully understood. It is also easier to get hold of busy people for a telephone interview than for a personal interview. Often telephone interviews are used to clarify points raised in a postal or personal interview.

The disadvantages of telephone interviews are that they may often be handed down the organization. Visual material cannot also be used in the interview and the replies received are instantaneous, as for personal interviews, and may have changed on further reflection.

There are several causes of errors in surveys, such as:

- Sampling errors.
- Bias in the interviewer: researchers may draw questions one way, depending on their belief of what the outcome should be.
- Impromptu replies from the interviewees, the replies not being carefully considered and therefore likely to change in real-life situations.
- Misunderstanding of the questions.
- Deliberately misleading replies, such as may be given by a potential competitor.
- Distortion caused when collecting the data.
- A sample is a snapshot of events as they occur at a particular time. The wrong time may have been chosen for the sample, for example at the occurrence of an unusual event which biases the replies.

15.3 Market segmentation

15.3.1 Defining a segment

A segment is an area of the market which reacts in a similar way to specific marketing actions (Berringan and Finkbeiner, 1992). For example, soft drinks are more likely to appeal to young consumers and wine to older ones. In marketing terms a segment is said to exist when the following conditions are fulfilled:

- The population (people or organizations) within the segment react in a similar way to specific marketing actions.
- The segment can be differentiated from other segments in the whole market-place.
- It is worth selling into the segment; that is, it is of sufficient size and sales potential.
- The segment can be accessed by the means available to the marketing organization.

Marketing effort can be targeted at the total population if the market is fragmented with no special characteristics between different parts, or if the product has universal appeal. These conditions do not normally occur; the products or services offered by an organization are not usually bought in equal amounts by all groups (Morris, 1996).

Under these circumstances market segmentation enables the marketing mix to be varied to match the requirements of the different segments more exactly. Customer requirements within individual segments, since they are similar, can be identified more easily and precisely. If the segmentation is done on a geographical basis distribution is easier. Generally the efficiency of the marketing effort is improved since it can be designed to match the segment requirements, such as the choice of sales message or advertising media.

Although market segmentation is a very useful marketing tool, it should only be used as a guide. It is not a precise science and one cannot always explain the buying behaviour of consumers on the basis of segmentation alone.

The case study 'Home florolls' illustrates the use of market segmentation.

Case study

Home florolls

Jack Fox, Marketing Director at the FloRoll Manufacturing Company, had been asked by Jake Topper, Chief Executive, whether the company should look at entering the consumer market with florolls.

'I don't suppose there is much of a market for home florolls', Jake had said, 'but it won't do any harm taking a look at it.'

Jack Fox asked Jill Search, his Marketing Information Manager, to propose a market segmentation which could be used. After some research Jill decided that florolls would appeal to four groups of users: young graduates, upper middle class, skilled working class and retired couples.

'I have also considered some of the basic characteristics of florolls which would appeal to each of these market segments', said Jill, placing her findings in front of Jack Fox (Table 15.2). 'Generally the price should be low, although I believe the upper middle class segment is not very price-sensitive. High performance is only likely to appeal to young graduates, although they are not very interested in high reliability. It is unlikely that they or the upper middle class segment will keep the product for long. The skilled working class and retired segments, however, are unlikely to change their florolls on a regular basis, so for them reliability is an important requirement.'

Table 15.2 Market segmentation for the case study 'Home florolls'

	Segment characteristics						
Market segment	**Low price**	**High performance**	**High reliability**	**Physical appearance**	**Readily available**	**Home delivery**	**Local advertising**
Young graduates	•	•					
Upper middle class				•		•	
Skilled working class	•		•		•		•
Retired	•		•			•	•

'I see', said Jack, studying the chart. 'What do you mean by physical appearance?'

'Exactly that', replied Jill. 'You must admit florolls are not very sexy to look at, although they provide excellent value for money. Generally appearance is not an important parameter, except for the upper middle class. A bit of snobbery here, I'm afraid.'

'And readily available means?', asked Jack.

'The skilled working class would expect to find florolls available in their local store; they are unlikely to have the time or the motivation to go searching for them. However, they would not want home delivery, as the upper working class and retired are likely to want. Finally, local advertising can be used to reach the skilled working class and retired segments.'

'Thank you, Jill', said Jack Fox. 'I'll talk this over with Jake Topper tomorrow and see what he thinks. 'Certainly the segments and their characteristics can be met by us, although the size of the market, as discovered by you earlier, is not substantial.'

15.3.2 Segmentation methods

It is usual in market segmentation to differentiate between consumer and industrial markets. Consumer markets are those in which end customers buy goods or services for their personal or household use. In industrial markets the goods or services are bought for producing other goods or services, which are resold.

Figure 15.12 illustrates some of the most commonly used market segmentation techniques. Demographic characteristics include age, sex, socio-economic group (as in Table 15.3) and household size.

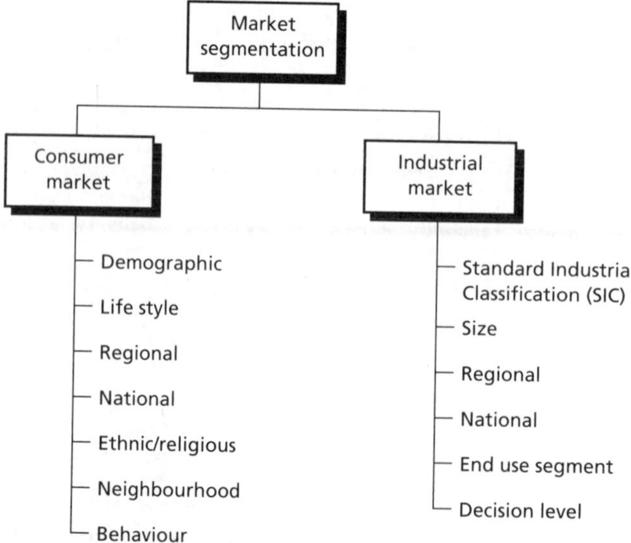

Figure 15.12 Some market segmentation methods.

Table 15.3 Socio-economic groups

Social grade	Social status
A	Upper middle class
B	Middle class
C_1	Lower middle class
C_2	Skilled working class
D	Working class
E	Those at the lowest level of subsistence

Demographic characteristics can often be linked to lifestyle, such as interests, leisure activities, food and membership of clubs. The assumption made in this segmentation method is that people buy according to what they do rather than what they are.

Regional and national segmentation are often combined under the geographical segmentation classification. Regional segmentation differentiates the population according to tastes, customs, income, and so on in the various regions of a country. It also differentiates the composition and density of industry in different regions. Regional differences also exist in sales outlets, such as wholesalers and retailers, and in environmental parameters, such as the weather, which will affect some products sold (for instance warm clothing).

National segmentation groups countries with similar characteristics into the same segments. These characteristics include user behaviour, such as culture; economic factors; political and legal restraints; and infrastructure availability. Ethnic and religious differences and similarities can occur within regional and national segments, or they can form the basis of separate segmentation methods.

The residential neighbourhood segmentation methods define segments into residential neighbourhood regions having similar characteristics relating to social class; age structure (such as level of elderly, students or very young); level of unemployment; level of immigrants; number of two car families and so on. It has been found that certain products and services sell better in certain types of neighbourhood, for example the sale of luxury goods, beer or cigarettes.

Behavioural segmentation classifies groups according to how they respond to marketing variables. Two main parameters are usually considered: benefit sought and usage rate. Benefit sought determines the benefit which people are seeking from their purchase. For example, toothpaste may be bought by consumers seeking whiter teeth, to have fresher breath or to prevent tooth decay. Clearly the advertising message will vary depending on which segment one is selling to.

Usage rate classifies users according to their usage of a particular commodity. Generally they can be grouped as heavy users, light users and non-users. Usually the 80/20 rule applies: 80 per cent of a given brand is bought by 20 per cent of the population which buys that type of product. Clearly, a company will want to focus its marketing effort on heavy users, but it must be remembered that light users may provide a niche market, which is free of competition, and non-users can be converted into users, so growing the market.

The Standard Industrial Classification (SIC) is a system used to classify companies into categories such as manufacturing, service and distribution. Within each category there are sub-categories for different products and services. This enables a very detailed classification on the basis of industrial segment. Organization size is also an important segmentation parameter for industrial marketing. Size can be measured by revenue, number of employees, product usage and so on. For example, a company may wish to sell directly, with its own sales force, to large companies, but to sell via distributors or agents to smaller organizations.

End user segmentation determines where the product is eventually used. For example, an electronic component may form part of a radio set, but if the set is to be used in the battlefield the component will need to have very different characteristics from when the radio is designed for use in a normal household.

For industrial markets the decision level is also a useful segmentation method. It determines the level within an organization at which a given buying decision is approved. Often the higher the level of spend the higher the level within the organization at which it needs to be approved. Based on this information the marketing effort can be targeted more accurately at the decision makers.

15.4 Consumer and industrial markets

15.4.1 Classification of consumer goods

Consumer goods are those goods that are bought for personal use or use by a household. The ultimate consumer is the person buying the goods. Consumer goods are usually classified on the basis of the buying habits of users. There are three main classifications:

- **Convenience goods**. These represent frequently bought items, which are purchased without much thought or search time. Examples include household supplies and commonly eaten foods. Usually the same brand of goods is bought (habitual buying), such as the same make of detergent or baked beans. If the usual brand is not available then a substitution will be made with very little extra thought. In an emergency the buyer will go to the nearest outlet to make a purchase. Convenience goods must be distributed widely; that is, sold through many retail outlets. Very little product differentiation is possible and price and packaging remain the most effective marketing weapons. These products must be promoted heavily in order to capture users.
- **Shopping goods**. These are infrequently bought items, such as cameras and refrigerators. The buyer will spend time shopping for these goods, making

careful comparison between alternatives and buying on the basis of the product's attributes rather than on brand. The buyer will do some research before visiting shops to make a purchase (discussion with friends, reading in magazines and so on). At the shop the buyer will test products and features, and items such as value for money are just as important as brand name. The buyer will not go to the nearest outlet to buy these products, but will shop around for the best (perceived) product. Shopping goods are sold through fewer outlets than convenience goods, and these outlets often specialize in the type of product concerned, such as camera shops. Personal selling is important, and the salesperson can influence the purchase. Brand advertising is needed for product differentiation, although advertising is less important than in the case of convenience goods.

- **Speciality goods**. These are goods which have unique characteristics, prized by the buyer. These characteristics could be taste (speciality foods) or prestige (prestigious brand of clothes), for example. Buyers will be willing to search for the product, to get exactly what they want and substitutes will not be accepted. Speciality goods therefore do not need to be widely distributed and often they are sold through a few prestigious outlets. The name of the store is now often as important as the product brand. Selective promotion is used to reach the small but select target market.

Goods can often fall into different product groups, depending on circumstance. For example a housewife may normally buy any loaf of bread (convenience goods), but if she is entertaining she may go through considerable trouble to find a special type of bread (speciality goods).

15.4.2 Population and income

Population and income are important factors when considering consumer goods marketing. Population defines:

- The number of people and whether this is increasing or decreasing.
- The age distribution of the population, which determines whether people are retired, students and so on.
- The geographical distribution of the population. Various factors affect this distribution: urban and rural; regions; countrywide; and so on.
- Household composition, such as married or single; number of children; or single parents.

Factors involved in income considerations include:

- Personal income. This is the total annual income (money) that a person earns.
- Disposable personal income, which is the money left over after taxes and other obligatory payments have been made.

- Discretionary income. This is the money left after essentials (such as food, clothing, housing and transport) have been bought. Clearly discretionary income will depend on the amount of these essentials bought and how expensive they are.

It has generally been found that as income increases a person spends less, as a percentage, on food, and the percentage spent on recreation and luxury goods, and on medicines and education, increases.

Generally the percentage of income spent on housing, heating, lighting, and clothing remains constant.

15.4.3 Classification of industrial goods

Industrial goods are normally those which are bought by an industrial buyer for use in a work context and not for personal or household use (Chisnall, 1994). There are three types of user:

- Users who buy goods for use within the business, such as trucks for transport or office supplies. The goods will not be changed or incorporated into other products.
- Original equipment manufacturers (OEM) who buy goods for incorporation into their own products, which are subsequently sold. An example is the purchase of tyres for incorporation into cars.
- Resellers, who buy goods and then resell them without making any change to the product. Examples are retailers and wholesalers. Value may be added to the overall service by providing a repair facility, home delivery and so on.

Industrial goods may be classified in several ways, such as:

- **Raw materials**. These are sold in the original state in which they are obtained, for example cotton, wool, tin and oil. The product goes from the primary source (which may be located in remote geographical areas) to the first processor.
- **Fabricated parts**. These are goods which are incorporated without change into a buyer's product. For example, semiconductor devices that are incorporated into a television set sold by the equipment manufacturer.
- **Capital goods**. These are high-cost items, which are bought to improve the efficiency of a company. There are few buyers for such goods, which are sold after lengthy negotiations and evaluation. The value of the product is not written off against the profit for the first year, but is usually depreciated over several years.
- **Accessory equipment**. These are items such as small tools, office equipment and computers. They do not become part of the buyer's finished goods.
- **Operating supplies (consumables)**. These items are needed to run the business. They are used up during operation and do not become part of the buyer's finished products. Examples are office stationary, fuel and office rent.

- **Badged products**. These are items which are bought in by the company and then resold as their own equipment, without any change. Often these goods are badged with the buyer's name so that it is perceived to have been made by them.

15.4.4 Industrial market characteristics

The industrial market has several distinctive characteristics, which are not shared by consumer markets:

- The demand within an industrial market depends eventually on the consumer goods used. Therefore the demand for industrial goods is known as a derived demand. For example, the demand for steel may depend on the sale of steel bolts, which are used in cars, and therefore on the eventual sale of cars within the consumer market. Therefore the marketing mix for industrial goods must often be applied down the line of users.
- Price elasticity is low since the cost of the material or tools sold is often a small part of the overall product cost. A small change in the material cost will have little impact on the overall cost of the finished product. Also, because of derived demand, a reduction in price is unlikely to generate much higher demand unless the demand of the finished product also increases. Therefore industrial products are often sold on quality or benefits to the user. rather than on pure price. However, long-term, price is important, since a manufacturer will often swap out expensive materials if cheaper alternatives are available.
- A small change in consumer spending can become magnified when translated back into the industrial demand. This is due to the delay in demand feeding back along the chain, and also due to the inability to increase or decrease production along the line in small units, to meet fluctuating consumer demand. Therefore a small increase in production may require installation of another machine, which will give over capacity.
- The number of buyers is much lower than in consumer markets, but they have considerably higher purchasing power.
- Industrial buyers are more discerning than consumer buyers. They know what they want and require detailed information about the product and the supplier. Personal selling is essential. The reliability of the supplier is as important as the quality of the product. Buyers often look for support from suppliers. Buyers seek to reduce buying risk and will often do this by buying from well-known suppliers or from those with whom they have dealt in the past.
- Buying responsibility is spread throughout the organization. Different people often produce the product specification, negotiate the purchase, approve the supplier, authorize the spend, place the order and so on. It is not always clear who the key decision maker is.

- Geographical distribution of customers is smaller than for consumer markets. Similar industries and their suppliers are often clustered together over a relatively small area.

Summary

The key aim of marketing is to develop products or services that customers want, to get them to the customers at a price they are willing to pay, and to make a profit in the transactions. Within an organization marketing operates at three strategic levels: at the corporate level, where it leads in the definition of the business mission and strategy; at the business level, where it is active in determining how the organization should compete in its chosen market areas; and at the operational level, where decisions are made about issues such as the promotion of goods and services.

The marketing organization covers many different disciplines, such as: Product Line Management (PLM) responsible for managing the product business case and for meeting profit and budget targets; Sales, which is responsible for closing the sales with the customers and providing them with the first level of contact; Marketing Information and Marketing Communications, which normally work together in collecting market-related information, such as market size and competitor activity, as well as leading the company's publicity and advertising activities; and Distribution, which is responsible for storage, handling and transport of products to customers.

An important element of marketing is determining the combination of marketing mix to use in any circumstance. The four elements of the marketing mix are product, which represents the physical parameters of the product or service being sold; price, which is the price put on this product or service, and includes all aspects of the price, such as discounts and credit terms; promotion, which is the advertising associated with selling the product, and includes sales promotions and sales staff; and place, which primarily refers to the distribution of the product and includes sales outlets such as retailers and wholesalers.

The strategy to be adopted when selling a product will depend on several factors, such as whether it is a new product or an existing product, and whether the market being sold into is new or existing. Promoting existing products into existing markets is aimed at increasing market share or expanding the market. Promoting a new product into an existing market is likely to be successful if it builds on the existing brands and the established reputation of the company in this market. An existing product in a new market is more difficult, since the image of the company needs to be established, although the success of the product in other markets can be stressed. The most difficult is selling a new product in a new market, since neither the product nor the organization is known. This is the highest risk strategy, but may be needed for diversification and growth.

The need for good marketing information has been driven by changing market conditions, such as the increased risks in the market-place, caused by competitive forces, and the high penalties associated with product failure; the decrease in product life cycles, caused by the increasing pace of technology change; changes in market structures, which have often separated the producer and the user, making it difficult to obtain good market intelligence; and the need to make decisions well in advance, with the risk that these can become obsolete unless continuously monitored and modified.

There are many sources of data, which form the basis of marketing information. Some are generated internally within the company, such as yearly financial performance or the analysis of bids made and lost. Some can be collected externally, such as government statistics and data published by trade associations or professional bodies. Some data can also be obtained from external marketing research companies on payment of a fee. Often, however, the company will decide to do its own market research. When this is done the objective of the research must be clearly defined to begin with and the method of data collection agreed. The data then need to be collected and analyzed, and the findings acted on. Several techniques can be used for data collection, such as point of sale research, which uses statistical sampling techniques to cover a large number of retail outlets; a consumer panel, where a group of consumers keep a diary of certain events in their lives; an *ad hoc* survey, which may be based on random or quota sampling; a census, which covers 100 per cent of the population and is therefore clearly limited to a small market segment; test marketing, which surveys a small area and applies this to the larger population; and experiments, which set up controlled situations to measure certain parameters.

It is usual to differentiate between consumer and industrial markets. In consumer markets the customers purchase goods for their own consumption, or for use by their household, while in industrial markets buyers usually buy on behalf of their employers. Industrial products are also often tailored to meet the requirements of a few large customers, and these may be incorporated into other products by these original equipment manufacturers (OEMs) and sold on to end users, who would be the final consumers. Long supply chains may also exist in industrial markets, for example the company's products going through several OEMs, wholesalers and retailers before reaching the end consumer. This means that marketing information is difficult to obtain and there can be a lag between consumer demand changing and the original supplier learning about it, resulting in adverse trading.

Consumer goods can usually be classified on the basis of the buying habits of users. These are convenience goods, which are frequently bought items, such as detergents and baked beans; shopping goods, which are infrequently bought items, such as cameras and refrigerators; and speciality goods, which are goods having a unique characteristic prized by the buyer, such as speciality foods and prestige brands of clothes.

A key consideration in marketing is the segmentation of the market such that the population within a segment reacts in a similar way to specific marketing actions, while at the same time the segments can be differentiated from each other. Techniques for market segmentation differentiate between consumer and industrial markets. Consumer market segmentation methods include demographic, such as age, sex and socio-economic group; lifestyles, such as leisure activities or membership of clubs; regional and national, which highlight differences in items such as environment parameters, sales outlets, user behaviour and culture, based on regional or national variations; ethnic and religious; neighbourhood, where neighbourhood regions have similar characteristics; and behavioural, such as how certain groups will respond to marketing variables.

Industrial market segmentation methods include the Standard Industrial Classification (SIC), which is a system used to classify companies into categories such as manufacturing, service and distribution; size of the organization; location of the company, such as the country and region; end-user segmentation, which is where the product is eventually used; and decision level, which determines the level within an organization at which the buying decision is made.

Case study exercises

15.1 Jack Fox, Marketing Director at the FloRoll Manufacturing Company, was considering his options. The company had a well-established range of florolls with a healthy market share and several loyal customers. A new competitor had entered the market and was selling equivalent products at a much lower price. There were several options open to Jack, such as lowering the price as well, or improving the product specification (for example by extended warranty) to increase product differentiation. Can you help Jack Fox, by outlining the options open to him and the factors that he should consider when choosing between them?

15.2 Jack Fox has been given an assignment by Jake Topper, Chief Executive of the FloRoll Manufacturing Company, to develop a business plan for a new floroll aimed at the high-end consumer market. Jack believes that the potential customers for the product cannot be easily identified and that they are widely distributed. He is considering what strategy to adopt for distribution and for advertising (or promotion). Can you advise him on the factors that he needs to take into account?

15.3 The FloRoll Manufacturing Company is about to launch a new product which has a specification far in excess of anything currently on the market. The product will be sold primarily to existing customers. The Board is considering the pricing strategy to be adopted. Jack Fox, Marketing Director, would like to sell the product at a large premium over their current range, since there are no competitors at present, and in order to recoup the heavy development expenses. Peter Sell, Sales Director, would like to keep the price comparable with the existing range, in order to retain customer loyalty. Discuss the two options. Are there any others which the company should consider?

15.4 The FloRoll Manufacturing Company is planning an extensive marketing campaign. They wish to sell their existing product range into a new market sector and at the same time to launch a new range into the existing market. What are the strategies they should adopt for the two scenarios? Should they be the same?

15.5 Jack Fox, Marketing Director for the FloRoll Manufacturing Company, has segmented the market for their new low-cost floroll range into three: mass market applications; applications requiring a quality product but still buying on price; and critical applications where quality and reliability are important. Can you advise Jack on the marketing mix which could be applied to these three sectors?

Endnotes

1. See also the following references: Renshaw, 1994; Shaw and Semenik, 1989; Slywotzky and Shapiro, 1993; Smith, 1993; Webster, 1984.

References

Baatz, E.B. (1991) The buyer's guide to market research, *Electronic Business*, 23 September, pp. 50–1.

Berringan, J. and Finkbeiner, C. (1992) *Segmentation Marketing*, HarperCollins, New York.

Boyd, H.W. and Walker, O.C. (1990) *Marketing Management: A Strategic Approach*, Richard D. Irwin, Homewood.

Cespedes, F.V. (1994) Industrial marketing: managing new requirements, *IEEE Engineering Management Review*, Fall, pp. 34–46.

Chisnall, P.M. (1994) *Strategic Industrial Marketing*, Prentice-Hall, Englewood Cliffs NJ.

Crask, M. *et al.* (1994) *Marketing Research*, Prentice-Hall, Englewood Cliffs NJ.

Doyle, P. (1994) Marketing Management and Strategy, Prentice-Hall, Englewood Cliffs NJ.

Frain, J. (1983) *Introduction to Marketing*, Macdonald and Evans, Plymouth, UK.

Grant, A.W.H. and Schlesinger, L.A. (1995) Realize your customer's full profit potential, *Harvard Business Review*, September–October, pp. 59–72.

Houston, F.S. (1986) That marketing concept: what it is and what it is not, *J. Marketing*, April, pp. 81–7.

Hutt, M. and Speh, T. (1989) *Business Marketing Management*, The Dryden Press, Chicago.

Jachson, B.B. (1985) Build customer relationships that last, *Harvard Business Review*, November–December, pp. 120–8.

Keller, K.L. and Staellin, R. (1987) Effects of quality and quantity of information on decision effectiveness, *J. Consumer Res.* , September, pp. 200–13.

Lele, M.M. (1994) The lessons of strategic leverage, *IEEE Engineering Management Review*, Spring, pp. 39–42.

McKenna, R. (1991) Marketing is Everything, *Harvard Business Review*, January–February, pp. 65–79.
Mitchell, A. (1994) Marketing's new model army, *Management Today,* March, pp. 41–9.
Mitchell, A. (1995a) The ties that bind, *Management Today*, June, pp. 60–4.
Mitchell, A. (1995b) Missing measures, *Management Today*, November, pp. 76–80.
Morris, J. (1996) Segment your market to survive, *Communications International*, June, pp. 131–2.
O'Brien, L. and Jones, C. (1995) Do rewards really create loyalty? *Harvard Business Review*, May–June, pp. 75–82.
Renshaw, P. (1994) *Marketing Plan Development Guide*, Prentice-Hall, Englewood Cliffs NJ.
Rice, V. (1991) The art and science of customer surveys, *Electronic Business*, 7 October, pp. 141–4.
Shaw, R.T. and Semenik, R.J. (1989) *Marketing*, South-Western Publishing Co., Cincinnati.
Slywotzky, A.J. and Shapiro, B.P. (1993) Leveraging to beat the odds: the new marketing mind-set, *Harvard Business Review*, September–October, pp. 97–107.
Smith, I. (1993) *Meeting Customer Needs*, Butterworth-Heinemann, Oxford.
Sowter, C. (1993) Marketing is too important to be left to the marketing department, *IEE News*, 4 March, p. 14.
Webster, F.E. (1984) *Industrial Marketing Strategy*, John Wiley & Sons, New York.
Webster, F.E. (1993) The changing role of marketing in the corporation, *EMR*, Fall, pp. 48–60.

16

Product management, sales and distribution

Introduction

This chapter introduces the key activities of product management, sales and distribution, which form part of the overall marketing operations of an organization.

16.1 Product management

One of the most important tasks within marketing is the management of the company's products. Generally there would be a Product Line Manager (PLM) for one or a range of products. It is the responsibility of the PLM to ensure that the products are managed so as to generate profits and ensure the continued profitability of the company.

This will mean: specifying the product initially so as to ensure that it meets market requirements; guiding the product through development and manufacture; managing the successful launch of the product, ensuring that it is rapidly adopted by the industry; managing the product life cycle so as to identify the requirements for replacement products; and managing the product portfolio for the company.

16.1.1 Product specification

In specifying a product all the attributes that a customer is likely to look for must be considered, both tangible and intangible. For example, tangible attributes of a car would be its wheels, steering, engine, brakes and so on. Intangible benefits, which are often more important, are comfort, fuel economy, prestige and so on. To many customers, side issues, such as warranty, service arrangements or packaging, are often more important than the main ones.

Generally there are three key factors to be considered: product performance; time to market; and cost. Technical products often rely on superior performance to gain competitive advantage, but time to market is equally critical, since if a market window has been missed the product will not be successful. Costs are important to the company if it aims to make a profit, since selling price is often fixed by competitors and by customer expectations.

One of the most significant causes of new product failure is lack of integration of R&D and Marketing early into the innovation process (Gupta and Rogers, 1992). However, Marketing must drive the R&D agenda and not the other way around (Kodama, 1992). R.W. Emerson said in the mid-19th century: 'If a man write a better book, preach a better sermon, or make a better mouse-trap than his neighbour, tho' he build his house in the woods, the world will make a beaten path to his door'. Unfortunately this is no longer true; a better mousetrap will not sell itself in the present competitive environment, especially if what the customer really wants is a cage in which to keep a pet mouse.

Trade-offs are often required between performance, time and cost, and this is why the best product specification teams consist of Marketing, Engineering and Project Management (Figure 16.1).

Improvements in communication between the three functions lead to more successful products (Stewart, 1994), and a better understanding of the problems involved. For example, Engineering usually wants Marketing to provide it with a clear-cut specification of customer requirements, which remain fixed during the period of the development, which Marketing is rarely able to do. On the other hand Marketing expects Engineering to develop a perfect product in a very short time and at minimum cost, which is again difficult to achieve. Working in a team ensures that each is aware of the other's problems.

The buyer's perception of the product is vitally important and should be considered when drawing up the specification. Generally buyers want a product which meets their needs, at an affordable price and with low perceived risk. Key questions are: will the product do what is wanted? Will it be reliable? Will it be easy to use? The newer the product or supplier the higher the perceived risk.

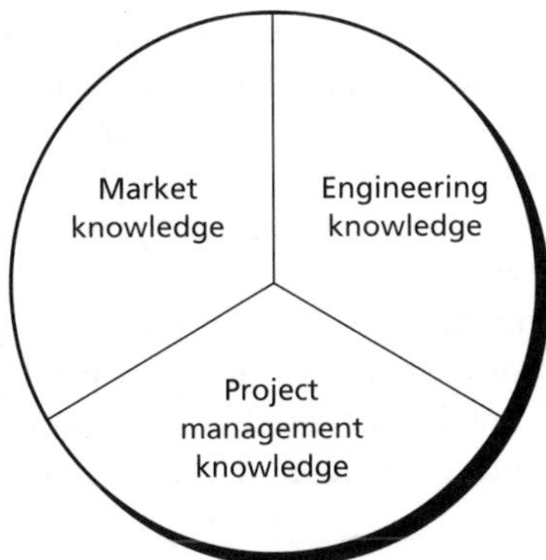

Figure 16.1 The information needed for effective product specification.

Figure 16.2 Buyers' perception stack.

Figure 16.2 illustrates a buyer's perception stack. Low price may attract potential buyers, but if the other parameters are missing then competitive pricing can work against the seller. For example, the low price may be perceived as indicating a poor quality product, with low reliability, and the warranty offered may be considered to be worthless. In these circumstances it is often better to keep price relatively high but to support the buyer in other areas, such as training or replacement unit service, so as to minimize the perception of risk.

The key to successful product specification is to be market guided (Ayal and Raban, 1990; McKenna, 1995). Markets are also continually changing and it is a business decision whether the organization chooses to react to change or to initiate it (Johne, 1992).

Scientific or engineering advances should be used to meet market needs, leading to commercializing of technology (Nevens, *et al.*, 1990). New technology should also be adapted to what the company is itself able to do with it, for example within its R&D and Manufacturing organizations (Iansiti, 1993).

In the current technological market-place there is real danger of innovation overload (Herbig and Kramer, 1993). Consumers have too many innovative options to choose from and too many detailed technical factors to consider in making their choice. The wide choice of complicated, sophisticated, expensive, high-tech products often results in the consumer being frightened off and not making a purchase. The risk factor is too high: the risk that the product will not meet requirements, will be too difficult to use, or may be unreliable.

This innovation overload can be countered by ensuring that new product developments are on the basis of small improvements to previous products and that new features and benefits are added to meet the target market requirements (Gaynor, 1993). This will also ensure that older versions of the product become obsolete gracefully as new products take over. Products should also be designed so that they are easy to use. A good product should not need an instruction manual to operate it. The more reading that consumers have to do the less likely they are to use the product. Products should not be technology platforms but machines designed for human use. The benefits from using the product should be obvious to the consumer.

In specifying a product the target market should be kept in mind (Slater, 1994). Markets are generally characterized by several features:

- **Size and phase**: This defines the overall market potential for the product and whether it is growing, remaining steady or is in decline. Most companies favour entering markets which will show growth over several years, although such a market is likely to attract other competitors.
- **Entry barriers**: The barriers to market entry may be high, such as the technical development needed; a new process in manufacturing start-up; a large number of strong competitors; or entrenched customers (those happy with their current suppliers). If a company is early into such a market, or has the resources to enter the market at a later date, then it would find the entry barriers an advantage as it would deter other competitors from entering.
- **Exit barriers**: This is the cost to a company if, after entering the market, it wished to withdraw at a later date. Generally companies favour a low exit barrier, such as the ability to sell its old product line to another company if it wants to concentrate its resources on another product or market. Low exit barriers would also allow competitors to exit the market more easily and leave the company with a larger share, if it decided to stay.
- **Capacity**: This is the capacity which the operators in the market segment have in meeting the requirements of the market. This consideration would usually arise when the market is in its growth stage. Generally a company entering a new market would like to see it operating in an under-capacity mode, so it can establish a significant market share. It is also important that the market does not later head towards over-capacity, which would result in price cuts and a reduction in the number of suppliers.
- **Market dominance**: It is important to the new entrant into a market that it does not have dominant suppliers who can dictate the agenda. It is preferable

to have several suppliers with equal market share. Dominant suppliers are acceptable, however, if the new entrant is seeking to establish itself in a market niche. Equally important is the absence of dominant consumers who can affect the price and profit margins of suppliers.

- **Market importance**: The market should not be contributing to a large part of a company's revenue, since this would make the company too dependent on the whims of a few customers. Also if competitors are critically dependent on any one market segment for a large part of their revenue, they are likely to defend it and put a considerable amount of resources selling into it. Therefore it could be difficult for a new company to break into such a market.

16.1.2 Product development

It is essential for a manufacturing-based organization to have a continuous stream of new products being developed (Lurban and Hauser, 1993; Webb, 1991; Zahra *et al.*, 1994), in order to:

- Replace existing products, which may be becoming obsolete.
- Extend an existing product's life.
- Broaden a product range, either to increase market share or counter competition. However, there are dangers in having too many product lines (Quelch and Kenny, 1994) so product extensions must be clearly matched to market demand.
- Reduce the cost and improve the performance of a product by incorporating the latest technology.
- Enter new markets.
- Meet changed customer needs.

As illustrated in Figure 16.3, products tend to have a shorter life now than they did over the past few years, and this is decreasing with time. The amount of revenue obtained by a company from new products is also increasing every year.

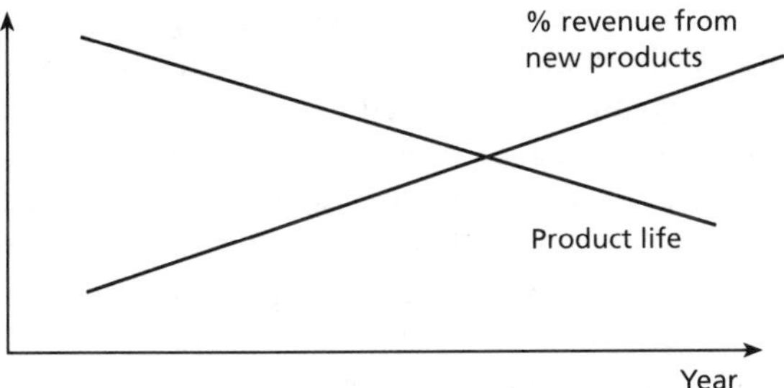

Figure 16.3 The importance of new products.

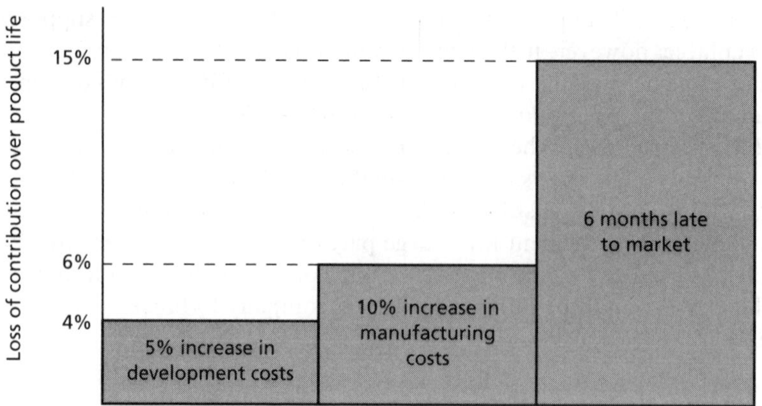

Figure 16.4 The importance of early market entry.

Time is therefore of utmost importance in new product development, especially the reduced time to market caused by shortening product development times (Rabino and Wright, 1993). Figure 16.4 for example, illustrates the effect in one market segment of three development factors: a 5 per cent increase in development costs, a 10 per cent increase in manufacturing costs, and a six month delay. Clearly product delay has the greatest impact on profitability and should normally be avoided, even if it requires an increase in development or manufacturing costs.

However, there is also danger in accelerated product development, since the risk of mistakes increases and the company's internal processes may not be able to cope with the faster development requirements and may break down (Crawford, 1993). Figure 16.5 shows the impact of engineering changes at various stages of a product. Accelerated product development may lead to poor designs and inadequate testing at early stages of the programme. This may mean that design faults are not found until late in the programme, increasing rectification costs. The problem of a slower development time, however, is that customer requirements may change during the development phase, forcing a change on the product, and this too can be expensive.

The effect of shortening development times on product risk is illustrated in Figure 16.6. The risk of failure is highest at the conceptual stage and decreases as the product enters the market. Shortening the development time also compresses the risk curve and reduces the overall risks involved.

Several techniques can be used to reduce product development times and still maintain product quality (Kopelman, 1994). Incremental development is an obvious solution where products are released at regular intervals with improved features and targeted at specific markets. Technology available from other projects or outside companies can also be used, the organization adopting the policy of incremental innovation rather than invention.

Figure 16.7 shows a concept used by many large organizations. A common

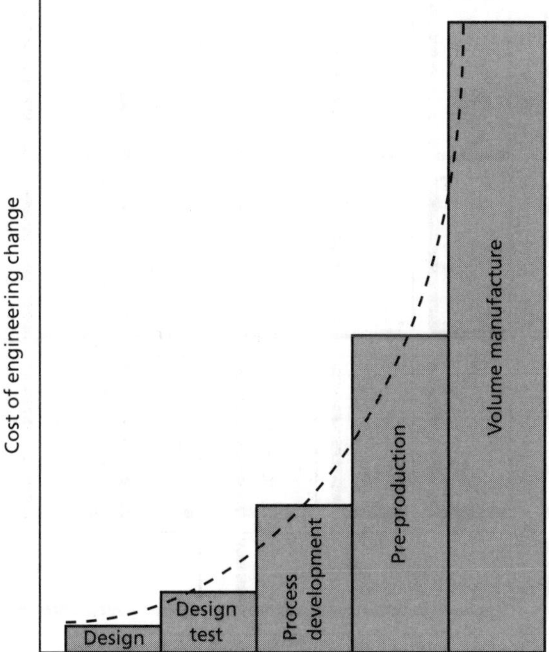

Figure 16.5 The impact of engineering change at various stages of product development.

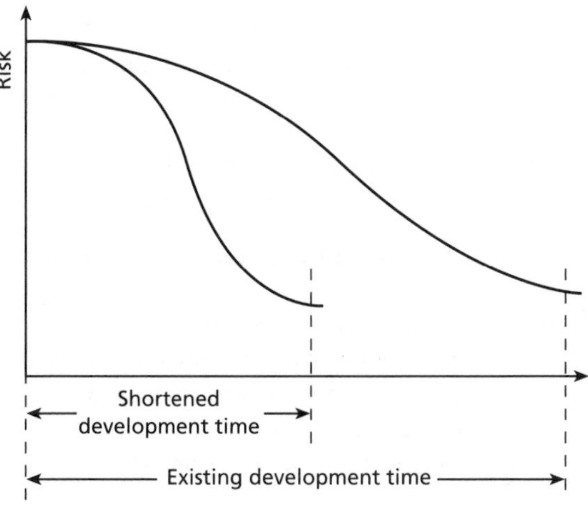

Figure 16.6 The effect of shortened development time on product risk.

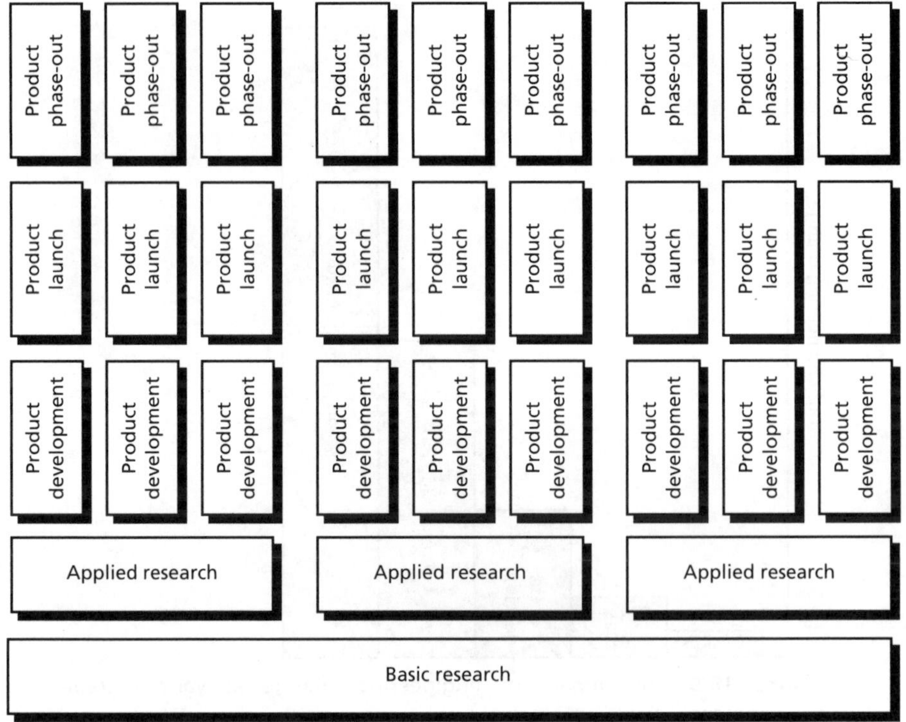

Figure 16.7 Product development from a common research base.

basic research base provides the base technology for all products. Several applied research projects are then developed off this base, each leading to products which are launched and eventually phased out. This enables the company to spend a considerable amount of time developing its base technology and at the same time launching a succession of products, to meet specific market needs, with relatively short development times.

Products may be acquired, rather than developed in house (Werther *et al.*, 1994). Acquiring a product or technology has several advantages:

- The uncertainties of development time and cost are avoided, and the product can be brought faster to market.
- If the product, or a similar product, is already on the market then its success is known and the cost of an unsuccessful product launch is avoided.
- Buying a product means that internal technical skills, which would have been used in its development, can now be used on other projects.
- Cash flow becomes positive earlier in the product cycle.

The disadvantages of buying in a product are:

- The company may have to pay over the odds for the product.

- The market reputation gained by the existing product may hamper its entry into other more profitable market segments which the company may wish to enter.
- There could be problems of supporting the product and of developing it further, since the company will not have built up internal expertise, as it normally does when a product is developed in house.
- Good internal engineers will leave, since most engineers prefer to do original development work rather than to modify an existing product. This will further hamper the organization's ability to develop future products in house.
- Internal processes used for new product development will not be used if a product is bought in, and these processes will be weakened. This further erodes the company's ability to develop products in future.
- The company's reputation as a innovator will be weakened in the marketplace if it regularly buys its products rather than develops them.

Generally companies tend either to buy in a product and badge it, or to develop a product from scratch. However, in a manufacturing environment many options are used, as shown in Figure 16.8. In development, too, an organization should explore all the alternatives, such as getting non-critical parts designed outside and doing essential developments in house (Strassberg, 1992).

Another product decision needed is whether to lead or follow a technology. Leading (being first into the market) provides a competitive advantage, since, in the absence of competition, higher prices can be charged and customer loyalty built up before serious competition enters the market. The disadvantage of being early into a market, however, is that the market must be educated, especially in the case of a technology product, and there is a higher risk of failure. The risk of product unreliability is also increased due to shortened development times, and an unreliable product will give the organization a poor reputation in the marketplace. An organization which enters a market early must ensure that it has sufficient resources to stay in the market while it develops slowly (Cooper and

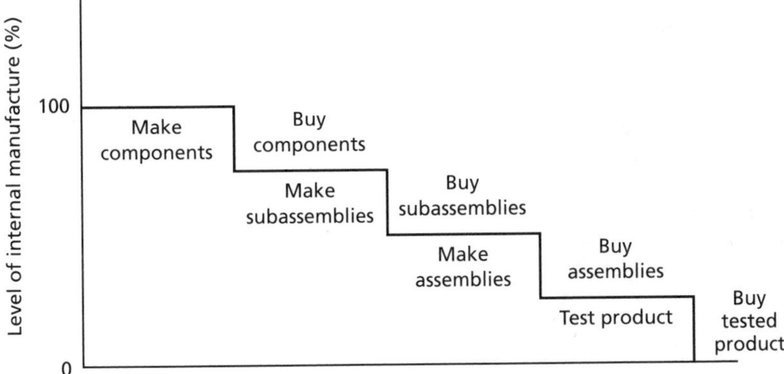

Figure 16.8 Manufacturing options.

Smith, 1993). The only advantage of slow market growth is that it will discourage competitors and so enable higher prices to be charged and larger profits to be made. Greater resources will also be needed if the market develops quickly, since the organization must strive to build up its market share in line with market growth, to prevent competitors obtaining a strong position from which they would be difficult to dislodge later.

Following technology allows an organization to learn from competitors' mistakes and to enter a market with a better engineered product. This will enable costs to be kept low and prices to be such as to provide a competitive advantage and so take market share. This strategy may not be successful if the early entrants to the market have built up strong customer loyalty and the market does not grow rapidly enough to support a new entrant.

16.1.3 Product launch

The stages leading to product launch are illustrated in Figure 16.9, along with the subsequent activities of selling, distribution and support which make a launch successful.

The aim of product launch is primarily to ensure a successful start to the

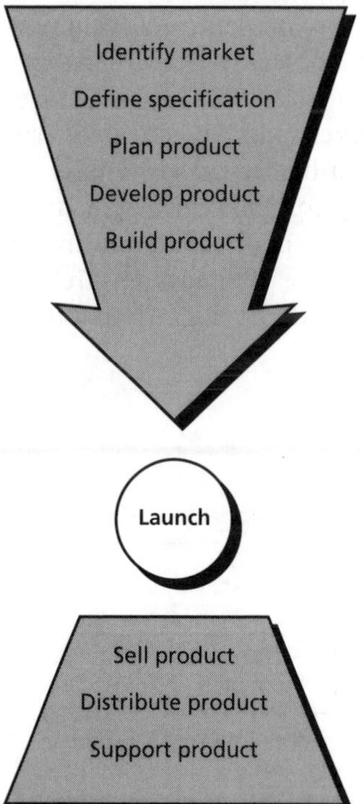

Figure 16.9 Product launch.

product's life, so that it quickly overcomes the early market development stages with rapid build-up of sales volumes.

Product launch can vary from an expensive public event to a quieter affair where a few key customers are introduced to the product by salesmen. The second method can be used when the new product is planned as an enhancement or replacement of an existing product, since a comprehensive list of customers is then available.

Successful product launches can enhance the reputation of a company. However, an unsuccessful launch can have a detrimental effect on the company's image and can affect all its existing products. Several guidelines exist for a successful launch:

- The product should first be launched only to those segments that have the most need or potential, in order to maximize the chance of success.
- The product should only be launched after it has been fully tested to ensure that it is reliable. Market launch is not the time to treat customers as beta sites.
- The launch must be effectively planned, with items such as promotional material, briefing packs for salespersons and training courses available. A product launch may be timed to coincide with another event, such as an exhibition, but this may in some instances dilute the effect of the launch.
- Sufficient resources must be allocated to make the launch successful. Product launches can be expensive but are a vital part of every business.
- The dangers of information overload must be avoided (Herbig and Kramer, 1993). Constant exposure to advertising can result in an almost unconscious rejection by the consumer of the message received. Careful attention should be paid to the quality of the information provided. The dangers of information overload apply to individual consumers as well as to organizations.

Products can fail after launch due to several reasons (Barclay, 1992; Wheelwright and Sasser, 1989):

- The product does not meet market needs, either because poor market research was conducted initially when the product specification was drawn up, or because customer requirements have changed while the product was in development and this was not tracked.
- The new product has not been differentiated enough from existing or new products launched by competitors.
- There are unexpected technical problems with the product after launch, which give it (and the organization) a poor reputation in the market.
- The product may also be technically inferior to others on the market.
- The product has been incorrectly positioned in relation to price, functions, target market and so on.
- The product has missed the market window; for example, it is too early and there is insufficient long-term demand, or it is too late and the market has been saturated by competitors.
- The product does not match the internal strengths of the organization.

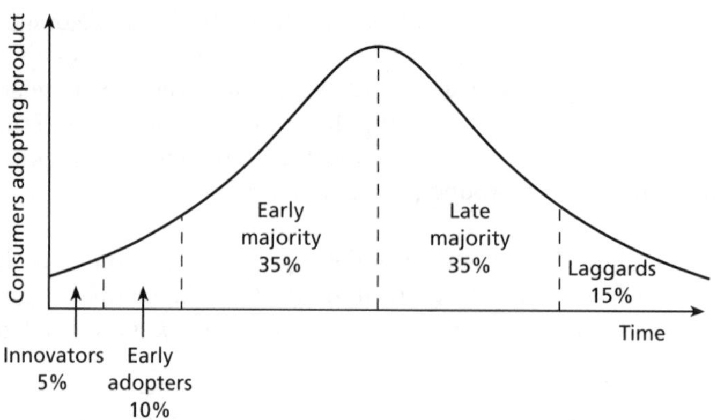

Figure 16.10 Product adoption stages.

A key to the success of a product launch is knowing how it is adopted by consumers. Figure 16.10 shows the stages a product goes through when it enters the market, based on the activity of classes of consumers. In this figure the percentage of the population in each category is approximate for a typical target market segment.

Innovators are those with high disposable income and high standing in society, who are confident of their own requirements and judgements. They have higher education than the average. They are willing to experiment and have the money to do so. They will be the first people to try out any new product, which they will probably hear of through the media.

Early adopters also have high income and a high social class. They rely on personal information sources to learn about the product, often from the innovators. The early adopter group have an active position within the community and are a key group which every successful product launch must reach. Once the early adopters have adopted a product it is likely to be successful, since they are an excellent recommendation for it.

The early majority have above-average income, education and social class. They are influenced in their buying by several sources of information, including early adopters. Adoption by the early majority means that the product has now entered the mass market.

Once the product has been established the late majority are likely to buy it. They have lower than average education, social class and income, so they are likely to wait until the price of the product has fallen, reliability has been proven and product improvements introduced before they buy it.

Laggards are those who are sceptical of innovation and do not belong to any one social or income group. They have a narrow range of experience and interest and are always late in adopting any new product. They constitute an important new market segment at the later stages of a product life cycle, as defined in the next section.

The product adoption stages are important to a product manager. Knowing the distribution of adopters it is possible to target them at each stage of the product life cycle. It is also key to locate and target the early adopters, as explained earlier, since they largely determine the success of a new product.

16.1.4 Product life cycle

The length of a product life cycle will vary depending on the type of product. The cycle for a house will occupy a long period of time, whereas that for technological and fashion goods will occur over a short time-scale. Irrespective of the actual time span, however, all products go through approximately the same stages, as illustrated in Figure 16.11. Knowing the stage that a product has reached enables the product manager to determine the best marketing strategy, as well as planning for the product's eventual replacement.

The market development stage can occupy a relatively long time for a highly technical product, since consumers need time to adopt it. Marketing effort is high during this stage, with test marketing, education of the market and so on. The key is persuading early adopters to buy the product. Investment will need to be built up as the market develops. Effort is expended on growing the market, and competitor activity often helps in this since it gives consumers confidence, because very few users are happy to go with a single supplier. Prices are high and so are costs, due to the low volumes being made. Many product failures can occur at this stage, and the product may go through several changes as market experience is gained.

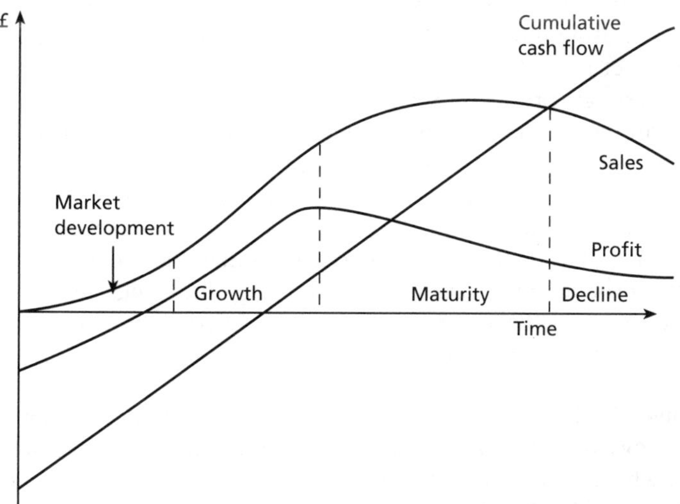

Figure 16.11 The product life cycle.

During the growth stage investment is maximized in order to build up capacity to meet the sharp take-off of sales. The product has been accepted by the early adopters and is likely to be successful. Marketing emphasis is on building brand loyalty and of maximizing market share. Pricing may still be kept to high levels and the product sold on its benefits and value to the customer. It may be appropriate at this stage to look at broadening the product range. Heavy marketing is needed to continue to grow the market and to look for new users.

At the product maturity stage selective investment is needed in specific product areas which show good potential and also to maintain market share. Pricing will have stabilized and may even start to decline, so that profits will start to fall, although cash flow will still be increasing. To be successful products must be differentiated from competitive offerings, and investment must go into cutting costs in order to reduce prices while still maintaining margins.

In the final stage, decline, the product starts to reach obsolescence. Cash flow is maximized. Very little new investment is made and cash can be increased by disposals, if necessary. The aim at this stage is to maximize profits and cash rather than to aim for increased market share. Prices can increase as competitors drop out of the market due to falling volumes. Cost reduction is important, where this can be done without excessive new investment. Marketing effort and expense is considerably reduced.

Many companies are reluctant to exit a market or product. It is important, however, that this is done before a company is forced out due to disastrous trading conditions. However, every effort must be made to protect existing customers who still need the old product. Sometimes it is possible to persuade them to stock their future requirements, or to get another company to take over the manufacture of the old product.

Resources saved can now be applied to more profitable product lines. Good product planning, using information regarding the product's life cycle, will ensure that as one product nears the end of its useful life another is available to take its place. For technology products, which have a long development phase and a short product life cycle, this may mean that the development of a new product starts before the previous one has been launched.

Figure 16.12 shows the successive product life cycles. As can be seen, each new product builds on the reputation of the previous one and so reaches a higher and higher level of sales. This is the ideal scenario in product planning.

Allied to the product life cycle is the technology life cycle, and it has been suggested that this may be more relevant for industrial markets (Popper and Buskirk, 1993). There are six stages in the technology life cycle:

1. **Cutting edge**. At this stage, more research than product development is done. The technology is well ahead of any practical application and its main output is patents, which may be marketable.
2. **State of the art (SOTA)**. During this stage cutting edge technology is applied to specialized applications. Very little marketing is carried out, since the

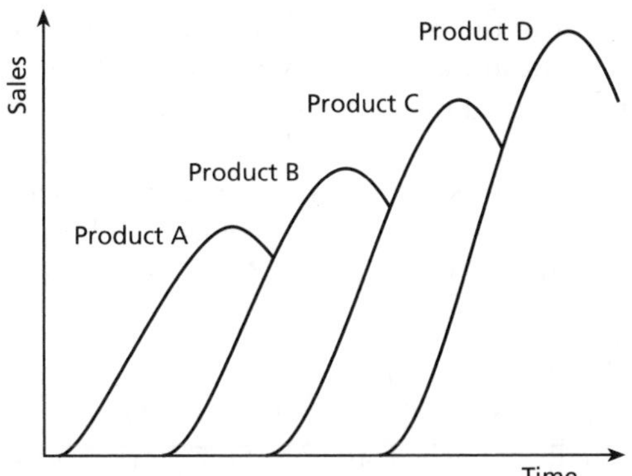

Figure 16.12 The importance of producing successive products.

market is small and contains sophisticated users who are aware of the product's benefits. There is limited competition.

3. **Advanced stage and market shake-out**. The market is larger now and has many less sophisticated customers. The company now needs to move from being a technology-led to a marketing-led organization and many find this difficult to do, and fail. Competitors begin to appear and marketing effort is needed to hold market share. Company activities shift from technology development to product development.
4. **Mainstream**. The market is now fully developed and the company's activities move towards production research. Products have become fully standardized, so competitive advantage can usually only be gained by efficient production, such as by automation or moving to areas with low labour rates.
5. **Maturity**. The products are low cost and competitive advantage can only be gained by better customer service.
6. **Decline**. Other technologies are displacing the old one. The life of the old technology can be prolonged by pricing it well below the new one, since the old capital has been fully depreciated, resulting in low-cost production.

16.1.5 Product portfolio

Products are not normally marketed as individual items, but as part of a product line or portfolio (Cardozo *et al.*, 1995). For example a company which is in the leisure business will not just market golf clubs, but may also manufacture items for other sports. In determining its product portfolio an organization needs to decide:

- How the products should be positioned relative to each other and to competitive products. This will vary depending on the target market segment. It is also the perceived difference which the consumer has regarding different products, rather than their actual difference, that is important.

- The level of coverage aimed for in the portfolio. For example, if a company wishes to be in the soccer business it may just make the football, or include boots and shorts, or also include goal posts, corner flags, whistles for the referee and so on.
- The changes in the product mix that is being aimed for. Products can be added and deleted depending on opportunities or changed conditions. Products can also be traded down (moved a model down) or traded up, relative to each other.

The well known Boston matrix (Boston Consulting Group, 1993) is a useful technique for analyzing a product portfolio for changes. This is shown in Figure 16.13. Products are normally launched as 'question marks' into markets which have high growth but in which the organization has low market share and is therefore in a relatively weak competitive position. Heavy investment in marketing and product development will be needed, especially as the technology in the growth market is likely to be changing rapidly, resulting in a short product life. The aim is to move around the matrix as indicated by the arrows. However, 'question marks' can move to the 'stars' box in Figure 16.13, if they are successful or to 'dogs' if they are unsuccessful.

If the product succeeds, a greater market share is obtained and the product becomes a star. High profits can be made on it, but a high level of investment is also needed in order to gain further market share and maintain the company's position as the market grows.

As the growth in the market begins to slow down the product moves from being a star to becoming a cash cow. Low market growth has resulted in a reduced number of competitors. The product is at the end of its maturity phase, so very little new investment is needed. Since the company is in a dominant market position, having a high market share, the product is giving good profits.

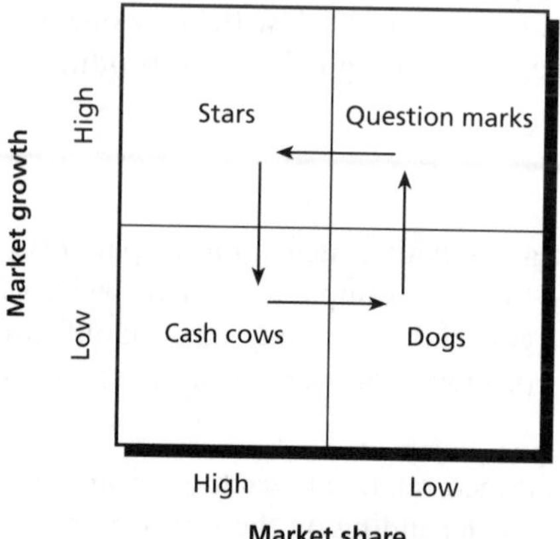

Figure 16.13 Portfolio analysis using the Boston matrix.

It is a cash cow which is generating the cash needed to fund other product developments and to support the question mark and star products, which will be the cash cows of tomorrow.

Finally the product runs out of steam; the market has declined completely and the product is phased out as a dog. An unsuccessful product at the question mark stage will move into the dog box and need replacing immediately. Such a product loses money and a company cannot afford too many dogs at one time.

At any one time the company is likely to have products in all four boxes of Figure 16.13, and knowing the mix of products helps the product manager to plan the product portfolio.

Figure 16.14 illustrates a products and markets analysis, and it should be compared with Table 15.1. The numbers in this figure correspond to the following:

1. New products are being developed for new markets. This is known as diversification and has high risk, since neither the product nor the company is known.
2. New products are developed for existing markets. This has lower risk since the company can build on its reputation with other products.
3. Existing products are upgraded for new or existing markets. Therefore the product can move two ways, as shown in Figure 16.14. The company can now build on its reputation and that of the previous product; that is, on the brand image.
4. Existing products can be sold without any modification into new markets or new applications. This is not usually successful since most market segments have their own special requirements.

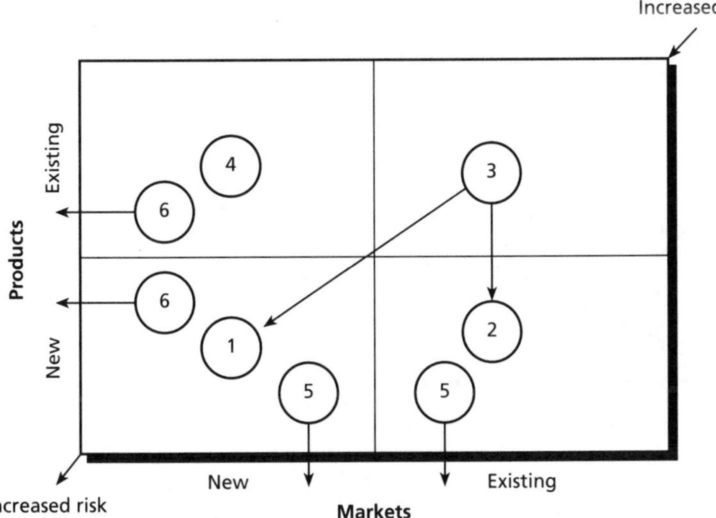

Figure 16.14 Product and market evaluation.

5. Exit markets.
6. Make products obsolete. Generally old products will reach the end of their lives, although if a new product is not successful then it too must be removed before it consumes further resources.

The level of risk increases, as shown in Figure 16.14, when moving from existing markets and products to new markets and products.

16.2 Pricing

16.2.1 Supply and demand

Prices for goods and services are usually determined by factors of supply and demand. However, the definition of supply and demand is very wide, and is influenced by many factors (Nagle and Holden, 1994). For example, a greengrocer may have only a few bunches of grapes left, but if this is on a Saturday evening and the grapes are not likely to last over the weekend, the greengrocer will be willing to sell them at a low price. Similarly, if a company has a cash flow problem it may be willing to sell its stock of goods at below cost in order to obtain cash immediately.

Price is not the only factor affecting demand for goods. However, if these other factors are assumed to remain constant then the demand curve for goods or services can be as shown in Figure 16.15. The higher the price the lower the demand, until at a given price demand will fall to zero, since no buyers will be willing to pay the asking price. As prices fall demand will also increase. Usually there is a maximum that demand can reach no matter how low the price falls (for example, free leaflets distributed in the street) since the number of consumers who perceive the product as a benefit will be limited.

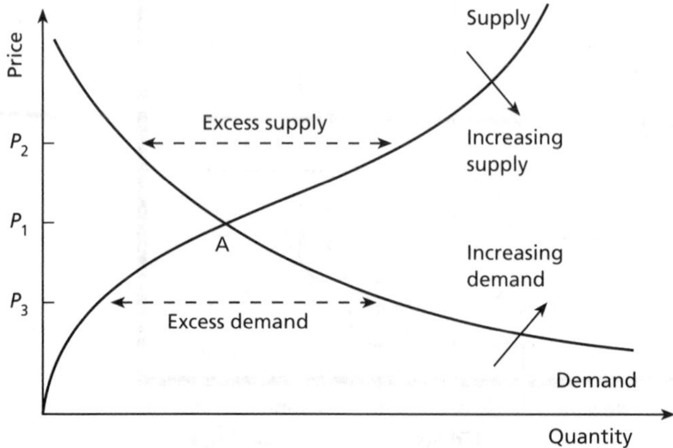

Figure 16.15 The supply and demand curves.

A typical supply curve is also shown in Figure 16.15. When prices are low there are very few suppliers and the quantity available is low. As prices increase the quantity of goods produced also increase, and so do the number of suppliers. There may be discontinuities in the curve due to special circumstances. For example, suppliers of seasonal goods may produce a limited amount only, and an unexpectedly high demand may cause the goods to run out.

With the supply and demand curves shown in Figure 16.15 the market will eventually settle at a price P_1 (point A), where supply and demand match. Below this price, such as P_3, there is an excess of demand over supply and the prices will be pushed upwards towards point A, as consumers compete with each other to obtain limited stock. Above price P_1, such as at P_2, there is excess of supply over demand, resulting in unsold stock, and prices will be forced downwards, as suppliers cut prices to reduce their inventory. As always there are special circumstances which may cause deviation from these curves. For example, antiques dealers may be willing to ride through a temporary lull in demand knowing that their stock is probably appreciating.

The curves for supply and demand are continually changing as market supply and demand change. For example, the demand for a product may increase due to an increase in consumer disposable income, or it may go down due to the arrival of a substitute product. The supply curve can also change as competitors enter or leave the market, and as automation is introduced into factories, enabling more goods to be made faster and at lower cost.

16.2.2 Pricing considerations

Pricing goods and services is not easy and many factors need to be considered (Dolan, 1995). Figure 16.16 provides a simplified flow diagram of the process.

Price must first be considered in relation to the total product portfolio and how the product is positioned in a given market segment. This will highlight all the elements of the marketing mix which need to be applied, of which product price is one element.

It should be remembered that price is the only one of the marketing mix factors that contributes to revenue; the others (product development, promotion and place) are costs and need to be paid for out of the cash generated by pricing. The price charged will also affect the other marketing mix elements. For example, the higher the price (and profit) the more the amount of revenue available to spend on product development, promotion and distribution.

Often prices are set on a 'cost-plus' basis: a percentage is added to the cost of the product, to allow for profit, and this is fixed as its price. Cost-plus pricing is commonly used, especially when a product is developed and made in conjunction with customers, to their specification.

In cost-plus pricing it is important that the cost covers all aspects of the product such as selling, distribution, packaging, etc. The margins of middle men

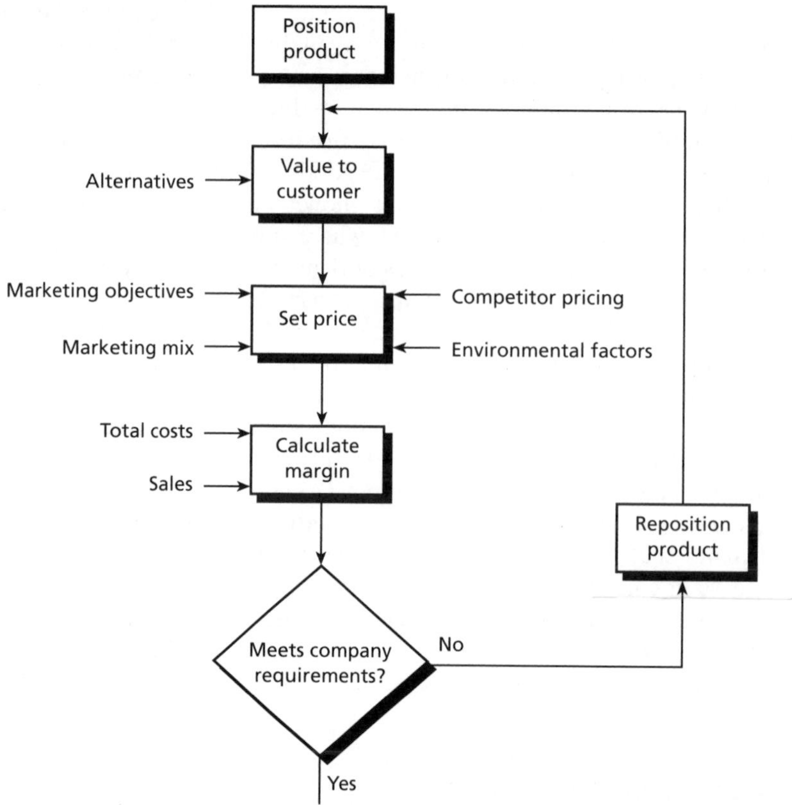

Figure 16.16 Simplified procedure for setting the price of a product.

such as wholesalers and retailers, if any, should also be included, as should any commission paid to agents. However the best method to use in determining costs is not always clear; for example, standard costing or marginal costing, as described in Chapter 11 and later in this section.

Cost-plus pricing should only be used in special circumstances, since it involves too much inward looking at costs and not enough outward looking at demand and the psychology of the buyer.

Generally market pricing methods should be used, as illustrated in Figure 16.16, which considers what the customer is prepared to pay; that is, the perceived value of the product to the customer compared with the alternative ways the customer has of satisfying needs (Rayport and Sviokla, 1995). It is important to remember that user perception of value is influenced by many factors, such as product packaging, promotion and distribution, and these can be varied in order to make a given price more acceptable.

Having determined the value to the customer other factors need to be considered, such as the marketing objectives of the company. These include:

- The effect that pricing has on the image the organization wishes to project. A low price could signal a value for money image, whereas a high price could indicate a quality company, the product being better than competitors' lower priced offerings.
- Obtaining maximum revenue with a new product. This is usually known as skimming pricing, where a high initial price is set for a product which is unique and well differentiated from others on the market. This should only be done where competition is not expected soon, either because the product is difficult to copy or the market segment is not large enough to attract many competitors.
- Increasing market share. This is know as penetration pricing, when prices are kept relatively low, as would normally be reached during the mature stage of the product life cycle. It may be applied to a new product if the aim is to build up a large volume quickly, in order to reduce costs by economies of scale, or to establish a dominant position in the market.
- To discourage competitors from entering the market. This is usually done if a company with a large market share keeps its prices low, since the risk for a new entrant is now high.
- To meet competitors' pricing; that is, to maintain the *status quo*. The organization is now a price follower rather than a price setter.
- To survive. This can occur, for example, in times of recession, when the prime aim of the company is to survive, with all its skilled workforce intact, until an upturn in the economy occurs. In these circumstances pricing can be very competitive and margins very low. This is known as survival pricing.

Environmental factors which need to be taken into account when setting price include:

- Government policy, such as maximum amount of price increase or price fixing.
- The economic climate, such as the effect of inflation on costs and customer expectations.

Once the price has been fixed the profit margin obtained can be calculated. This is when product costs need to be considered, and they are discussed later in this section.

If the margins are acceptable then the price is fixed. However, all prices must be reviewed at regular intervals to ensure that changes in market conditions and costs are being taken into account. Continuous market research is essential to provide the information for this. If margins are too low then the product is repositioned and a new price set.

Several other factors need to be considered when fixing the price of a product. For example, discounts are normally given on the list price, and these can be varied according to a number of parameters, such as the volume bought; the mix of products; the total yearly spend; the importance of the customer; or an inducement for prompt settlement of bills.

Transfer pricing, usually between divisions of a company, also needs special consideration (Kovac and Troy, 1989). These can be on the basis of allocation of a fair proportion of production costs, including overheads; generating profits for the originating division, which may be a profit centre; competitive pricing, if the selling division does not have a monopoly; or regulation, if the transfer price is controlled by government legislation, such as the price charged by an equipment manufacturing division of an organization to its public service division.

Pricing of technology is often difficult. It usually involves some up-front payment followed by royalty fees. These fees could be fixed on the basis of the cost savings to the licensee by use of the technology; the profits which the licensee is likely to generate by use of the technology; or the impact which the licensee is likely to have on the licensor's market, since they could become potential competitors.

The case study 'Pricing the image' illustrates that there are no hard and fast rules when setting product prices.

Case study

Pricing the image

It became clear to the Board of the FloRoll Manufacturing Company that, if the company was to succeed in the industrial market with their defeatured florolls, then it was essential that the product was priced correctly.

'I believe that the product will be quite sensitive to price', said Jack Fox, Marketing Director, at Jake Topper's Executive Staff Meeting. 'Based on market research and on the cost figures provided by Adrian Elton (Manufacturing Director) I believe the cost and total profit figures will be as on my slide' (Table 16.1).

Table 16.1 Illustration of the case study 'Pricing the image'

Price (£000)	Cost (£000)	Total profit (£000 000)
10	19	1.5
12	10	1.8
15	12	1.6
18	14	1.4
20	15	1.2

'I see', said Jake Topper. 'Looks like sales volumes are falling as price increases from £10 000 to £20 000 and this is giving an increase in manufacturing costs.'

'Mainly caused by the material costs', interrupted Adrian Elton. 'Unfortunately, we are using a different mix of materials from that used in our standard floroll product and the quantity discount we get on this falls with low volumes.'

'I see', said Jake. 'Profit clearly peaks at a price of £12 000. Is this the selling price you would like Peter?'

'Yes', said Peter Sell, Sales Director. 'This is well below the list price for our standard floroll so it gives us considerable competitive advantage and, as you can see, the profits are also the highest at this price.'

'I see', said Jake. 'How much do we charge for the standard product?'

'£30 000 list price, although most customers have a negotiated discount of between 10 and 15 per cent.'

'I see', said Jake. 'So the lowest price to any of our current customers is about £25 000. What do you think, Jack?'

Jack Fox smiled. He was pleased that Jake Topper had reached the same conclusion as himself. 'Unfortunately, the analysis I showed you does not indicate the impact which the selling price would have on our standard range of florolls, which is, after all, our bread and butter at present. If the price of the industrial product is set at less than half that of the standard product there will be suspicion among our current customers that we have been making excessive profits out of them. I think you will have a problem selling the standard floroll at the current price, Peter. Not only that but there is a real danger that some customers may think that the low-cost product is inferior in quality, because it is so cheap, and we will get a reputation for producing shoddy goods, which will be very damaging in our military markets.'

After some discussions it was agreed by the Board that the price of the industrial floroll would be set at £20 000. This would make it 20 per cent cheaper than the standard product and still give an adequate profit.

16.2.3 Elasticity of demand

Figure 16.15 illustrated one curve of variation of demand with price. This variation of demand is known as the price elasticity of demand. If a price reduction gives more than a proportionate increase in demand, or if a price increase gives more than a proportionate decrease in demand, then the demand is said to be elastic. If the price change results in a less than proportionate increase in demand then demand is said to be inelastic.

If demand is elastic then pricing is an important marketing consideration. If demand is inelastic then pricing is not as important and the other elements of the marketing mix should be used for product differentiation.

Figure 16.17 illustrates constant elasticity. For the perfectly inelastic curve, once the price has fallen below the basic price at which consumers are willing to buy, the actual quantity bought is not related to the price. This could be the case of people buying a car; once the price falls below the figure they have set they will buy a car, but are unlikely to buy more and more cars if the price falls further.

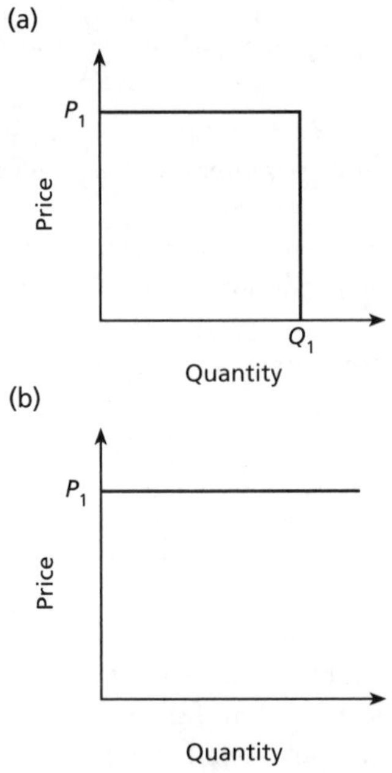

Figure 16.17 Constant elasticity: (a) perfectly inelastic; (b) perfectly elastic.

In the perfectly elastic case the consumers buy an unlimited quantity at the present price, although a small increase in price causes the quantity sold to be zero. This would be the case in a perfectly competitive market. As expected, constant elasticity rarely occurs in real situations, the price demand curve usually being as in Figure 16.15.

Several methods exist for calculating elasticity. In the arc elasticity method the price elasticity of demand (PED) is given by Equation (16.1).

$$\text{PED} = \frac{\text{Proportionate change in quantity demanded}}{\text{Proportionate change in price}} \tag{16.1}$$

This is called arc elasticity, since it calculates the average elasticity between two points on the arc of the demand curve. It does not consider changes between these two points. The smaller the price changes considered the closer the elasticity represents that along the demand curve. Arc elasticity also varies depending on whether price rise or price fall is being considered, since a different base is used for calculating proportionate change in the two cases. (See the case study 'Calculating the elasticity'.)

Point elasticity, illustrated in Figure 16.18, measures the tangent at the point on the demand curve being considered. It is given by Equation (16.2), where Δ represents an incremental change.

$$\text{Point elasticity} = \frac{\Delta Q/Q}{\Delta P/P} = \frac{\Delta Q}{\Delta P} \times \frac{P}{Q} \tag{16.2}$$

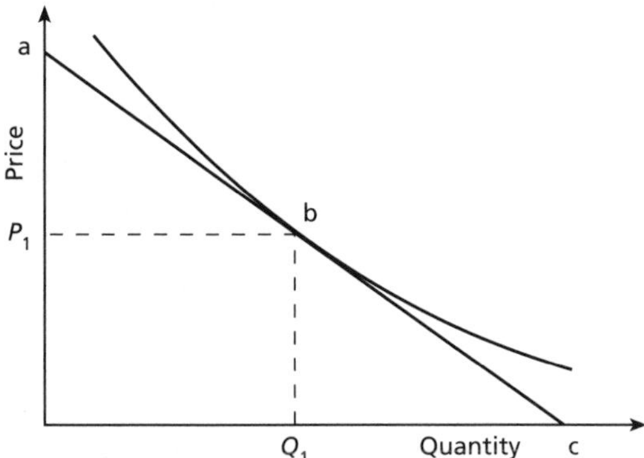

Figure 16.18 Point elasticity.

The point elasticity therefore depends on the slope of the demand curve at the point being considered and on the original price and quantity at the point. For Figure 16.18 it is given by Equation (16.3).

$$\text{Point elasticity} = \frac{\text{bc}}{\text{ba}} \tag{16.3}$$

If the arc and point elasticity of demand do not change appreciably as price changes then the product is said to be relatively inelastic. If they change significantly then it is relatively elastic.

Several factors affect price elasticity of demand (Maile, 1985):

- Whether there are substitution goods available. If there are, then demand is likely to be elastic, since consumers can switch to other goods if the price is considered to be too high.
- Whether there is consumer loyalty to the product or if there is habitual buying, for example consumers always buying the same brand of soap powder. If any of these conditions exist then there is likely to be demand inelasticity, consumers being unlikely to change unless the price change is significant.
- The frequency of purchase; for example, the purchase of goods such as televisions and freezers. These are likely to have considerable price elasticity, since consumers will buy them infrequently.

Elasticity of demand can be calculated with respect to other factors apart from price. For example, Figure 16.19 illustrates a typical income demand curve. The elasticity of demand (YED) (the effect of changes in income on the quantity bought), is given by Equation (16.4).

$$\text{YED} = \frac{\text{Proportionate change in quantity demanded}}{\text{Proportionate change in income}} \tag{16.4}$$

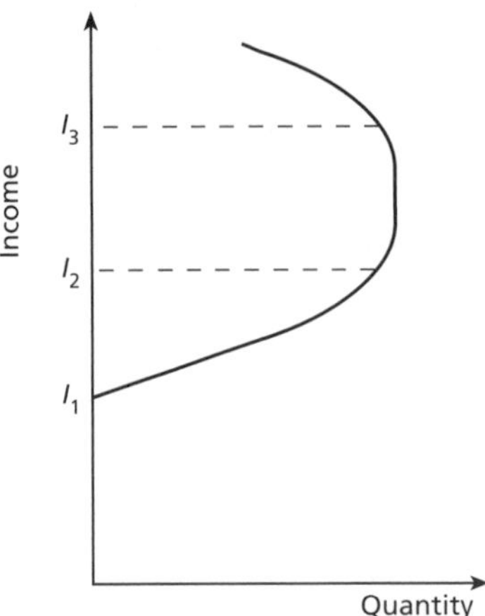

Figure 16.19 Income elasticity of demand.

A negative YED means that fewer goods are bought when income increases; that is, they are inferior goods. A positive YED means that more goods are bought when income increases: it is a normal product. In Figure 16.19, below income I_1 there is insufficient income to buy any of the product. Between incomes I_2 and I_3 the demand for the product is relatively flat, while above I_3 it starts to decrease, as consumers switch to substitutes which may be more expensive but are perceived to be better.

Case study

Calculating the elasticity

The Roller Company makes bearings which go into several of the critical parts of florolls. There are substitution products and the market therefore has considerable price elasticity.

Joe Ball, Marketing Manager of the Roller Company, is studying the market research data, as in Table 16.2, to determine the price elasticity for the product. He decides that the low prices are unrealistic, since they would give poor profitability. However, prices in the range £10 to £15 are economic and worth further consideration. Joe calculates the arc elasticity for a price rise from £10 to £15 as in Equation (16.5).

$$\text{PED} = \frac{(10-15)/10}{(50-20)/50} = -0.83 \qquad (16.5)$$

If a price fall from £15 to £10 is considered then the arc elasticity is given by Equation (16.6).

$$\text{PED} = \frac{(15-10)/15}{(20-50)/20} = -0.22 \qquad (16.6)$$

Table 16.2 Illustration of the case study 'Calculating the elasticity'

Price per box (£)	Quantity (000)
20	0
15	20
10	50
5	100
3	140

This shows that the arc elasticity differs depending on whether a price rise or a price fall is being considered, as Joe Ball expected.

16.2.4 Costs and pricing

As seen earlier, although costs are not the foremost consideration when fixing product price, they are important when determining the return obtained at any price. Usually prices are related to costs, and an example is shown in Figure 16.20. When the product is first launched the cost of the product is high, and although the product price may also be kept high the price may be below cost.

The cost of the product decreases rapidly after launch, as it moves down the learning curve, with full production ramping up. However, prices can be maintained at their original level, generating profits from time t_1 onwards.

At time t_2 it is assumed that competitors have entered the market, and the price is reduced in order to maintain market share. There may be one or more step

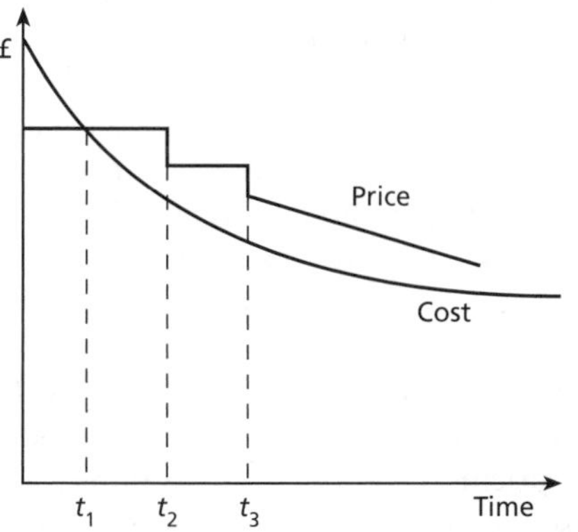

Figure 16.20 A typical cost–price strategy.

decreases in price, for example at time t_3, before the price of the product tracks the cost.

A variety of pricing techniques can be used based on costs. Three of these are discussed here: absorption pricing, rate of return pricing and marginal pricing.

Absorption pricing is concerned with absorption costing. In this method the cost of making the product at the expected volume of output is calculated and a profit margin is added to this to arrive at the selling price.

The advantage of the absorption pricing method is that it is straightforward and easily understood. However, as discussed in Chapter 11, this method of costing does not provide guidance on such items as the increase in cost or profit from making an incremental increase in volume, since the costs include fixed costs. In a competitive situation it does not provide information on how much prices can be changed and still ensure that a contribution to fixed costs is made by the product. The case study 'Pricing with absorption' illustrates the use of the absorption method of pricing.

In the rate of return pricing method the requirement is to cover costs and to earn a certain rate of return on the capital employed. This is illustrated in the case study 'Pricing with return'.

Rate of return pricing is a refinement of absorption pricing and, as in any absorption costing method, the costs and prices are based on expected volume and fixed overheads. The results are therefore only true once that volume is reached. The method is not flexible enough to give management the guidance it needs to price products in a competitive environment.

The marginal pricing method does not fix the product price to recover all the costs, but instead uses marginal costing principles; each product is considered to earn a contribution towards total fixed costs and profit. Some products may make a loss in cost terms, but provided they make a positive contribution they may still be useful products to make. (See the case study 'Pricing with margin'.)

Marginal pricing ignores fixed costs, which are historical and unchangeable. It concentrates on future costs and capital outlays, with the aim of maximizing the total contribution to costs and profits. In pricing decisions one is interested in changes in costs which result from these decisions. This is reflected in marginal pricing, which does not take fixed costs into account for decision making since these are not affected by management decisions. The danger of marginal pricing is that fixed costs may get ignored completely, although eventually all these costs have to be absorbed into the product lines. Generally a company uses full costs for pricing in a stable situation and marginal pricing in special situations, such as in competitive markets.

The discussions so far have ignored inflation, since in these circumstances the profit earned must be such as to maintain the capital of the company. Other considerations are also important in pricing decisions. For example, a company may reduce prices in order to increase the volume of sales, and the profit/volume ratio and breakeven point calculation will indicate whether it is worthwhile doing so. When entering a new market a company may also initially sell well below cost in order to get to a monopoly position in which it can usually eventually make bigger profits.

Case study

Pricing with absorption

Pack-U Ltd makes packaging material for florolls. It has three product lines, known simply as Package P, Package Q and Package R. The direct costs and sales volume in the next year are expected to be as in Table 16.3.

Table 16.3 Costs for the case study 'Pricing with absorption'

Item	P	Q	R
Direct material cost per unit (£)	100	300	400
Direct labour cost per unit (£)	100	100	200
Sales volume in units	3000	2000	1000

The fixed overheads are predicted as £700 000 during this period and the labour rate as £10 per hour. The company absorbs overheads in proportion to labour hours. The Marketing Manager calculated the selling price of P, Q and R, assuming a profit margin of 20 per cent, using absorption pricing, as in Table 16.4. The fixed overhead absorption is taken as a proportion of the total labour hours worked in the factory during this period, which is predicted to be 30 000 + 20 000 + 20 000 = 70 000. The fixed overhead absorption for Package P is therefore given by Equation (16.7), and those for Packages Q and R can be similarly calculated.

Table 16.4 Calculation of selling price for the case study 'Pricing with absorption'

Item	P	Q	R
Total labour hours	30 000	20 000	20 000
Fixed overhead absorbed	300 000	200 000	200 000
Fixed overhead per unit	100	100	200
Total cost per unit	300	500	800
Profit (20 per cent of cost)	60	100	160
Selling price	360	600	960

$$\textbf{Fixed overhead absorption} = \frac{3}{7} \times £700\,000 = £300\,000 \qquad (16.7)$$

Case study

Pricing with return

The Wrap-It Packaging Company also makes packaging materials for florolls. It has a single low-cost product, WrapSafe, which competes with Package P produced by Pack-U Ltd. The company uses the rate of return pricing method to set its prices and the Marketing Manager is evaluating the price which should be set for their product for the coming year.

Table 16.5 gives the costs per unit, fixed costs for the plant and the sales volume expected over the next year. Like its competitor the production volume is targeted to meet the predicted sales volume.

Table 16.5 Costs for the case study 'Pricing with return'

Item	WrapSafe
Cost per unit (£)	200
Fixed cost of plant (£)	100 000
Sales (production) volume in units	5 000

The mark-up on total costs is given by Equation (16.8).

$$\frac{\textbf{Profit}}{\textbf{Total cost}} = \frac{\textbf{Profit}}{\textbf{Capital employed}} \times \frac{\textbf{Capital employed}}{\textbf{Total cost}} \qquad \textbf{(16.8)}$$

The total costs for 5000 units are given by Equation (16.9) and the mark-up on total costs by Equation (16.10).

$$\textbf{Total costs} = \textbf{5000} \times \textbf{200} + \textbf{100 000} = \textbf{£1 100 000} \qquad \textbf{(16.9)}$$

$$\frac{\textbf{Profit}}{\textbf{Total cost}} = \textbf{20} \times \frac{\textbf{1 000 000}}{\textbf{1 100 000}} = \textbf{18.2 per cent} \qquad \textbf{(16.10)}$$

Table 16.6 shows the calculation used to arrive at the selling price, to achieve the level of profit given by Equation (16.10).

Table 16.6 Calculation of selling price for the case study 'Pricing with return'

Item	WrapSafe
Cost per unit (£)	200
Fixed costs (£)	20
Total costs (£)	220
Mark up on costs (18.2 per cent)	40
Selling price (£)	260

Case study

Pricing with margin

It is becoming clear to the Pack-U company that its product Package P is losing ground to its competitor's WrapSafe product. The main problem is the price differential of £100, which the company has not been able to overcome by using other elements of the marketing mix, since customers perceive both products as being essentially similar.

The Marketing Manager of the Pack-U company has decided that the only way that Package P is going to maintain market share is to reduce its price to match that of the competition. A price level of £250 per unit is required.

Based on the figures given in Tables 16.3 and 16.4 this means that each unit, in total cost terms, will be making a loss of £50. This indicates that if absorption pricing methods continue to be used then Product P must be discontinued, which would increase the price of the remaining product lines unless fixed costs can be significantly reduced.

However, using marginal pricing it is clear that the direct material and labour costs of Product P are £200, so that even at a selling price of £250 each unit makes a contribution of £50. Since the spare capacity freed up if Product P ceases production cannot be put to better use the Marketing Manager decides to continue with Product P, selling it at the lower price of £250. This would give a profit of $3000 \times £50 = £150\,000$ more than if the product were discontinued.

16.3 Marketing communications

Marketing communications can cover several activities, but its main aim is to ensure that the company and its products are effectively presented to existing and potential customers. Several methods are used for this, such as advertising, direct mail, trade exhibitions and promotions, as discussed in this section.

Also included within the remit of marketing communications is publicizing the product and the company through such items as sponsorship, where the event chosen must match the image that the company wants to project for its product and the company as a whole. Although sponsorship of sporting events, such as motor racing, is common, many companies have discovered that they can also benefit by sponsoring other charitable and fund-raising events.

Public relations (PR) is an important consideration for the marketing communications function. Generally the aim is to get media coverage to publicize the company and its products. Every assistance should be provided to busy editors, such as the preparation of press releases and tailoring the information to meet the needs of certain media. The press releases should be chosen to be of interest and clearly presented. Obtaining press coverage is important, since it enhances the image of the company when customers and potential customers see that the company is able to influence the press.

However, it is important to remember that public relations is all about the public and potential customers; the press is only a vehicle for reaching the public.

Many companies spend a large amount of money on marketing. For example, a company's sales and marketing expenditure may represent 15–20 per cent of its total revenue (Slywotzky and Shapiro, 1993). The marketing communications budget is a significant part of this and may be fixed in several ways, for example as a percentage of sales; as a rate of return, to try to obtain value for money; as a proportion of the level of competitor activity; on the basis of what is left over from other expenditure, while still maintaining the required profit margins; or as a percentage of the total marketing budget, which is tied into the product life cycle.

No matter which method of budgeting is chosen, the problem is measuring whether this money is being spent effectively. Unfortunately, measuring effectiveness of marketing communications is not easy, unless it is based on response to specific advertisements or exhibitions. Even then it has been found that although sales are affected by specific sales campaigns, there is always a lag factor, and successive promotions reinforce the sales message. For example, consumers questioned during a particular sales campaign may remember the message promoted in the previous campaign or the one previous to that, rather than in the present campaign.

16.3.1 Advertising

Some industrial markets contain a relatively few, easily identifiable, consumers, and in these instances direct sales visits by the company's internal sales force may be a more effective way of reaching the market than advertising. However, if the company is a newcomer to the market, or if its products are bought by a cross-section of the industry, then advertising is essential (Felton, 1994). Advertising is a one-way communication mechanism, from the seller to the buyer. Its main aims are to:

- Generate interest in the product and to get consumers asking for more information. This will then provide the sales lead needed to sell the product.
- Position the product, differentiating it from competing products on the market.
- Enhance the company's image with its present and potential customers, remembering that, especially in industrial markets, people like to buy from reliable suppliers, as a means of reducing their perceived risk. Image improvement also helps to improve staff morale and reduce turnover, since people generally like to work for well-known companies. Therefore advertising is sometimes targeted at internal audiences.

- Keep the product and company names constantly in front of the consumer. Often if a new product is advertised heavily it can create an impression in the customer's mind that the product has been established for some time.
- Support the sales effort. Advertising may be slanted so as to help the internal sales team within the company, or to support distributors, retailers or agents, some of these often being mentioned by name.
- To reassure a buyer, who has recently bought the product, that a sound choice has been made. Buyers who are thus reassured are likely to make further purchases.

Companies carry out their own advertising or employ an advertising agency. However, the procedure used in advertising, as illustrated in Figure 16.21, is often the same:

1. The target audience must first be identified and its characteristics defined.
2. The advertising medium then needs to be selected. Clearly this must be such as to reach the target audience. There is no point, for example, in advertising baby foods in a magazine targeted at the retired. Television was a very effective medium when a handful of TV stations commanded vast audiences (Mitchell, 1994). Now this medium is fragmented, which means that it is often prohibitively expensive to reach mass audiences through TV advertising. Posters are relatively cheap, but are generally more effective for consumer, rather than industrial, goods. They give regional cover, unless a national poster campaign is launched. The location of posters also has an impact on the type of audience reached, for example in a busy shopping centre or on the underground. Posters are best when used for general appeal products, rather than specialized ones.

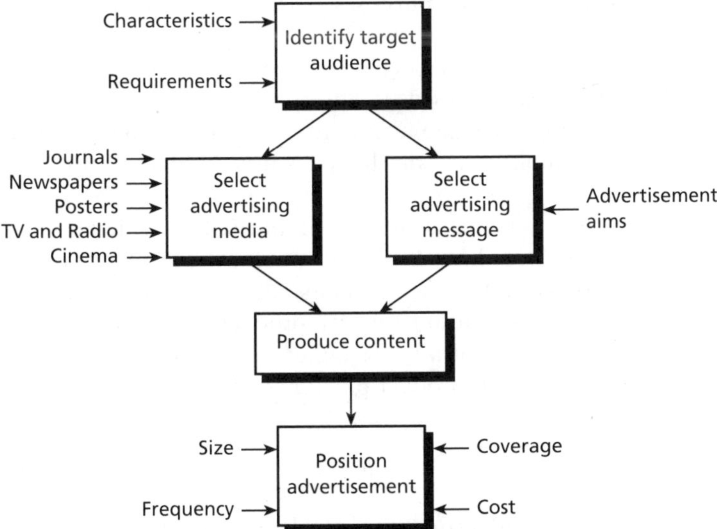

Figure 16.21 The basic steps in the production of an advertisement.

3. The advertising message has also to be chosen carefully. The aims of the advertisement and the target audience must be kept clearly in mind when the message is being formulated.
4. The content of the advertisement is then produced. If this is for TV, radio or the cinema it can involve considerable amounts of time and money. The presentation of any advertisement is clearly important, since it will affect not only the sales of the product concerned but also the image of the company. A quality product and image require a quality advertisement in a quality medium. The key message of the advertisement must be clearly brought out.
5. The positioning of the advertisement must then be considered. Factors include the size of the advertisement (for example whether it is full page, half page and so on); the frequency of appearance, such as daily, weekly or every other issue; and the coverage (local, regional, national). All these factors will clearly be linked to the budget limits set for the advertisement campaign. Often there are critical trade-offs, such as whether to take a full page advertisement in every alternate issue of a magazine, or to go for a half page advertisement in every issue.

An important area of advertising which is effectively used on consumer goods, but not so well on industrial goods, is packaging. The package should carry the brand name and the company name. When it sits on shelves, whether at home or in the customer's storeroom, it should reinforce the sales message. It should be eye-catching, to be noticed when it is on supermarket shelves or when it is being transported.

Apart from protecting the goods inside and making it easy to transport and display goods, packaging also enables product differentiation. Unique packaging designs can be used for several purposes, such as easy to dispense packages; reusable packages, such as those which can be used as storage containers; packages which can be used to prepare the product, such as boil in a bag packages; child safety packages; or decorative packages.

Varying the packaging also allows different markets to be targeted with the same product, for example different size packages for family use or for single persons.

A well-chosen product name is also an important consideration in product advertising. The name should be chosen to reinforce the image which is to be projected about the product and the company, and it should be used for product differentiation. A key consideration for any product is whether to name it as a standalone product or as part of a wider family of products. The eventual choice must be in line with the overall product portfolio aims of the company.

16.3.2 Direct mail

Direct mail is a form of advertisement which is seen by a limited number of people. It is sent to a select audience, who would presumably be the potential customers most likely to buy the product. Unlike more general advertisements it will not reach any other potential users, so its coverage is limited. However, direct mail shots are normally cheaper and if the target audience is small and known then it is a more effective mechanism.

The problem is obtaining the names and addresses of the target audience. This information may come from marketing communications' own files, or it can be hired or purchased from external agents, who determine and maintain such lists for different categories of products.

The direct mail shot material must be carefully designed if it is to avoid the fate of most 'junk mail': going straight into the bin after a cursory glance. It must be eye-catching and get the message over in the first few seconds. The recipient must be made to read on.

Brochures and catalogues are the most common form of direct mail, and these are also expensive to produce. They may be sent as part of the initial mailshot or as a follow-up, once the customer has asked for further information. Presentation is once again very important, since the reader is judging the products and the company from the brochures. They must be high-quality, eye-catching and, above all, informative.

16.3.3 Exhibitions

Exhibitions are used by companies within consumer and industrial markets as an effective way of reaching a target audience. A company needs to consider many factors when preparing for an exhibition, such as the design and location of its stand; its size; the number of company personnel on the stand and their composition (technical, marketing, sales); how to handle competition which comes onto the stand; and stand management. The main advantages of an exhibition as a market communications method are:

- It is a two-way communications medium. Not only do producers tell buyers about their goods, but buyers can also feed back their likes and dislikes.
- It enables the seller to show actual products and therefore to influence potential buyers more effectively than can be done in an advertisement. The buyer's risk is also reduced since the product can be seen, not just a picture of it, and buyers can operate the equipment and ask questions about aspects which may be causing them concern. Exhibitions often turn interest created by an advertisement into potential sales.
- Exhibitions often provide useful sales leads, which can be followed up later. They also provide an opportunity for meeting and entertaining potential and existing customers. This can considerably reinforce the sales message, and salespeople often invite their key customers to visit the company's exhibition stand.
- Exhibitions can enhance a company's image. For this it is clearly important that the company has a prestigious stand and that it is well

designed and effectively managed throughout the exhibition. Foreign companies wishing to enter new markets use exhibitions to gain credibility with new customers.
- Exhibitions are often used to launch new products, since feedback can be obtained from potential customers, which can be used to modify the product if necessary. Exhibitions also present a glamorous setting for new product launches, although if many companies launch new products at the same exhibitions the impact of each launch is diluted.

The exhibition chosen must clearly match the marketing aims of the company (Williams *et al.*, 1994). Considerations which need to be taken into account are:

- Whether the company should participate in a general exhibition with other suppliers, even competitors, or whether it should organize its own exhibition, perhaps only for itself or in conjunction with a few other suppliers. Generally one-company exhibitions are expensive to arrange, even if they take the form of a travelling trailer, and they reach fewer people, although they can be carefully targeted.
- The coverage of the exhibition; that is, whether it is international, national, regional or local. Clearly, the exhibition coverage must match the aims of the product (such as whether it is aimed at international audiences or whether the customers are clustered around a given region).
- The content of the exhibition, such as whether it is general or specialized. Generally the attractiveness of either type of exhibition to the exhibitor depends on the product range being exhibited. Most exhibitions usually have a theme, such as telecommunications or business equipment. Sometimes it is good for a company to exhibit in a related market area, since it enables it to meet fringe customers, which it would not normally have done, and so increase market size. However, such moves may not be cost-effective and should not be done often.

Many exhibitions have conferences associated with them and technical papers presented at these conferences often enhance the reputation of the company. Conferences are also organized by professional bodies and papers may be published in the technical press and in conference proceedings. These have the benefit of reaching the technical decision makers within the buying organization and often do much to enhance the image of the selling company. Many companies obtain reprints of papers presented by their staff and send these out to customers as a very effective supplement to the normal colour brochures and catalogues.

16.3.4 Sales promotions

Sales promotions usually consist in offering incentives to buy a given product (Helme, 1996; MacVicar, 1996). They may be targeted at the trade or the consumer, as shown in Figure 16.22. Trade promotions are aimed at

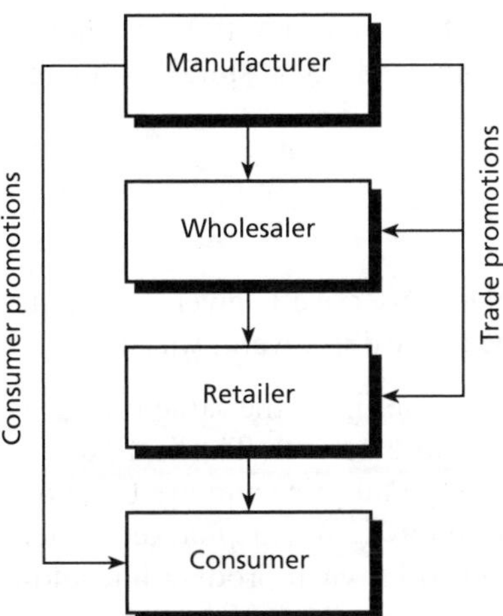

Figure 16.22 Types of promotion.

intermediaries (wholesalers, retailers and agents), and they may take the form of discounts or gifts. The aim is to get the intermediaries to stock the company's goods, and to sell them aggressively to their customers, in preference to competing goods.

The aims of consumer promotions are:

- To attract new users to an existing product. This can be done by offering price reductions or by selling goods on a sale or return basis. Small quantities of the product may also be given free, such as by mailshots or by inviting readers to send off for samples following an advertisement. The aim of such promotions is to reduce consumer risk by enabling them to try out the product at low cost to themselves.
- To introduce a new product to a target market segment. The aim is to provide publicity for the product, and this can be done by holding contests and games in which the product is given away as prizes; providing a heavy price discount for a limited period; giving free samples; and so on. A new product may also be given free when an established product is bought, for a limited time, or trial offers can be given.
- To encourage the consumer who has bought a product to make a repeat buy. This is often done by including coupons on the product's package which gives a price discount for the next purchase.

Promotions enable a manufacturer to buy sales on a once and for all basis, although this may not generate repeat purchases (Jones, 1990). Promotions usually mean price reductions, so they sacrifice income. As such they are an expensive method of advertising and generally devalue the image of the promoted brand in the consumer's eyes. Promotions have the tendency to bring forward sales from a later period, when prices are again raised, and this also weakens the brand. Another side effect of promotions is that they often cause competitors to retaliate, which can lead to a spiral of price cuts and diminishing returns.

16.3.5 Branding

Branding is the term used when a product or service is given a definite name or identity (Maruca, 1995). Branding can be applied in several ways:

- **Family branding**. In this all the product lines have the same brand, as done, for example, by Campbell's, the food manufacturer. This is a good technique to use if all the items have similar quality and price and are targeting the same market segment. It provides an umbrella for the marketing activities.
- **Separate branding**. This is usually applied for each product line and is necessary if the products are targeted at separate market segments.
- **Flanker branding**. In this method the brand name of an established product in one area is used to launch a new product into another area. For example, the brand of Levi jeans was extended to cover Levi shoes. The common brand provides goodwill carry-over from one area to another and is cheaper and more certain of success, although an unsuccessful launch can weaken the original brand.
- **Intermediary branding**. In this method the product is not branded by the manufacturer but by the retailer. An example is the brand of goods sold by many supermarket chains as part of their own products. The advantage of this to the retailer is that it ties the customer into their outlets, although the retailer and not the manufacturer now has to support and market the brand and carries legal liability for it.
- **Generic branding**. In this the product does not have a brand name, and is sold as a low-cost item which has minimum money spent on it for packaging and advertising. The key selling factors are low price and consistent quality. Generically branded products are generally sold to price-conscious consumers and are not sold in industrial markets.

Branding a product or service achieves several marketing objectives:

- It allows the product or service to be differentiated in the market-place and links it to the seller. The brand name, if it is well known, such as Heinz baked beans, often causes the consumer to choose the goods in preference to other products.
- Branding allows the consumer to relate past experience with using the product, good or bad, when making a purchase decision. For example, it would be very difficult to choose chocolate from a confectionery stall if all the brand names had been removed. Branding allows past buying experience to be carried forward to the next buy.
- Different market segments can be targeted with different brands of the same product, such as is done with toothpaste.

- Branding enables easier market introduction of a new product if the brand name is carried forward between products, such as is done for Xerox copiers or IBM computers.
- Branding reduces the price elasticity of a product; with brand loyalty built up the product is likely to be less sensitive to small price changes.
- There is strong correlation between profitability and quality (Chernatony, 1993). A powerful brand offers the customer an assurance of consistent high quality, since a well-known brand would be supported strongly by its manufacturer.

There are three directions in which a branded product can move (Jones and Ramsden, 1991):

- To reinforce the product by strengthening the brand within its traditional product category.
- To extend the brand into new categories which fit its concept. For example, confectionery suppliers such as Mars extended their chocolate bars into ice cream chocolate bars.
- By exploiting a discontinuity in the market, usually technology-based, to bring about a transformation of the market itself.

16.4 Sales

16.4.1 The aims of selling

Sales activities are fundamental to any organization, since they result in income being received by the company. The aims of any sales operation are:

- To make a sale, converting interest in the product into positive buying intention.
- To understand the customer's needs and to position the product so that it meets this need. Salespersons must be skilled in differentiating the product and creating a competitive advantage for it when dealing with customers. For this they must have knowledge of competitor products and also what the buyer values most, such as price, service or credit facilities. The customers should be influenced such that their needs match the strengths of the company and product. Generally this is done by assisting the customers, when they are preparing their product specification, by offering to discuss the draft of the proposals with the supplier's technical staff.
- To ensure that buyers are satisfied with their purchases. Buyers must not feel that they are being sold to. Rather, they should be completely satisfied that the sales process is of mutual benefit to buyer and seller and appreciate that the salesperson has a genuine desire to recommend the right product to meet the customer's needs. The relationship between buyer and seller should be

regarded as a long-term partnership and not as a one-off sale. The reputation of the seller, and not that of the buyer, is at stake if the customer is not satisfied with the purchase. Often the seller needs to help the customers understand their true requirements, so that they can make a correct purchase. The salesperson must ensure that full support is provided to the customer both before and after the sale.

Personal selling has several advantages in meeting these aims in an industrial environment. For example, the selling message can be tailored to the individual needs of the customer. Instant feedback is also received during the sales interview and the message can be modified to suit the customer. Body language can be studied and a rapport established between buyer and seller, which is important for a long-term relationship. Personal selling also allows the selling message to be illustrated by samples, demonstrations and so on.

16.4.2 The sales organization

Many different methods are used by an organization in selling, such as the use of its own direct sales force; the use of distributors or agents; and selling through retail outlets (Moriarty and Moran, 1990). Personal direct selling is probably the most widely used within industry and is effective, since many industrial products require a detailed knowledge of the product and of the customer in order to make a successful sale. However, a hybrid system is usually better (Moriarty and Moran, 1990). The demarcation between the various methods of selling can be made on three bases: customer characteristics, geographical boundary and product boundary.

Differentiation based on customer characteristics usually involves customer size or the potential order value. For example, large customers may be serviced by the company's own internal sales force, while smaller ones could be reached by distributors or retailers. This provides increased market coverage at lower initial cost and also has the advantage of reaching new customers.

Another customer characteristic, which is often used to differentiate between the various types of sales effort, is the industrial sector in which the customer operates. For example, an organization's products may be primarily targeted at one market sector and the company may wish to use its own trained sales force to sell into this market. Other market sectors may be covered by other methods, such as by agents or distributors.

Geographical boundaries are another differentiation technique, such as using one's own sales force to reach those areas which are close to the manufacturer's home base, while operating through local distributors and agents for remote areas and overseas.

When differentiating on the basis of product boundaries a company usually sells its high-valued complex products using its internal sales force. These products require technical selling, and the customer is usually more comfortable dealing directly with the manufacturer for high-valued items. However, for low-valued or routine items, especially when the major sales activity consists in order taking (for example repeat orders) third parties, such as distributors or wholesalers, can be used to sell the product.

The internal sales organization reflects the differentiation method used by the company in its selling process. Usually the sales force is organized along one of four lines, although frequently a mixture is used depending on the market or product:

- **By customer**. Usually the company will allocate several key accounts to its top salespersons. These accounts may be large organizations, such as multinationals, or organizations which have considerable influence in the area in which the company operates.
- **By industry or type of customer**. For example, a salesperson may be responsible for customers in the telecommunications industry, another cover customers in the transport industry and so on. This method is used if the message the company wishes to give varies by industry, or if its products are mainly sold into a few industries. This organization structure allows the company to understand customer problems within a given industry more effectively. Industries which are not targeted directly may be reached by other methods, such as distributors and agents.
- **By geography, such as by region or country**. This method is used if the customer base is widely spread and the product sold is relatively standard. Usually the sales force will also be distributed geographically, either working from home or from regional offices. Selected areas may also be targeted by the company, using its internal sales force, with others being reached by distributors and so on.
- **By product line**. This is used when products are complex. It is a common method for industrial products, where different divisions of the same company make a given line of product. This organization enables the salespersons to specialize in their product lines. The danger of this organization is that it can result in duplicated calls on a customer by salespersons from the same company, leading to customer confusion and annoyance. Long travelling distances may also be involved when customer calls are spread over a geographically wide area. One way to overcome this problem is to have non-specialist salespersons, who cover a variety of products, but are backed up by specialist technical support teams, which can be called on as necessary.

16.4.3 Sales process

The first step in the sales process is to identify customers. These may be prospects, such as potential customers who have replied to advertisements or visited the company's exhibition stand. The salesperson's aim is to generate enough interest in these prospects to enable a call to be made.

Prior to the call research must be done so that the product or service being sold is fully understood by the salesperson, regarding its technical performance and applications as well as price flexibility and associated items such as support and delivery. Advertising, brochures and so on can also be used to provide the customer with advance information. Apart from research into the product the salesperson also needs to know the customer and the products being offered by competitors. The customer's requirements and any objections to the product being offered should be determined, if possible, in advance. (Figure 16.23.)

The first call on the customer is important, since its main aim is to smooth the way for subsequent calls. The call will establish the importance of the prospect. The customer's requirements must be thoroughly understood during this and other calls. As the sales process develops the salesperson must show how the

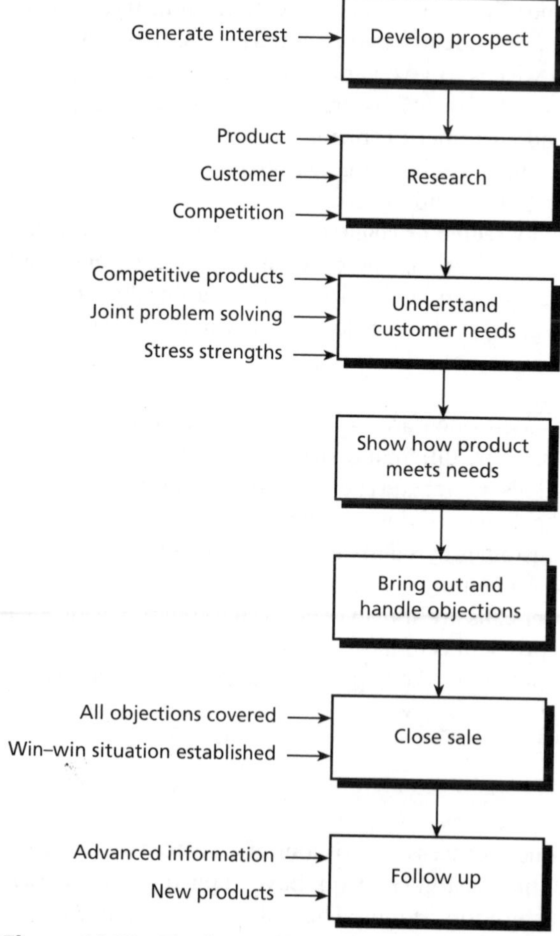

Figure 16.23 The basic steps in the sales process.

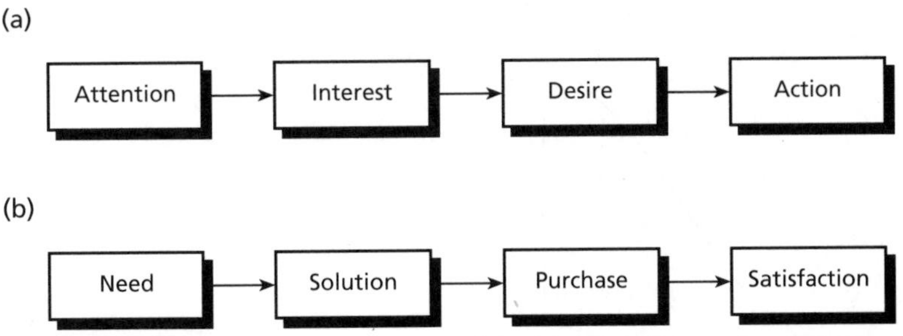

Figure 16.24 Buying behavioural models: (a) AIDA; (b) need–satisfaction.

product or service being offered meets the customer's needs. An atmosphere of joint problem solving should be established so as to gain the confidence of the customer. The strengths of the product offering should be stressed and competitor weakness exploited, although it is bad practice to directly attack competitors or their products.

A key part of the sales process is bringing out and handling objections. Customers' concerns with the product should not be glossed over; if the customer is unhappy or has any niggling doubts the sale will not be successful. Salespersons must show that they have genuine regard for the customer's objections and concerns, and must show why these are not important in the present situation.

Before the sale can be successfully closed all objections must have been covered to the satisfaction of the customer and a win–win situation should have been established between seller and buyer. Each must benefit from the transaction.

Several models exists to explain buying behaviour, two being illustrated in Figure 16.24, and these need to be kept in mind during the sales process.

In the AIDA model (Attention, Interest, Desire, Action) the first step involves catching the attention of the buyer, by the product as well as by the salesperson. Once attention has been caught the next step is to arouse interest in the product being sold. This is followed by a desire for the product, which leads to action to buy the product.

The need–satisfaction model starts with the customer's need, the desire to solve a specific problem. The salesperson's task is to show how the product or service being offered solves this problem. This will lead to the purchase action, resulting in customer satisfaction.

Follow-up actions are a vital part of every sales process. The customer must be kept informed of new products and given advanced notice of any price changes. Customer orders must be monitored and the customer helped to prepare for and use the product when it is delivered. The buyer must be assured that a good buying decision was made if repeat orders are to follow and competitors are to be kept out.

One of the problems in any sales call is determining the length of time which should be spent on the call. For example, Figure 16.25 shows how the effectiveness of calls, measured in sales per call, can vary depending on the time spent on a call. Clearly, as this time increases the sales will also increase, but

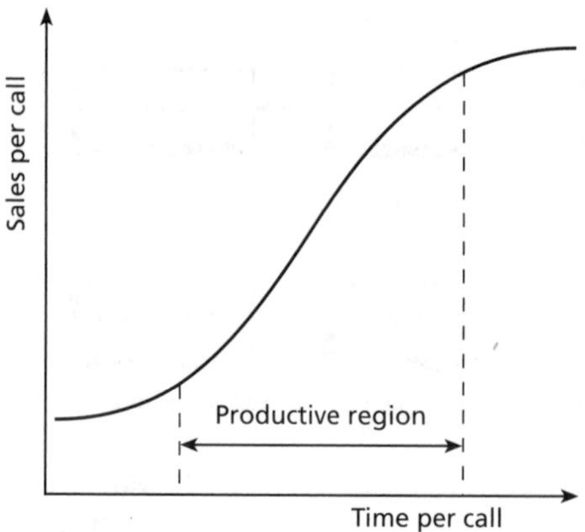

Figure 16.25 The effectiveness of time spent on a call.

eventually the gain made is very small. The key is to operate within the productive region in Figure 16.25, although often salespersons operate well outside this region, spending long periods of time with established customers who have become friends, and neglecting other potential customers.

16.4.4 Sales force compensation

Salespersons have very special qualities and the compensation method chosen must match this. The key qualities are:

- Ability to learn quickly. Industrial buyers expect sales representatives to be technically knowledgeable about the products they are selling (which are also continually changing), to know the buyer's business, and therefore to be able to recommend the best product fit to meet their needs.
- Good at personal interaction. Establishment of empathy with customers is essential if long-term relationships are to be developed. Salespersons have traditionally had a reputation for being the 'party-going' type, ready to entertain the customer on every occasion. Modern buyers do not expect this, and many would be suspicious if they were being continually entertained. The salesperson must ensure that the relationship developed between seller and buyer is much more subtle and long-lasting.
- Persistent and with a strong competitive spirit. There will be many failures, but the desire to win through in the end must be strong. Salespersons must never take 'no' as the final answer, nor must they be put off by indifferent or rude customer behaviour. They must look for ways in which to turn a competitive disadvantage into an advantage, always remembering, however, that the sale must not be forced onto an unwilling buyer.
- Self-motivated and disciplined. Salespersons very often work at their own time and set their own agenda. They must be able to work on their own for

long periods of time, often from home, while still keeping their long-term goals clearly in mind.

The compensation package offered to salespersons is important, since it can, as in the case of other employees, act as a demotivator if it is not satisfactory. Three main methods are used for determining sales compensation: straight salary, commission, or a combination of the two.

The salary method is chosen when high-cost items are being sold, or where a long time is needed to develop prospects and to close a sale. In these circumstances the salesperson also requires the support of a relatively large team of people, dealing with commercial matters, providing technical information and so on, so the eventual sale depends as much on team effort as that of the salesperson. In these circumstances the role of the salesperson is primarily to represent the customer's interest within the seller's organization and to coordinate the overall sales effort. The level of salary will usually be fixed according to the importance of the contracts handled by the salesperson and on performance.

The commission method of sales compensation is used for less technical and lower cost products, where the sales are determined primarily by the salesperson, with little support from others within the seller's organization. A fair system for determining the commission must be used, such as setting a quota for the year based on prospects' size, product qualities and so on, and then providing a scale of commission depending by how much the quota is exceeded.

The commission method provides an incentive for salespersons and also reduces the short-term costs within the company, since there are no sales costs until sales revenue is received. The method has its shortfalls, since it encourages the salesperson to close a sale even if it is not suitable for the customer's application (Oliver, 1996). Long-term customer relationships may be sacrificed in the interest of short-term gain and the company's reputation in the market-place will suffer. (See the case study 'A long-term relationship'.)

Most organizations adopt a mixture of salary and commission as the method for compensating their sales force. This provides a measure of security to salespersons, since their company pension rights are often based on their salary and do not take commissions into account, and also provides an indication of sales success and incentive for increasing sales.

Case study

A long-term relationship

Jane Warm, Sales Manager for the Southern Region of the FloRoll Manufacturing Company, decided that it was time to have a talk with John Bright, Sales Executive.

'Sit down, John', said Jane, watching as John slid awkwardly into the seat in front of Jane's desk. Seeing him, Jane thought of herself, many years ago, sitting nervously in front of the then Sales Manager, whenever she had been summoned into his office. John was very young, barely five years out of college, but already he showed tremendous potential.

'As you know, I had lunch with old Billy Westby from Westby and Wringle Ltd. Generally Westby was satisfied with the service he is receiving from us, but he did say that he felt that the Floroll ZB range you sold him last month was not right for their application. Of course I offered to take back the whole batch at once and credit them with the amount.'

'But it took me a long time to close that sale!', protested John.

'I know', said Jane, not unkindly. 'Perhaps that is the problem. You were so intent on making the sale that you forgot the number one rule: customer satisfaction takes precedence above all, even over a sale.'

John groaned silently. 'Let me tell you a story', continued Jane. 'When I first joined the FloRoll Manufacturing Company, some years ago, I was allocated the Kingsley account. As a new girl I was determined to exceed my quota and prove myself. Besides that, we worked on the basis of commission only at that time, and since I had a large mortgage it was essential that I closed as many sales as possible.

'Soon after I joined the company, Kingsley put out a major request for a quotation for a two-year supply of florolls. We were just introducing our new range of florolls at that time and had only made a relatively low level of sales to Kingsley in the past, so it was important that we won this order. Our rivals, Winger Inc., were thinking of entering the UK market and were also going for Kingsley Ltd; therefore there was much to play for. By winning the order not only would we establish our new product range, but we would also keep a potential rival out of the market.

'It was evident to me after the initial negotiations that our florolls were not a good fit for Kingsley's requirements. However, I managed to get our technical boys to bamboozle Kingsley's engineers with facts and figures so that they were thoroughly confused. Then I went in with attractive credit terms, which I knew Winger Inc. could not match. It took quite a lot of persuading before Kingsley finally signed and we had the two-year supply contract.

'Unfortunately, things started to go steadily bad after the first few months. Our florolls just did not fit Kingsley's application and although we made several modifications it was still not very effective. As you know, our relationship with Kingsley began to go sour, since they blamed us for selling them a product which we knew would not do the job.

'Although we had the order we made a loss on it, and after two years Kingsley left us and moved to Winger Inc., who had a product which matched their requirements much more closely. We have not been able to sell anything to Kingsley since that time and I believe it also damaged our reputation in an industry where customer service is very important. Our stocks fell and I am sure that it was that which precipitated our takeover by RGU International.'

Jane paused and looked at the young face staring quietly at her. 'John', she said, softly, 'I won the sale but lost a valuable customer.'

16.5 Physical distribution

Physical distribution is primarily concerned with ensuring that the product or service is available in the right place at the right time. It adds value since it saves the customers the cost, time and effort if they had to do distribution themselves.

Overall, the cost of distribution is included within the price of the product, but usually this is less than if each individual customer were responsible for their own distribution. For example, Christmas trees are sold in shops and market stalls, but they may be grown some considerable distance away, even in another country. Clearly it is cheaper for the seller to transport these in bulk, and increment the price of each tree marginally to cover transportation costs, than to have each buyer travel to the place where the trees are grown and transport their own trees back.

The Christmas tree example illustrates the importance of 'place' in distribution. 'Time' is equally important, since these trees are required before Christmas. Distributing these to market after Christmas is unlikely to sell many.

Distribution is different from production. For example, goods may be made in a distant factory, in large economical batches, and then transported to market and sold at different times, in smaller batches. Stock can be carried within the factory itself, in a warehouse, or in the channel of distribution, such as with wholesalers or retailers.

Distribution is one of the major factors in determining customer satisfaction. This is usually measured as a customer service level: the percentage of customer orders which can be delivered within a given time of receiving an order. The lower the service level the higher the stockholding is likely to be, so it should be tailored to what the customer expects and is willing to pay for. Clearly if the customer is in the business of making goods to order for others, then fast delivery of piece parts is required. However, if goods are made for stock then the actual delivery time is less important, but the reliability of the delivery period is much more critical. If a delivery of four weeks is promised then the customer can order goods and plan production four weeks in advance. If the goods are delayed by a week it will mean production delays; if goods are delivered early then it will increase the customer's inventory costs.

Apart from distribution time the order must be accurately fulfilled. One wrong item may make all the other delivered parts unusable.

16.5.1 Distribution channels

Several types of distribution channel can be used, such as direct delivery from manufacturer to user, or going through an intermediary such as a wholesaler or retailer. A wholesaler buys goods and resells them to retailers, end users or to special outlets such as government agencies. Wholesalers carry the risk of holding goods in bulk and incur their storage and distribution costs. Retailers sell the goods to end users. Usually this is done via a store, vending machine, mail order and so on.

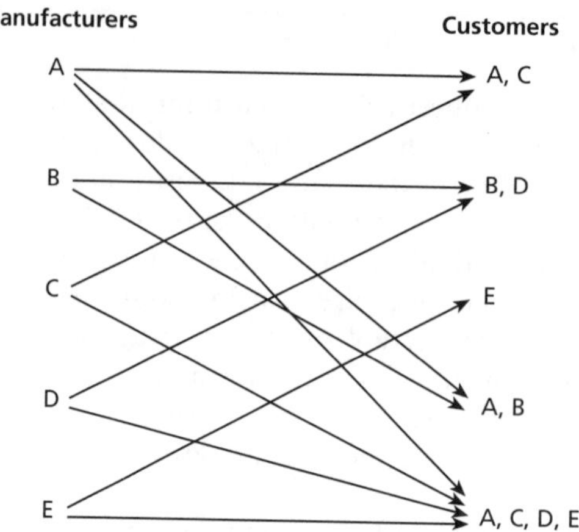

Figure 16.26 Physical distribution from five manufacturers direct to their customers.

Figure 16.26 illustrates the direct distribution channel between manufacturer and customers. It is assumed that there are five manufacturers (A to E) and that the five customers each want goods made by one or more of the manufacturers, as shown. The manufacturers are responsible for their own distribution and make individual deliveries. This can be costly, especially if the customers are located in geographically dispersed areas. It can also be annoying for the customers, since they are receiving many different deliveries.

An alternative is to use intermediaries, as shown in Figure 16.27. It is assumed that the five manufacturers all supply their goods to a common wholesaler or distributor. Therefore only a single distribution needs to be made by each manufacturer. The wholesaler holds stock in bulk and will break this down into smaller units and supply retailers situated in different geographical areas. It is assumed that Retailer 1 is located close to the first three customers and Retailer 2 is near the other two customers. Therefore the supply pattern is as in Figure 16.27, from which it is seen that the overall cost of distribution is much less and the customers each have a single distribution from their retailer.

The distribution channel is often dependent on the type of product and the market segment. In this, distribution uses the marketing channels. For rapid turnover consumable products, such as food and drink, the usual distribution channel is through retail outlets, or through wholesalers and then to the retail outlet. However, wholesalers may go direct to trade customers, such as providing food items to restaurants, or the manufacturer may sell direct to the retailer, such as breweries who provide beer to the pubs. For goods which have a limited life, such as perishable foodstuffs or fashion goods, the lines of distribution must be kept short, so that the number of intermediaries is kept small.

Durable consumer products, for example cars, TV and washing machines, are usually distributed through specialized outlets, such as car dealers or electrical shops. The manufacturer often distributes direct to the retailer since technical

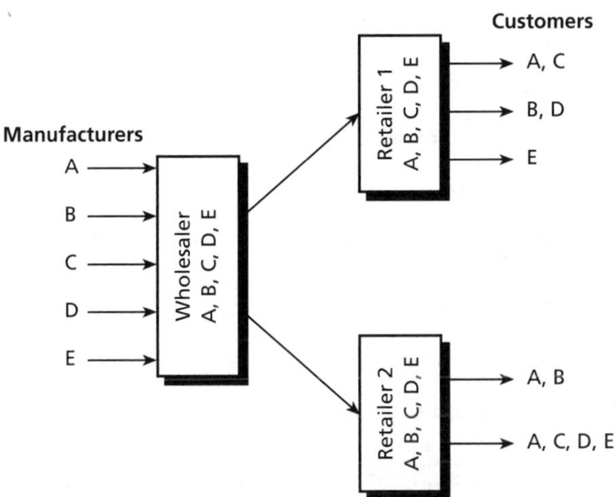

Figure 16.27 Physical distribution from five manufacturers using intermediary distribution.

support and after-sales service will need to be provided. Mail order is also used for distribution, and this can be done by the manufacturer, wholesaler or large retailer. Goods may be distributed to the customer by public post or by a private delivery service.

Industrial consumable products, such as chemicals or heating oil, may be distributed direct to the end customer if there are only a few large users, or if the manufacturer wishes to keep close control on quality, or if technical selling is required. Otherwise distributors may be used who buy in bulk and then break the bulk down for selling in smaller quantities.

Industrial capital goods, such as machine tools and vehicles, are sold direct to the end customer, since they involve a long period of negotiations before sales occur and a considerable amount of technical support is needed. Agents may be used for overseas markets, the agent being trained in the product.

Figure 16.28 illustrates some of the considerations when determining the length of distribution channel for any product or market. The channel can be kept short (with few intermediaries) if there are a limited number of customers, especially if they are geographically clustered, or if the product is complex, needing technical selling. Channels should also be kept short if the product has a short life or if the manufacturer wishes to maintain close control over the quality delivered to the customer. High valued transactions also favour short distribution channels.

Often the distribution channel can change depending on the stage the product has reached in its life cycle. When first introduced it may need intense technical support and a short distribution channel. As it matures, and becomes more common place, the distribution channel can be lengthened. For example, when small photocopiers and personal computers were first introduced they

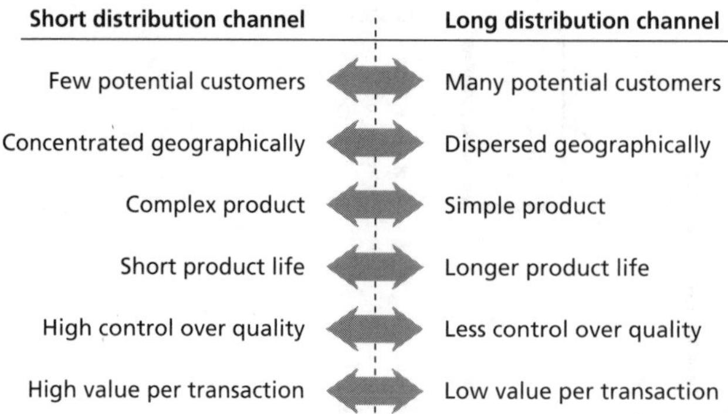

Short distribution channel
Long distribution channel
Few potential customers
Many potential customers
Concentrated geographically
Dispersed geographically
Complex product
Simple product
Short product life
Longer product life
High control over quality
Less control over quality
High value per transaction
Low value per transaction

Figure 16.28 Factors that determine the length of the distribution channel.

needed to be supported and were sold direct from the manufacturers. Now that they are more established they are sold through retailers, by mail order and so on.

16.5.2 The distribution process

The stages in the distribution process are illustrated in Figure 16.29. The overall process is one of storage, packaging, transportation, storage, packaging, transportation, and so on, depending on the number of intermediaries the product goes through. Products which are made to order (to a customer's requirements), would be delivered to the customer from the factory. The process illustrated in Figure 16.29 assumes that the product is made for stock.

Products go from manufacturing into storage. This could be at the manufacturer's site, at a wholesaler or retailer, or on the customer's premises. Warehousing may be privately owned or a public warehouse may be used, which is owned and run professionally by a third party. A public warehouse can involve relatively low cost if it is only used occasionally, since only the actual usage has to be paid for. The location of the storage space may be close to the manufacturing plant or to the customer site. It may be centralized or distributed.

Storage of goods is required for several reasons:

- To be able to hold stock to meet unexpected surges of demand. Competitors could step in and take sales if demand cannot be met.
- To enable economies of scale in production where the product is manufactured in large batches but sold in much smaller quantities.
- To smooth out seasonal and cyclical variations in demand.
- To improve the quality of some goods, which mature with age, for example wine or cheese.

Order processing is the process of interpreting the customer's order into the instructions needed to dispatch the correct goods to the customer. Speed and

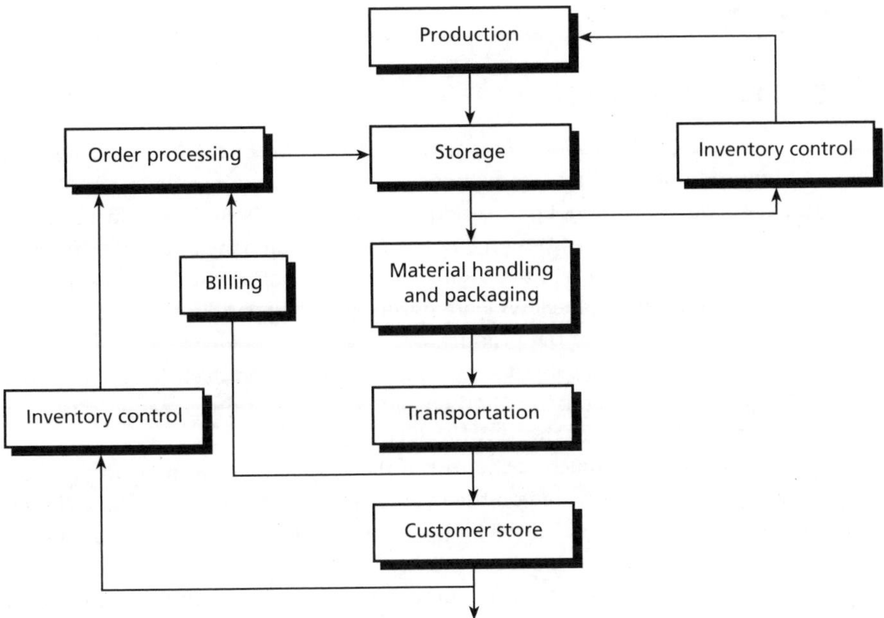

Figure 16.29 The distribution process.

accuracy are important and most of the order-processing activities are now computerized. Systems also exist for obtaining direct computer to computer orders from customers, so further shortening the order-processing time.

Material handling and packaging involves moving the products from storage and packaging them in sizes convenient for customer delivery and use. Material handling can be expensive, and automation is often introduced to reduce costs. There are also trade-offs in items such as packaging. For example the size and weight of the packaging can be reduced to lower the cost of transportation, but now there is a higher risk of the product being damaged in transit.

Inventory control is important due to the high cost of holding stock and the equally high cost if the stock runs out and halts production. A mathematical analysis of inventory control was given in Chapter 8.

There are many types of transportation, both private and public, and it includes electronic transportation: electronic transfer of software and services. Examples of conventional transportation include air, sea, road and rail. Often the method chosen is determined by the bulk to be transported. Costs are not always easy to determine. For example, air transport may be expensive compared to that by sea, but the faster delivery could mean that capital is tied up for a shorter period, smaller batches can be delivered more frequently (so reducing customer inventories), and payment is received sooner by the manufacturer.

The case study 'Small is beautiful' illustrates the trade-off which often occurs in distribution.

Case study

Small is beautiful

Conference Room 3 was full to capacity. It was the day of the Design Review for a cost-reduced floroll, which was planned to be introduced into the market over the next year. Micro Dowling, Engineering Director of the FloRoll Manufacturing Company, had introduced Design Reviews into the company as a means for ensuring that all the functions within the organization were happy with the way the design of the product was going. Reviews were normally held at various stages during the development life cycle of the product.

The electrical and optical aspects of the new floroll had been reviewed without any major issues being raised, when Jill Tait, Systems Design Manager, rose to cover the mechanical aspects of the new product.

'It is surprising how much of the cost of the product is associated with its frame', said Jill, placing her first slide on the overhead projector. 'The problem is the material used, which has to withstand the high speeds and temperatures generated when the floroll is operating at full speed.'

Jill placed her second slide on the projector. 'We have also had to introduce fans to cool the product and keep its size small. However, as you can see, most of our current market does not need the smaller size. Therefore, provided we can dissipate the heat generated, we could dispense with the fans and so save a considerable amount of money.'

Jill went through several slides, explaining the problems associated with the current floroll design, which required the use of fan cooling and special high melting point material for its casing. She then reached her last slide and paused for maximum effect, before continuing.

'The key to the new design is finding a material which dissipates heat quickly. And, ladies and gentlemen, we have found it! Kover 33, produced by Bell Agnus Company, not only dissipates twice the heat per unit area as our present casing material, but also costs a third less! By using this material we also do not need to use fans, making a saving of over 5 per cent in the material cost of florolls.'

Jill Tait paused, surveying the audience with a smile. It was Heather Bates, Customer Service Manager, who was also responsible for physical distribution within the company, who broke the silence.

'Very good; an excellent achievement. Have you had to change the dimensions of the product to do this?' asked Heather.

Jill Tait gave her a disbelieving look. 'Of course! That is the whole point. To dissipate the extra heat the frame of the floroll has had to be increased, but since it costs much less I am sure none of the customers will mind.'

Jill Tait placed one of her back-up slides onto the overhead projector, which showed the dimensions of the new floroll. Heather Bates studied it for a few minutes before speaking.

'That is quite an increase! The length is about the same as before, but the height has been increased by almost 50 per cent. Do you know what that means?'

'It is bigger?', volunteered Jill Tait, blinking.

'Yes', agreed Heather Bates. 'Which means that we will only be able to pack half the number of florolls into our trucks in each delivery. So, at a stroke, we have doubled our transportation costs! As an estimate transportation costs account for about 8 per cent of the overall costs of the product, so although the new design saves 5 per cent on material costs it adds 8 per cent due to distribution cost increases.'

The review ended soon after that. It was agreed that Engineering would return to the drawing board and would work with the Distribution Department, to see if the mechanical shape of the product could be modified so as to give the required heat dissipation while not having any significant effect on transportation costs.

Summary

A key function of product management is to specify the product or service that the company should make or provide. To do this effectively requires a knowledge of marketing, engineering and project management. It is important that the buyer's perception is fully understood and the target market, into which the product is to be sold, is always kept in mind. In developing the product speed is important, since being late into the market can cause a large loss of market share. However, this must not be done at the expense of quality, since the cost of engineering changes increases exponentially the later in the product development stage that they are made. Product development times can be reduced by acquiring products, or by developing a technology base from which different products are evolved.

An effective product launch is important for its eventual success and for the reputation of the company. Products should be launched first into those market segments that have the most need, and only after their reliability has been proven. Careful planning is required and resources must be allocated to ensure a successful launch. However, it is important to guard against the problems associated with information overload, which can confuse the consumer and negate the impact of the launch. Early adopters form an important segment of users who need to be targeted for success.

After launch the products need to be monitored throughout their life cycle, since the marketing mix will vary depending on the stage reached. A high degree of marketing effort and investment is required during the market development stage, when the product is first introduced. This must be continued during the market growth stage, when sales starts to increase quickly. Brand image and customer loyalty need to be built up before the entrance of major competitors. At the product maturity stage the emphasis is on maintaining market share by selective investments. Prices will have stabilized, due to competitor activity, and profits may fall. Product differentiation is important for continuing success. At the final product decline stage the market is in decline and the emphasis is on maximizing cash flow. Very little new investment is needed and it may even be possible to increase prices if the exit of competitors results in a scarcity of the product. Just as timing was important when entering a new market, so the organization must ensure that it exits a declining market in a timely fashion.

Several techniques are available for managing product portfolios, such as the Boston matrix. This classifies products into four segments, and it is possible that at any time a company will have products in all four categories. These are question marks, when the product is first launched by the company into a market which has high growth but low market share, so its future is unclear; stars, when the company secures a high market share in a high growth market; cash cows, when high market share is maintained but the market is in decline; and finally dogs, when the market growth and market share are both low, at which stage the company should exit from the market with this product.

The key factor determining product price is supply and demand, and the price would normally stabilize at a level at which demand is met by the level of supply. Several aspects need to be considered when determining the price of a product, including: product positioning in a given market segment, both in relation to other products supplied by the company and by its competitors; the marketing objectives of the company, such as company image or increasing market share; environmental factors, such as Government policy or the economic climate at the time; and the cost of the product.

Although pricing is a key marketing mix factor, some markets may be relatively inelastic, which means that changes in price will not significantly affect sales.

The benefits of a product need to be communicated to the market-place, usually by the Marketing Communications function. This can be done by placing advertisements in journals and newspapers, by taking stands at selected exhibitions, by direct mail, and by sales promotions. Branding is also important, and several types of branding can be applied, such as family branding, separate branding, flanker branding, intermediary branding and generic branding. Branding a product achieves several objectives, such as allowing product differentiation in the market-place; allowing consumers to relate to past experience with use of the product; allowing different market segments to be targeted with different brands of the same product; easier market introduction of a new product if it uses the same brand name as a previous product; reducing the price elasticity of a product by building up brand loyalty; and providing consumers with an assurance of consistent quality, which is normally associated with a strong brand name.

The key function of the sales force is to close the deal with the customer, but this must only be done on a win–win basis, so that the product also meets the needs of the customer. Because the sales force has close contact with the company's customers it also has several other aims, such as understanding customer requirements and feeding these back to the marketing organization, and building up customer relations and loyalty.

The sales organization can be structured in different ways depending on the company's products and structure. Customer segmentation is useful if the company has key customer accounts which need careful attention. Segmentation by industry type is used when the company's message varies by industry and its products are sold into a few selected industries. Segmentation by geography is useful for covering a widely scattered customer base, which is then usually served out of regional offices. Finally, segmentation by product line is used for technological products when it is important that the sales force learn about the product and effectively sell this to the customer.

A very important marketing consideration is physical distribution, which involves getting products to the customers at the right time, in the right place and in the packaging required. Products can be delivered direct from the manufacturer to the customer, perhaps going from the shop floor to storage in the process, or they may go from the manufacturer to several intermediaries, such as wholesalers and retailers, before reaching the customer. The shortened distribution channel, from manufacturer to consumer, is preferred when there are a few large customers, or they are concentrated geographically, or the product is complex, or a high level of control is needed over the quality of the delivered product, or each transaction has a high value. Longer distribution channels may be used in other cases.

Case study exercises

16.1 Floroll QCK was introduced some time ago, and although its sales have fallen off over the past two years, it is a well-known, established brand. The company has discovered a much more effective formula for propulsion, and has developed a product which gives a greatly enhanced performance at a nominal increase in costs. Graham Host, Product Line Manager with the FloRoll Manufacturing Company, is considering two strategies: launch the product under a new name, or launch it on the existing name (flanker branding). Discuss both alternatives and make a recommendation.

16.2 The existing range of florolls, which was introduced about a year ago, is selling well into its target market segment. There are several competitors, and more are expected to enter the market, but no one company has captured a significant market share. The FloRoll Manufacturing Company has been making investments in new manufacturing plant and Adrian Elton, Manufacturing Director, has stated that he now feels that the cost of the product can be reduced by about 10 per cent. Jack Fox, Marketing Director, has been asked to recommend to the Board the policy that should be adopted regarding pricing. What factors would Jack need to consider, and should the price be reduced in line with costs?

16.3 Jake Topper, Chief Executive of the FloRoll Manufacturing Company, would like to bring out a special floroll to mark the 50th anniversary of the floroll. He believes that, although only a limited number of these will be sold, they could result in a high profit margin. It is unlikely that any competitors would bring out a similar product. Table 16.7 shows the cost and number of units that are expected to be sold. Calculate the selling price based on this information, using the absorption and marginal pricing methods, assuming that the company wishes to make a 25 per cent profit. Would you use costs as a method of pricing in this instance?

Table 16.7 Costs for case study exercise 3

Item	Value
Total direct material cost (£000)	35
Total direct labour cost (£000)	8
Total overhead (£000)	12
Units made	100

16.4 The FloRoll Manufacturing Company is considering introducing a new product, and Jack Fox, Marketing Director, is preparing the business case. He needs to determine the best distribution channel and considers the relevant facts, which are: (a) the product is a modification of an existing product and so will need a limited amount of technical support for a short period only. (b) There are likely to be only a few customers for the product and they will be clustered in a few industrial areas of the country. (c) The product is not considered to be a mainstream product as far as the company is considered. It provides an opportunity to capitalize on some work done on an earlier project which was never completed. (d) This is a relatively low valued product, by company standards.

What are the different distribution channels open to Jack Fox, and what strategy should he adopt?

References

Ayal, I. and Raban, J. (1990) Developing hi-tec industrial products for world markets, *IEEE Transactions on Engineering Management*, August, pp. 177–84.

Barclay, I. (1992) The new product development process: past evidence and future practical application, *R&D Management*, **22**(3), 255–63.

Boston Consulting Group(1973) The product portfolio, *Perspective No. 66*, Boston Consulting Group.

Cardozo, R. *et al.* (1995) Product-market choices and growth of new businesses, *IEEE Engineering Management Review*, Summer, pp. 10–16.

Chernatony, L. de (1993) The seven building blocks of brands, *Management Today*, March, pp. 66–8.

Cooper, A.C. and Smith, C.G. (1993) How established firms respond to threatening technologies, *EMR*, Fall, pp. 30–9.

Crawford, C.M. (1993) The hidden cost of accelerated product development, *EMR*, Summer, pp. 21–8.

Dolan, R.J. (1995) How do you know when the price is right? *Harvard Business Review*, September–October, pp. 174–83.

Felton, G. (1994) *Advertising*, Prentice-Hall, Englewood Cliffs NJ.

Gaynor, G.H. (1993) Exploiting product cycle time, *EMR* Spring, pp. 30–43.

Gupta, A.K. and Rogers, E.M. (1992) Internal marketing: integrating R&D and marketing within the organization, *EMR*, Spring, pp. 29–36.

Helme, S. (1996) Winning customers with CDs, credit cards and new Toyotas, *Mobile Communications International*, June, pp. 45–8.

Herbig, P.A. and Kramer, H. (1993) The phenomenon of innovation overload, *EMR*, Fall, pp. 21–9.

Iansiti, M. (1993) Real-world R&D: jumping the product generation gap, *Harvard Business Review*, May–June, pp. 138–47.

Johne, A. (1992) How to pick a winning product, *Management Today*, February, pp. 72–4.

Jones, B. and Ramsden, R. (1991) The global brand age, *Management Today*, September, pp. 78–80.

Jones, J.P. (1990) The double jeopardy of sales promotions, *Harvard Business Review*, September–October, pp. 145–52.

Kodama, F. (1992) Technology fusion and the new R&D, *EMR*, Spring, pp. 6–12.

Kopelman, O. (1994) Streamline your design process with QRPD, *Electronic Design*, 27 June, pp. 94–104.

Kovac, E.J. and Troy, H.P. (1989) Getting transfer prices right: what Bellcore did, *Harvard Business Review*, September–October, pp. 148–54.

Lurban, G.L. and Hauser, J.R. (1993) *Design and Marketing of New Products*, Prentice-Hall, Englewood Cliffs NJ.

MacVicar, D. (1996) How to create cost-effective promotional plans, *Laser Focus World*, April, pp. 105–8.

Maile, R. (1985) *Economics*, Charles Letts, London.

Maruca, R.F. (1995) How do you grow a premium brand? *Harvard Business Review*, March–April, pp. 22–40.

McKenna, R. (1995) Real-time marketing, *Harvard Business Review*, July–August, pp. 87–95.

Mitchell, A. (1994) Marketing's new model army, *Management Today*, March, pp. 41–9.

Moriarty, R.T. and Moran, U. (1990) Managing hybrid marketing systems, *Harvard Business Review*, November–December, pp. 146–55.

Nagle, T.T. and Holden, R.K. (1994) *Strategy and Tactics of Pricing*, Prentice-Hall, Englewood Cliffs NJ.

Nevens, T.M. *et al.* (1990) Commercializing technology: what the best companies do, *Harvard Business Review*, May–June, pp. 154–63.

Oliver, J. (1996) New, improved salesforce, *Management Today*, December, pp. 82–4.

Popper, E.T. and Buskirk, B.D. (1993) Technology life cycles in industrial markets, *EMR* Spring, pp. 44–50.

Quelch, J.A. and Kenny, D. (1994) Extend profits, not product lines, *Harvard Business Review*, September–October, pp. 153–60.

Rabino, S. and Wright, A. (1993) Accelerated product introductions and emerging managerial accounting perspectives, *EMR*, Fall, pp. 40–7.

Rayport, J.F. and Sviokla, J.J. (1994) Managing in the market space, *Harvard Business Review*, November–December, pp. 141–50.

Slater, S.F. (1994) Competing in high-velocity markets, *IEEE Engineering Management Review*, Summer, pp. 24–9.

Slywotzky, A. and Shapiro, B. (1993) Leveraging to beat the odds: the new marketing mind-set, *Harvard Business Review*, September–October, pp. 97–107.

Stewart, D (1994) Improve engineering's diaglogue with Marketing, *Electronic Design*, 18 April, pp. 85–92.

Strassberg, D. (1992) Increasing your effectiveness by bringing in outside help, *EDN*, 12 November, pp. 91–9.

Webb, D. (1991) Don't just lay an egg – hatch a profit centre, *Electronic Business*, 9 December, pp. 58–62.

Werther, W.B. *et al.* (1994) The future of technology management, *IEEE Engineering Management Review*, Fall, pp. 13–19.

Wheelwright, S.C. and Sasser, W.E. (1989) The new product development map, *Harvard Business Review*, May–June, pp. 112–25.

Williams, J.D. *et al.* (1994) Trade show guidelines for smaller firms, *IEEE Engineering Management Review*, Summer, pp. 59–66.

Witt, C.A. (1994) *Service Operations Management*, Prentice-Hall, Englewood Cliffs NJ.

Zahra, S.A. *et al.* (1994) Creating a competitive advantage from technological pioneering, *IEEE Engineering Management Review*, Spring, pp. 76–85.

17

Leadership and motivation

Introduction

This chapter introduces the concepts of leadership, the differences between leadership and management, the theories of leadership, and the importance of delegation in leadership and motivation. This is followed by a consideration of what is meant by motivation, the effect of motivation on performance, motivational theories and techniques that may be used to improve motivation.

17.1 The nature of leadership

What is a leader? How does a leader differ from a manager? Can leadership be learned?

As in all areas there are some people who are more able to lead than others, but the performance of these so-called 'born leaders' can be improved by training, and mediocre leaders can learn to become good leaders (Adair, 1983; Barnard, 1958; Bennis and Nanus, 1985; Bittel, 1984; Burns, 1978; Fiedler, 1967).[1] It is also important to appreciate that natural circumstances often create leaders. For example, at the onset of a fire the fireman first on the scene is the leader, controlling events and directing the action of people. Similarly the first-aider who initially arrives at the scene of an accident is its leader.

Industrial situations are different, since the activity of the leader must be sustained: it needs to carry on after the fire has been put out or the ambulance has arrived. Teamwork is also very important, since most modern tasks within industry can only be achieved by well-directed collective effort.

It is important to realize that the majority of teams want to be led, and they will do this by following the example, not the words, of a leader. Leaders can appear and disappear, in a similar manner to popstars. Leaders change their characteristics, team members and attitudes change, or the situation changes, all of which can result in the existing leader being replaced by a new one. Leaders, especially if they are also managers, are perceived in different ways depending on their position within an organization (Morris *et al.*, 1995). For example, the shop floor supervisor is seen as a different type of leader to the managing director, although both can be equally effective within their teams as leaders.

17.1.1 The manager and the leader

It is important that the manager is also the leader of the department or team (Caulkin, 1993; Crainer, 1995a; Zaleznik, 1992). If the manager is an ineffectual leader then an informal leader will spring up, who may challenge the manager's position, creating a team within a team. The manager then has four options:

- Get rid of the informal leader. This will usually only result in someone else being chosen by the team to be its informal leader, since the original problem which caused the informal leader to be created still exists.
- Undermine the informal leader's position, which is similar to the above and will also result in the rise of a new informal leader.
- Try to divide the group so as to diminish the power of the informal leader. This will be very difficult to do since the team will stay loyal to its chosen leader.
- Come to terms with the informal leader, to supplement the deficiencies in the manager's ability. This is by far the only sensible position since the formal and informal leaders must work together for success, and this will be recognized by the team.

Managers have the authority of their formal position within the organization. Their power is given to them by the company. Leaders have the authority of influence over the team and their power is given to them by team members.

Good leaders are not necessarily good managers, since leaders may be very poor at organizing and achieving the tasks set to the team, although they may be very good at inspiring the team. Good managers, on the other hand, are usually good leaders, since leadership skill is one of the fundamental requirements of management.

Leaders are part of the team (that is, they are insiders), although they are different from the other team members, being their nominated leader. Managers will be 'outside' their team if they are not also its leader, and it will be very difficult for them to manage the activities of the team. In one respect teams are similar to countries: they want to be led, not managed. After all, we refer to the president or prime minister of a country as its leader, not as its manager!

Leaders do not always set out consciously to be the leader of the team; they are elected by the team members. Managers, on the other hand, become managers by design. They wish to manage as part of their professional learning and skill. Management is a profession; leadership is not.

A leader inspires the team to achieve its goals; a manager can only help the team by traditional methods, such as providing resources and training. Figure 17.1 shows the stages in the relationship of new team members. Initially they are fully dependent on the team leader and other established members of the team. They are learning new skills and getting used to their environment. They learn by following others. (Buchholz and Roth, 1987.) After this phase they reach the independent stage, where they have acquired sufficient knowledge and experience to work by themselves. Good managers and leaders will now ensure that their team members continue to learn and to develop, and will direct their work to that end.

Finally, interdependent stage is reached. In this stage a good manager becomes

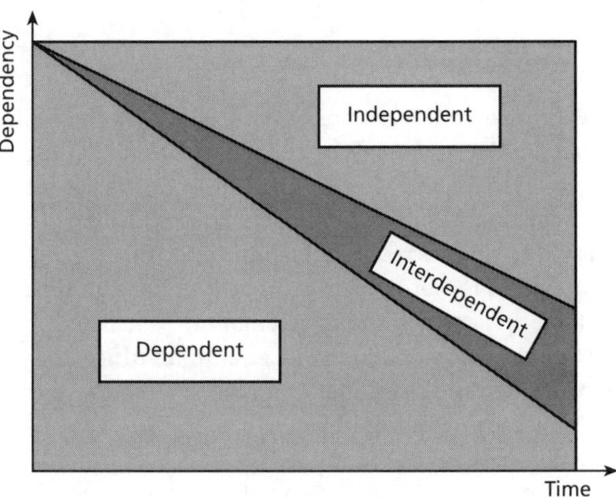

Figure 17.1 Stages in the relationships of new team members.

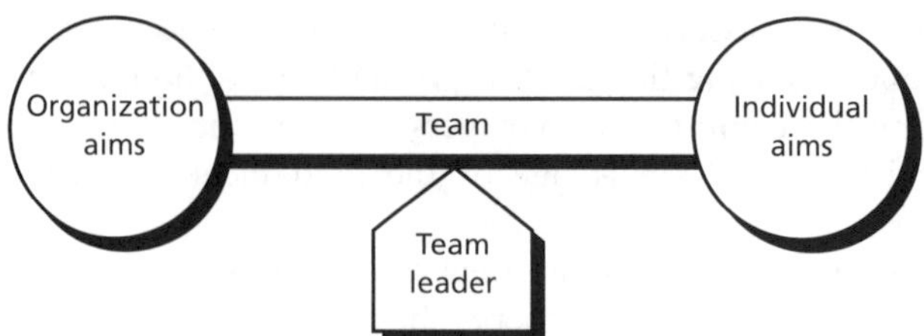

Figure 17.2 The 'aims dumb-bell'.

the team's leader. The manager enters the team as a collaborator and draws on the strength of the team to achieve joint goals. There is a sense of community, of belonging, and synergy is said to exist in the team. (See the case studies 'The purpose of the exercise', and 'Packing them in'.)

Managers concentrate on achieving goals or tasks, which are relatively short-term targets with a defined beginning and a measurable end result. Leaders, on the other hand, give the team its purpose, enabling it to understand why it exists as a team.

The purpose or mission of the team must also be clearly defined (see the case study 'The purpose of the exercise'). However, it has no measurable beginning or end. It defines a general direction and it may (and often must) change with time. The purpose enables team members to share the same vision. It puts individual goals into context, giving the full picture.

A good leader, like a good manager, works through the team to achieve its goals or aims. The organization's aims need to be balanced with those of the team, as illustrated by the aims dumb-bell of Figure 17.2.

Case study

The purpose of the exercise

At Jake Topper's monthly Executive Staff Meeting, Adrian Elton, Manufacturing Director at the FloRoll Manufacturing Company, stated that it was taking too long for Engineering to design new systems and the cost of manufacturing products from these design was too high. Smarting from these criticisms, Micro Dowling, Engineering Director, returned to his office determined to 'show them', to take the initiative and improve productivity. He set himself a target to cut design time by 10 per cent and reduce manufacturing costs by the same amount.

Micro called Jane Furley, his hardware design manager, into his office and reviewed all the work on hand. He discovered that the design for a new spin mechanism, used on florolls, was just about to begin.

'Who have you got to do this design?', asked Micro.

'It needs three people. I was going to give it to Connelly and her team', replied Jane.

'Good! I want you to set up two teams: Connelly's team and also another team with your best designers. I want to compare how long both teams take to complete the design and the manufacturing costs from both designs. Let's see if we can learn from the two teams' mistakes and so improve our designs.'

Jane Furley left Micro's office and told Elizabeth Connelly about the plan. Word soon spread amongst the hardware designers: Micro Dowling wanted to measure designers' performance. What would happen to the poor performers? Was redundancy planned in the department? Morale fell. Connelly led a deputation to Micro's office, to inform him that all the designers were unanimous; they would not duplicate designs, since it was wasteful of effort.

Micro realized his error! He was concentrating on the goal: reducing design time and manufacturing cost for the spin mechanism by 10 per cent. He decided to define the purpose to his department. Calling them together he said, 'I am sorry; I blew it! Let me tell you guys the problem. The market out there is getting competitive. We are getting too late to market with new designs, allowing our competitors to take valuable market share, and our manufacturing costs are increasing, eroding margins. Our job is to produce designs which make money for the company. Any ideas how we can hack this one?'.

The team had plenty of ideas. They were intelligent engineers and knew that they could do better. They agreed to set up a working party to make recommendations. Design rules were examined and improved. A relatively low-cost computer-aided design tool was introduced. Manufacturing was asked for and produced a guide on 'design for manufacture'. Within six months the goal of 10 per cent reduction in design time and manufacturing cost had been reached. But the team continued with its purpose; costs decreased steadily after that as new design tools and techniques were introduced, always at the suggestion of the team members themselves. Micro was able to show the savings which were being made by these improvements and so obtain backing from the Board for the increased investments.

Case study

Packing them in

Pauline Packer is manager of the Packaging and Shipping Department within the FloRoll Manufacturing Company. It is a small department, with four full-time packers, a clerical assistant and a secretary. Generally the department copes well with shipments, but recently there has been a problem within the Assembly Shop which has resulted in its output being delayed. This backlog is now being cleared and pressure is being applied to the Packaging and Shipping department to get the orders out to customers before the close of the financial year, so that they can be billed.

Pauline Packer considers her dilema. For the past month her staff have been working a seven-day, eighty-hour week, and with four weeks to go to year end she will have to ask them to work through the coming weekends again. She knows that her staff resent working excessive overtime and are on the point of rebellion. She considers her options:

- Ask the team to come in again this weekend. Point out that they will be paid overtime and that a nice tidy sum was building up for each person. Pauline knows that the answer to this would be a firm no; the staff have had enough and want to be with their families. By taking this course she risks being snubbed as a leader and losing her authority. The staff have already complained that Pauline is not being strong enough in standing up to pressure from higher management and refusing to drive them so hard.
- Threaten her staff by pointing out that if the orders did not go out on time then company profits would be reduced, resulting in possible redundancy for some of them. Pauline dismisses this approach; it is out of character for her to threaten her staff and they were hardened people who would not be affected by it anyway. She considers bribing them with a special bonus, in addition to the normal overtime. Pauline feels confident that she could persuade Adrian Elton, her boss, to agree to this payment. Unfortunately, she knows that her staff will consider this also as a threat: no bonus in future if they do not come in over the weekend.
- Try to gain the sympathy of the team by showing that she is one of them, part of the team. She could point out that she has argued with her boss against the team coming in. She has told Adrian that her team has already worked too hard, and that they were now being asked to cover for inefficiency in another department (the Assembly Shop). Pauline dismisses this approach; she has always been honest with her staff and does not want to spoil their relationship.

Eventually Pauline decides to talk to her staff about the problem. She calls them into her office and explains the importance of getting the orders out on time, even though the delay is not their fault in the first place. The staff understand. Together the leader and the team came up with a solution. The packaging process is time-consuming and laborious, but does not require a high degree of skill. If the company can take on two agency staff to cover weekend working over the next four-week period, then two of the permanent staff could have alternate weekends off, so that they would not need to come in for all four weeks.

Pauline thinks that this is a very workable scheme. She speaks to Adrian Elton, who agrees to the extra staff; the overall cost is less than the overtime the permanent staff would have worked.

17.2 Leadership theories

Because of the importance of leadership in all aspects of human endeavour, such as industry, commerce, government and the armed forces, a considerable amount of work has gone into determining the methods which can be used for identifying leadership qualities. If such a method can be found then one can use this as a selection process for future leaders, and also determine whether leadership can be taught to those who do not possess this quality. Several leadership theories have been put forward, a few of these being discussed in this section.

The leadership traits theory was one of the earliest theories on leadership, and stated that people are born with leadership qualities, and that leadership cannot be acquired. This is therefore also known as the Natural Leader Theory. If one could identify the key leadership traits, then this could be used for the selection process of future leaders.

Trait theory states that these characteristics or traits are essential for good leadership and that they are situation-independent. Therefore to find a good leader for every situation one only needs to find a method for measuring these traits.

Peterson and Plowman (1957) defined four groups of traits for a leader: physical, personality, character and intellectual.

Stogdill (1974) catalogued 40 traits, collected into five groups as follows: five traits relating to physical ability; four traits relating to intelligence and mental ability; six task-oriented traits; sixteen personality traits; and nine social traits.

Researchers soon found that there was very little correlation between traits and leadership (Jennings, 1961). Many non-leaders were found to possess these leadership traits, and many of these traits were missing from good leaders. However, some correlation was found between leadership and traits such as initiative, intelligence and self-confidence (Ghiselli, 1963).

The situational leader theory says that leaders are selected as a result of the situation rather than the personality of the leader. For example, Hitler was chosen by Germany as its leader after the First World War, and the start of the Second World War established Churchill as the leader in the UK. People follow a leader if they perceive that by so doing they will achieve their personal aims in any given situation.

The situational leader theory generally confirms and accounts for the fact that managers' performance as leaders depends on the environment as much as on their personal qualities, but it is not very helpful in assisting with the selection process for new leaders.

Several theories were postulated which tried to relate success in leadership with the style adopted by the leader. These theories can be considered to be based on one-dimensional, two-dimensional and three-dimensional models.

One example of the one-dimensional model is whether the leader adopts an autocratic or a democratic style. In the autocratic style the leader is outside the group and leads it from that position, while in the democratic style the leader acts like a team member, although separated from it by virtue of being its leader. Actual styles can vary between these two limits.

The best known of the one-dimensional models was that proposed by Douglas McGregor (1960). He postulated the Theory X and the Theory Y style of leadership, X and Y being chosen to indicate that they are not intended to indicate 'good' or 'bad' styles. The Theory X manager thinks that all people work only for monetary gain and that they are motivated by fear. It is also assumed that people do not want any responsibility, that they want to be given detailed instructions and monitored and guided continually.

Theory Y managers believe the opposite about the people they manage. They think that people are motivated by achievements and want to have the freedom to act according to their own judgement and resent close supervision. McGregor's theories are considered further later in this chapter in relation to motivation.

The two-dimensional model was popularized by Blake and Mouton (1964). They defined a Managerial Grid which could be used to identify and measure leadership behaviour. Figure 17.3 shows a simplified layout of the Grid. Concern for people and for production (or for the organization, or for tasks) are shown along the two axes. The higher up the axis a style appears, the greater the coverage of this particular style by the leader. Five basic styles are identified:

- The 1,1 style, where the manager has very little interest in people or in production. These managers are often considered to be 'time servers', expending the minimum amount of energy needed to survive within the organization. Often these managers act as messengers between their boss and their subordinates, relaying instructions and conveying progress.
- The 9,1 style, in which the manager is mainly interested in the efficiency of the organization or the team, and not in the people within it. All work is structured such that the human element is kept to a minimum.

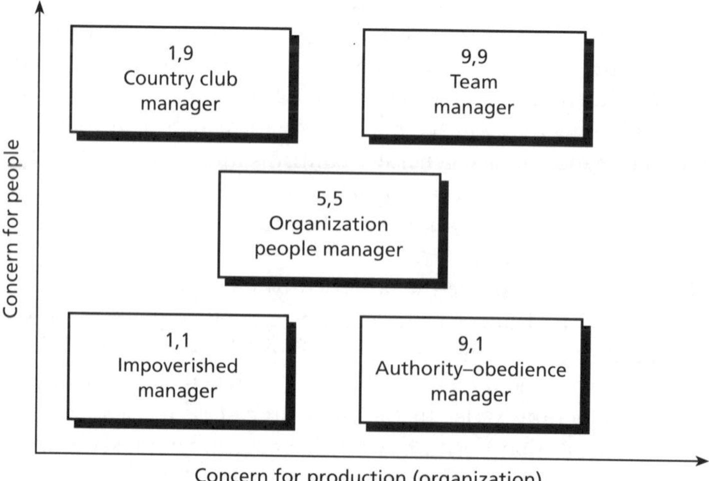

Figure 17.3 The Managerial Grid (simplified).

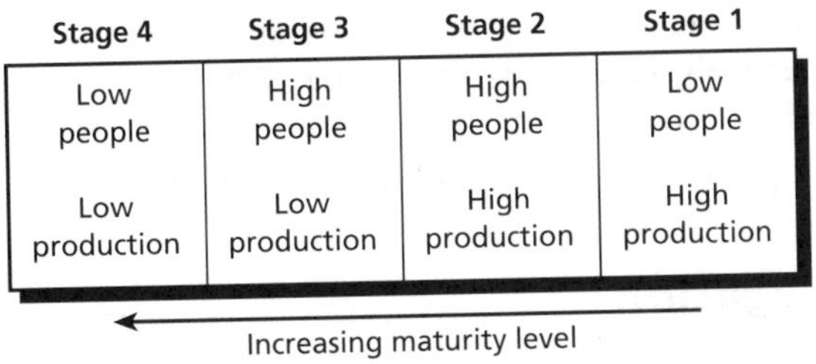

Stage 4	Stage 3	Stage 2	Stage 1
Low people	High people	High people	Low people
Low production	Low production	High production	High production

← Increasing maturity level

Figure 17.4 The Hersey–Blanchard model.

- The 5,5 management style, which is a balanced style. The leader shows concern for the organization and for people, but not to a very high level. This means that the performance of the team will also be mediocre. This is usually the average style adopted by managers within large organizations.
- The 9,9 management style, where the manager shows the highest concern for both people and for production. Considerable leadership skills are needed to match the requirements of the two. Blake and Mouton believed that this style produces the best results, motivating the team to work towards a common mission or purpose, and with mutual trust and synergy between its members.
- The 1,9 management style, where the manager is mainly concerned with the well-being of the team rather than getting tasks done. Needless to say, productivity is low.

Hersey and Blanchard (1977) believed that leadership style varied with the maturity of the task being done by the team. They added a 'maturity level' to the Blake and Mouton Managerial Grid, as in Figure 17.4. Stage 1 represents the start, when the leader shows high concern for production or for the task (planning, budgeting and so on). As the task progresses the leader's concern for people increases, until at Stage 2 concern for the team and for production are equally high.

By Stage 3 the task starts to become fairly well established; there are very few unknowns and the tasks become relatively routine. The leader is now mainly concerned with people issues, to maintain motivation and to ensure that team members are working towards a common goal. Finally, by Stage 4, the team becomes mature and self-sustaining. The leader can now relax and concern for people and tasks is low. However, leaders must still mix freely within their team, to see and be seen, so that they can act if problems arise.

17.3 Delegation

Delegation is one of the key activities performed by managers and leaders. It must not be viewed as a discrete event; it is a continuous process which occurs all the time, leaders often not being aware that they are delegating something.

Delegation is not easy, but it is essential if the leader is to operate efficiently

(Fisch, 1973, 1976). Most managers spend a considerable amount of time on tasks which could be effectively delegated, so freeing them to carry out other duties which add more value to the organization. Managers who cannot delegate will often be heard to say: 'if you want it done properly, do it yourself'.

17.3.1 Definition of delegation

When a task is usually allocated to a subordinate the manager retains responsibility and requires frequent interaction with the subordinate as the task progresses. The subordinate has less freedom to act, needing to agree the method of tackling the job with the leader.

In true delegation subordinates should have a high level of autonomy to act as they think best, to achieve the results agreed with the leader. Monitoring of the activity by the leader would normally occur at agreed points only, unless the subordinate requested help from the leader earlier.

In delegating a task leaders are carrying out the following:

- Giving to subordinates some of their own responsibility, usually for specific tasks and relevant decisions.
- Giving subordinates sufficient authority to match the responsibility which has been delegated, to help them achieve the agreed targets. Boundaries may be set on the authority, for example spending limits.
- Ensuring that subordinates accept accountability for success or failure of the delegated task, although leaders carry the ultimate responsibility to the organization for all activities under them.
- Being available to help with advice if called on by subordinates. It is important to appreciate that managers have not abdicated their responsibility, only delegated it.

17.3.2 The delegation process

Before leaders delegate tasks they need to consider what tasks to delegate and the level to which they are to be delegated. Examples of some levels are:

- Leaders instruct subordinates to do the task, but to check with them before taking any decisions. This is not really delegation but task allocation.
- Leaders instruct subordinates to do the task and to keep them fully informed.
- Leaders request subordinates to carry out the task, holding weekly reviews with them.
- Leaders delegate the task and say 'let me know if I can help'.
- Leaders abdicate the task by saying: 'Here you are; do this and let me know when finished'.

The steps a leader needs to follow when delegating a task are:

- Analyze all the jobs that need to be done and decide on which can be delegated. Usually if a task can be clearly defined then it is a candidate for delegation.
- Out of the jobs that can be delegated, decide on which tasks will be kept and which will be delegated and to whom.
- Analyze the subordinates who will receive these tasks, to determine whether they require any special training or coaching. Since delegation is a long-term activity this investment in training is worthwhile.
- Agree the content and expected results of the delegated task with the subordinate and also the delegation process to be used. Often delegation can be in stages, with reducing intervention from the leader.
- Delegate the task and trust in the subordinate to achieve the agreed objectives. However, be available so as to discreetly monitor progress and to provide help if asked. It is important that no decisions are made by the leader and that subordinates do not feel that the leader is constantly looking over their shoulders. Considerable tact is needed to help without interfering (see the case study 'Service with tact').
- If the subordinate achieves the results then provide reward, such as praise, promotion, bonus or a bigger assignment. If the subordinate fails treat it as a learning exercise rather than as an occasion for censure, remembering that the leader should have monitored progress so that failure should not have been allowed to occur on a critical programme.

Case study

Service with tact

One of the first tasks planned as part of the automation programme in the Dinsher Shop was to increase the efficiency of the stores and manufacturing control departments. A large number of IBM personal computers were to be bought from a recognized distributor for this.

Normally an order of this size would be closely supervised by the Materials Manager, James Silver, but he decided to delegate it to a promising young buyer, Jill Getter. After discussing the assignment with Jill, and agreeing her level of responsibility and authority, James asked that Jill come to him for any help or advice she needed.

After a thorough round of negotiations with several recognized distributors, Jill Getter decided to buy from PC Service Inc. She consulted with James Silver, who agreed with her choice and congratulated her on the work she had done so far.

A few days later James Silver met Micro Dowling at a meeting and after the meeting he mentioned to Micro that they were proposing to buy quite a few IBM personal computers.

'Who are you getting them from?', asked Micro.

'PC Service Inc. Do you know them?'

'Yes, my neighbour works at Willaby Engineering and he told me that PC Service Inc. are pretty hot on price but not so hot on service. There are quite a lot of companies around which can service personal computers these days; might be better to go to them.'

On his way back to the office James Silver stopped off at Jill Getter's desk. 'Say, Jill', he said. 'I just ran into Micro Dowling in Block H. Micro told me that his neighbour works in Willaby Engineering and they also buy their personal computers from PC Service Inc. because they give the best prices. Looks like you made a good choice there! I don't know if there is anything in it, but Micro also mentioned that their servicing is pretty lousy.'

Jill decided to look into it further. She telephoned some of her contacts in other companies and found that PC Service Inc. definitely had a very poor service record. She also discovered that, although PC Service Inc. bundled the service price into the cost of the machines, several of her contacts who made large purchases had got them to unbundle this and had taken out service contracts with other companies.

By the end of the following week Jill Getter had agreed a contract with PC Service Inc. for supply of the computers and had also negotiated a very good contract with another company for service. It was a task successfully completed and James Silver congratulated her on it, while at the same time deciding to try her with a bigger task to see if she was ready for promotion.

17.4 Defining motivation

The 'new company' syndrome is well known; after all most people within industry have been through it. New entrants walk into the company buildings on their first day. They are highly motivated, having chosen to join the company, although they may be a little apprehensive. However, it is often only a matter of a few months before they join the rest of the staff in muttering the familiar cliches: 'The only way to progress around here is to marry the boss's daughter'. 'Reward for doing a good job is getting to keep it'. 'Quality is something to talk about and not to practise'. 'If you want to know what's going on around here read the daily papers'.

What caused this change, this fall in morale and motivation? Several factors:

- Unfair treatment of employees; favouritism.
- Excessive emphasis within the company on status; the 'haves' and the 'have-nots'. For example, different canteens for staff and for managers; different toilets; managers travelling first class and others economy class.
- Non-recognition of achievements. All employees receiving the same rewards, irrespective of performance.

- Petty regulations, for example disallowing small expenses, such as purchase of morning newspapers when travelling on company business.
- Poor communications; employees having to discover important facts affecting them from rumour and hearsay.

A few dictionary definitions of motivation-related words are:

Motive: A consideration or motion that excites to action.
Motivation: A motivating force; incentive.
Drive: To urge along; to hurry on; to furnish motive power to.

Motivation is therefore a force which excites and drives a person to action. Motivation results in drive. It is subjective and qualitative, rather than objective and quantitative. It cannot be measured.

Motivation varies with time (for example the age of the person, mood at the time, immediate history) and with the person's individual characteristics. It is therefore a psychological and not a logical entity. It causes action which is done subconsciously and emotionally, with the reasons behind it not always clearly definable. There are two main types of motivation driver:

- Primary motivation drivers, which are instinctive, such as hunger, thirst, pain avoidance.
- Secondary motivation drivers, which are learned, for example that certain behaviour gives pleasure (for example, writing a good report will get praise from the boss).

17.4.1 Motivation and performance

Every action has a positive and a negative factor associated with it. Generally a person will carry out the action if it is perceived that the positive factors exceed the negative factors (Taylor, 1996; Nicholson, 1997). For example writing a report requires work (negative factor) but if done well it will get praise from the boss (positive factor). The positive and negative factors associated with an action will vary from person to person.

Performance on a job is determined by several factors (Gellerman, 1963; Gomersall, 1971; Kornman, 1974), the key ones being shown in Figure 17.5. These are:

- Ability related to the task to be done. This can be influenced by training, for example.
- Availability of support tools. Once again this can be provided in order to increase the performance on the task.
- The organizational environment. Several factors are involved here, such as the communications within the company; the personal relationships within the team; the quality of the team leader (manager); the recognition received for good performance; and job security. All these factors can be influenced

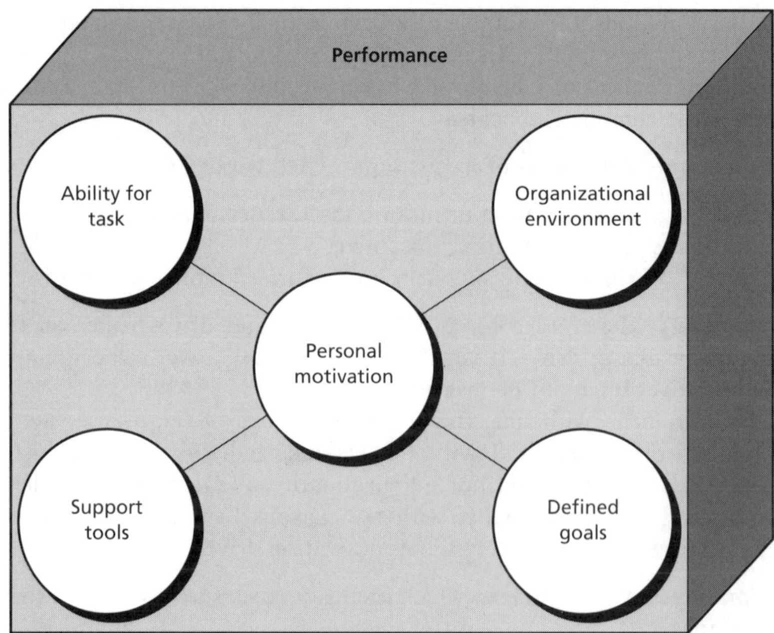

Figure 17.5 Motivation and performance.

by the team leader, although many of them are more difficult to achieve than training and providing suitable tools.

- Clearly defined missions and goals, so the person knows what is expected and can measure progress. This too is something which will be done by the good manager and leader.
- Motivation on the job. This is probably the most important factor, since if the person is not motivated to achieve then no amount of training or support tools will result in good performance. Motivation cannot be directly affected by the manager; it is determined by the other factors, as in Figure 17.5. Therefore motivation is a vital but indirect factor (Atkinson and Raynor, 1974; Crainer, 1995b).

17.5 Motivational theories

Early writings on motivation emphasized the importance of economic gain to the individual. It was this that was assumed to be the main motivation driver, so the emphasis was on incentives, such as work study and rates for the job. Research work on motivational theory proved that his was not the case (Smale, 1996). Although much of this work was carried out some time ago, it is still relevant in today's industrial environment.

Over the period from 1924 to 1932, Elton Mayo, Professor of Industrial

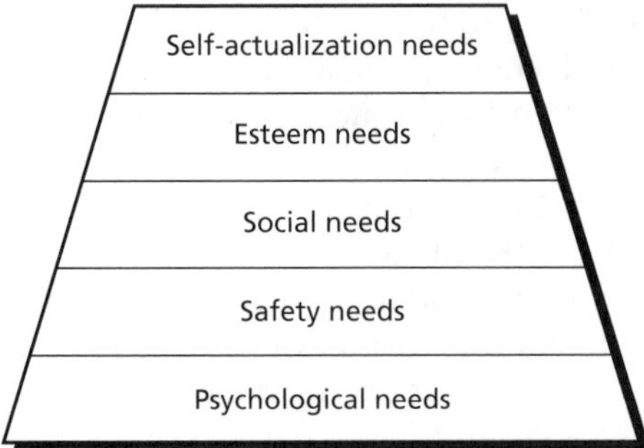

Figure 17.6 Maslow's need–hierarchy model of motivation.

Research at the Harvard Graduate School of Business, and his team carried out a series of experiments at the Hawthorne works of the Western Electric Company in Chicago, which proved conclusively that monetary gain was not a major motivational factor (Mayo, 1971).

Abraham Maslow proposed the Need-Hierarchy model of motivation (Maslow, 1943, 1954) as shown in Figure 17.6. He postulated five levels of need:

- Psychological needs, which are the basic needs, such as hunger, thirst and the need for shelter.
- Safety needs, such as the need for protection from threats and danger. This may be physical danger or emotional danger, such as a verbal attack on a person at a meeting.
- Social needs. This can take many forms, such as the need to belong to a group; acceptance by one's peers as their equal; and the giving and receiving of affection.
- Esteem needs, which are often also known as ego needs. These can be of two forms, self-esteem and the esteem of others. An example of self-esteem is the need to develop self-confidence and to achieve, while an example of esteem of others is the need for status and recognition from others.
- Self-actualization needs. This is the need for continual self-development, and the realization of one's own potential.

Regarding the hierarchy of needs, Maslow further proposed that:

- A higher level need only arises when the ones below it are satisfied.
- A satisfied need no longer dominates an individual's behaviour; the next higher need takes over. (See the case study 'Bright young thing'.)
- An unsatisfied need acts as a motivator. Lower level needs can be satisfied but higher level needs (such as the need for continual self-development) are never completely satisfied.

It is important to appreciate that in reality needs vary with individuals, and one can have instances where higher level needs may become important and require satisfaction even if lower ones have not been satisfied.

Fredrick Herzberg and associates promoted the two-factor theory of motivation (Herzberg *et al.*, 1959; Herzberg, 1968), also referred to as the motivation hygiene theory. He stated that human needs can be grouped into two levels: lower and higher. To some extent, therefore, Herzberg may be considered to have simplified Maslow's hierarchy of needs into two levels or two factors.

Examples of lower level needs (also referred to as extrinsic or hygiene factors, or as dissatisfiers) are pay and working conditions. Examples of higher level needs (also referred to as intrinsic factors, motivators or satisfiers) are doing a challenging job, having responsibility for the task being done, getting recognition for work well done, and belonging to a group.

Lower level needs, if absent, create dissatisfication and if present do not create satisfaction. Higher level needs, if absent, do not create satisfaction but also do not create dissatisfaction. Therefore two processes are needed to minimize dissatisfaction and to maximize satisfaction.

It should be noted that hygiene factors are controlled by the manager within an organization, for example by giving subordinates salary increases and determining their working conditions. Managers usually erroneously believe that it is these factors which motivate their staff. Motivators are controlled by subordinates themselves, for example job satisfaction. True motivation therefore arises from doing things which can be controlled by oneself rather than having it bestowed from above. However, the manager must ensure that the climate within the organization is such that this is possible.

People are often classified as hygiene seekers and motivation seekers. Their characteristics are given in Table 17.1. However, it must be kept in mind that very few humans operate in this black and white way, and most exhibit a combination of both characteristics, although there may be a predominance of one.

Vroom (1964) postulated that a person's motivation to achieve a goal is dependent on the value which the person places on achieving the goal and on the expectancy of attaining it (Vroom and Yetton, 1973).

If people perceive that there is very little chance of success then they will not be

Table 17.1 Characteristics of hygiene seekers and motivation seekers

Hygiene seekers	Motivation seekers
Low interest in type of work or standard of performance	Very high interest in type of work and the standard to which it is performed
Very little satisfaction obtained from succeeding in the job	Very high level of satisfaction obtained from succeeding in the job
Concerned with the environment in which the job is performed rather than the job itself	Motivated by the job itself and less concerned with the environment in which it is performed
High reaction to changes (improvements or deterioration) in hygiene factors	Low reaction to changes in hygiene factors

motivated to do the task. For example, employees may want promotion, but if they perceive that the situation is such that promotion is not possible (for example the company is shrinking) then they will not be motivated to work hard towards it. If, however, they believe that hard work will result in promotion then they will work for it.

Maximum motivation is obtained when people want to attain the goals and feel that they control the means for achieving these; that is, they control the job method so that there is a higher expectation of success.

As stated earlier, Douglas McGregor put forward two extremes of management style, which he referred to as Theory X and Theory Y (McGregor, 1960). These styles of management treat subordinates in set ways and therefore affect their motivation. Although these extremes were postulated, McGregor accepted that actual behaviour lies somewhere in-between, although most people have a bias to one side or the other. The Theory X manager believes that:

- The average person dislikes work and will do everything possible to avoid it.
- Because of this dislike for work subordinates must be controlled and closely supervised. One cannot bribe subordinates into doing work; they must be threatened with punishment if they don't conform.
- The average person is lazy and dislikes responsibility. Their prime wish is for job security and they are happy to be directed in all they do.

Theory X can be equated to the carrot and stick motivation technique. It can work in certain instances and for short periods of time, for example by threatening with unemployment or loss of wages. Subordinates who respond to a Theory X management style are usually hygiene or maintenance seekers.

Theory Y managers believe that:

- The average person does not inherently dislike work. The use of physical and mental effort in work is as natural as play or rest.
- Employees will use self-control and self-direction in meeting organizational objectives to which they are committed. (Note that they need to be committed to these objectives, which cannot be imposed from above.) External control and threat of punishment are not the most effective ways for meeting objectives.
- The average person wants responsibility. If responsibility is avoided it is because of a direct result of some previous unpleasant experience and is not inherent in human character.
- Most people are capable of applying a high degree of creativity to problem-solving in their daily work.
- The intellectual capabilities of the average person is underutilized in the normal work environment.

Subordinates who respond to a Theory Y management style are usually motivation seekers.

Theory X gives management an excuse for poor organizational performance, since this can be blamed on the generic nature of workers, their dislike for work and responsibility. Theory Y, on the other hand, places responsibility for performance squarely on the shoulders of management. Since the workers are inherently good and motivated, any sign of laziness or unresponsibility on their part is due to the poor management methods used to motivate them.

Although no modern manager or team leader will admit to ever using Theory X practices, this is the most frequent method in use, and managers often practise it subconsciously. Managers assume that they know what is best for subordinates and for the organization (see the case study 'All together now'). It is assumed that in exchange for money and benefits an employee will be willing to be directed to meet the company goals. Theory X managers also take for granted that the aims and needs of the organization come before the individual's needs and aims.

Theory Y is not easy to implement in practice, and that is probably why managers and leaders do not follow its style more often. It is not easy, within a large team, to allow individuals to control their own work and still ensure that common goals are met. However, if one accepts the principles of Theory Y, then it is management's prime task to look for ways in which this can be achieved (see the case study 'Lucky Jim').

David McClelland (1961) began his studies into motivation in the 1950s. He postulated that each person is motivated by one of three needs:

- Need for power, where a person is motivated by influencing others. For this person exercising power is more important than achieving goals. Power is desired for power's sake, not as a means for achieving an aim.
- Need for affiliation, where the person is motivated by the need for social intercourse and the need to belong to a group.
- Need for achievement, where the person is motivated by the need to meet goals. No motivation occurs on receipt of money, praise and so on, except where it is taken as a sign of successful achievement of a goal.

People do not fall neatly into one or other of these three groups. The degree to which these three needs, or motives, affect a person's behaviour varies with the person's personality and with past experience.

Case study

Bright young thing

Sarah Jones shone at university. She majored in economics and was President of the Student Union and head of the Debating Society.

Sarah was ambitious. She planned to join a large organization where she could exercise her many talents, and progress up the corporate ladder as fast as she could. She applied to, and received offers, from four companies, but decided to accept that from the FloRoll Manufacturing Company because she was impressed with the discussion she had with Philip Elkins, the HR manager.

'We are a young progressive company', Philip had said. 'We believe in promotion on merit. Young recruits are encouraged to develop quickly so that they can take over positions of responsibility. We believe in a lively environment with working conditions second to none.'

After six months with the FloRoll Manufacturing Company, Sarah found that she had very little to complain about regarding the working conditions. She had an office to herself, which was unusual for a newly recruited graduate, and she was well paid. Everyone was also very friendly and helpful.

However, although her reputation had preceded her and she suspected that she was sometimes looked on as the 'bright young thing', she had been given very little responsibility. Most of the work she did was mundane and the sort of jobs which new graduates usually received. She was bored and felt her talents were being wasted. When she told her boss he was friendly and sympathetic.

'You must have patience', he told her. 'The work is not the same as what you learned in university. It needs quite a lot of practice and experience.'

'But how will I get this practice and experience unless I actually take on the responsibility of doing some of it on my own?'

'You will learn, you will learn. Be patient! The company has a good working environment, give yourself time to enjoy it.'

Sarah gave herself another six months to enjoy the excellent working conditions. She then decided that enough was enough and left the company.

Case study

All together now

Things were getting bad in the Development Tools group of the FloRoll Manufacturing Company. Projects were running late and customer deliveries were being threatened. David Laith, Manager of the group, was told by Micro Dowling that he needed to look closely at his organization to see if productivity could be improved.

'That's going to be difficult', moaned David. 'You know what a lazy lot of engineers I have! They all struggle in at 8:30 a.m. and disappear promptly at 5:00 p.m. Unless I am there to continuously supervise and motivate them nothing gets done.'

'Maybe you need to look at how you are managing them,' suggested Micro.

David Laith decided to prove that he was right. He reorganized his group, eliminating intermediate levels of supervision, so that he could directly monitor the work of everyone himself. Most of the day David spent in the laboratory, looking over the shoulders of senior and junior engineers, issuing orders and handing out criticism, in order to improve performance. He insisted that people stay late if they had a task to complete. David himself came in to work very early each morning and left late at night, to set an example to his team and to catch up on paperwork, which was piling up.

Six months later morale in the Development Tools group was at rock bottom, projects were running later than ever, and David Laith suffered a nervous breakdown and was off work for an extended period.

In David's absence Micro Dowling set up a study group consisting of the three section heads, Angela Code, Stew Long and Maude Klinger, to come up with recommendations on how the group should be run. They spent a week consulting with their teams and then recommended that the three sections heads run a joint management team reporting to Micro. Micro agreed and gave the group full autonomy, asking for weekly reports on progress.

David Laith returned to work six months later. By then morale in the team was the highest it had been for many years, no further slips had occurred in the programmes and several had been completed ahead of schedule.

Case study

Lucky Jim

Jim Twiddle, Manager of Manufacturing Engineering at the FloRoll Manufacturing Company, was very good at his job. He was a recognized authority on assembly techniques and had considerable experience in floroll manufacturing techniques. Jim supervised a small team of engineers and he loved his job and his team. He was a family man and a good people manager.

Jim's reputation was known to the parent company, RGU International, and they decided that Jim should be promoted to their head office in Chicago, into a staff function where he could act as a consultant to the other manufacturing plants belonging to the organization.

'This is a wonderful opportunity for you', said Jake Topper, when he first proposed the move to Jim. 'Of course, Adrian Elton does not want to lose you, but then we have to accept that it is best for the wider good of the company. You are a lucky man to be offered this move.'

Jim did not want to go. It was inconvenient to move his family to Chicago and he knew that he would miss the line responsibility, no matter how humble it was. But Jake brushed aside all objections; it seemed that he just could not believe that Jim was serious in objecting to a move which was obviously so good for his own personal development and for the good of the company.

Jim thought long and hard about the move. He spoke to his family and they all agreed. After a week he decided to quit the company and handed in his notice.

Jake Topper was dumbfounded. He listened carefully to Jim's reasons and then asked that he be given a week to sort things out. In that week Jake spoke to the President of the company and they agreed that Jim Twiddle was too valuable a man for the company to lose. A meeting was set up with Jim and they agreed to an alternative strategy. Jim would be assigned a deputy, whose task was to take much of the mundane work off Jim's shoulders. This would free Jim to do a limited amount of consultancy work. It was also agreed that Jim did not need to move to Chicago, since the consultancy work could be done just as effectively out of his present office. This arrangement pleased everyone, including Adrian Elton.

'If only they had consulted me first', Jim Twiddle told his wife that evening, 'rather than assuming what was good for me, we could have arrived at this arrangement in the first place and saved ourselves a lot of heartache.'

17.6 Defining needs

The motivational needs of professional people, such as scientists and engineers, include many of those which are common to other workers, such as white collar and manual workers. However, some factors, such as pay and benefits, although still important, have less impact on professional employees.

As mentioned in the last section the goals that are set must be challenging but must also be achievable. Figure 17.7 shows how motivation varies with success probability. If the task is impossible to achieve (zero probability of success) then motivation to do the task is low. If success is certain, no matter what is done (unit probability of success) then again motivation is low. In-between these two extremes there is an optimum probability of success (P_m) which results in a maximum motivation factor (M_m). The values of these two variables, P_m and M_m, vary depending on the characteristic of the individual.

Figure 17.7 indicates that if an employee's motivation is low it could be on either side of the optimum line. Managers usually assume that their subordinates are automatically operating to the left: that they are perceiving the goals to be too difficult to be attained. Their first reaction is therefore to reduce the amount of responsibility and work they delegate, so as to make the task easier. Partly this is done to avoid the risk of failure on a critical task. However, it could be that subordinates have low motivation because they perceive the task as being very easy to complete, and in this case the manager should be increasing the level of the task, not decreasing it.

Professionals, in common with most other employees, also desire variety in their work. In addition there should be scope for self-expression and creativity. It is important that the employee learns and acquires new skills in performing tasks.

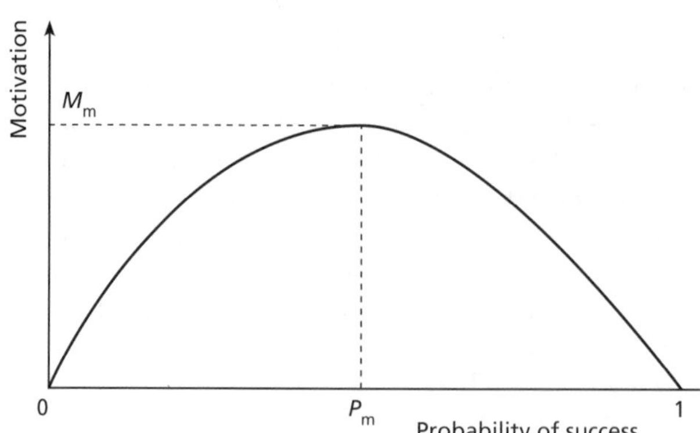

Figure 17.7 Variation of motivation with success probability.

Most professionals like to make their own decisions and to choose the work methods for achieving their tasks based on their experience, skills and knowledge. They require independence to act, with minimum supervision and controls. The higher the qualifications or intelligence of the person concerned, the more will be the need for individuality.

All professionals are usually very proud of their profession and take pride in adopting a professional approach in their work, and in associating with other professionals. They recognize that they have been through extensive learning and training periods and they are anxious to put this to good use. Motivation is highest when this is being done as part of their work.

Recognition of achievements is important to all employees, and professionals are no exception. They are proud of their achievements and require recognition from peers, their managers and from other professionals within industry. Professionals usually work very hard and recognize that their effort will result in enhanced performance and results (as in Figure 17.8). Rewards are usually required as a sign of recognition for work which is well done, and this recognition will lead to further effort and higher performance. Therefore the cycle is repeated and can spiral to peak performance.

17.7 Motivation techniques

In order to motivate staff effectively, a manager or team leader must understand the needs of subordinates and ensure that these are matched to the requirements of the organization in which they work. Several considerations must be taken into account in motivating staff (Cole, 1993).

Financial reward is usually the first to be considered, although, as has been stated earlier, it is not the most important motivational factor. The financial package must be enough to meet hygiene needs (bringing up a family, holidays and so on) and to prevent dissatisfaction.

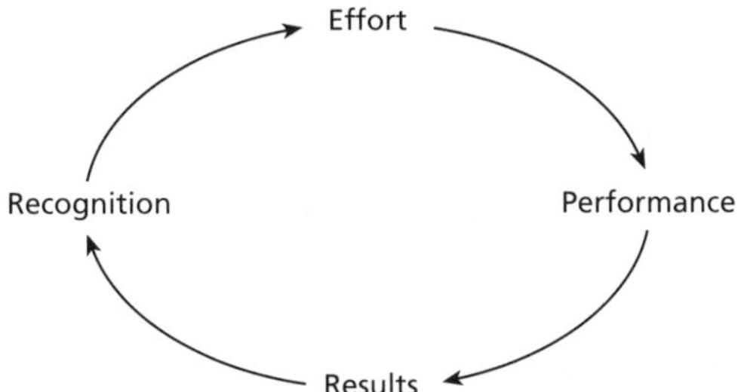

Figure 17.8 The recognition–performance spiral.

However, financial remuneration is often considered as a mark of status, meeting the esteem needs of employees and also acting as a recognition of success. It need not be large, but it should be applied selectively and on merit. Usually this is best done soon after the event, perhaps as a cash award for a job successfully completed. Usually its impact is lost when such rewards are reserved for the annual incremental award, when it is also received by other employees.

Several financial techniques have been applied to promote motivation, such as incentive payments, profit sharing schemes (Pickering, 1994) and bonuses. These have limited success since very soon they are considered to be part of the normal financial package of the employee.

Working conditions are another hygiene factor which needs to be considered, but which can act as a dissatisfaction preventer rather than as a motivator. It is essential that fairness is used in allocating working space, since this is one of the most visible signs of status.

Job security is an important demotivator, although in certain circumstances it can act as a powerful short-term motivator. To threaten people with loss of employment is to provide a negative incentive, which does not pull them to the desired goal (as a positive incentive does) but pushes them away from an undesirable goal.

No manager should consciously apply the threat of unemployment as a motivator, since it can devastate the morale of staff and many will be dissatisfied and leave even when the threat is removed. Instead, managers should normally do all they can to assure their staff of job security. This can be done by reassuring subordinates, praising them for a job well done and giving them a view of how the task they are currently doing is likely to develop in the future.

The working environment must meet the social needs of employees. The manager can help by assigning team tasks and calling team meetings. The need for social contact should be recognized; too many managers frown on subordinates who spend time talking to each other over coffee, or go out occasionally for a lunchtime drink. Provided targets are being met these contacts are to be encouraged.

Communications at all levels of the organization must be effective, both up and down the chain. Managers and team leaders must make time to listen to grievances and to talk to their staff. In particular they must be equitable in dealings with subordinates, providing equal opportunities for benefits such as promotions, and be fair in enforcement of company regulations.

Managers must provide recognition for high performers, as this is an important part of their esteem requirements. They should ensure that peers know about the achievement, and this can be done by making awards, or giving praise, in public, and by publishing the information in the company newsletter. Their family should also be involved, and this can usually be done by providing a family award, such as a holiday or a dinner paid for by the company.

Recognition or reward, in the language of transactional analysis, are known as strokes. A positive stroke can be unconditional ('That was an excellent report you prepared, John; well done!') or conditional ('That was an excellent report you prepared, John. If only all your reports could be as good!). Negative strokes indicate criticism ('That was a very poor report, John.').

As part of recognition the manager must be prepared to help subordinates, even if it means promoting them out of the department. The manager must help promising subordinates find positions in other departments, if it will further their career aims, and must ensure that they are adequately trained for their next assignment (Waterman *et al.*, 1994).

Subordinates must also be encouraged to publish professional papers and take part in conferences. They should be helped with filing of patents and be given time to participate fully in their professional institution activities, if they so wish.

Many professionals seek power to influence others. The manager can satisfy this need in several ways, such as making them responsible for a task force on a specific topic, or getting them to lead an important technical project. Often subordinates can satisfy this need by being part of a team making important decisions on company policy, where they will be sharing information with senior executives. Managers must also be willing to delegate meaningful decision-making tasks to subordinates, showing trust in their capabilities, and ensure participation in goal setting and decision making.

To professionals, work and its content are important. The work should be challenging and the subordinates should be making use of their professional skills in problem solving (Cooper, 1977). Tools used in this work are very important and professionals are aware of what constitutes a good tool and demand the best. The manager's task is to enrich the job as much as possible, making subordinates responsible for meaningful tasks and giving them total responsibility.

Summary

Teams want to have a leader. Managers have power over their team given to them by the company, while team leaders have the power given to them by the team members. Managers generally concentrate on achieving goals or tasks, which may be relatively short-term with clearly defined beginnings and measurable end results, while leaders give the team its purpose, enabling it to understand why it exists as a team. It is desirable that the team manager is also its leader, but if the manager is an ineffectual leader then an informal leader will spring up within the team, who may challenge the manager's position. Rather than fight the informal leaders, managers need to be skilled to use the informal leaders to supplement their own ability, so that the informal and formal leaders are working together for the good of the team, and this will be recognized by its members.

Several theories have been proposed for identifying the characteristics of good leaders. The leadership traits theory, also known as the Natural Leader Theory, postulates that people are born with leadership qualities and that leadership cannot be acquired. It then sets out to identify these traits, such as physical characteristics, personality, character and intellect. The situational leader theory, on the other hand, assumes that leaders are selected as a result of the situation rather than the personality of the leader. It therefore assumes that managers' performance is related to the environment as much as their own personal qualities.

Other leadership theories have been postulated which relate success in leadership to the style adopted by the leader. Examples of these are the one-dimensional model proposed by Douglas McGregor (Theory X and Theory Y) and the two-dimensional model by Blake and Mouton (Managerial Grid). Hersey and Blanchard introduced a 'maturity level' to the Managerial Grid to account for their belief that leadership style varied with the maturity of the task being done by the team.

It is important that managers delegate tasks in order to improve their own efficiency and the motivation within their teams. True delegation is a difficult process, since it involves giving subordinates some of the manager's own responsibility along with the authority needed to see the task through. Managers must ensure that subordinates accept accountability for the success or failure of the delegated task, but that they are available to help and advise if called on by subordinates. Leaders need to go through several steps in delegating tasks, such as analyzing all the jobs they have to see which ones can be delegated; deciding on which subordinates to delegate tasks to and whether they need special training or coaching to succeed; agreeing the delegated task with the subordinates and the method of working such as how often to meet to review progress; delegating the task and then trusting the subordinate to get on with it, but at the same time being available to help and advise if needed; and providing reward and praise if the subordinate succeeds, or treating it as a learning exercise if the subordinate fails.

Motivation is the force that excites and drives a person to action. It is subjective and qualitative, rather than objective and quantitative, so it cannot be measured. There are two main types of motivation driver: primary drivers, which are instinctive, such as hunger and thirst; and secondary drivers, which are learned, for example that certain behaviour provides reward or pleasure.

Performance on a job is critically determined by the motivation of the person carrying out the task, but motivation cannot be affected directly by the manager; it is determined by many other factors. Examples of these are the ability of the person relative to the task being done; the availability of support tools to do the job; the environment in which the task is being done, for example the communications within the company, personal relationships between team members, job security, and recognition received for good performance; and clearly defined missions and goals, so the subordinate knows what is expected and can measure progress. Early motivational theories emphasized the importance of economic gain, but later research showed that this plays a small part in motivating employees, and motivation is related to many more complex issues. Abraham Maslow proposed the need-hierarchy model of motivation, which has five levels of need: psychological, safety, social, esteem and self-actualization. Higher level needs only arise when the ones below them are satisfied, and it is an unsatisfied need which acts as a motivator. Although lower level needs can be satisfied, the higher level needs, such as those for continual self-development, are never completely satisfied.

Fredrick Herzberg and associates promoted the two-factor theory of motivation, also called the motivation–hygiene theory. In effect it simplified Maslow's hierarchy of needs into two levels, or factors. The theory postulates that the lower, or hygiene, motivators, such as pay and working conditions, if absent, result in demotivation, but in themselves they do not result in motivation. Higher level, or motivation, factors, such as doing a challenging job and having responsibility for the task being done, have to be met in order to result in motivation.

In his theory on motivation, Vroom postulated that a person's motivation to achieve a goal is dependent on the value that the person places on achieving the goal and on the expectancy of attaining it. Douglas McGregor's Theory X and Theory Y styles of management can be extended to the theory of motivation. People who respond to the Theory X style need to be driven by management to achieve tasks, being associated with the carrot and stick technique of motivation. Those responding to Theory Y need to have minimum supervision and given maximum responsibility, and they are known as motivation seekers. David McClelland postulated that people are motivated by three needs: for power, for affiliation and for achievement. The degree to which these three needs, or motives, affect a person's behaviour varies with the person's personality and with past experience.

To improve motivation within their teams managers must first remove the demotivators; that is, satisfy the hygiene factors. These include providing adequate financial reward, improving working conditions and making employees feel secure by removing the threat of redundancy. The motivation factors then need to be met, such as recognition of the ability and professionalism of staff, challenging and interesting tasks, improved communications up and down the organization, and good team atmosphere.

Case study exercises

17.1 Jane Furley, Hardware Design Manager, has been asked by Micro Dowling to attend a meeting the next day, to agree the way forward for the special modification to be made to the current floroll design for Saudi Arabia. Micro apologized for the short notice; it seemed that the Commercial Department had been asked to provide a quotation for the modification and had, as usual, left it until the last minute before telling anyone.

Jane Furley realized that this presented an excellent opportunity to introduce the new analogue design which the whole team had been talking about and so decided to delegate the task to Peter Wave, Section Head of the Analogue Design section. It was 5:00 p.m. when Jane spoke to Peter and most of the analogue design team had gone home. Peter therefore decided to work on the proposal himself, since he had all the information required. He worked through most of the night and had sketched out a proposal by the morning.

Peter Wave called his team together first thing next morning and ran through his proposal with them. He had an hour to go before Micro's meeting, so only minor changes were possible at this late stage. Most of the team agreed to go along with the scheme except for Mary Stick, who was recognized by the team to be the best designer among them. Mary's main objection was that the design was very complicated and they needed to work on it for at least a week before they could be certain that it would work. Commercial had to put the proposal in to Saudi Arabia by the following day. Peter Wave therefore decided to overrule Mary Stick, but, in order to pacify her, he invited her to come to the meeting. He hoped that once she saw the pressure which was being put on getting this proposal off on time she would drop her objections. Unfortunately, at the meeting, Mary Stick raised the same objections which she had raised with him at their team meeting.

Peter Wave was furious and decided to tackle Mary after the meeting. He suspected that she was objecting in such a large gathering in order to attract attention and was angling for promotion. He considered her behaviour to be disloyal to the team.

Discuss the above. Should Jane, Peter and Mary have acted differently. What should Micro do in these circumstances?

17.2 The automation programme for the Dinsher Shop was in danger of being delayed due to a disagreement between Jenny Pett, the Automation Team Leader, and 'Big' Joe Bull, the Line Supervisor. Two new trial machines had been installed within the plant over the past month, one made by Trader Inc. and the other by a German company, Fender GmbH. Big Joe wanted to standardize on the machine supplied by Trader Inc. He believed that he knew best; after all he had the line experience and his men would be the ones using the machines. Jenny thought that Fender's machine would result in the greater productivity. Although she was relatively new to the Dinsher Shop, being brought in as a specialist in automation, she believed that a radical change in the way things were done was needed if productivity was to be significantly improved.

John McCaully, Manager of the Dinsher Shop, called a meeting to discuss the proposals and to agree a way forward. As he now sat watching the two protagonists glare at each other across the conference table, he wondered whether he had done the right thing in leaving it completely to the 'experts' to carry out the automation programme. Perhaps he should have kept more in touch with the technical side of developments so that he could now arbitrate on the best way forward.

Discuss John McCaully's dilemma. What should he have done? What should he do now?

17.3 Micro Dowling liked to delegate; it gave him more time to contribute technically in areas of his speciality and it also trained his direct reports and motivated them. The problem was that Jake Topper, the Chief Executive, had an annoying habit of calling Micro to his office at short notice and wanting an update on some programme issue or another; and Jake liked to go into detail. It had now got to the state where Micro did not like to walk past Jake's office door in case it flew opened and Jake pounced on him for another status report. Micro was reluctant to start holding daily meetings with his staff, so that he could be closely briefed, but he did not see how else he could satisfy Jake's demands.

What should Micro do?

17.4 The annual rate of inflation was running at 10 per cent but, owing to poor sales the average salary increase at the FloRoll Manufacturing Company had to be kept to 8 per cent. Each department was allocated 8 per cent to share out as it thought best. Micro Dowling was not sure what he should do. Should he give all his staff an 8 per cent increase? This was below the inflation level, but they would understand that it was the best that could be obtained and it was after all only 2 per cent below the inflation rate. However, it might be better to treat this as a merit increase, giving high performers an increase above the inflation rate and poor performers well below the rate of inflation.

What do you think Micro Dowling should do?

17.5 A key engineer from David Laith's group handed in her notice, giving the reason for leaving as a much higher salary offer from a competitor. David told his boss, Micro Dowling, that they should top the competitor's offer in order to keep the engineer within the company, even though this would mean that she would be earning well above the maximum for her grade. 'We can always give her smaller increases in subsequent years to bring her salary back into line,' David said.

What should Micro Dowling do?

17.6 Micro Dowling asked his direct reports if they had any nominations for promotions from within their teams. At most only two could be promoted from the whole Engineering group, so Micro asked them to be selective and only make nominations if performance had been outstanding. David Laith though that his whole team had been outstanding. He recommended all ten of them for promotion, and told them that he had done so. 'It is better for the morale of the team that they know that I support them all fully', he thought. 'If Micro Dowling now only promotes one of them or none at all, then they will not feel so bad knowing that at least I have thought them all capable of promotion.'

Do you agree with this?

17.7 Motivation in the Machining Section of the Assembly Shop was high. However, the supervisor on the section, Bill Hale, discovered that although daily productivity was well above average for the company, it fell into a dip for the first hour after the lunch break. The problem was that the operators tended to go to lunch together and then carry on socializing afterwards, so that they were not in full production mode until well after the lunch break. In a bid to increase productivity Bill staggered the break over a two hour period, so that groups of operators could not go to lunch together. After a month motivation began to fall and productivity dropped dramatically.

What do you think was the problem? What motivation principles did Hale break?

Endnotes

1. See also the following references: Geigold, 1982; Gibb, 1954; Gretton, 1995; Guest, 1962, 1967; Kouzes and Posner, 1987; Margerison and McCann, 1985; McCall and Lombardo, 1978; Pagonis, 1992; Wright and Taylor, 1994; Yukl, 1994.

References

Adair, J. (1983) *Effective Leadership – a self-development manual*, Gower/Pan, Aldershot.

Atkinson, J.W. and Raynor, J.O. (1974) *Motivation and Achievement*, Holt, Reinhart & Winston, New York.

Barnard, C. (1958) The nature of leadership. In *Organization and Management* (ed. C. Barnard), Harvard University Press, Cambridge MA.

Bennis, W. and Nanus, B. (1985) *Leaders: the Strategies for Taking Charge*, Harper & Row, New York.

Bittel, L.R. (1984) *Leadership – the Key to Management Success*, Franklin Watts, New York.

Blake, R.R. and Mouton, J.S. (1964) *The Managerial Grid*, Gulf Publishing Co., Houston.

Buchholz, S. and Roth, T. (1987) *Creating the High-Performance Team*, John Wiley, New York.

Burns, J.M. (1978) *Leadership*, Harper & Row, New York.

Caulkin, S. (1993) The lust for leadership, *Management Today*, November, pp. 40–3.

Cole, L. (1993) Winning a moral victory, *Computing*, 28 January, pp. 26–7.

Cooper, R. (1977) *Job Motivation & Job Design*, Institute of Personal Management, London.

Crainer, S. (1995a) Have the corporate superheros had their day? *Professional Manager*, March, pp. 8–12.

Crainer, S. (1995b) Re-engineering the carrot, *Management Today*, December, pp. 66–8.

Fiedler, F. (1967) *A Theory of Leadership Effectiveness*, McGraw-Hill, New York.

Fisch, Gerald (1973) Do you really know how to delegate? *The Business Quarterly*, Autumn, pp. 17–20.

Fisch, Gerald (1976) Towards effective delegation, *CPA Journal*, July, pp. 66–7.

Geigold, W.C. (1982) *Practical Management Skills for Engineers and Scientists*, Lifetime Learning Publications, Belmont CA.

Gellerman, S.W. (1963) *Motivation and Productivity*, American Management Association, New York.

Ghiselli, E.E. (1963) Managerial talent, *American Psychologist*, October, pp. 631–41.

Gibb, C.A. (1954) Leadership. In *Handbook of Social Psychology* (ed. G. Lindzey), Addison-Wesley, Cambridge MA.

Gomersall, E.R. (1971) Current and future factors affecting the motivation of scientists, engineers and technicians, *Research Management*, **XIV**(3), 43–50.

Gretton, I. (1995) Taking the lead in leadership, *Professional Manager*, January, pp. 20–2.

Guest, R.H. (1962) *Organizational Change: The Effect of Successful Leadership*, Dow Jones-Irwin, Homewood IL.

Guest, R.H. *et al.* (1977) *Organization Change Through Effective Leadership*, Prentice-Hall, Englewood Cliffs NJ.

Hersey, P. and Blanchard, K.H. (1977) *Management and Organizational Behaviour*, Prentice-Hall, Englewood Cliffs NJ.

Herzberg, F., Mausner, B. and Snyderman, B.B. (1959) *The Motivation to Work*, Wiley, New York.

Herzberg, F. (1968) *Work and the Nature of Man*, Staples Press, *New York.*

Kenneth, J.D. *et al.* (1992) Perceptions of an innovative climate: Examining the role of divisional affiliation, work group interaction and leader/subordinate exchange, *IEEE Trans. Engineering Management*, August, pp. 227–36.

Kornam, A.K. (1974) *The Psychology of Motivation*, Prentice-Hall, Englewood Cliffs NJ.

Kouzes, J. and Posner, B. (1987) *The Leadership Challenge*, Jossey-Bass, San Francisco.

Margerison, C. and McCann, R. (1985) *How to Lead a Winning Team*, MCB, Bradford.

Maslow, A.H. (1943) A dynamic theory of human motivation, *Psychological Review,* **50**(4), 370–3.

Maslow, A.H. (1954) *Motivation and Personality*, Harper & Row, New York.

Mayo, E. (1971) Hawthorne and the Western Electric Company, in *Organization Theory*, (ed. D.S. Pugh), Penguin Books, London.

McCall, M.W. and Lombardo, M.M. (eds.) (1978) *Leadership: Where Else can we Go?* Duke University Press, Durham NC.

McClelland, D. (1961) *The Achieving Society*, Van Nostrand, New York.

McGregor, D. (1960) *The Human Side of Enterprise*, McGraw-Hill, New York.

Meyer, C. (1994) How the right measures help teams excel, *Harvard Business Review,* May–June, pp. 95–103.

Morris, J. *et al.* (1995) Effective leaders at a premium, *Professional Manager*, March, pp. 26–8.

Nicholson, J. (1997) Employee demotivation – the bottom line, *Call Centre Focus*, January, pp. 6–7.

Pagonis, W.G. (1992) The work of the leader, *Harvard Business Review*, November–December, pp. 118–26.

Peterson, E. and Plowman, E.G. (1957) *Business Organization and Management*, Richard D. Irwin, Homewood IL.

Pickering, B. (1994) Made for sharing, *VAR World*, April, pp. 30–4.

Smale, R. (1996) Does money motivate? *Call Centre Focus*, May, pp. 4–9.

Stogdill, R.M. (1974) *Book of Leadership*, The Free Press, New York.

Taylor, M. (1996) Exploring motivation, *Call Centre Focus*, May, pp. 14–17.

Vroom, V.H. (1964) *Work and Motivation*, Wiley, New York.

Vroom, V.H. and Yetton, P.W. (1973) *Leadership and Decision-Making*, University of Pittsburgh Press, Pittsburgh.

Waterman, R.H. *et al.* (1995) Towards a career-resilent workforce, *Harvard Business Review*, July–August, pp. 87–95.

Wright, P.L. and Taylor, D.S. (1994) *Improving Leadership Performance*, Prentice-Hall, Englewood Cliffs NJ.

Yukl, G. (1994) *Leadership in Organizations*, Prentice-Hall, Englewood Cliffs NJ.

Zaleznik, A. (1992) Managers and leaders: Are they different? *Harvard Business Review*, March–April, pp. 126–35.

18

Team building

Introduction

This chapter describes the characteristics of teams, how their members can be recruited, and how their teams should be built up, trained and managed for maximum effectiveness.

18.1 Defining an effective team

A department within a company is usually focused on achieving its own aims. Managers of departments believe that their main job is to ensure that their departments run as smoothly and as efficiently as possible, and meet all their targets. Similarly, individuals within a department are set personal goals and are measured against these. Very little credit is given to individuals who achieve a team goal, but miss a personal goal in doing so.

Teamwork is essential within modern industry (Belbin, 1982; Bursic, 1992; Ford and McLauchlin, 1992; Handy, 1997; Kernaghan and Cooke, 1990; Stott, 1994). The key to this is the support provided by individual team members to each other. However, supporting team members does not mean doing their share of the work for them. Instead it is helping them attain their goals for the mutual good of the whole team.

Animals hunt in packs and fighter pilots fly in formation, each supporting its neighbour. No two members of the team are alike; they complement each other and work as one entity in the joint task which they are performing. Every page of a telephone directory is different, and whereas a child can easily rip a page in half, it would require considerable strength to tear a bound telephone directory.

Defining a team is sometimes not easy, even when a team has been specially assembled for a specific short-term task. This is because each task requires support from many ancillary functions, who may be considered to be part of the team. Examples are the secretary who types the minutes of team meetings and makes all the administrative arrangements, such as coffee, and other professional functions, such as procurement and model shop, who carry out tasks for the team. However, these are not strictly part of the team, as defined here, but part of a much wider group.

Katzenbach and Smith (1993) defined a team as: 'a small number of people with complementary skills who are committed to a common purpose, set of performance goals, and approach for which they hold themselves mutually accountable'. A team is not just a group of people working together, for example a committee or a task force. Groups do not become teams just because someone refers to them by that name.

Table 18.1 compares some of the differences between a working group and a team. Although the output from a working group is due to the individual contributions of its members, that from a team is equal to both individual contributions as well as collective contributions.

The key difference between working groups and teams is that a working group is driven by individual goals and measures, whereas a team focuses on team goals and mutual accountability. No member of a working group would be measured on the performance of other members, whereas a team member accepts that failure of the team task will mean that each member has also failed, no matter how significant their contributions.

Teams must have measurable goals, whereas this is not essential for working groups (Meyer, 1994). For example a working group may be asked to look at methods for reducing the wastage of material in a given manufacturing process, whereas a team would be tasked to reduce the wastage in the manufacturing process by, say, 10 per cent over a six-month period. The methods of arriving at solutions also vary between a team and a working group. In a working group, harmony is usually important, whereas conflict may frequently occur in a team which is striving to arrive at the most optimum solution. However, once this agreed solution has been obtained conflict is removed and the team commits to the result.

Table 18.1 Comparison between a working group and a team

Working group	**Team**
Output determined by individual contributions	Output determined by individual contributions and collective work
Focus on individual goals and accountability	Focus on team goals with individual and mutual accountability
Individual commitment. No responsibility for results apart from own	Common commitment. Team responsibility
Share information and knowledge and help each other do their jobs to achieve individual goals	Share information and knowledge and joint contributions of members achieve collective goals
Performance achieved equals the sum of that of individual members	Performance achieved exceeds the sum of that of individual members
Purpose of group clearly defined by outside authority and does not change	Broad purpose of group defined by outside authority but is then shaped by team members. This can change during the life of the team
Need not have measurable goals	Must have measurable goals
Emphasis on harmony and avoidance of conflict	Emphasis on questioning and probing to arrive at best agreed solution
Actions agreed and then delegated	Actions agreed and then implemented

18.1.1 Stages of development

A team goes through three distinct stages. This applies whether the team has grown naturally or has been brought together for a specific task (Figure 18.1).

In the first or drifting stage, individuals come together and get to know each other. The official role which each individual plays within the team is defined, usually by the team leader. Team members may also size each other up for unofficial roles. There is very little synergy within the team, or sense of belonging and purpose. The interests of individual team members are placed before those of the team as a whole.

In the second stage, which can be referred to as the gelling stage, like-minded individuals form into small groups, as in Figure 18.1. Each group starts to develop its own identity under an unofficial leader. This is a dangerous stage and one which needs to be passed through very quickly if the total team is to survive as an entity under the official leader and not split into smaller teams. Once again, self-interest comes first, although there is some loyalty to the group and its unofficial leader.

In the unison stage the whole team is behaving as a single, highly organized body, under a single leader. All team members work towards a common goal and pull in the same direction, with a common purpose. The interests of the team match those of the individuals and there is no goal conflict.

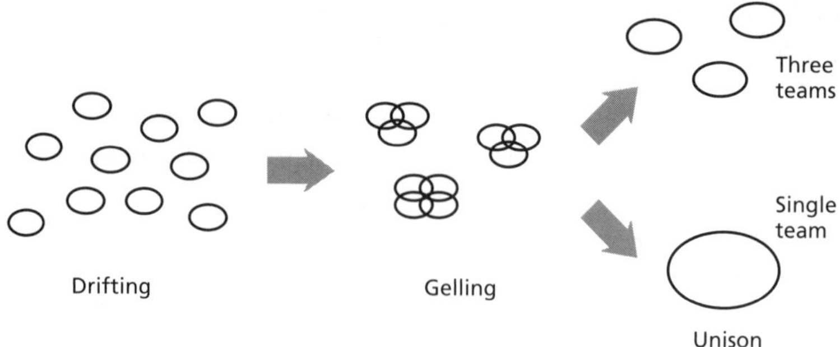

Figure 18.1 Stages in the formation of a team.

18.1.2 Team characteristics

An effective team has several key characteristics:

- It is highly efficient and results-oriented. The focus is on achieving results, and accepting and creating change (Glass and Saunders, 1992).
- There is a high level of energy and enthusiasm within the team. Morale is very good and all team members are committed to the tasks being performed (Leavitt and Lipman-Blumen, 1995).
- There is synergy between individual members of the team, who act in unison to meet team goals. The personal objectives of team members coincide, and there is interdependency between individual members in achieving common goals.
- There is a sense of purpose; the team knows why it is there and what it has to achieve.
- There is an excellent working atmosphere of trust and mutual support. Communications are good, and although conflicts exist these are resolved and lead to better solutions.
- The team has a strong leader who uses a participative leadership style to gain commitment and share responsibility. The individual strengths of the team members are used to meet team aims.

18.1.3 Team membership

A common mistake made by managers when setting up a team is to recruit clones of themselves (see the case study 'The lop-sided team'). People play several roles within a team, depending on their personality. The number of people within any one of these roles, required by any team, will vary depending on the task which is to be performed. Ten roles can be identified (Francis, 1987):

- Process Managers, who are usually managers or leaders of the team. They organize the teams, set their goals, know the characteristics of the team members, draw on their strengths, and generally keep things moving along.

- Conceptual Thinkers, who are the team's source of new and original ideas. They have imagination and vision and can think logically. Conceptual Thinkers test and develop ideas and see the impact of alternatives. They are not only theoreticians but also aim to develop practical solutions.
- Radicals, who do not accept conventional thinking and solutions. They bring an unusual perspective to problem solving, and propose new approaches to the team.
- Technicians. These are usually the specialists on the subject being considered by the team. They are important members of teams, since without them the teams would not have the knowledge to complete their task efficiently. For technically oriented teams it is usual to have several Technicians, although many leaders make the mistake of filling their team with Technicians and ignoring other team types.
- Harmonizers, whose main aim is to ensure that there is good feeling and a sense of harmony between team members. They are concerned with human behaviour within the team, and create an atmosphere of cooperation and support, helping to resolve conflicts and ensuring that team members enjoy each other's company.
- Planners or Implementors, who drive for completion of team goals. They may be autocrats and very inflexible, but they bring method to the team's tasks and ensure that they strive for quality in their work.
- Facilitators, who are ready to provide help and support wherever needed. They are hard-working, adaptable and flexible in their approach and a bit of a jack of all trades.
- Critical Observers stand back and observe the team, with the aim of judging their activities. They look for problems, often advising caution. Critical Observers may sometimes be considered to be a bit of a 'wet blanket', since they question and challenge every new idea. However, they are very objective in their criticism and their activity is essential in keeping the team's feet firmly on the ground, and stopping it from pursuing misguided objectives.
- Politicians or Power Seekers, who believe that they are always right and aim to influence other team members into their way of thinking. They are usually responsible for shaping the team's views and moving them towards their common goal.
- Salespeople or Diplomats, who provide the link between the team and other teams and groups. They develop contacts which are useful to the team, selling the team's accomplishments, and obtaining vital information for the team.

Case study

The lop-sided team

Peter French is still remembered by his colleagues at The FloRoll Manufacturing Company. He was responsible for a major marketing flop and achieved notoriety in so far that the French Marketing Case is now used in most books on business studies.

Peter French joined the company as Marketing Director when florolls were still a relatively new industrial product. He decided that the market for florolls extended well beyond industrial applications. 'Every home should have one', was his slogan. Unfortunately everyone else within the company disagreed with him.

'It is too noisy for use in the home', the designers told him, 'it has not been designed for home use.'

'It is not suitable for mass manufacturing', Manufacturing told him, 'it will be too expensive.'

'It is not the sort of thing people want in their homes, and we don't have the right distribution channels for direct selling', he was told by the Sales Department.

Peter French decided to ignore them all and set about recruiting a task force to spearhead the sales of florolls into the domestic market. He recruited an entirely new team who shared his views, but knew very little about florolls or the company.

'This way we will make sure that we think alike and act as one team', Peter told his team. 'I don't want anyone in the team who has the least doubt about the success of the project.'

After nine months it was evident that the project was going to fail. Money was poured into advertising but consumers showed very little interest. The few who bought it soon complained that the product was too big, too noisy and too expensive for what it did.

'Can't we redesign it to a smaller size and make it quieter?', Peter demanded of Engineering. The engineers shook their heads; if only Peter had involved them earlier they could have worked with his team to redefine the product for certain niche domestic markets.

'Why can't we make it cheaper?', he shouted at Manufacturing. It could be made cheaper, they replied, but that would take time and would affect some of its qualities. If only Peter had involved them earlier ....

Peter French left the company a few week later and the FloRoll Manufacturing Company withdrew from the domestic market.

18.2 Recruiting team members

Many organizations pay lip service to the fact that their strength lies in their people, but people are a company's greatest asset. It is important that managers get the best people to fill the positions in their teams and that they make the effort to keep them once they have been recruited.

All teams need a mix of skills. For example, a football team needs a striker to score goals, a midfielder to make opportunities, a full-back in defence, and so on. The number and type of people chosen will depend on the tactics the coach wishes to adopt for the game, such as attack or defence.

A project team also needs a mix of skills and experiences. If the task requires a high amount of original research then the team will contain more conceptual thinkers. A team in development engineering or manufacturing is likely to be biased towards implementors.

It is wrong to assume that the best team consists of that staffed with 'like-minded individuals'. This is often extended to apply to age groupings as well, a team staffed by young engineers being assumed to be more creative and better motivated than one which has a mix of skills and ages. Every team needs a mix of personality types and experiences, and this includes a mix of ages.

Teams should be built from volunteers. Often members are seconded from other departments, sometimes against their will. The team leader should ensure that the team is composed of suitable members, and not people who have been moved from their present jobs because they were misfits or could be spared by their current functions.

18.2.1 Defining the job

Defining the content of the job, and from that arriving at a description of the ideal candidate, is probably the most important task of recruitment. Too many managers do not have a clear idea of the job requirements and tend to mould them around promising candidates. Tasks are related to the organization's needs, and it is important to define each job to achieve these tasks, not to expect the job to change to meet the candidate. The manager should, however, correctly structure the team so that the job meets two important criteria:

- It is not so difficult to perform that it can only be done by the top 5 per cent of employees. All jobs should be defined so that they can be performed by the average, well-motivated, person with the right qualifications and skills. Managers sometimes over-specify jobs, because it boosts their own ego to think that this job will report to them, and also in the mistaken belief that an over-qualified person will do the job more effectively. In reality a person who is overqualified for the job will get bored and demotivated, and could cause problems for the team. Such a person will also be more expensive to recruit, leading to mismatches in salary within the team.
- The job is big enough to stretch the job holder. This means that the job should be enriched, allowing scope for creativity, and not just be made more difficult. For example, the job of the people responsible for washing dishes in a restaurant could be made more difficult by giving them more dishes to wash in a given time. Alternatively it could be enriched by making the job

holders responsible for other activities, such as selecting the washing-up liquid, organizing the bench layout and so on.

As a minimum a job description should cover the following:

- The purpose of the job and how it fits into the overall organization and meets organization goals.
- The scope and limits of authority within the job, such as the control on spend.
- The relationships within the job and how they fit into those of the team and the organization. The interfaces needed to line and staff within the organization, and with suppliers and customers outside the organization, must be defined.
- The duties and responsibilities within the job, with performance measures. If a measure cannot be set for any task then it should not form part of the job description. Examples of duties are delivery of a product to a given budget and time-scale, with a specified amount of rework.
- The salary associated with the job. This should be in line with the company structure and that of other team members doing similar jobs. If the salary is too high then it will demotivate others in the team and will affect the candidates if the plan is to bring their salary in line with others by giving smaller salary increases after recruitment. If the starting salary is too low then it will demotivate candidates, once they have joined the team and discovered this fact, and it could also demotivate others in the team who will look on this as the going rate for their job.

  Candidates should be attracted to the job by the challenge of the job itself and by factors such as working conditions, rather than by salary alone. Problems such as the extra housing costs, when a candidate moves from an area of lower cost housing to one of higher cost housing, can be taken into account by relocation allowances or other one-off payments.
- The title of the job. This could be a generic title, such as the use of 'member of scientific staff' for all persons below the department head, or it could be such as to indicate seniority, such as engineer, senior engineer, principal engineer. There are pros and cons to both approaches and titles should be in line with the company's structure and the way it likes to operate (for example with or without external signs of hierarchy). However, it must be remembered that titles are important and there should be uniformity. If one department calls its first line supervisors by the title of section leader, then another department should not refer to an equivalent job holder as manager.

The description of the suitable candidate will include the amount of experience needed in specific areas and qualifications. It is usually a good idea to divide the job requirements into 'musts' and 'wants', since it is unlikely that a candidate will be found who meets all the requirements of the job.

18.2.2 Attracting candidates

Candidates may be recruited internally, from within the company, or externally. Most companies have the policy that all jobs are first advertised internally before they can be advertised externally. This is a good idea, since it allows internal staff to apply for promotion or a career move, and they would be demotivated and leave if they thought that all good jobs were filled from external recruitment even when suitable candidates existed internally.

The advantage of internal candidates is that they know the company, its people and processes, and so can become effective team members sooner than external candidates. The disadvantage is that they are likely to leave a hole in some other part of the organization, which will need filling. It is also sometimes desirable to bring new ideas into the company, and an ideal team would consist of a mix of existing staff and new recruits.

Several factors need to be consider when recruiting from outside the organizations. There is usually no hard-and-fast rule in many of these, and it depends on the situation at the time. These factors are:

- The composition of the job advertisement. For example, should this contain an indication of the salary for the job? The problem with salaries is that they should be high to attract suitable candidates, but since the advertisement will be seen by internal staff it must not be out of line with the salary structure within the company. Also, how precisely should the advertisement describe the candidate, including range of skills, age and so on? Generally, if there are likely to be many suitable candidates available then one can reduce the number of applicants by being precise; otherwise it is best to keep the description relatively open.
- The method for advertising the job, such as in national newspapers or trade journals. Also, one needs to consider whether it is acceptable to openly target a competitor. Examples of this are placing advertisements in a paper which is local to where the competitor is sited, or holding informal 'open days' in a hotel close to a competitor's site.
- The use of recruitment agents and head-hunters. Often an advertising agent is used if the company wishes to remain anonymous, so as not to let its competitors know that it is working in a given field. Recruitment agents may also have suitable candidates on their books, or can head-hunt suitable candidates from other organizations. It is usually a good idea to define the job very closely when using external recruitment sources and to allow the organization to short-list suitable candidates for interview.
- The number of new graduates to be recruited. Usually most companies carry out campus recruitment, which takes the form of one- or two-day visits to the university concerned. To get the best recruits, however, a company should be in constant touch with the university's placement bureau, so that they are aware of the company's needs and will have the company's promotional literature available to hand to suitable candidates. Close links should also be

maintained with the teaching staff, who exert influence on candidates. Lecturers and professors should be invited to visit the company's plant and research facilities, and the company should encourage their staff to give occasional lectures to undergraduates.

18.2.3 Selection criteria

During the selection stage it is important to have a set of defined selection criteria against which to measure candidates. Six factors have been suggested (Munro Fraser, 1971):

- General health and physical fitness. This is of more relevance in manual jobs. It is generally assumed that for an office position a candidate will be fit enough not to have excessive time off from work due to sickness, and to be able to work the required number of hours per week.
- Impact on others. This relates to how other people are likely to react to the candidate. It is determined by factors such as manner, social behaviour, speech and appearance. It also determines how confident candidates are in company and how they interact with others, such as whether they are domineering or persuasive.
- Acquired knowledge and skills. This can usually be determined relatively easily. It relates to how well candidates can perform the technical elements of the job. It can be determined from their educational achievements; work experience, both general and specific to the job being considered; and achievements outside work, such as membership of committees and service with professional institutions.
- Innate abilities. This is a measure of the special aptitude which candidates may possess. Examples are quickness in adapting to new situations and how quickly they can learn new tasks.
- Motivation. This is a measure of the drive which candidates exhibit to achieve results. It is the effectiveness with which they apply their abilities to meet those goals which they set out to complete. It determines the types of target that the candidates set for themselves and their determination to see them through. It is also an indication of their initiative in overcoming obstacles. People with motivation are often referred to as self-starters.
- Emotional adjustment. This is indicated by several factors, such as the ability of candidates to stand up to stress at work; their outlook on tasks (whether they are optimistic or pessimistic); their ability to adjust to difficult people at work; and whether they are continually moaning about things, such as the organization in which they work.

The manager should determine where the job fits into the above six areas and choose the person who fits these necessary areas most closely. If the job is that of a project manager, with considerable interaction with people and the requirement to influence others, then clearly the impact of the candidate on others is an important consideration. If the candidate is to set up a new division in another country, then motivation, with the ability to work on one's own initiative, is important. If the work is that of an operator, and is relatively routine, then the level of motivation needed is lower because of the higher supervision involved.

For all jobs the manager must staff from strength, filling positions on the basis of what a person can do. Maximize strengths within a team rather than minimizing weakness (Drucker, 1967). Look for what the candidates excel in. Everyone has weaknesses as well as strengths; where there are peaks there will also be valleys. If one is looking to build a team which has the minimum amount of weakness then one will end up with a mediocre team. One good apple is more enjoyable to eat than two mediocre ones.

The team should be structured such that the weaknesses of individuals are no longer relevant, and only their strengths are used. For example, if some members are very good technically but cannot mix with people, then they can be given a defined work area and encouraged to become the thinkers of the group, so maximizing their output while minimizing the need for them to mix with others.

18.2.4 The selection process

Selecting a candidate is still a relatively hit-and-miss affair, with no guarantee that the most suitable person will be chosen for the vacancy (Downes, 1980; Wheatley, 1996). If an unsuitable candidate is chosen this will be evident after a few months on the job. If a suitable candidate has been rejected the manager is never likely to find out.

The selection process is usually in two stages. The first stage consists of sifting through application forms to short-list the most suitable candidates. Some companies have their own application forms, which ensure that all relevant information is provided and presented in a uniform way, to make comparison easier. Other companies prefer to have the candidates provide their own CVs or résumés, on the assumption that the style and presentation of the CV will in itself tell them something about the candidate.

Several techniques are available to help the manager in the selection process, as defined in the following sections (O'Neill, 1993).

The application form

This is a useful vehicle for providing basic factual information on the candidate, such as name, address and family status. Unfortunately, most other information is not very useful on its own, and needs to be explored further using other techniques. For example, having a technical degree or a business qualification is no guarantee that the candidate has the required technical knowledge to do the job, and it is certainly no indicator of management success. If candidates have recently graduated then their qualifications probably indicate that they have acquired a level of knowledge, but it still gives no indication of how well they will apply this to an industrial situation.

Most jobs, unless they are highly technical, can be learnt by the job holder after some time in the position. So in addition to raw qualifications, other things need to be considered, such as, for example, how the qualification was obtained. If this was done by part-time study, in the evenings and weekends, then it shows that the candidates can be motivated and have the dedication to carry out tasks which they believe to be worthwhile.

Selections are also often made on the basis of a candidate's past experience and career progression. The assumption is made that if candidates held jobs and did these well, then they will be likely to do equally good jobs in similar positions in other companies. However, past performance is no guarantee of future performance; much depends on the environment in which the candidates operated and their own temperament and circumstances, which will change with time.

The accuracy of selection based on hobbies and outside interest is also open to debate. If candidates have extensive outside interests, such as being on the local council, chairman of their professional institute's local branch, and members of several sports clubs, it may indicate that they are excellent team players and leaders. On the other hand it might mean that they are not motivated by work and seek their fulfilment outside of the work environment. This will need to be determined by the use of other selection techniques, such as interviewing.

Psychological testing

Various forms of testing have been used in the selection of candidates (Boothroyd, 1996). It is important that all these tests are administered and interpreted by trained personnel, and that they are only used as indicators; they must be checked out during the interview stage. Intelligence testing was very widely used, but has now largely fallen out of favour. Although it was found to be reliable in predicting academic abilities, it did not predict how well candidates would perform on the job (Platz *et al.*, 1959). Similarly, tests used to measure creativity were found to be unreliable (Wallack, 1976) primarily because, as will be seen later, creativity is not an easy parameter to define or measure.

Psychological tests are becoming more reliable and are more widely used today in candidate selection. These provide an indication of the candidates' personality types and the roles which they are most likely to adopt within a team. Once again, however, these tests must only be used as an aid to selection, and the results obtained are open to interpretation and must be verified during the interview. The results of the tests should therefore be available to the interviewer.

Assessment groups

The basic principle in this method of selection consists in putting groups of candidates through simulated work conditions and using trained assessors to observe their performance.

The assessment is typically held in a hotel and runs over two to five days. During this time the candidates go through group and individual exercises. Group exercises are designed to measure abilities such as decision making, persuasion, leadership and competitiveness. Individual exercises include making a presentation and the in-basket, where the candidate goes through a pile of mail and separates the critical from the trivial.

Although assessment groups provide more information on candidates than a simple interview, for no other reason than the fact that they are being observed over a longer period, they have several disadvantages. They are expensive to run, both in terms of money spent and assessors' time. They are also conducted in an artificial environment and do not necessarily indicate how well candidates will perform in real-life situations. Generally it is found that they do not provide much better information about a candidate than the more traditional paper and pencil type of psychological tests.

Selection interview

Interviewing, in conjunction with the application form, remains the most widely used method of candidate selection. The application form provides facts about the candidate; the selection interview should be used to verify these, check for inconsistencies and to determine the candidate's temperament and attitude. In this the interviewer is helped by the results of psychological tests.

It is important to remember that the candidate is also assessing the interviewer and the company during the interview. All candidates should feel that they are being treated fairly, which means that interviewers should show that they have done their homework by reading the candidates' application forms before the interviews and preparing for them. Even though a candidate may not be selected for this particular job, the candidate may be suitable for another job in the future; candidates also talk to others and this may affect the reputation of the company as a suitable employer. This applies equally for candidates from within the company. Personnel records can be studied before the interview and the person's manager consulted, although it is important to appreciate that people perform differently depending on how they are managed, and all views given are subjective.

The 'old school tie' effect must be guarded against. So often one can build up a bias about a candidate, based on one's past experience, such as the fact that the candidate's mannerism reminds the interviewer of someone else. First impressions are important; discard them! Get to know the person behind the bright T-shirt, sandals and earrings!

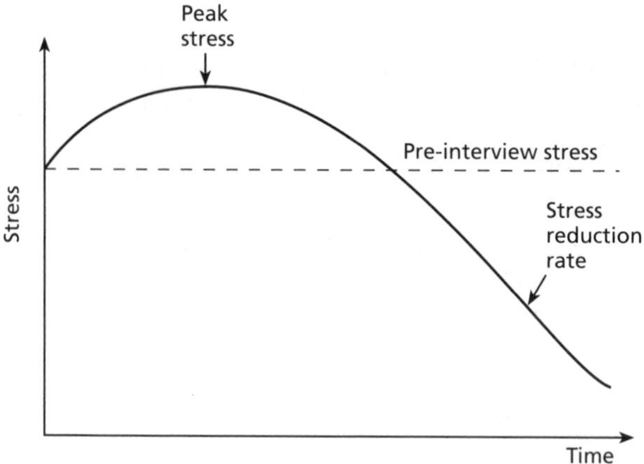

Figure 18.2 A candidate's stress curve.

It must also be remembered that the applicant's response will depend on the environment in which the interview is held and on the person doing the interviewing. Allow enough time for the interview and ensure that there are no outside interruptions.

'Gestapo' techniques should be avoided. They are often used in the mistaken belief that they will determine how candidates performs under stress, or that they will force them to reveal their 'true selves'. Instead, interviewers should do all they can to reduce stress during interviews. Figure 18.2 shows a typical stress curve through which a candidate goes during an interview. Before the interview stress has built up to the pre-interview level, and during the early stages of the interview it will rise rapidly, reaching a peak. After that the stress level will decrease, as the candidate gets to know the interviewer and a rapport is established between the two. A good interviewer will be aware of this curve and will endeavour to minimize the peak stress and to increase the stress reduction rate.

Listening is the key to good interviewing. Encourage the candidate to speak by asking open questions (questions to which the replies are not a simple 'yes' or 'no') and listen to the reply, not forgetting body language (see Chapter 19). Some interviewers like to speak first, telling the candidates about the job and then asking them to say how they meet its requirements. Others prefer to let the candidates speak first about themselves and their past work and then they provide information about the job. This way candidates cannot bias their backgrounds to match that of the job.

Usually candidates are interviewed by a representative from the human resources department and then by the person who has the vacancy. Sometimes candidates may be interviewed by others, but it is important that they do not go through a rapid succession of interviews. If necessary it is better to short-list candidates at the first interview and then to bring them back for further interviews. Panel interview techniques are also used, where the panel may consist of two or more people. It is important in such situations that the candidates know the roles of the individuals on the panel and they are not overwhelmed by questions fired in quick succession from different directions.

Often candidates may have come from competitors and in these situations the confidentially of their current employers must be respected. Candidates must not feel that they have been invited to interviews for the sole purpose of being grilled to get competitor information. Clearly such situations are difficult, since potential employers must determine the suitability of candidates by knowing what they have done in the past, and candidates must get some information on the sort of work they will be doing if they were to join the new employer. This becomes more difficult if it is planned that, as part of the interview, the potential candidates would be taken on a walk around the laboratory, so that they can be observed in the environment in which they will eventually work. It is usually best to recognize this problem at the start of the interview, as it will put the candidates at their ease and build trust between interviewee and interviewer.

After the interview the interviewers should make careful notes about the candidates, while the facts are still fresh in their minds. They should compare the candidates against their 'musts' and 'wants' list of job requirements, since this will help them later when they are selecting from a number of potentially suitable candidates, who may have been interviewed over a number of days.

18.3 Building teams

Teams are usually formed by edict, with membership, titles and goals defined by an external authority. However, it is after the team has come together that the real team-building work begins, with the definition of team goals, roles which members play, and selection of a leader (Classe, 1994; Rapaport, 1993).

Being chosen to be part of a team is usually looked on positively by employees. It is often an opportunity to do stimulating work, which meets the self-actualization aims of engineers. It represents an opportunity to be noticed by higher authority and it may be considered to be a reward and recognition for past good performance. The official leader must recognize this and build on it when developing the team.

The manager who is to lead the team must ensure that traditional good management practices are carried out, such as strong leadership, effective communications and high team motivation.

In addition the manager should do the following:

- Select team members to meet the task requirements, and not base selection on a consideration of personalities (see the case study 'The angry man'). The team should be balanced and it must be remembered that members are not there simply to provide another pair of hands; each person should bring some special skill or strength to the team.

- Break down early barriers when the team is being formed. Pass through the team gelling stage quickly. Ensure that during the formation stage every opportunity is taken to get people to work together and to mix socially. Set up joint activities; have frequent briefing meetings. Ensure contact between team members, even if it is by telephone or electronic mail.
- Adopt a participative leadership style wherever possible. There are two areas where this is required, in goal setting and in decision making. Decisions made by the team will be accepted by them and there will be a much higher level of commitment. However, the manager should be aware that, at the early stages, not all staff are able to talk freely in front of their manager. It is important to build trust first.
- Define the purpose of the team and its broad goals. Work with the team in formulating detailed goals and tasks. Although managers will allocate tasks to individuals they must be prepared to change these if this is required by the team (see the case study 'The reporter'). Ensure that all team members know how their individual tasks fit into those of the team and the organization.
- Ensure that all team members have the same interests and aims. Understand differences and resolve them.
- Avoid role conflict. Jobs may overlap but responsibilities must be defined. Team structures must be clarified and matrix relationships understood. The workload in the team must be evenly distributed.
- Identify group norms, which is what the group accepts as normal behaviour. Build on those norms that are considered to be good; influence those considered to be bad so as to modify or change them. The best method of changing a norm is to show the team that a change would be beneficial to the team and to the organization.
- Ensure that there are measures of success associated with each task. Show progress towards goals, with rewards and recognition for success.
- Show the importance, to the organization, of the work being done by the team. Publicize the work of the team to senior managers within the organization. Obtain funding and resources needed by the team. Ensure smooth handover of tasks from the team to other groups within the organization, if necessary.

Case study

The angry man

John Eldridge was a good analogue designer with the FloRoll Manufacturing Company, and excelled in component technology. When Jill Getter, the Buyer, was tasked with setting up a team to reduce the cost of procured components by 10 per cent in the next year, she wanted to have John on her team. However, she was aware of John's reputation: he was outspoken and continually argued with his manager and with other team members, behaviour which had earned him the title of 'the angry man'.

Jill though about the problem for some time. She knew that, on technical grounds, John was by far the best choice for the team, but would he be a disruptive influence?

At last she decided to talk with John. She interviewed him for the job and was quite open with him. She explained that it was a vitally important position and she believed that he was the ideal candidate for it. She, however, also told him about her fears for his disruptive influence on the team.

'John, I don't expect you to be a saint. I want to ensure that all the members on the team are frank and speak their views openly. However, you are a relatively senior engineer and may have an adverse effect on some of the less experienced members of the team. Can I ask, please, that you consider this before accepting the job.'

John accepted the job. He was outspoken and forceful but fitted in well with the team. They recognized his special talents and made allowances for him, which tempered his behaviour somewhat. The team achieved all its goals and was very successful. Jill was pleased with her choice of team member.

Case study

The reporter

When Fred Doyle first became manager of the Network Software group he planned to run the group on democratic lines. Realizing that administration work was a necessary evil, but was disliked by engineers, he decided that it would be spread equally around the team. One such task was the preparation of technical reports for the Marketing department, which they used in customer briefings.

Fred soon discovered that there was very little enthusiasm among his staff for these reports. They thought of every possible excuse to avoid doing them, usually because they were fully immersed in their programming work, which they found more interesting.

The exception was Craig Wallace. He considered himself to be a mediocre programmer, compared to the others, and quite liked writing reports. In spite of Fred's edict on equal administration burden between his staff, the team soon found that it was better for them to do the programming, taking on Craig's tasks as well, while Craig produced all the reports. This worked very well, so much so that Fred Doyle decided to rescind his earlier order on task allocation. Craig was appointed to be the formal interface to Marketing and the author of all reports coming from his department. The team had clearly allocated tasks more efficiently than the leader had done on his own.

18.3.1 Creativity

Every team requires creativity from its members, especially when this is a technical function. However, it is difficult to predict or measure creativity (Buchholz and Roth, 1987). Should it be measured on the basis of the number of patents held? Or the number of technical papers published? Or the number of difficult tasks completed on time? If these are the measures of creativity, then how can it be measured in a new graduate who has not had the time to obtain patents or publish papers? Should it also be related to time, for example the number of papers published within a year, and if so does creativity vary if this number changes from year to year?

Creativity is normally considered to be the ability to produce new and useful results. These may be new developments or novel applications of known facts. Creative people normally have the following characteristics:

- They are non-conformists. This does not, however, mean that all non-conformists are creative. Generally this needs to be recognized and allowance made for it, since creative people are sometimes considered to be difficult to manage. However, all team members must be treated equally by the manager. The team will recognize creative members and will often relax its norms to make allowances for them.
- They want to be original and deliberately look for different solutions to problems.
- They are able to think laterally around problems.
- They like problem solving and will often seek out problems. They approach all problems with curiosity.
- They have lots of ideas, many of which may not be practical.
- They have a high level of confidence in their ability. They are usually loners and prefer to work on their own with minimum supervision.

Creative people must be managed. It is important that they are set targets and these are monitored. However, the level of supervision should be less than for other staff. To encourage creativity in the team, a creative environment is essential, as follows:

- A flat organization should be used to ensure that the level of supervision is reduced.
- The organization should be loosely structured to allow a high level of autonomy. Creative members should be given the opportunity to spend less time on activities such as administration.
- The team should be encouraged to accept those that do not closely conform to its norms, such as those who are loners.
- There should be good facilities for study and research, such as a library or patent searching.
- Individual creativity should be encouraged, recognized and rewarded. Ideas always start with an individual and are then developed by the team (Nonaka, 1991).

18.4 Managing for results

A key task of every manager or team leader is to obtain the highest level of performance from the team (Schaffer, 1991). To achieve this it is important that each team member has the following information:

- Clear targets, so that they know what is expected of them and how this activity fits into those of other team members and of the corporation. In setting targets any constraints must be taken into account and the levels of responsibility clearly defined.
- Clear standards of performance, along with the methods used to measure the targets.
- A system of feedback, so that team members know the progress which is being made.

Most companies now use some form of target-setting and measurement technique to manage teams. One such technique, which meets these requirements, is known as Managing By Objectives (MBO). Odiorne (1965) was one of the earliest writers on the subject and he defined MBO as a process in which the superior and subordinate jointly identify the subordinate's major areas of responsibility in terms of results expected, and use these for operating the unit and for assessing the contribution made by each of its members.

18.4.1 Target setting

Any target-setting operation, such as MBO, has the objective of moving the corporate strategy to the individual level, so that individuals can see their contributions in the overall scheme (Reddin, 1971; Seyna, 1986). This is shown by the goal-assessment staircase, as in Figure 18.3.

There can be many more levels within the staircase than those shown in Figure 18.3, such as corporate, divisional, departmental, section, group, team and individual. The aim is, however, the same: to ensure that individual targets are derived from corporate objectives and that the appraisal of individual targets leads to a review of corporate strategy (Bartlett and Ghoshal, 1994). An individual's target is changed by going up and then down the staircase, although targets can be shaped to meet the requirements at different levels, such as for teams. Linking individual targets to corporate objectives ensures that all the employees are moving in the same direction and also provides 'purpose' and a sense of 'belonging' to the organization and the team. It shows the value of the contribution being made by individuals.

Figure 18.3 illustrates the theory; in practice very often there are so many levels between corporate objectives and individual targets in a large organization, that it is not easy to relate the two. It also assumes that there is a common overall goal, but the interpretation of this common goal may be very different depending on the level on the staircase at which it is being considered.

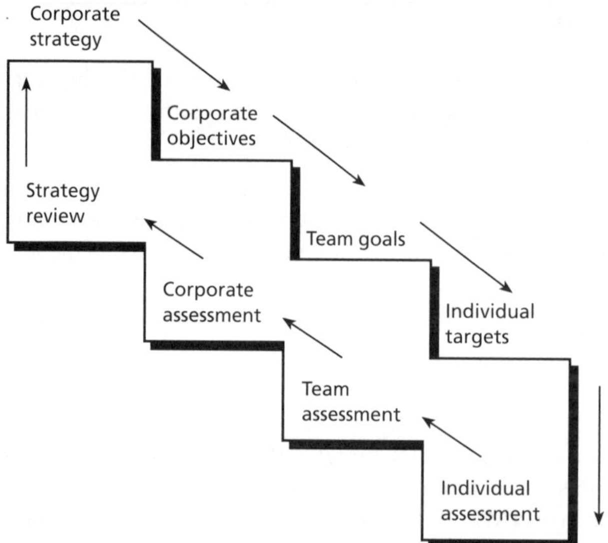

Figure 18.3 The goal-assessment staircase.

The aim of moving up and down the goal-assessment staircase is to ensure that there is a year-on-year improvement in targets and goals. This is usually achieved by changes in technology, training and perception. For example, it was many years before the magical four minute barrier was broken in the one mile race, although all world-standard runners can now regularly beat that time.

Targets, or objectives within the MBO, must have the following characteristics:

- They should be significant and not trivial. They should pick on the essential elements of a job or task.
- They must be results oriented so that they are done in order to achieve a specific objective and not just for the sake of being done.
- They should be very clearly specified and understood by both the manager and subordinate. There must be no misunderstandings when the targets are later reviewed, perhaps a year later and by another manager. (See the case study 'The moving target'.) Targets should be documented and signed by the manager and the subordinate.
- The targets should be committed to by the subordinate. This is usually achieved if subordinates have been involved in developing their targets.
- The targets should be attainable within the time frame specified and with the resources available, but they should also be challenging and stretch the subordinate.
- The targets should be linked to those of the organization and the team, and subordinates should be able to see how their effort will support the wider aims of the corporation.

- The targets should be measurable, so that subordinates know whether they have succeeded and how well they have done. It is generally felt that all targets should be quantifiable. However, putting numbers on everything is not always a good idea, especially if this results in subordinates playing the numbers game rather than aiming to meet the spirit of the target (see the case study 'The moving target'). The method for assessing the targets should also be developed jointly by the manager and subordinate.

It is unfortunate that many targets tend to have a short-term focus, even at high levels within an organization. In large corporations Divisional Managing Directors often concentrate on profitability in the two to three year span, sacrificing long-term investment for short-term profits. This way they hope that their targets are met, and if successful they will be promoted out of their current positions within a few years, before the problems of lack of investment become evident.

Case study

The moving target

When John Heston was Sales Manager of the Northern Region within the FloRoll Manufacturing Company, he agreed a sales target for the year with his Sales Executive, Tony Shaw, of £20 million. Unfortunately, John had to retire, due to ill health, half-way through the year, and when Matthew Swift succeed him, he was too busy learning the job to have time to revise individual targets. Therefore at year end Tony Shaw was appraised against the original targets by Matthew. Tony believed that he had met his target: he had taken sales of £21 million for the year.

'But these were all of Floroll LC', protested Mathew. 'Not only does this have the lowest profit margin, and is therefore easier to sell, but I note that you sold many of these well below the recommended list price.'

'That was the only way I was going to get into these new accounts,' explained Tony.

'But the profit contribution made by your sales is less then a third of that from the other salesmen', said Mathew. 'I can't agree that you have performed well or that you have met the spirit of your target.'

'You are being unfair', protested Tony. 'When John Heston set my target he did not say anything about profits. I assumed my job was to sell as many florolls as possible, irrespective of how much profit they made, in order to capture market share.'

Mathew realized his error in not reviewing the targets of his staff when he took over the group. At the appraisal meeting he now worked closely with Tony to ensure that they defined a new set of targets for the following year which they fully understood and both accepted as being fair.

18.4.2 Reviews

There are two types of reviews: those that deal with tasks or activities and those that are concerned with individual and team performance. Often these two are combined into a single review. Reviews also occur with varying frequency, and may be differentiated on this basis, as:

- The day-to-day informal review. This usually occurs either accidentally, when the manager and subordinate meet over lunch, coffee or a social event, or when the manager or the subordinate seeks the other out for a specific query. The discussion may be related to the task or to a personal matter. It is important that the manager takes these opportunities for continual feedback and appraisal of the subordinate and that these are not only done at formal reviews, which are usually held annually in most organizations. A key management technique is MWA, Management by Walking About, since not only does the manager see and is seen by staff on these occasions, but it also presents an opportunity for informal feedback on tasks and personal performance.
- Formal reviews held regularly at frequent intervals. This is a weekly or monthly review and usually deals with task-related matters only. It may be an entirely written review occurring one-way, such as a weekly progress report produced by a subordinate. Such a report may lead to a face-to-face meeting between manager and subordinate if the manager wishes for clarification, and at these meetings personal performance issues may also be discussed. Alternatively, the formal meetings may be conducted as presentations on task progress, and personal issues may be discussed following the meeting.
- Formal reviews held regularly at infrequent intervals, such as quarterly. These are more likely to be personal reviews and would occur on a one-to-one basis between the manager and subordinate. The meeting would review progress on personal objectives and carry out a performance appraisal. Targets would be reviewed for the following period and modified if necessary. It is recommended that, wherever a management by objective system is used, targets are reviewed at least quarterly to ensure that they are still current. These reviews also provide valuable feedback to subordinates on their performance. Nothing could be worse than for subordinates to go through a full year and then discover, at the annual review, that the manager considered their performance to be inadequate.
- Formal annual review. This is usually a personal appraisal of the subordinate's performance. It is a vitally important review and is discussed in the next section.

Personal appraisal

Conducting a personal appraisal can result in a considerable amount of stress for both the subordinate and the manager. Little wonder, therefore, that many managers think of every excuse to avoid appraising their subordinates. Yet an appraisal, if conducted in the right spirit, is probably the most important task which a manager can do and it will build a strong bond between the manager and the subordinate (Orpen, 1995).

Appraisals can be conducted on the basis of ability, or on the performance against specific targets, or a combination of the two. Whichever method is used, it is important that the results of the appraisal are recorded. Most large organizations have a formal form which needs to be completed at the appraisal interview. This standard form ensures that a uniform process is used in appraising all staff, irrespective of department or function. It also allows the manager's views to be seen (and confirmed) by others, such as the personal department and the manager's manager. Written appraisals also provide a record for future promotions and protects the company against a charge of unfairness or discrimination. It can be given to subordinates as their personal targets for the next period.

Figure 18.4 shows part of a typical appraisal form which uses ability as the basis of assessment. The manager records the aptitude shown by the subordinate in each of these categories over the past year. The problem with this is that the assessment is very subjective and may not be directly related to the work actually

Name: ____________________

Department: ____________________

Date: ____________________

Tick one of the boxes below (1 indicates lowest performance and 5 indicates highest performance).

Communications, oral	1	2	3	4	5
Communications, written	1	2	3	4	5
Time-keeping	1	2	3	4	5
Attitude to work	1	2	3	4	5
Judgement	1	2	3	4	5
Leadership	1	2	3	4	5

Figure 18.4 Part of a formal appraisal form using ability as the basis of assessment.

done. It is difficult to measure and can also be vague and open to interpretation. Terms such as judgement and attitude to work mean different things to different people and one person's views regarding what is good or bad in these may be different to another's. The use of numbers as a measure can also generate emotion and put subordinates on the defensive if they find that they are receiving a low mark.

The advantage of an appraisal measure based on ability is that it is quick to complete and provides a direct comparison between other personnel within an organization. It also aims to measure those qualities which may be important for promotion to the next level, such as leadership and communication abilities.

Figure 18.5 shows part of an alternative appraisal form in which targets are used as the basis of assessment. In this the bulk of the appraisal meeting is spent in discussing the subordinate's performance against each of the targets which were set for the previous period and only in the end is a final overall rating allocated. This overall rating is usually included, since it forms the basis of rewards, such as salary increases, and although it is still an emotive action it is reserved for a small part of the meeting. The problem the manager faces, however, is how to allocate the final number, since is it is an accumulation of the performance against individual targets, and these will need to be weighted. It could be that all the targets have been performed exceedingly well, but performance on one crucial target has been poor, resulting in a low overall rating.

Because everyone within the organization, especially in different functions, will have had very different targets, direct comparison between employees, using the target method of assessment, is difficult. Even if the targets are the same,

Name: ________________

Department: ________________

Date: ________________

Targets	Actions completed	Comments on performance

Overall rating: (5 = highest) [1] [2] [3] [4] [5]

Figure 18.5 Part of an alternative formal appraisal form using targets as the basis of assessment.

circumstances may result in unequal comparisons. For example, two ice-cream salespeople may have had a target to sell a certain amount of ice-cream in a period. Freak hot weather in the first salesperson's area and an unexpected cold spell in the second's might distort the amount of ice-cream sold by the two, even though the second person is a better salesperson. Another problem is that the appraisal method based on targets may indicate how well subordinates have done their present job, but it does not show whether they have the qualities for promotion to another job.

Figure 18.6 illustrates the main activities at the appraisal meeting. Both the manager and the subordinate usually receive notice of the meeting and can therefore prepare for it. One of the main activities is to note all the tasks that were done by the subordinate during the appraisal period, which is usually the past year.

During the appraisal meeting these tasks are recorded on the appraisal form. The inputs for this are the targets that were agreed with the subordinate at the previous appraisal meeting, but they should also include any other major tasks which were done.

The next stage is to assess the subordinate's performance against these tasks. How well were they done? Did they meet or exceed the standards set? If the targets were clearly defined and measurable the problem of deciding whether they have been attained is easier, but, as mentioned above, meeting targets may not necessarily mean that the subordinate has done a good overall job, as illustrated in Figure 18.7.

Figure 18.6 The key activities in an appraisal process.

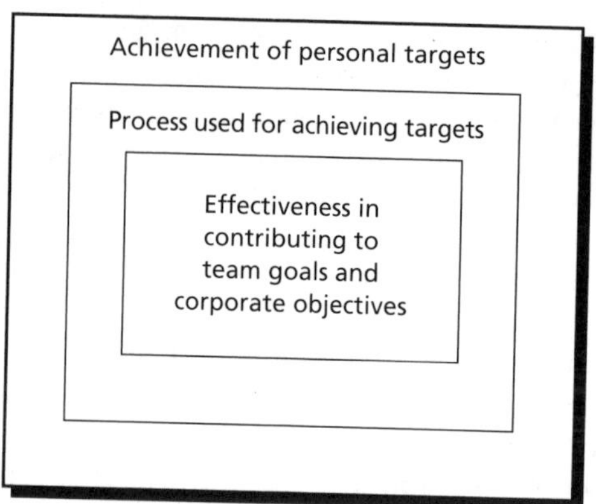

Figure 18.7 Hierarchy of performance measures.

The subordinate may have achieved personal targets, but the process used may not have been good and the achievements may not have contributed effectively towards team goals or corporate objectives. In this instance the fault usually lies with the fact that the targets were specified too tightly, with too many numbers as measures, which could not be effectively linked into the overall organization aims. Assessment of individual targets is usually an objective process, although assessment of the process used and the effectiveness of the contribution to team and organizational goals is subjective.

It is important that subordinates know how well they have performed against targets, and if they have not totally achieved them then how close they have come to meeting them. Nothing can be worse than shooting at a lighted bulb in the dark, where one can tell if the bulb is hit or not, but not have any indication of how close one has come to hitting it.

Following performance assessment the opportunities open for the subordinate are discussed. These include training, coaching and assignments to widen the subordinate's experience. Rewards, such as promotion or salary increases, will be implied depending on the performance assessment, and it may be discussed at the meeting. Actual salary increments, however, usually depend on the overall rating of other staff, and the total pot of money available within the organization, and cannot be determined at the appraisal meeting.

The final stage of the performance meeting is for the manager and subordinate to agree the subordinate's targets for the next period. In doing this the goals of the organization and of the team need to be considered, as well as the personal aspirations of the subordinate.

Counselling

Counselling is usually considered to be part of the appraisal meeting, in which the manager provides the subordinate with feedback on performance and discusses strengths and weaknesses. Most counselling sessions conducted in this way are not successful. Three methods may be used in counselling:

- The tell and sell method. In this managers give their views on the subordinates' performances and set out the plans they have in mind for the future. This is then 'sold' to the subordinates, who have no opportunity for commenting on the assessment. This method reduces the counselling time, but it leaves the subordinate with a sense of frustration and injustice. It can demotivate the subordinate, who will consider the appraisal meeting as a vehicle for fault finding.
- The tell and listen method. In this method managers give their views to the subordinates and then ask the subordinates to comment. The inference is that the subordinate is expected to agree with what the manager has said. It puts the subordinate on the defensive; the manager has been the judge and passed judgement, and the subordinate must now defend his or her actions. Once again, this is not a satisfactory method.
- The ask and discuss method. In this managers adopt the role of helpers, and help the subordinates to decide on their own level of performance and further development needs. The manager is sharing ideas with the subordinate and is not just giving advice. The manager lets the subordinates bring out the aspects of the work they did well and those they did less well and so together they identify the subordinates' strengths and how these can be better utilized. The question of any weakness must only be considered in the context of arranging work and targets such that these weaknesses are not relevant, and for organizing a training plan for the subordinate.

  The manager and subordinate are problem solvers working together, and the subordinate feels in control of the discussion. This method of counselling takes much longer and requires advance preparation by the subordinate and manager, but it builds a strong bond between them and provides a very effective counselling technique. It is successful because the manager does what is very important in any counselling: listen to the subordinate.

The counselling discussion should concentrate on major issues and the manager must avoid the danger of being seen to nit-pick. Behaviour patterns should be mentioned only if they have a cumulative significant impact on the performance of the subordinate, and only if they can be clearly identified. It is important to concentrate on behaviour rather than the person. For example, a statement such as 'being forceful with Joe Bloggs seemed to have got the meeting off to a bad start' is better than saying 'you are a very aggressive person and upset Joe Bloggs before the meeting.'

All suggestions or feedback must be considered from the perspective of the subordinate rather than that of the manager: what value does it have to the subordinate? Also, all suggestions must be actionable so that the subordinate can do something about them.

Managing conflict

Conflict within industry is caused by an incompatibility of goals, interests and ideas, which leads to mental strife between the participants. Occasionally this conflict takes the form of physical activity, such as shouting and hand-waving, but most of the time it is passive, such as avoidance of each other's company or over-politeness.

Conflict can occur between individuals or between groups, such as teams or departments. Often this is caused by competition, which can be beneficial, although it is sometimes as a result of arguments over demarcation of responsibilities or attempts to lay blame for failure of a task. Conflict also occurs in appraisal or counselling situations between the manager and subordinate.

Many managers believe that conflict is disruptive and must be prevented at any cost. However, conflict must be accepted as a fact of life within an organization. It is part of the process of change, both personal and organizational. It can often be very useful in generating new ideas or simply 'clearing the air' and forming a bond between the participants.

Managers usually do not know how to handle conflict within their organizations, and ignore it if it occurs. Some attempt to stop it by using their authority. However, stopping conflict between two people in this way results in a lose–lose situation, since the issue has not been resolved to either party's satisfaction and so neither one is happy. The manager may take sides, but this creates a win–lose situation, resulting in one disgruntled employee. It is much better to look for a win–win situation to resolve a conflict (Covey, 1994).

Conflicts are usually resolved between two parties in the following ways, and the method chosen is dependent on the characteristics of the people involved:

- Avoidance. In this the two parties recognize that conflict is about to occur and pull away from the brink. It is clearly important that both parties act together, since if one party seeks to avoid conflict while the other actively pursues it, hoping to gain advantage, then conflict will occur.
- Negotiation. In this both parties look for compromise. They aim to reduce the differences that separate them and to build on the areas of agreement. Again this method will only work if both parties follow the same plan.
- Confrontation. In this conflict actually occurs, the two parties tackling each other head-on. It is important that this is only done if the two are equally matched. In nature two bull stags will walk up and down sizing each other up before a fight. If one is clearly much stronger, the other will withdraw gracefully. Only if they both rate their chances equally will battle occur.

  Similarly, confrontation cannot succeed if one of the combatants is much stronger than the other, for example one is a senior manager and the other a junior member of staff. However, for equally matched parties confrontation helps to define the problem and the differences sharply. Following confrontation the two parties can either adopt an avoidance mode or a negotiation mode (see Figure 18.8). Alternatively, if they both share the same

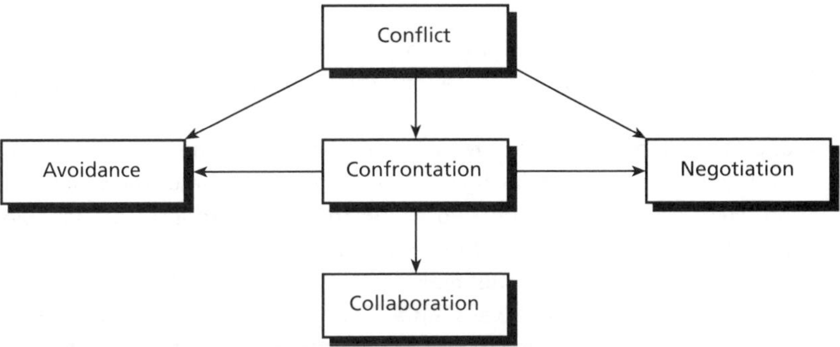

Figure 18.8 Methods for resolving conflict.

goals, as between a manager and subordinate or between team members, they will collaborate to come up with a joint solution. This is the best resolution of conflict, since it produces a solution much better than that propounded by either party, leading to a true win–win situation (see the case study 'Collaborators in conflict').

Case study

Collaborators in conflict

It was clear to Adrian Elton, during his monthly staff meeting, that conflict existed between Joe Plant, Manager of the Assembly Shop, and Jim Twiddle, Manager of Manufacturing Engineering. The problem was relatively simple, but had resulted in some heated arguments. One of the tools used within the Assembly Shop needed to be modified to fit a new variation of floroll. Usually this would be carried out within Joe Plant's Shop. Unfortunately, an accumulation of urgent tasks had meant that the Assembly Shop was overloaded and critically short of anyone to supervise the tool modification task.

'It won't take us long to do', said Joe, 'once we get the Cyprus job out of the way.'

'But that won't be finished for another ten weeks', Adrian Elton pointed out. 'Perhaps we should look at doing the modification this time in the Manufacturing Engineering department, as Jim seems to have suggested.'

Joe scowled at Jim. Was this the thin edge of the wedge?

'The manufacturing engineers don't fully appreciate our requirements', he said. 'Anyway, if they do the work it will take a long time first to explain to them clearly what we want and then to transfer the knowledge of the modified tool back to my department.'

'I think we understand your requirements very well', Jim scowled back at Joe. 'The problem I have is that we are currently very short of staff. I still have two vacancies to fill and one of my engineers is off sick.'

'It's a joint problem', Adrian told Joe and Jim. 'Why don't you two talk it over after the meeting and see if you can't agree something. Perhaps you could both brief me later.'

Joe and Jim stayed on in the conference room alone together after the meeting. The first hour was spent in 'speaking their minds' and, in spite of the sound-proofing, Mary Maid, the tea lady, swore later that she heard a cry and saw blood oozing out from under the closed door when she was passing by.

It got a bit quieter as both parties realized that they needed to brief Adrian soon. The solution they came up with was simple. Since Joe was short of supervisors he had three experienced operators who were not fully occupied and causing him a problem. Similarly Jim had an excellent supervisor who was short-staffed and not fully employed. The operators would temporarily transfer from Joe's department to Jim's. They would work on the modification to the tool under the supervisor, taking with them their knowledge of requirements. After the design had been completed they would transfer back, bringing their knowledge of how the new tool worked with them.

Both Joe and Jim had to agree that this not only solved the tool modification problem, but also solved the problem of how they were going to gainfully employ some of their staff. It was a win–win situation.

18.5 Training and development

18.5.1 Why training and development?

Sceptical managers often use the phrase 'good managers are born and not made'. They claim that people cannot be trained or developed to become good managers, and use this as an excuse for not bothering with training and development for themselves or their subordinates.

Generally managers associate training with courses (Gretton, 1995). They go to courses or send subordinates for various reasons:

- To improve their performance in their present job. This is usually referred to as training, and is very specific.
- To ensure that they are ready for some future position. This is known as development and is more educational (see Figure 18.9).
- To ensure that all team members share a common vocabulary.
- To ensure that team members have the same basic knowledge.
- Because the advertising leaflet on the course looks good and it is felt that someone from the team should go to it. Usually this is the person who can be spared, and may not be the one needing the course.

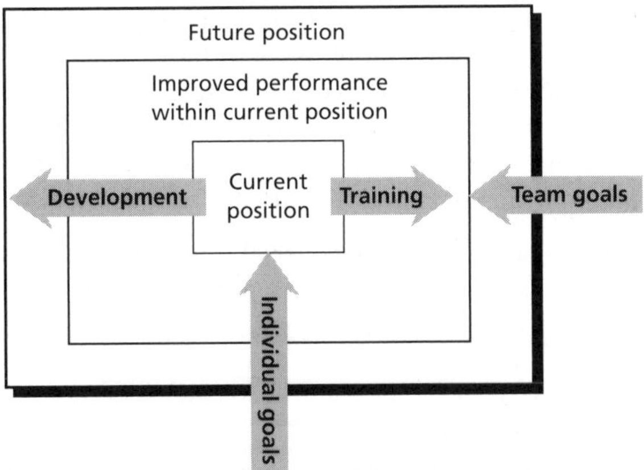

Figure 18.9 Training and development.

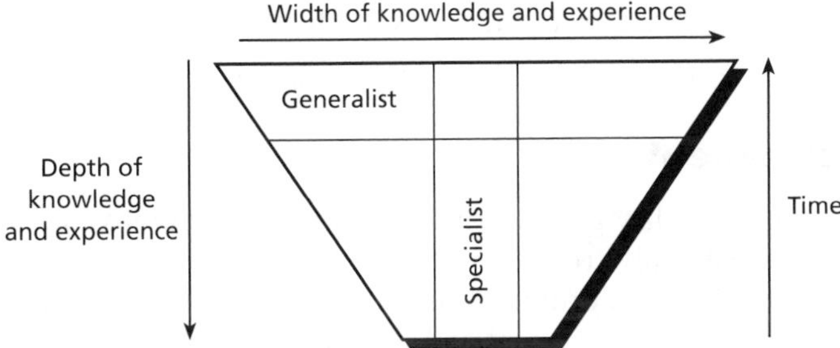

Figure 18.10 The knowledge and experience of a generalist and a specialist.

The aims of all training or development programmes are threefold:

- To provide the student with knowledge. This can be width of knowledge, resulting in a generalist (Figure 18.10), or it can be very narrow, producing a specialist. Usually people move over time from being specialists to generalists. A specialist knows more and more about less and less. This can be useful if the speciality is in demand, but will leave the person on a limb if it is no longer needed.

  A generalist knows less and less about more and more. It is dangerous to have a team without specialists to carry out the detailed tasks, which are usually technical in nature.
- To develop specific skills in the student.
- To affect the student's attitudes and values. This is difficult to do, since attitudes have been build up over many years and are often a part of the student's characteristics. Usually, also, attitudes are affected more by those of the student's boss and the organization than by what is taught on the course.

Courses which aim to change attitudes are best held away from the normal work environment. The person is then more likely to move away from established practices and accept new ones. However, this is likely to take a relatively long time and there is a danger that the person will drift into old practices when back at work. Returning to a new job helps maintain what has been learnt, especially if this is on a different site from the previous job.

Much has been written about technical and managerial obsolescence and the need to combat this by training and development (Argyris, 1991). Usually obsolescence is not related to age but to interest level, and interests change with time, such as moving from things to people. Managerial obsolescence can be combated more effectively by management experience rather than by training; taking part in the decision-making process is the key factor.

Organizations also suffer from obsolescence, and this affects the individuals who work in it. For example, the structure and control systems can age, resisting change. Such organizations are characterized by rigid structures. People in it stay static, there being few leavers or joiners. Promotion is by filling dead men's shoes. The vitality and vision needed from the top is not forthcoming.

18.5.2 Self-development

People cannot be taught; they must learn. A person must be motivated to train or develop, and this is especially true where behavioural change is required. It is the manager's job to motivate subordinates to want to develop by creating the right atmosphere, such as by encouragement and reward.

Apart from motivating staff to train and develop, managers must also take responsibility for their own development (Dent, 1995). Several activities need to be carried out:

- Planning, to determine the areas which need improvement. The areas must be clearly identified, whether it is enhanced performance in the present job, or to prepare for the next position, or merely for interest and curiosity.
- Selection of the training or development methods. Examples of this are attendance at courses; private reading; hands-on experimenting; and 'sitting by Sally', for example shadowing a senior manager.
- Making time for self-development, whether this is at work or outside of working hours.
- Seeking feedback, for example from manager, subordinates and peers, on the effectiveness of the self-development (Carter, 1993).
- Creating opportunities for self-development. Examples of this are seeking new experiences, such as assignments, especially in functions different from one's own, working overseas, and joining task teams. Deputizing for one's manager is valuable experience. Contact with customers is important for self-development, such as visits to other companies, and making customer presentations.

Interests not directly related to one's work are also useful for self-development. Examples are joining the local committee of one's professional institute; presenting papers at conferences; and acting as a visiting lecturer in a local college.

18.5.3 Learning theories

Probably the most popular of the learning experiments is associated with Ivan Petrovich Pavlov. His studies dealt with conditioned behaviour, and he showed that a dog could be conditioned to perform certain involuntary actions, such as salivating or twitching, at the expectation of food. The dog learned to associate items such as a light or bell with the presence of food.

John B. Watson postulated that all human behaviour can be defined in terms of two items: a stimulus (any change in the environment) and a response (behavioural reaction to the stimulus). Some responses to a stimulus are inborn, although most of these need to be learned. Therefore Pavlov's dog learned to salivate at the sight of light or on hearing a bell, in anticipation of food.

B.F. Skinner postulated the operant conditioning learning theory, in which behaviour is obtained, eliminated or maintained depending on the outcome of the behaviour carried out. For example, if every time engineers make suggestions they receive a negative response from their managers, then they will learn not to make suggestions.

18.5.4 Learning and development methods

Education teaches the student to be confident in an uncertain environment. It must be learned and cannot be taught. Training is usually applied to a much more specific aim and environment, and it can be taught. Most industrial courses provide training, while a university degree leads to education. In training a person to move from engineer to manager three parties can effect success: the engineer, the employer and the organization carrying out the training, such as a college (Hall *et al.*, 1992). Engineers often feel that employers can do more to help them move into management, by greater encouragement and recognition of advancement in their training, by helping to relate their training to their job responsibilities, and by career counselling. Many companies still send people on training courses out of a sense of duty or because it is common practice to do so, rather than because of a genuine need on the part of the organization.

In management training one is dealing with adults. It is important to learn by participation, and to ensure that training examples are related to those normally encountered at work. Methods used should also be varied, in order to reinforce the message. Common techniques for management training are:

- Classroom lectures. These should only be used to present basic principles, since they are similar to reading a book. Questions must be posed throughout, in order to get the students to think for themselves and not just sit passively absorbing facts. Senior management from the company may be invited to give short talks in order to relate the training to the work context. Questions and discussions should be encouraged.

- Programmed learning. This can take many forms, such as the use of video and computer-based training. It is usually quite successful in teaching specific skills, and can be accompanied by kits which provide 'hands-on' experience.
- Case studies. These usually form part of group discussions. The cases cover only an incident and the students analyse what happened, recording problems and recommending alternative solutions. Cases used should always relate to people's work.
- Role playing. This is an extension of the case study, in which the students play the part of individuals in the case. For example, one may play the role of an interviewer and interview several candidates, the rest of the class watching, with the interview recorded on video. This can then be played back to the class, who comment on the interview.
- Business games. These usually take the form of role playing and organizing. Students take the part of different functions within an organization and teams compete with each other. Business games need to run over an extended period of time for maximum effect, and they are often used on a residential course lasting for a week or more.
- Sensitivity training, or T-Groups. This teaches the students how people react to each other and how the student reacts within the group. Basically the group is left alone for some time in an unstructured environment when they first come together. Then the trainer asks the group to analyze their behaviour. T-Groups can be emotionally disturbing for unadjusted people.
- Job rotation. This is a good method for gaining wide experience, but it is very difficult to implement. It is also risky moving out an established person and introducing someone with less experienced. The problem also remains of ensuring that everyone is accounted for when the jobs are rotated. This method is usually best carried out by subordinates looking for vacancies as they arise in different functions and then applying for these in order to broaden their experience.
- Job shadowing. In this the student is temporarily assigned to a more senior person, usually as an assistant. This is an effective on-the-job training method, but it is very dependent on the effectiveness of the trainer and how much time the trainer is willing to devote to the trainee. For best results the assistant must act as a true assistant, but must also take every opportunity to learn.

18.6 The HR function

The Human Resources, or Personnel, function has a very specific role to play within an organization. Line managers are responsible for personnel management, where it applies to those reporting to them, and it is the aim of HR to help line management in this and to ensure that uniform practices are followed across the company. It is important that the HR function works through the line management organization within a company, and does not by-pass any layers of management in going from senior management to staff.

HR personnel often feel that they have divided loyalties, between the employees within the company and the management of the company. However, it is clear that HR is part of the management structure of the company and therefore represents the company first, although they must deal fairly and sympathetically with employees, whose interests they also represent.

HR is usually responsible for the following within an organization:

- Being aware of legislation, such as employment and trade union laws, and making sure that the company works within these in all its actions.
- All industrial relations matters, such as negotiations with unions and dealing with strikes and disputes.
- Setting policies on terms and conditions of employment, such as holiday entitlement, and fixing pay scales, depending on market conditions, for classes of jobs within the company.
- All health and safety matters, again ensuring that, as a minimum, they meet government regulations.
- Taking the lead in organization matters and ensuring that the company exhibits organizational effectiveness.
- Administering the company's grievance and disciplinary procedures, ensuring that both of these meet legislative guidelines and are fair to the employee.
- Internal communications within the organization, such as operating the company suggestion scheme and carrying out employee attitude surveys.
- Assisting the line manager on personnel matters, as required, such as in recruitment, staff development, counselling, promotions and new starter induction.

The responsibilities of the HR function vary depending on the organization size and structure. For small companies the Chief Executive may take on some of the above responsibilities and delegate a few specific tasks to an HR officer. In other organizations the HR function may have an enhanced role, and include other areas as well, such as site services.

Summary

A team is normally considered to be a small number of people with complementary skills who are committed to a common purpose, set of performance goals, and approach for which they hold themselves mutually accountable. The key difference between teams and working groups is that a working group is driven by individual goals and measures, whereas a team focuses on team goals and mutual accountability. Members of a working group are not measured on the performance of other members of the group, whereas a team member accepts that failure of the team task will mean that each member has also failed, no matter how significant their contribution.

Teams go through three stages when forming: drifting, gelling and unison. In the final unison stage the whole team is behaving as a single, highly organized body, under one leader, and the interests of the team match those of the individuals, so that there is no goal conflict.

A team has many types of member, all of whom have an important role to play. Examples are Process Managers, Conceptual Thinkers, Radicals, Technicians, Harmonizers, Planners or Implementors, Facilitators, Critical Observers, Politicians or Power Seekers, and Statesmen or Diplomats. The actual composition of the team will vary depending on the task to be performed, but for maximum effectiveness all these types of member should be represent.

The first step in building a team is to define the jobs that are to be done by the team members. In breaking the total work down into these jobs it is important to ensure that they are not so difficult to perform that they can only be done by the top 5 per cent of employees. The jobs should, however, be big enough to stretch the job holders. The jobs should therefore be enriched, allowing scope for creativity, and not just be made more difficult. Job descriptions should be prepared, and these should include the purpose of the job and how it fits into the overall organization; the scope and limits of authority within the job; the relations within the job and how they fit into those of the team; the duties and responsibilities within the job; the financial benefits, such as salary, associated with the job; and the title of the job.

Having prepared the job description it needs to be advertised, internally or externally, to attract candidates. Internal applicants are more likely to know the company and its structure, and internal movements help provide opportunities for promotions. However, it is useful to have new people on teams, sometimes recruited from outside the organization, since this ensures the generation of new ideas.

Several techniques can be used for selecting from the applications received for a job, such as assessment of the application forms, psychological testing, assessment groups and the selection interview. Generally, six criteria are used for selection: general health and physical fitness; effect on others; acquired knowledge and skills; innate abilities; motivation; and emotional adjustment.

Having recruited the team it is important that they are managed effectively for results. It is the key task of every manager or team leader to obtain the highest level of performance from the team. This is done by ensuring that every team member has clear targets and knows what is expected of them, and how these activities fit into those of other team members; team members have clear standards of performance, along with the methods used to measure the targets; and a system of feedback exists so that team members know the progress that is being made.

Several techniques are used for target setting and appraisal, such as Managment By Objectives (MBO). In setting these targets the aim is to move the corporate strategy to the individual team level, so that individuals can see their contributions in the overall scheme, as shown by the goal-assessment staircase. Targets or objectives within the MBO need to have several characteristics, such as: they should be significant and not trivial; they must be results-oriented; they should be clearly specified and understood by both the manager and subordinate; they should be committed to by the subordinate; they should be attainable within the time frame specified and with the resources available, but they should also be challenging and stretch the individual; they should be linked to those of the organization and the team; and they should be measurable.

Reviews can be held at periodic intervals to progress targets. These can be day-to-day informal reviews, which are usually task-oriented, or formal reviews, which are held regularly at frequent or infrequent intervals. These can cover task- or personal-related matters, although those reviews held at frequent intervals are usually team-related and cover task matters. Formal annual reviews primarily deal with personal matters and form the annual appraisal for the individual concerned. These personal reviews can be ability-based or targets-based. Apart from agreeing on the performance of the individual, the reviews should also determine personal training and development issues.

A manager is often called upon to counsel staff, and three techniques can be used for this: the tell and sell method, the tell and listen method, and the ask and discuss method. This last method is the most successful, since manager and subordinate work together to determine the subordinate's performance and how it can be improved.

Conflicts can arise in many situations and these need to be carefully managed. Usually they are resolved between the two parties in one of three ways, depending on the situation and the characteristics of the people involved: avoidance, negotiation or confrontation. To be successful confrontation must lead to collaboration and a win–win situation for all the parties concerned.

It is important that managers and their staff are continually upgrading their skills and ability. This can be done by training and development, such as specific skills courses or more general further education courses. Courses which aim to change attitudes are best held away from the normal work environment and have a greater chance of success if the student is moving to a new job soon afterwards. Common techniques for management training are: classroom lectures; programmed learning; case studies; role-playing; business games; sensitivity training; job rotation; and job shadowing.

The HR function has an important role to play, but it should always ensure that it works through the line organization rather than around it. The key responsibilities of the HR function are: being aware of legislation and advising the line managers; taking the lead in industrial relations matters; setting policies on terms and conditions of employment; health and safety issues; administering the company's grievance and disciplinary procedures; and assisting line managers on personnel matters, such as recruitment, staff development and counselling.

Case study exercises

18.1 Jenny Pett, Automation Team Leader within the Dinsher Shop, worked well with all the people within her small team. Mike Mold was the oldest and was recognized by the team to be Jenny's second in command. He had been there the longest and had come up through the line, having worked on all the processes within the shop. Jenny tended to depend on Mike for day-to-day running of the team.

When the automation programme moved on to the computer integration phase Jenny decided that she needed to recruit a new member to the team, who was a specialist in computer systems. Because Mike was not familiar with computers Jenny did all the recruitment and selection herself, and John Smythe joined the team from an outside organization. John was young, bright and ambitious. Jenny decided to work very closely with John, in order to ensure that this part of the programme went smoothly, and they spent long hours together, often working overtime on a particular problem. Mike Mold saw less and less of Jenny, although she still depended on him to carry out the normal functions within the team. He resented being shut out from the work being done by Jenny and John. He did not complain, but after six months he resigned from the company and went to a competitor. Jenny was shattered, realizing what a loss this was to the team and to herself. She suspected why Mike had left but she did not know how else she could have acted, since Mike was obviously not familiar with computer systems, and it would have taken too long to train him.

What would you have done if you were in Jenny's position?

18.2 David Laith was an autocrat and his reputation was well know. He met Jane Wilmore, the HR Director, in the corridor and grumbled once again about the attitude of his staff towards work. 'They are lazy', he moaned. 'If I don't constantly keep on them no work will get done.'

Jane persuaded David to send his team on a new interactive skills course which the company was running. David felt that it would be a waste of time ('what they want is a course to teach them how to get off their backsides and do some work'), but eventually agreed.

After the course the team members realized the shortcomings in David's management style and openly rebelled against it. David was furious. Following a confrontation with one of his staff, he stormed into Jane's office and shouted, 'I told you the course was a waste of time and would only make matters worse'.

Discuss the above. Was the course beneficial? Should Jane have made the suggestion to send David's staff on the course? Should David Laith have gone on the course as well?

18.3 The new Ministry of Defence contract was running late and there was intense pressure on the design team to deliver the program. Micro Dowling, the Engineering Director, was eventually told by Jake Topper, the Chief Executive, to 'deliver or else'. Micro therefore decided that they should ship the product, even though it had not gone through all its testing. The members of the design team were angry: 'We don't know whether the software is stable or not', they cried. 'We will just have to take that chance', was Micro's reply.

Should Micro go against the wishes of the team? What actions should he take to minimize the risk? What would you do in his position?

18.4 Mary Stick was an excellent hardware designer, but she wanted to move into management in order to broaden her experience. Peter Wave, her manager, encouraged her to go to evening classes on management, and Mary worked hard for two years, coming first in her year. Soon after that a management vacancy arose in the procurement department. Mary applied for it, but the position was filled by outside recruitment. Mary was turned down because she did not have sufficient experience in the procurement function, which was a key requirement of the job. 'But I have better technical knowledge on components than all your procurement engineers put together', she protested, 'and I learned about procurement on my course. The only way I will get experience in it is to work on the job.'

Should Mary have been given the job? How would you have handled this situation?

References

Argyris, C. (1991) Teaching smart people how to learn, *Harvard Business Review*, May–June, pp. 99–109.

Bartlett, C.A. and Ghoshal, S. (1994) Beyond strategy to purpose, *Harvard Business Review*, November–December, pp. 79–88.

Belbin, R.H. (1982) *Management Teams – Why they Succeed or Fail*, Gower, Andover.

BIM (1985) *Improving Management Performance*, BIM, London.

Boothroyd, D. (1996) Psychometric selection: Not so sinister, *New Electronics*, 27 February, pp. 20–2.

Buchholz, S. and Roth, T. (1987) *Creating the High-Performance Team*, John Wiley & Sons, New York.

Bursic, K.H. (1992) Strategies and benefits of the successful use teams in manufacturing organizations, *IEEE Trans. Engineering Management*, **39**(3), 277–89.

Carter, S. (1993) Developing and organizational mentoring scheme, *Professional Manager*, May, pp. 15–16.

Classe, A. (1994) All pulling together, *Computing*, 13 October, pp. 52.

Covey, S.R. (1994) *The Seven Habits of Highly Effective People*, Simon & Schuster, New York.

Dent, F. (1995) Taking Control, *Professional Manager*, November, pp. 28–9.

Downes, C.W. (1980) *Professional Interviewing*, Harper & Row, New York.

Drucker, P. (1967) *The Effective Executive*, William Heinemann, England.

Ford, R.S. and McLauchlin, F.S. (1992) Successful project teams: A study of MIS managers, *IEEE Trans. Engineering Management*, **39**(4), 312–17.

Francis, D. (1987) *Organizational Communications*, Gower , Andover.

Glass, K. and Saunders, L. (1992) Managing organizational handoffs with empowered teams, *AT&T Technical Journal*, May/June, pp. 22–30.

Gretton, I. (1995) Build people – build business, *Professional Manager*, March, p. 24.

Hall, J.L., Munson, Michael J. and Posner, B.Z. (1992) Training engineers to be managers: A transition tension model, *IEEE Trans. Engineering Management*, **39**(4), 296–302.

Handy, C. (1997) No place for the prima donna, *Management Today*, January, p. 27.

Katzenbach, J.R. and Smith, D.K. (1993) The discipline of teams, *Harvard Business Review*, March–April, pp. 111–20.

Kernaghan, J.A. and Cooke, R.A. (1990) Teamwork in planning innovative projects: Improving group performance by rational and interpersonal interventions in group process, *IEEE Trans. Engineering Management*, **37**(2), 109–16.

Leavitt, H.J. and Lipman-Blumen, J. (1995) Hot groups, *Harvard Business Review*, July–August, pp. 109–16.

Munro Fraser, J. (1971) *Introduction to Personnel Management*, Thomas Nelson, London.

Nonaka, I. (1991) The knowledge-creating company, *Harvard Business Review*, November–December, pp. 96–104.

O'Neill, B. (1993) Going beyond the interview, *Professional Manager*, March, pp. 20, 25.

Orpen, C. (1995) Making appraisal schemes even more effective, *Professional Manager*, May, pp. 26–8.

Platz, A.C. *et al.* (1959) Undergraduate grades and the Miller Analogies test as predictors of graduate success, *The American Psychologist*, **14**, 285–9.

Rapaport, R. (1993) To build a winning team: An interview with head coach Bill Walsh, *Harvard Business Review*, January–February, pp. 111–20.

Reddin, W.J. (1971) *Effective MBO*, HIM, London.

Seyna, E.J. (1986) MBO: the fad that changed management, *LRP*, **19**, 116–23.

Schaffer, R.H. (1991) Demand better results – and get them, *Harvard Business Review*, March–April, pp. 142–9.

Stott, K. (1994) *Teams, Teamwork and Team Building*, Prentice-Hall, Englewood Cliffs NJ.

Wallack, M.A. (1976) Tests tell us little about talent, *American Scientist*, January–February, pp. 57–63.

Wheatley, M. (1996) The talent spotters, *Management Today*, June, pp. 62–4.

19 Effective communications

Introduction

This chapter describes the process used for communication and how this can be improved. Presentations and meetings represent special instances of communication and are also considered.

19.1 The communications process

No modern manager follows the principle of mushroom management: keeping subordinates in the dark and feeding them manure. However, many fail to realize that passing information on down the line is only a small part of the communications process. Communication is much more than that; it is about conveying information, feelings and ideas to others and getting them to accept and understand what the sender meant, and to make that communication a part of their own thoughts (Barczak and Wileman, 1991; Berlo, 1975; Bolton, 1979; Francis, 1987; Haney, 1973; Keefe, 1970).[1]

Communication is not easy, because people are involved and people are variable and unpredictable. Engineering problems can usually be communicated more exactly because of the use of standardized symbols and drawings, which form part of the common vocabulary of engineers. Since managers spend about 90 per cent of their time in communications (Luthans, 1973), it is important that they communicate well.

Information and ideas can be communicated by three main methods:

- Written, such as by letters, reports, specifications, proposals and published articles.
- Spoken, as in face-to-face discussions, presentations and meetings.
- Visual, which is often used in conjunction with one of the above, such as body language used in face-to-face discussions or meetings; slides used in presentations; and videos (Fast, 1971).

The message being communicated usually consists of two parts, the report, which gives the information, and the command, which indicates how the message should be received and defines the relationship between the sender and the receiver. For example, the message 'it is raining' may be considered to be only a report, a statement. However, it is usually delivered much more emphatically, as: 'It is raining!'. This is intended to tell the receiver of the message that it, the rain, is undesirable, perhaps because they will need to postpone their game of tennis or will get wet on their way to lunch. Similarly, the innocent statement 'this is expensive' carries the implied command 'agree with me that this is expensive'.

The communication process can be shown simply as in Figure 19.1. The sender formats the message before transmitting it to the receiver. Formatting can take several forms, as discussed above, such as the spoken word or body language. The formatted message is transmitted and must be unformatted by the receiver before it can be understood. The response from the receiver to the sender is important feedback, since it indicates to the sender whether the message has been correctly received.

Errors or misunderstandings can occur in the communication process, primarily due to the following:

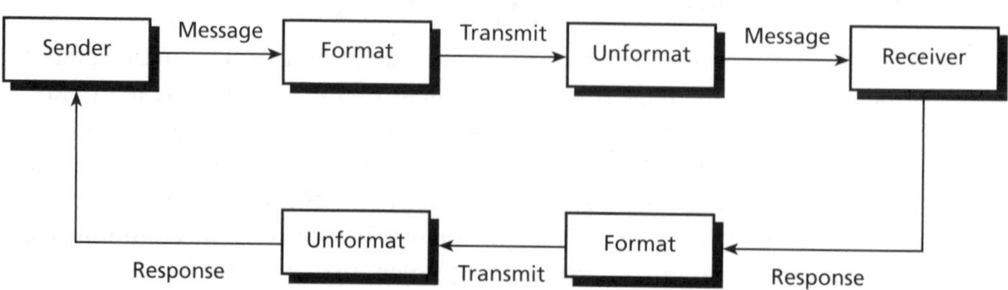

Figure 19.1 The communication process.

- The characteristics of the sender, such as background, education, past experience, personal values, and history of dealing with the receiver.
- The characteristics of the receiver. Clearly the closer the characteristics of the sender and receiver the lower the probability of errors and misunderstandings.
- The environment, such as time factors; whether the receiver is an individual or a group of people with mixed values; the characteristics of the organization (whether it is open or whether most information is communicated by rumour and via the grapevine); and the expectations of the receiver, since people often hear what they want to hear.
- The content of the message, such as its complexity to the sender and the receiver; and the importance of the message to the sender and receiver, since greater effort is usually taken by both parties to communicate messages which are of mutual interest.
- The format chosen for the communication process: whether it is written, spoken or visual. Clearly the format must suit the content of the message, the environment and the style of the sender and the receiver.

It is important to remember that in communications 'actions speak louder than words', as illustrated by the case study 'Austerity measures'.

Case study

Austerity measures

Improvements in the design of florolls meant that the amount of raw materials required was being steadily reduced. Although this was good news for the FloRoll Manufacturing Company, it was not so good for some of its suppliers. Two of these, Kelvin Inc. and Meyer Inc., decided to cut ancillary costs within their companies.

The Chief Executive of Kelvin Inc., Hal Jansen, called a meeting of his directors and then gave them an hour long pep-talk on the need for cost cutting in such areas as expenses and travel. The meeting ended when his secretary interrupted to say that the chauffeur was waiting to take him to his private jet; he had arranged to spend the weekend, at company expense, at an exclusive hotel working on his speech for the annual shareholders' meeting.

Needless to say the cost-cutting exercise within Kevin Inc. was not very successful.

The Chief Executive of Meyer Inc., Freddy Smith, adopted a different strategy, communicating by example. He told his secretary that from now on she was to book him into middle-priced hotels, and all his flights were to be arranged on scheduled airlines, economy class. Freddy assumed, correctly, that his secretary would soon communicate this information to the secretaries of his directors and that they would follow his example.

19.1.1 Communications flow

Communications must flow up and down the organization as well as sideways. Before communication takes place several questions need to be clearly answered: What is to be communicated? When is communication to take place? Where is communication to occur? Why is it necessary to communicate? How is communication to occur and feedback obtained? Who should be involved in the communications process?

The most common methods for downward communication within an organization are:

- Down the organization line.
- On a one-to-one basis between the manager and subordinate.
- Through representatives, who then pass on the message to their groups.
- Simultaneously to all the parties involved, for example by means of reports, posters or in meetings.

Common methods for upward communication are:

- Up the formal organization chain.
- On a one-to-one basis between the subordinate and manager.
- Through surveys of the group concerned, which may be anonymous.
- Via a formal system such as a suggestion scheme.

Communication has four main objectives:

- To share a vision and to ensure that everyone is clear about the goals and the purpose. This may be directed at people within the organization as well as outside, such as shareholders and customers.
- To ensure that all are pulling in the same direction, with coordinated effort. The aim of communication is to break down barriers, such as exist between teams and departments. Informal communications, such as occur around the coffee machine, play an important role in this.
- To motivate individuals and teams, to influence attitudes, and to build trust between team members and their leader (Quinke, 1996).
- To ensure that errors in decision making are reduced. The quality of a decision is dependent on the information it is based on. Upward and downward communications is essential to ensure that the best possible information is available at all levels for effective decision making.

There are several patterns for communications flow within teams, as shown in Figure 19.2. In the mesh pattern communications flows from every member of the team to every other member. For the star pattern the team members communicate through one member, usually its leader. The linear pattern passes the communications on from team member to team member, while for the tree messages flow up and down as if along the branches of a tree.

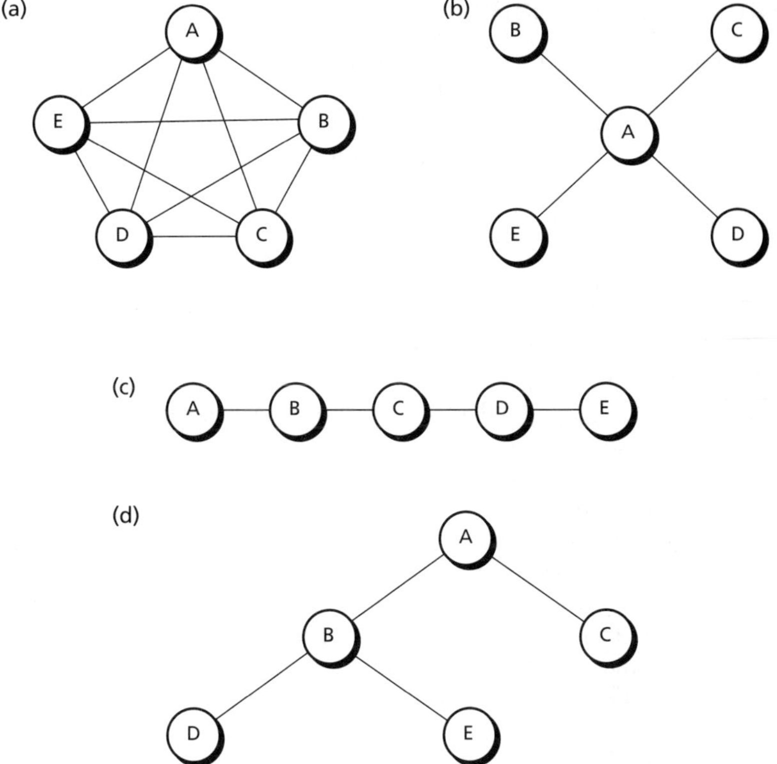

Figure 19.2 Patterns for communication flow: (a) mesh; (b) star; (c) linear; (d) tree.

Each one of these communications patterns establishes its own mode of behaviour on the team members and is used in different situations. It is important that the correct pattern is used since it determines several factors:

- The creation of a leader, since some patterns, such as the star, are more conducive to creating a leader than others, such as the mesh.
- The effectiveness of the team and the motivation and job satisfaction of individuals within the team.
- The accuracy of the information being communicated within the team.
- The speed of understanding of the task and the readiness of team members to accept it.

Generally the positions that individual team members occupy within the communications pattern will determine the extent of the above factors. For example, the accuracy of the information in the line pattern is likely to be greatest at the start of the line, getting progressively inaccurate as the distance from the sender increases. Similarly, a person in position B within the tree pattern is more likely to be elected its leader than one in position D.

19.1.2 Transactional analysis

Transactional analysis (TA) is often used in the analysis of the communications process (Ernst, 1973; Novey, 1976; Wortman, 1981). This theory states that people have three parts to their personality, which are referred to as ego states. These are parent, adult and child, as in Figure 19.3. At any time people operate from one of these states, although they can change rapidly between states, and often do so subconsciously.

The parent ego state can be subdivided into critical or controlling parent and nurturing or sympathetic parent. The critical parent is extremely demanding and operates in the mode of giving commands. For example, managers may say to subordinates: 'I want this report completed by today', and they will not accept any excuses from the subordinates regarding how busy they are.

Sympathetic managers tend to be overprotective towards subordinates. They may say: 'I want this report by today', but when the subordinates reply that they are very busy the manager may reply: 'Well don't worry, do your best, and let me have whatever you can by then'.

The person using an adult ego state in communications is following an adult thought process. Actions are based on facts and data, and follow an organized approach. Usually adult communication is used to resolve problems. For example, the manager may ask a subordinate for the report by the end of the day, and to the reply that the subordinate is very busy the manager may say: 'Yes, you probably are. Let's review your priorities and see if we can delay some of the other tasks'.

The child ego state involves emotional behaviour, rather than reasoned thought. There are three types of child: natural, little professor and adapted. The natural child is uncontrolled, aggressive and selfish, acts in an impulsive and carefree manner, and is affectionate.

The little professor child is creative and inventive, and likes to show off. It wants to impress its peers and superiors with its abilities. For example, at the first

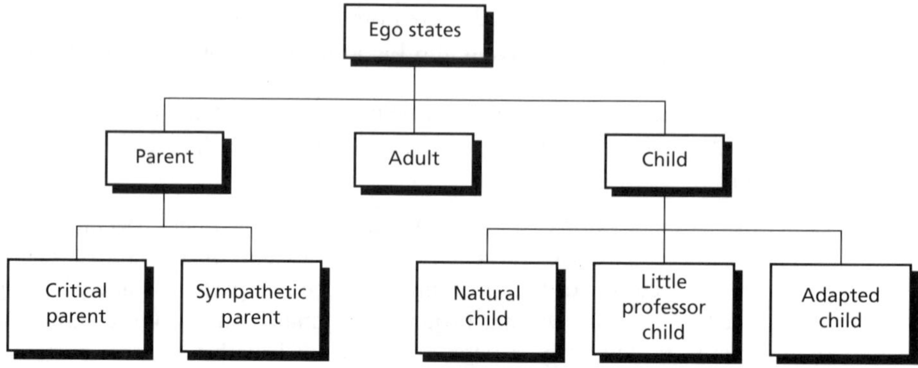

Figure 19.3 Ego states defined in transactional analysis.

hint of success the junior salesperson is likely to run to the manager with: 'I did it! I did it! I think I am going to close the sale!'.

The adapted child is well-behaved and disciplined, and well versed in social niceties. It conforms to expectations, is shy and retiring and depends on others for support. For example, the salesperson may approach the manager with the statement: 'I think I may be successful in making the sale. Can I go though the information with you, just to be sure?'.

Transactional analysis uses diagrams to show communications flow, as illustrated in the case studies 'The case of the EV meter' and 'Brass monkey weather'. Note that communications can also be direct and crossed. In direct communications the reply from the receiver is as expected by the sender. For crossed communications the receiver's reply is not as expected by the sender.

Case study

The case of the EV meter

The hardware design team was under intense pressure to complete the design of the new spin mechanism for the floroll, and the whole team was a bit on edge. Peter Wave went into the laboratory early one morning and found that the EV meter, which he wanted to use, was not at its usual place in the instrument storage cupboard.

'Mary', he asked of Mary Stick, who was at an adjoining bench, 'do you know where the EV meter is?' (Note: Peter Wave is using an adult to adult mode of communications, shown as 1 in Figure 19.4.)

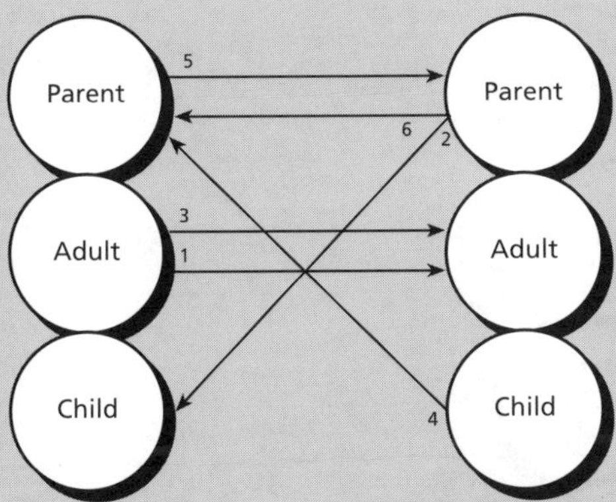

Figure 19.4 The transactional analysis process for the case study 'The case of the EV meter'.

'It's where you left it last evening', snapped Mary. (Mary is adopting the parent to child mode, as in 2 in Figure 19.4. The communication is also crossed since Mary's reply is not as Peter expected it to be.)

Peter felt his temper rising. 'That's not good enough, Mary. You know that we have a rule here that all instruments have to be put away in the cupboard last thing at night, and that the last person out must do so. You were the last to leave last evening. It was your job to make sure all instruments were put away.' (Peter is replying to Mary by using the adult to adult path, as in 3.)

Mary realized that she was in the wrong. 'I am sorry', she said. 'I was very tired and forgot.' (Mary is now using child to parent, as in 4.)

Peter was also sorry for having lost his temper with Mary. He realized that they were all under pressure. 'That's OK, Mary', he said. 'Maybe this rule we agreed about last person out putting away instruments needs to be looked at. We are all forgetting to check for instruments before we leave. Perhaps we should have a nominated person each week.' (Peter has switched to parent to parent mode, as in 5.)

'Sounds a good idea', replied Mary. 'Shall I bring this up at our next staff meeting?' (Mary has also switched to parent to parent as in 6, and the problem can now be resolved.)

Case study

Brass monkey weather

Jane Furley, Hardware Manager, and Jill Tait, Systems Manager, were in the staff restaurant having lunch.

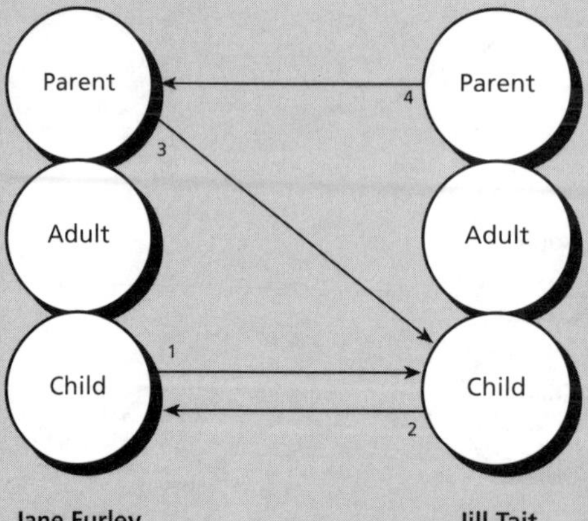

Figure 19.5 The transactional analysis process for the case study 'Brass monkey weather'.

'I have to go to Moscow again next week', said Jane. 'How I hate going there this time of the year. Last time I went the temperature was minus twenty with a wind chill temperature of minus forty.' (Jane is operating in child to child mode, as in Figure 19.5, path 1.)

'I know! It's ridiculous! Last time I went it was just as bad', replied Jill. (Jill is also operating in child to child mode, as in Figure 19.5, path 2. Also the conversation is direct, since Jill has replied as Jane would have expected.)

The conversation between Jane and Jill continued for some time, child to child, and can do so indefinitely while each complains about the weather in Moscow and having to go there in winter.

'Say!', said Jane at last. 'The problem really is that we don't have the right clothes for Moscow. What we need is thick coats and fur hats. The company should really buy these for us.' (Jane has now moved to parent to child mode, path 3.)

'I agree', said Jill. 'Let's go and see Micro. He is usually very fair about these things.' (Jill has also moved into parent mode and the conversation is between parent and parent, as in path 4. The problem can now be solved.)

19.2 Establishing communications

Several techniques and considerations are involved in setting up good communications within an organization. These can be grouped into environmental considerations, attitudes and emotions, questioning, and listening.

19.2.1 Environmental considerations

Enviromental considerations are primarily related to contact between the communicating parties. This is manifest in the following ways:

- Geographical location. It is well known that communication flows more freely between staff located next to each other than those on different sites or even in different buildings on the same site. This partly accounts for the popularity of open plan offices. One of the main reasons for better communications is the opportunity that geographical location presents for improvements in social contact, and the building of trust between people. This is one of the reasons why the manager should take the opportunity to collocate multifunctional teams and to bring people together, such as at lunch time or coffee breaks.
- Time separation. This can occur when teams are working in different countries on the same project, when even telephone contact becomes difficult and the best means of communication is by electronic mail. The most common form of time separation is shift working, where mistrust builds up between shifts, each considering the other to be inferior. Sometimes this can

get so bad as to lead to one shift deliberately sabotaging the work of another shift.

- Functional separation. This is often seen as the rivalry and mistrust between different departments, such as marketing and engineering, or engineering and manufacturing, each believing that the other department is not pulling its weight. Once again this can, in the extreme, result in departments working against each other rather than pulling in the same direction against a common competitor. Line and staff management are also frequently at loggerheads, the staff managers being looked on suspiciously by line management as an interfering organization who have no grasp of reality ('those ivory-tower people'), while the line manager is considered by staff management to be inflexible and hidebound.
- Lack of opportunity for communications to occur. This is especially true for upward communications between subordinates and managers. Managers can call meetings, put up notices and send official letters to subordinates, but very few opportunities exist for subordinates to communicate upwards, unless this is actively encouraged by the manager. Most companies recognize this and set up formal review processes where the manager and subordinate meet for review. Unfortunately, this usually only occurs once a year and is a formal process, with the wrong emphasis, since it is often linked to the annual salary review.

  Managers must not assume that subordinates will automatically come to them if they have a problem. They must adopt policies which create frequent opportunities for formal and informal contact. Unfortunately, managers are busy people and often discourage such contact. For example if the manager is rushing to prepare for a meeting and a subordinate comes into the office, it is so easy to say: 'Not now, John, I'm busy'. The subordinate will leave and probably never return with that, or any other, problem. Much better for the manager to listen for a few minutes and then say: 'Sounds like we need more time on this one John. I have to dash to a meeting. Can we meet at 4:30 today to discuss it further?'
- Faulty communications structures. As discussed, with reference to Figure 19.2, there are many communications structures, most being suitable for different occasions. The wrong structure can result in poor communications. The most common problem is long communications lines, where the information gets distorted by the time it takes to reach distant groups, resulting in misunderstanding and mistrust.

19.2.2 Attitudes and emotions

Probably the most important consideration for good communications is the attitudes and emotions of the parties involved (Argyris, 1994). Several factors can represent barriers to good communications:

- Prejudices, such as those already mentioned between line and staff organizations and between different departments. It is the responsibility of the manager to ensure that these prejudices are recognized and broken down, by providing the opportunity for the staff concerned to meet frequently, often on an informal basis, so that they get to understand each other.
- Differences in background, education and values between the people concerned, resulting in lack of empathy with each other's views. This will result in the receiver of the message putting a different interpretation on it from that intended by the sender. There is little the manager can do except to be aware of this and to compensate for it wherever possible. (See the case study 'Promotions for all'.)
- A general climate of low motivation and mistrust within the company, resulting in communications from management being disbelieved. This is often as a result of the management having been dishonest with employees in the past. (See the case study 'Promotions for all'.)
- Management failure to show that communication is required. A manager must actively encourage upward communications by being good at downward communications and by rewarding good communications practices from subordinates. Often a simple 'Thank you for letting me have the facts; that was very useful' will lead to greater upward communications.
- Communication insensitivity. It is important that both sender and receiver communicate at the 'feelings' level, to get behind the words and understand what the other party is really feeling when sending the message.
- A large number of management actions which indicate to the employee that 'the manager is the boss' and is to be obeyed. Examples of such behaviour are when the manager criticizes or judges the subordinate; acts in a superior manner and belittles the subordinate, usually before peers; threatens or implies a threat; talks down to the subordinate; acts with indifference towards the subordinate; displays a closed mind; uses the authority of position to command. (See the case study 'I am the boss'.)
- A manager should avoid assigning blame. Nobody likes to pass bad news upwards since staff usually fear authority. (Some people withhold information from the manager, as a form of punishment, but this happens very rarely.) It is difficult for the subordinate to tell the manager that the fault lies with the subordinate or with the boss himself. (Bartolome, 1989). Often the subordinate will remain quiet in the hope that the problem can be overcome, so the boss need never know. Usually the boss is told when things go very wrong and it is then too late for any action to be taken to correct it. (See the case study 'Too late to say sorry'.)
- Sometimes the subordinate is afraid that the manager will reject suggestions. Some may even worry that the boss will hijack their ideas and claim credit for it with higher management. (This has been known to happen!)
- Management belief that communication is not always a good thing. Many managers think that employees should not receive feedback on their

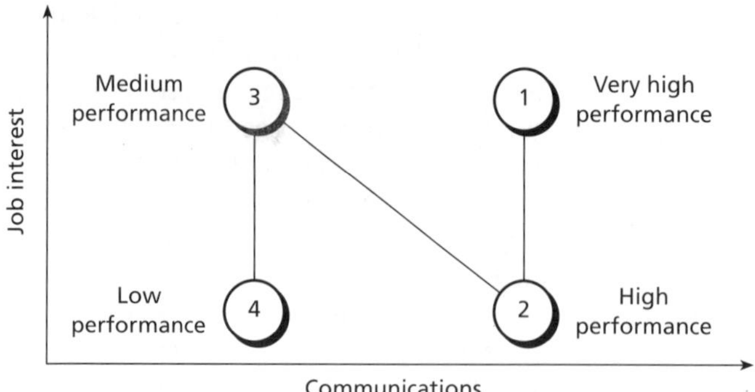

Figure 19.6 The N curve of performance.

performance. As one manager said: 'If I tell someone that he has done a good job he is likely to get conceited, sit back on his laurels and ask for a salary increase. If I tell him he has done a bad job he will become disheartened and demotivated and do even worse. Much better to keep them in the dark'. And feed them manure?

Feedback should always be given, but in a constructive manner. It has been shown (Buchholz and Roth, 1987) that performance in a job is related to job interest and communications. This can be illustrated by the N-curve of Figure 19.6. If job interest and communications are both high then performance on the job is very high (point 1). If job interest is low but communication is high then performance is high (point 2). For high job interest but low communications, performance is medium (point 3). Where job interest and communications are both low, performance is also low (point 4). Therefore, even when job interest is low a good measure of performance can be obtained by keeping communications high. One such vehicle is enabling employees to let off steam and to know that management is listening.

- Evaluating messages. There is a natural tendency to evaluate all messages from one's own viewpoint (Rogers and Roethlisberger, 1991). For example if someone says: 'I don't like people with bushy moustaches' the natural reaction is for us to evaluate the statement and either to agree with it: 'Yes, horrible, aren't they' or to disagree with it: 'Oh, I don't mind them'. Either way, if one is evaluating then full attention is not being given to receiving the message and seeing it from the sender's viewpoint.

Case study

Promotions for all

Mike Delling was due to retire as supervisor of the Machining Section in the Floroll Assembly Shop. This was the first time in two years that an opportunity had arisen for promotion within the Shop, and consequently there was much interest in who would be promoted.

The group members thought that Tom Cole would be chosen; after all he was the craftsman with the greatest skill among them. 'It'll be Tom', they all agreed over their cup of tea, 'and a good thing too. Tom is the best among us. Even Mike Delling thinks so. Doesn't he get Tom to show us how to use these new tools which are brought into the Shop?'

Joe Plant, manager of the Assembly Shop, had other ideas. He wanted to appoint someone with high leadership qualities. 'What the team wants is good leadership', he told, Adrian Elton, Manufacturing Director. 'They want someone who can set an example to them, build up team spirit and get them all pulling together. Morale is quite low in the Shop, as you know.'

Joe was right. Morale was low mainly because management had lost its credibility and was considered to be dishonest. When many of the operators had been recruited into the Shop, a few years earlier, they had all been told that the company policy was to promote strictly on merit, and there were plenty of opportunities for promotion. Unfortunately, although all the operators had worked hard and were good at their jobs, there had been no openings for promotion to supervisor level until now. Little wonder that motivation was low and management was not trusted.

Eventually Bill Hale was appointed as the new supervisor. He was promoted out of the Packing and Shipping Department and although he had very little knowledge in operating tools, he was known to management as a good leader. The members of the Machining Section were outraged and went on strike. Management had been insensitive to the fact that the operators' sense of values was different from that of the management, and had led them to expect different criteria in the selection of their next supervisor.

Case study

I am the boss

David Laith, Manager of the Development Tools section, was known to be an autocrat and was a bad communicator, as Jim Swail, a young graduate, was about to discover. The development of the new software tool was running late, and as this was part of Jim's project he was called into David Laith's office to explain the delay.

David was reading a letter when Jim entered and he ignored the interruption.

'You sent for me to explain the delay in the project', began Jim.

David continued reading. After a pause Jim repeated the statement, but David took no notice. As Jim turned to go David put down the letter.

'Where do you think you are going, young man?', he said. 'You can't escape that easily.'

'I'm sorry', said Jim. 'You are busy. I can come back some other time.'

'A good employee waits until his boss is free', said David. 'But then you wouldn't know anything about that. Looks like you are not a very good engineer either. You are certainly making a mess of the new software tools project.'

Jim stood silently by the desk.

'I told Angela Code that you were too inexperienced to take on the project. Looks like you bit off more than you could chew. When you get more experienced, like me, then you will know what you are capable of doing, which, on present showing, is not much.'

'The programme was going well until you transferred one of our workstations to the Embedded Software Group. We are short of development computers now.'

'So you have problems! We all have problems and have to overcome them.'

'But...'

'Don't but me! I know when someone isn't pulling their weight. There is no room in this department for slackers. Get out there and make sure the project is pulled back on schedule.'

Case study

Too late to say sorry

Further problems had hit the software development tools project. Jim Swail confided in Angela Code, Section Head of the Software Design Tools group. 'We are very short of workstations', he said. 'I estimate that we are currently one month behind schedule and if things continue as they are we will be slipping a week in a month.'

Angela Code decided not to tell David Laith, her manager. Every time she reported a slip David reminded her that Jim Swail had been her choice for the project. Anyway, Angela reasoned that she was being paid to solve problems herself and not run to her boss when things went wrong.

'We will be getting our workstation back next week, and John Frazer from the Embedded Software group has agreed to lend us one of his machines as well. This should enable us to move an engineer from testing to code writing, and with overtime working we should be able to recover lost time.'

The plan did not work. The workstations were late in being released by the Embedded Software group and the test engineer needed extra training before he could contribute to coding. The project slipped from being one month to five months late. Eventually David Laith had to be told since it was coming close to the date when the software had to be delivered to the customer. By then it was too late to recover the lost time. The customer invoked the penalty clause on the contract and although David Laith tried to blame Angela Code and Jim Swail, Micro Dowling reminded him that managers are always responsible for all the activities under them.

19.2.3 Questioning

To obtain more information it is important that the manager ask open-ended questions: questions to which the replies are not limited to 'yes' or 'no'. Using words such as 'what', 'why', 'when', 'where', 'how' and 'who' is essential. Avoid use of the word 'you' which can sound aggressive, for example: 'You wanted this...' or 'You should have thought of that...'.

Care is also needed to avoid digging for minute detail, otherwise the sender of the message is likely to feel under interrogation and mistrust will result. There is a fine line between soliciting for details and interrogation, and the manager must learn to know when this is being crossed. The style must be adjusted to suit the other party.

It is also important to ask searching questions (known as aggressive enquiry) and not to accept things at face value. For example, if the manager is told by the subordinate that the design time for a component has been cut down from ten weeks to ten days, then the manager should ask: Was such a drastic reduction necessary? How did the company survive in the past with ten weeks? What were the costs of achieving this reduction? Can these techniques be applied in other areas to reduce design times? (See the case study 'Personal service'.) Only the highlights of messages are often communicated; it is essential to use aggressive enquiry to get the whole story.

Case study

Personal service

Peter Sell, Sales Director at the FloRoll Manufacturing Company, decided to use the forum of Jake Topper's Executive Staff Meeting to gain some political advantage for his department. Towards the end of the meeting, when Jake asked if there were any other matters to discuss, Peter spoke up.

'You may be interested to learn, Jake, that our Northern Region Manager, Matthew Swift, has received a very nice thank you letter from Johnston & Company for the excellent service he provided them. They telephoned our Northern Office at six o'clock one evening last week to say that a floroll we had supplied to them the previous day was faulty. Fortunately there was a spare floroll in the office, so

Matthew drove through the night to get to the offices of Johnston & Company by the time they opened the next morning and deliver the new floroll. That's what I call really putting the customer first!'

Jake was impressed, but wanted to know more. 'Excellent!', he said. 'How much business are we doing with Johnston & Company?'

Peter looked embarrassed. 'Well, not too much. In fact they have only bought one floroll, but after all one or one thousand is not that important. It is the principle; a customer is a customer.'

'Is this extra special service going to get them to buy any more florolls?'

Peter coughed before replying. 'Well, actually, no. They only need one floroll for their business.'

'Was it important that the product be taken to them the next day and did the manager of our Northern Office have to do so personally? Couldn't it have been sent by our normal delivery service the following day?'

Peter turned a little red. He could sense that Jake Topper was no longer impressed. 'Well, yes, but getting it to them without any delay showed how much we value our customers.'

'While Matthew was visiting Johnston & Company which other customers were not being looked after?'

No reply from Peter Sell.

'And, by the way, why did we ship a faulty floroll to Johnston & Company in the first place? I thought you had put a test section into all our Regional Offices, so that they could carry out a final check before sending anything out.'

19.2.4 Active listening

On average people spend 45 per cent of the total communications time in listening (Nichols and Stevens, 1957). It is important, therefore, that managers learn to listen effectively. Yet most only prove the saying 'in one ear, out the other!' The problem is often developed from a young age. After all, many years are spent in school in learning to read, but very little time is devoted in learning to listen.

Effective listening requires active listening, being responsive and not sitting passively without any sound or reactions. The receiver of a message must not only hear but also listen. Hearing is the process by which auditory sensations are received by the ears and transmitted to the brain. Listening, on the other hand, is the psychological process of interpreting and understanding the significance of the sensory process (Drakeford, 1967). Therefore one can hear without listening.

Active listening is important since it improves the quality of the information being sent by the sender. Also it is difficult for the listener to sit for any length of time without doing something, as required by active listening.

There are a few simple ways to improve active listening:

- Show interest in what is being said and indicate that the message is being understood. One can respond by word and also by gesture and facial expression. Give undivided attention. Listen with understanding (Rogers and Roethlisberger, 1991).
- Pay attention. The mind can absorb information about five times faster than the speaker can deliver it. Therefore make sure that attention does not wander. Concentrate on other aspects of active listening, such as watching for visual communications.
- See the communications from the sender's frame of reference. Try to feel the sender's feelings. Ask the question of yourself: 'What does the speaker feel when saying that?'.
- Show empathy with the sender's feelings. About 80 per cent of all communications is non-verbal, so watch out for body language (Kitchener, 1996). Much of the sender's feelings can be determined from this. Furthermore a large part of the feedback that a listener provides must also be non-verbal.
- Listen without interruptions to start with, so as to get the full story. Don't lead the speaker with questions. (See the case study 'Was the car OK?')
- Adopt the correct posture. Lean forward slightly towards the speaker; have an open position (legs and arms uncrossed); use movement, but be careful of distracting movements or appearing to be nervous. Eye contact is important, but don't stare. No one likes talking to someone whose eyes roam all over the room but do not look at them.
- Paraphrase and provide reflective responses. A paraphrase is a concise response which summarizes the essence of the message from the sender. In a reflective response the receiver restates the feeling or content of the message from the sender in a form which indicates to the sender that the message is being understood and accepted. (See the case study 'Man of few words'.)
- A sender will often code messages without knowing that this is being done. The receiver must therefore evaluate the message carefully to determine whether coding has been applied before decoding it.
- Often the sender will have difficult identifying, handling or conveying feelings. The sender may also 'beat around the bush' and not get directly to the main aim of the message. The listener must be prepared to help and encourage by active listening.

Case study

Was the car OK?

Joe Penn, the mechanic in the garage at the FloRoll Manufacturing Company, walked into the office of Dave Simpson, Transport Manager.

'Peter McKelly was involved in an accident', said Joe, 'and unfortunately...'

'What!', interrupted Dave. 'Not another accident! The number of accidents these sales guys have is ridiculous! No respect for their company cars. They would be a lot more careful if they had to pay for the repairs. How bad is it?'

'Pretty bad', said Joe. 'The assessor will come to see it tomorrow, but what is worse...'

'Was it in good mechanical condition?'

'Should be, we serviced it a week ago, but...'

'Thank goodness for that. We have renewed the insurance on it, haven't we?'

'Yes, the insurance on all the cars is up to date, but...'

'Good. Well, I suppose we had better get Mr Peter McKelly to fill in an accident report.'

'Bit difficult to do that', replied Joe. 'He was killed in the accident.'

Case study

Man of few words

Ed Bramble, Sales Executive, shuffled into the office of his boss Jane Warm, Southern Region Sales Manager. Jane could see that Ed was despondent.

'I blew it!', moaned Ed. 'I lost the Liverpool contract! Those guys in manufacturing really let me down. I had been at them all week to bring forward the delivery date for the batch of florolls, and the best they could come up with was six weeks! Six weeks! Our competitors are quoting half that time! I should have known better! I should have cut the price as my first instincts told me to. So what if the profit margins would have been a little lower; at least we would have got the sale. I am going to do that for the next contract.'

Jane decided to practice reflective responding and paraphrasing. 'Oh, no!', she said, her face and posture reflecting Ed's despair. 'I am so sorry Ed, I know how hard you worked for this contract and how you must feel about losing it. Never mind, you lost this one on delivery; you will win the next one on price.'

19.3 Presentations

Presentations within industry are as common as reports, and in some organizations major reports consist of no more than a collection of presentation slides. It is therefore crucial that every manager becomes an effective presenter. It must be remembered that, when making a presentation, the manager is on display, being judged, consciously or subconsciously, by the performance put on while standing up in front of the audience.

There are three stages in presentations: preparation, delivery and follow-up. The follow-up phase is as important as the other two, if not more important, since it will ensure that the aims of the presentation are achieved. However, the style and procedure for follow-up will vary depending on what is to be achieved, and so this is not considered further here. Instead, the common aspects of the first two stages, preparation and delivery, will be described in further detail.

19.3.1 Preparation

Several steps need to be followed during the preparation phase:

1. The first step is to define clearly the aim of the presentation. There should only be one aim, for example obtaining approval for a new project or getting a customer to buy a new product.
2. The characteristics of the expected audience should be carefully analyzed. The aim is to clearly understand what each person is likely to be interested in and to find a common ground to which the presentation can be targeted. (See the case study 'Common ground'.)
3. The subject matter needs to be thoroughly researched. The audience will want to feel that it is being presented to by someone who knows the field, so that they can have confidence in what is being said. Although presenters can carefully limit slide material to areas with which they are familiar, audience questioning may cover wider topics.
4. Put down, in bullet form, all the facts connected with the subject. Then categorize these into three or four major headings with several sub-headings. Remember that too many facts can confuse; concentrate on the key ones so that they can be clearly presented and remembered.
5. The slides can now be prepared. Nothing is truer than the old saying; 'A picture is worth a thousand words'. Also it has been shown that people only remember 10 per cent of what they hear, but they remember 50 per cent of what they see and hear.

   The slides must not be too complex; they have to be assimilated in a very short time by an audience who may be unfamiliar with the topic. One idea per slide should be the rule. Start from what is known, and where there is agreement, before moving to more controversial areas. Aim to build understanding and empathy with the audience, so that they are thinking the way you wish them to.

Case study

Common ground

It was coming towards year end and, as always, budgets were tight. The Engineering Department urgently needed to buy a new software-based design tool in order to improve their overall performance, and Micro Dowling, Engineering Director, felt that this should be ordered now and not in the next financial year.

When Micro spoke to Jake Topper, Chief Executive, about the investment he was asked to make a presentation on the subject at the next Executive Staff meeting.

Micro sat down to prepare his presentation. He realized that he would be competing for capital money with the other department heads. It was therefore crucial that he found some common ground, some benefit which this investment would have for the whole audience. He considered the main people who would be present.

Jane Wilmore, the HR Director, would be primarily interested in the effect that the tool would have on staffing. Would it require more people? Would the skills mix change? Adrian Elton, the Manufacturing Director, was always grumbling about the defect level in the designs that he received from the Engineering Department and the impact this had on his manufacturing process. Jack Fox, the Marketing Director, and Peter Sell, the Sales Director, were constantly complaining about the length of time it took to get from the concept stage to having product ready for delivery to the customer. Johnny Bacon, the Financial Director, had the overall responsibility within the corporation for ensuring that any computer-based purchase met corporate guidelines and gave an adequate payback.

Micro decided that the common ground he would tackle would be financial. He prepared his slides, which showed the following:

- There would be no increase in payroll spend since the same people could be used to operate the new tool. A one-week training course would be required, but this could be given by the training department (Micro checked this with Vic Smith, the Training Manager).
- The defect level in the product would be significantly reduced, since the tool had an in-built test and check facility. This would show considerable savings in rework and defect correction when the product went into manufacture. (Micro estimated the savings and put these on his slide, after discussion with Adrian Elton.)
- The design time would be reduced by almost 40 per cent, providing an opportunity for much increased revenue by being quicker into the market, ahead of the competition. (Micro obtained figures for the increase in sales from Jack Fox and Peter Sell.)
- The new tool would run on the same computers that were recommended by Johnny Bacon and would therefore fit into the corporate strategy. It would also show a financial payback which was better than that obtained from most of the investments made earlier that year. (Micro got the payback calculation from Johnny Bacon.)

Micro Dowling's presentation was successful and he received approval to buy the new design tool.

19.3.2 Delivery

The moment speakers stand up in front of an audience they are on their own, and they are in command. Thorough preparation will give them confidence, but several factors still need to be kept in mind:

- Go slowly; better to risk offending the audience by being too slow than to lose them by going too fast. Start the presentation by saying what information is to be provided; then provide the information; then end by saying what information was provided. Ensure that the audience is concentrating on one subject at a time. Announce all the main headings at the start of the presentation and then announce all sub-headings at the start of the main sections. Ensure that the audience knows when the presentation moves from one heading to the next.
- Remember that the audience is more important than the presentation! It is the speaker's responsibility to ensure that the presentation style and information match the audience, to maximize understanding, absorption and retention. (See the case study 'Bow tie and sandals'.) The larger the audience the greater the problem of understanding and retention.

  The presenter must also be prepared to vary the style depending on feedback received from the audience. Often, for large audiences, presenters will not know their characteristics until the presentation has started, so they must be flexible.
- Remember that the aim of the presentation is to sell an idea to the audience as well as to sell oneself as an authority to be listened to and agreed with.
- Throughout the presentation remember that it is important to show how the proposals being made are benefiting each individual member of the audience.
- Develop empathy with the audience. Present information in terms that they understand and can relate to, so that they will make the ideas their own. Use examples in which the speaker and audience have acted successfully together.
- Anticipate resistance and deal with it during the presentation, rather than waiting for them to come up as questions from the audience. The speaker has an advantage over the audience since the speaker, in effect, controls the floor as long as required. It is a good idea at the start of the presentation to indicate whether questions should be asked during the presentation, to encourage audience participation, or saved for the end, to minimize interruptions to a prepared sequence of information flow.
- The whole posture of the presenter is important. The presenter must show enthusiasm for the proposal being made. After all, if the presenter is not enthusiastic then why should the audience be? Stand and not sit during the presentation, but avoid wandering about as it can be distracting. Body language is important, but a contortionist act is not called for! Smile but avoid looking like a half-wit with a perpetual grin, especially when giving serious information, such as a plunge in company profits.

- Don't look at the screen unless this is being used as a prompt. Focus on the audience. Eye contact is crucial if their attention is to be held. Pause long enough on each person so that they feel that you are talking to them personally. For large audiences divide them into groups; don't sweep rapidly from side to side.
- In a written report one only has punctuations for emphasis. Presentations have many more facilities, and these should be used to the full to maintain interest and to enhance the message. For example, one should pause briefly after introducing a complex idea to give the audience time to assimilate it. Volume and pitch can be varied for emphasis, as can the pace of delivery. The voice should be projected to be heard by the last row in a large audience, but beware of shouting.
- The last few minutes of the presentation are vital. The last point is usually the one best remembered by the audience. Summarize and reinforce the main message. Aim to get commitment and agreement on the next steps to be followed.

Case study

Bow tie and sandals

As part of a quality awareness programme within the FloRoll Manufacturing Company, Vic Smith, the Training Manager, was given the task of ensuring that all company employees went through a two-day training session on quality. Vic was aware that this was not likely to be popular with many of the employees, who often considered quality to be the responsibility of the quality department. The company culture had to be changed.

It was evident that a variety of styles had to be adopted due to the wide spread in background, education and attitudes of the employees, although the basic message would be the same. Vic and his team devised three separate courses, as follows:

- **Course 1**. Primarily for those on the shop floor. This was given to small teams, consisting of a mix of operators and supervisors. The courses were held in a corner of the shop floor which was screened off for the duration of the training. The aim was to ensure that the training on quality could be easily related back to the work environment. The training team ensured that their style of dress was close to other operators on the shop floor, in order to build empathy. The material was kept simple and short, with flip charts and whiteboard used primarily as the presentation media.
- **Course 2**. This was designed for junior and senior staff within the various functions. Larger audiences were involved and care was taken to ensure that managers were mixed in with their subordinates, and there were representatives from the different functions (such as marketing, engineering, sales,

manufacturing and support) at each session. The talks were held in a conference room, and were longer and more detailed. A variety of presentation styles were used, such as overhead projection, slide and video. Again the dress used matched that of the audience, one member of the training team even wearing sandals, a favourite among the engineering community during summer.

- **Course 3**. Designed for Jake Topper and his direct reports. Vic Smith considered this to be a very important course, and it was given first, to indicate the commitment of the board to quality. It was a one-day course, held in the Board Room. The material was kept relatively direct, with emphasis on the main messages. Formal presentation techniques were used, such as overhead slides, which was the favourite method for normal presentations within the company. All the trainers wore suits.

19.4 Communication at meetings

Many managers fail to appreciate that a meeting or conference is a form of presentation, and it calls for as much preparation in order to be successful (Churchill, 1993). During the meeting each time people speak they are holding the floor and are presenting, so very similar techniques must be used. The speaker is also creating an impression and being judged on performance at the meeting, as much as during a presentation.

Meetings should represent a 'meeting of the minds', although all too often they end up being nothing more than a meeting of bodies, an event where several people collect together in order to pass the time. Little wonder that many consider meetings to be a waste of time in which nothing is achieved, but which are a necessary evil perpetuated by the organization. This is usually as a result of the meeting being badly planned, ineffectively controlled and running too long.

Meetings should not be held if there are other more effective ways of achieving the same aims. Generally meetings are called in order to:

- Give information to several people at the same time. This is a presentation or briefing and not a true meeting.
- To pool knowledge and decision making in order to solve a problem. The contribution by the various members in the meeting, each usually having a different expertise, will often result in a more effective solution than could be obtained by any of the members on their own.
- To obtain commitment from the members at the meeting to a given course of action.
- To introduce people to each other, usually new members of a team to the other team members.

The frequency with which meetings are called depends on the topic being discussed. For example, a manager may call daily meetings to determine progress or to tackle problems on a critical project. *Ad hoc* meetings, called infrequently, may be more appropriate where a group is trying to formulate long-term strategy for a product development.

19.4.1 Behaviour in a meeting

Chairpersons need to be skilled in recognizing the various behavioural patterns that participants adopt within a meeting. This will enable them to control the meeting more effectively, and also adopt a style which is suited to the occasion and to the aim which they wish to achieve. Ten main behavioural patterns have been identified by researchers:

- Proposing new actions and making suggestions. In normal meetings this should be done by all members, although if participants feel threatened, for example if they are among very senior people, then their proposing behaviour may be low. Examples of proposing statements are: 'Why not start with item 2 on the agenda?' Or: 'I suggest that the project is delayed by three months'.
- Building on the proposal made by someone else. This is especially useful in some types of meeting, as will be seen later. However, in all meetings building on a proposal shows support for the original speaker and fosters good relationships. Examples of building statements are: 'Starting with item 2 on the agenda, let us move to item 3 and then item 1 and the other items'. Or: 'If the project is delayed by three months we should see if we can add the extra features which the customer wants and which we could not previously include in the development'.
- Supporting a proposal or an opinion put forward by another person. Again supporting action improves the atmosphere within a meeting. Examples of supporting statements are: 'I agree; we should start with item 2 on the agenda'. Or: 'Slipping the project by three months sounds like a good idea'.
- Disagreeing with a proposal or opinion, or raising obstacles and objections. Needless to say, people feel threatened when their opinions are disagreed with, and excessive disagreeing behaviour in a meeting can result in a poor atmosphere and low productivity. Examples of disagreeing statements are: 'The problem with starting with item 2 on the agenda is that we will lose the flow of the meeting'. Or: 'I disagree! If we slip the programme by three months the customer might cancel the order'.
- Attacking and defending behaviour. This is usually in the form of emotional statements, aimed at people or an idea. For example, an attack statement would be: 'Starting with item 2 on the agenda is the stupidest thing I have heard for a long time!'. The defensive statement to this could be: 'It makes more sense that way; I'm not to blame if the agenda was

drawn up wrong'. Another attacking statement would be: 'Just because you can't meet deadlines I don't see why we should slip the programme by three months'. A defensive statement to this could be: 'I am on time; it is the delay in the delivery of components from suppliers which is holding up the programme, and you are responsible for that'. Usually the spiral of attack and defence will destroy a meeting unless the chairperson steps in to nip it in the bud. Note, however, the importance of conflict, as described in Chapter 18.

- Testing understanding, to see whether the earlier discussions have been understood. For example, the chairperson may say: 'Let me just check here; by slipping the programme by three months you are proposing to deliver the product to the customer in April rather than in January?'.
- Asking for and giving information, which can be related to new facts, or more information or clarification on an existing statement. For example: 'If we slip the programme to April, will we still be able to deliver 100 florolls per month from that time?'. The reply could be: 'No, we will have to ramp up to that value over a two month period'.
- Summarizing earlier discussions in a much shortened form, with highlights. Summarizing is crucial in establishing clarity, structure and control and it is an essential behaviour of a good chairperson. For example, the chairperson might say: 'So we have agreed that we will slip by three months but add the new features, and ramp up production over two months'.
- Bringing in contributions from participants who have not been active. For example, the manager might say: 'I seem to be doing all the talking; what do you think of the proposal, John?'. Or: 'You seem to be lost in thought, Peter, what impact do you think the three month slip will have on the manufacturing programme?'.
- Shutting out, which is the opposite of bringing in, the aim being to reduce the contribution from a participant. Usually this is done unconsciously by members at the meeting, by one person interrupting another, although a chairperson may do this to prevent a dominant personality hogging the meeting. For example: 'Sorry to interrupt you, David, but before you go any further can I ask Fred what he thinks about the proposal to add the extra features?'.

19.4.2 Planning and running a meeting

Meetings usually have a chairperson, although frequently the manager who has called the meeting acts as its chairperson (Jay, 1976). This is usually not a good idea. A chairperson must be impartial, interested mainly in the process being adopted for the meeting and not the content of the meeting. The manager, on the other hand, is usually the most senior member at the meeting and will probably have very strong views about the topic under discussion, and will want to have these aired.

Although a chairperson needs to be completely unbiased, many do show bias and this is done subtly, such as by bringing in participants who share their views and shutting out those who disagree with them. A good chairperson is aware of this and avoids such behaviour. Chairpersons use the following behavioural patterns during a meeting:

- Proposing is done on procedural items rather than on content. For example, the chairperson might say: 'Seeing that we are running late, I suggest that we skip item 5 on the agenda and return to it later if we have time'.
- Building is done in order to combine contributions rather than to add a new contribution. For example: 'Joe, you suggested earlier that the programme is likely to slip by three months, and now John has stated that the customer wants new features added. Why not slip the programme and use the extra time to add the new features?'.
- Very little supporting or agreeing/disagreeing behaviour is exhibited, which is as expected, since the chairperson is neutral.
- The chairperson is constantly testing understanding and asking questions to see whether information provided by other participants has been understood. In addition, the chairperson usually has exclusive responsibility for summarizing at periodic intervals, which is often done in order to keep the meeting on track and to show progress.
- Bringing in and shutting out is done periodically and consciously, in order to bring in a low contributor or to stop any one person, or a group of participants, from dominating the meeting. In some types of meeting, where time is critical, the chairperson will also shut out participants if they are rambling off the main point or going over ground which has already been covered.
- The chairperson never attacks or defends, but is constantly on the watch for such behaviour, and stops it as soon as it occurs, before it can build into an attack–defence spiral.

Some of the chairperson's tasks can be delegated to other participants. For example, one of the members may be asked to summarize at periodic intervals. In this case the chairperson acts more as a meeting's manager, delegating tasks as appropriate.

Whether a formal chairperson function is required also depends on factors such as the size of a meeting. The larger the meeting the greater the degree of control required and the greater the need for a chairperson. This is shown in Figure 19.7. For small meetings, with about four to five participants, very little formal control is required and a chairperson's role is largely superfluous. As the size of the meeting increases the level of control needed also increases, although the increase reduces for very large meetings. For example, the style and control needed for a meeting with 200 participants will be very similar to that needed for a meeting with 300 participants.

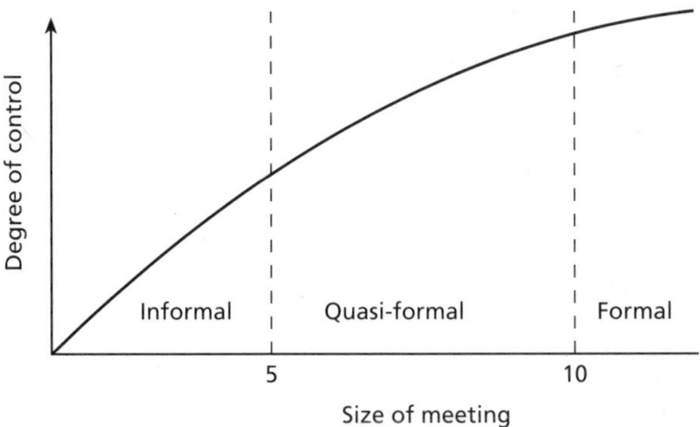

Figure 19.7 The degree of control required in meetings of various sizes.

Every meeting must have a clearly defined objective, and it is the chairperson's responsibility to ensure that all participants are aware of the objective when the meeting is called. Also, the meeting should have an agenda, with an indication of the time to be spent on each item. The chairperson is responsible for putting together the agenda, but does so after taking inputs from all participants, who must finally agree the agenda. The chairperson may edit the agenda, to vary the sequence in which items are taken or to delete items to fit the time allocated for the meeting, but all changes must be agreed with the participants, and the agenda must be published before the meeting. Once published the agenda should be stuck to unless there is a very good reason for deviating from it.

It is important that every meeting has a formal beginning and a formal end. Usually the chairperson starts the meeting by stating its aims and briefly going over the agenda, and ends the meeting by summarizing the main points, showing achievements. It is important that participants don't leave the meeting with the feeling: 'That was a waste of time; we did not achieve anything.'

The minutes of the meeting are an important document for recording agreement and actions. It must be taken by a secretary whose main task is to produce accurate minutes and who does not take part in the discussions. Although a competent shorthand secretary could fulfil this role, for highly technical meetings it is important that the secretary fully understands what is being said. The chairperson must ensure that when an action is allocated it is clearly defined to the secretary and a completion time put against it. Before the end of the meeting the actions should be gone through to ensure that the action holders accept them. The minutes should be action-oriented; their purpose is to record agreement and actions and not to act as a chronicle of every event that took place. The following style should be avoided: 'Joe Smith said that he thought that item two on the agenda should be taken first. Fred Bloggs asked him why. Joe Smith replied that this was a more important item than item one and should be discussed first. Mary Mince thought that item three was more important and should therefore be discussed first. Joe replied that he did not agree with this statement, since ...'. If it is important to record the change in agenda sequence it should be done by a simple: 'After discussion it was agreed to take agenda item two first'.

19.4.3 Types of meeting

Meetings can be classified into two major categories: filter meetings and amplifier meetings. A third type, brainstorming, is really a hybrid of the filter and amplifier meetings.

Filter meetings

In a filter meeting the participants put forward a large number of proposals. Usually these proposals have been formulated before the meeting and the purpose of the meeting is to consider them and to select the best few proposals. Each proposal is considered in turn and, following a brief discussion, is accepted or rejected. There is very little building behaviour, since the aim is to reduce the number of proposals and not to increase them. There could be a lot of agreeing and disagreeing behaviour, and attack and defence, as participants try to get their favourite proposals accepted.

Filter meetings are also referred to as win–lose meetings, since the winners are those who have their proposals accepted and the losers are those whose proposals have been rejected. Most of the work is done before the meeting, when the proposals are being formulated, and relatively little work is done during the meeting. The outcome of the meeting is also predictable, since one knows that no new ideas will emerge from the meeting. Although the meeting may agree on the proposals that are to be accepted, there is usually very little commitment to these from the other participants, who may have been supporting proposals which have been rejected. Morale can also be low, since proposals are often rejected with very little explanation.

Amplifier meetings

In an amplifier meeting only a few new proposals are put forward. However, there is a considerable amount of building activity, as participants build on these proposals until they arrive at the final solution, which is accepted by all. Because all participants have had a hand in developing the solution, there is total commitment to it. Very little attacking and defending takes place in an amplifier meeting, although there could be agreeing and disagreeing. The atmosphere is one of kinship and excitement, since the result from the meeting can rarely be predicted before the meeting.

An amplifier meeting is also known as a win–win meeting, since all participants win in the end, by having their joint proposal accepted. It is much more creative than a filter meeting and the quality of the decisions from the meeting is also higher, since it draws on the skill and knowledge of all its participants. Generally it is better if no preparation is done by the participants before the meeting, and all the work is carried out at the meeting. This prevents people from coming to the meeting with preconceived ideas, which they may find difficult to abandon at the meeting.

The problem with amplifier meetings is that they are time-consuming if a large number of proposals have to be gone through. They are also not favoured by many managers, who usually call a meeting in order to get their own proposals accepted.

Brainstorming

Brainstorming is a form of filter–amplifier meeting in which participants are encouraged to put forward new ideas and to build on other's proposals. It is an established technique, being described by Osborn (1984) as: 'Using the brain to storm a creative problem'. It has very few rules, but the few that there are should be closely followed for best results.

The venue for the meeting should be carefully selected. It should be free from outside interruptions and allow participants to relax. The people coming to the meeting will be excited, since they are aware that they have been chosen because of their creative abilities. The meeting is likely to be exhausting and so should be kept relatively short, an hour to an hour and a half being the maximum. Small groups of participants should be involved, from six to twelve being best. There should not be any dominant personality present, such as a very senior manager with junior subordinates, as this may inhibit the subordinates. Competition between participants should be encouraged, but this competition should be in thinking up new suggestions and not in having their suggestions selected.

A very strong chairperson is essential for the successful running of a brainstorming meeting. The objectives of the brainstorming session, and the procedure to be followed, must be clearly stated at the start of the meeting. For example, the objective could be to select a name for a product which is soon to be launched.

The chairperson would go round the group in turn and ask each person for a proposed name. (Alternatively participants may be allowed to shout out a name as they think of one, although this requires many more note takers and may allow some participants to be idle.) It is important that when asked the participant makes a proposal, no matter how poor. If they cannot make a proposal then the next person is asked.

All proposals are written down in full view of all participants. This is probably best done on flip charts which are pinned around the room. The aim is to get quantity of proposals and not necessarily quality. Wild suggestions are encouraged and no proposal is discussed or criticized at this stage; it is simply noted. Participants are encouraged to build on other people's proposals.

When no more proposals can be obtained the chairperson may go though the list of proposals in turn, seeking clarification from the contributor. At this stage some of the very wild proposals may be rejected. New ideas which arise can be added to the list. The aim is eventually to narrow down the proposals to a few which can then be taken away for further analysis. However, in some brainstorming meetings the task of narrowing down the proposals is done in a second meeting (which is really a form of filter meeting at which no new proposals are allowed) or by another group of people, where again no new proposals are allowed.

Summary

Communication is not easy, because people are involved and people are variable and unpredictable. Errors or misunderstandings can occur in the communication process due to the characteristics of the sender and receiver; the environment; the content of the message; and the format chosen for the communication process.

Communication has four main objectives: to share a vision; to ensure that all are pulling in the same direction; to motivate individuals and teams; and to ensure that errors in decision making are reduced.

There are four main patterns for communications flow: mesh, star, linear and tree. Each of these establishes its own mode of behaviour on team members and determines several factors, such as the creation of a leader; the effectiveness of the team, and the motivation and job satisfaction of individuals within it; the accuracy of the information being communicated within the team; and the speed of understanding of the task and readiness of the team to accept it.

Transactional analysis is often used to analyze communication processes. It refers to the three ego states that people have (parent, adult and child), and how these interact. Parents can be subdivided into critical or controlling parents and nurturing or sympathetic parents. The child ego state can be subdivided into natural child, little professor child and adapted child.

Several techniques and considerations are involved when setting up good communications, such as environmental, attitudes and emotions, questioning, and active listening. Environmental considerations are primarily related to contact between the communicating parties, and are seen in the following ways: geographical location of the communicating parties; time separation; functional separation; lack of opportunity for communications to occur; and faulty communications structures. Probably the most important considerations for good communications are the attitudes and emotions of the parties involved. Factors that represent barriers to good communications are prejudices; differences in background; climate of low motivation; management failure to show that communication is required; communications insensitivity; management actions which indicate that the manager is the boss; manager assigning blame; subordinates afraid to talk to their managers; and management belief that communications is sometimes to be avoided.

Questioning and active listening are important communication techniques. In questioning, open-ended and searching questions should be asked, but care should be taken to avoid digging for minute detail. Effective listening requires active listening in which the listener is responsive. This can be done by showing interest in what is being communicated; paying attention; seeing the communication from the sender's frame of reference; showing empathy with the sender's feelings; listening without interruptions to get the whole story; adopting a correct posture; paraphrasing and providing reflective responses; evaluating the message received to determine whether it has been coded; and helping the sender in conveying feelings.

Presentations are widely used within industry as a communications vehicle, and in making a presentation the manager is on show. Preparation is essential: clearly define the aims of the presentation; analyze the characteristics of the expected audience to understand their interests and frame of reference; thoroughly research the subject matter; note all the facts related to the topic; and prepare the slides, ensuring that maximum information is conveyed in the minimum space. In delivering the presentation several factors need to be considered: it is better to go slowly and risk offending the audience than to go too fast and lose them; constantly remember that the audience is more important than the presentation; keep in mind that the aim of the presentation is to sell an idea, as well as selling oneself as an authority on the subject; show how the proposals made benefit the listeners; develop empathy with the audience; anticipate resistance and deal with it during the presentation rather than wait for it to appear during question time; posture of the presenter is important; eye contact is also important, so look at the audience but don't sweep rapidly from side to side; use punctuation to maintain interest; and ensure that the key points are reinforced at the end of the presentation.

There are several types of meeting, such as filter meetings, amplifier meetings and brainstorming (which is a form of filter–amplifier meeting). Meetings can be very effective or a waste of time, depending on how the communication within it is managed. A skilled chairperson is essential. Members at the meeting will exhibit several types of behaviour: proposing new actions and making suggestions; building on proposals made by others; supporting a proposal or opinion; disagreeing with proposals and opinions; attacking and defending; testing understanding; asking for and giving information; summarizing; bringing in contributions from participants; and shutting out. Good chairpersons may propose, build on, test understanding, bring in and shut out, but being neutral they should rarely support, agree or disagree, or attack or defend.

Case study exercises

19.1 The FloRoll Manufacturing Company decided to move its Engineering centre to another site, since it was running out of space at its current location. This new site would have all the latest facilities and was planned as a showpiece for the company. As such, management felt certain that the engineers would welcome the move. Negotiations were started with several local authorities for suitable accommodation. In order to keep the workforce fully informed it was agreed that they would be told that a move was to be made, but that no site had as yet been chosen.

This led to discussions among the engineers as to the location of the new site, and rumours were rife. Many with small families decided to look for alternative employment, fearing that the new location would not be within commuting distance. Morale fell and productivity suffered. Negotiations took longer than expected and no suitable site had been found after six months. By then morale was so low that the company decided to abandon its plans and to overcome the space problem by splitting the Engineering group, putting a smaller team into another factory site close by.

Could the company have handled this situation better? Was poor communication to blame? Did the company do the right thing by telling the workforce about the plans, even though they were incomplete?

19.2 In order to reduce the amount of waste of materials which occurred in the production of florolls, Joe Plant, Manager of the Assembly Shop, decided to set up a working group to examine the problem. Because of its importance he decided to run this himself and selected two of his supervisors to work with him. The aims of the working group were well publicised and they met at regular intervals over a three-month period, the minutes of the working group being put up on the noticeboards for all to see.

Joe was disappointed that there was so little interest in the work of the group, since most of the people on the shop floor ignored the notices. Eventually the group came up with several very good recommendations. These required changes in working practices, but most of the workforce implemented them very reluctantly and some even secretly worked to ensure that the new practices failed. Eventually they were withdrawn.

Joe thought that he had communicated very well with his staff. Do you agree? If not, how could communications have been improved?

19.3 The project was running late and David Laith, Manager of the Tools Development group, thought it was time for a pep-talk with his staff. He realized that he was considered to be somewhat of an autocrat by his staff, but this time he thought that he would show them that he was 'one of the lads', and that they would work together as one team in order to succeed.

David Laith thought he made quite a good speech. He pointed out that the programme was running late and that, if they failed, the customer could cancel the contract. He explained that, as manager, he was responsible for the success of the project and so they were all equally to blame for the failure of the programme. He was disappointed therefore when, a few days later, a deputation came to see him to ask if they were all under threat of redundancy if they were late and the contract was cancelled by the customer.

Where did communications fail? Why did the engineers read something into David Laith's talk which he had not said? (See also the case studies 'I am the boss', and 'Too late to say sorry'.)

19.4 Jim Bishop, Sales Executive within the Northern Region sales branch of the FloRoll Manufacturing Company, suspected that one of his major clients was having financial problems. However, he was reluctant to confide in his boss, Matthew Swift. 'I could be wrong', thought Jim, 'and anyway Matthew will probably get at me, as he always does, for having recently taken that big contract from them.' Jim kept quiet for several months, continuing to take large orders and hoping that he was wrong. However, the customer went bankrupt and defaulted on payment of several large bills.

Do you believe the reasons Jim Bishop gave for not confiding his suspicions to his manager? What could have caused this communication breakdown to occur? What could his boss have done to discover the problem earlier?

Endnotes

1. See also the following references: Lindauer, 1974; Rockey, 1977; Scanell, 1970; Scott, 1984; Scott, 1986; Tubbs and Moss, 1974; Wolf and Aurner, 1974.

References

Argyris, C. (1994) Good communications that blocks learning, *Harvard Business Review*, July–August, pp. 77–85.

Barczak, G. and Wilemon, D (1991) Communication patterns of new product development team leaders, *IEEE Trans. Engineering Management*, May, pp. 101–9.

Bartolome, F. (1989) Nobody trusts the boss completely – now what? *Harvard Business Review*, March–April pp. 135–42.

Berlo, D.K. (1975) *Information and Communications*, Holt, Reinhart & Winston, New York.

Bolton, R. (1979) *People Skills*, Prentice-Hall, Englewood Cliffs NJ.

Buchholtz, S. and Roth, T. (1987) *Creating the High-Performance Team*, John Wiley, New York.

Churchill, D. (1993) Hangovers are out and cost effectiveness is in, *Management Today*, January, pp. 56–8.

Drakeford, J. (1967) *The Awesome Power of the Listening Ear*, Word, Waco TX.

Ernst, F. (1973) *Who's Listening? A Handbook of Transactional Analysis of the Listening Function*, Addresso, Vellejo CA.

Fast, J. (1971) *Body Language*, Pocket Books, New York.

Francis, D. (1987) *Organizational Communication*, Gower, Andover.

Haney, W. (1973) *Communication and Organizational Behaviour*, Richard D. Irwin Inc., Homewood IL.

Jay, A. (1976) How to run a meeting, *Harvard Business Review*, March–April, pp. 43–57.

Keefe, W. (1970) *Open Minds: The Forgotten Side of Communication*, Wadsworth, Belmont CA.

Kitchener, S. (1996) Body talking, *Call Centre Focus*, September, pp. 11–14.

Lindauer, J.S. (1974) *Communicating in Business*, Macmillan, New York.
Luthans, Fred (1973) *Organizational Behaviour*, McGraw-Hill, New York.
Nichols, R.G. and Stevens, L.A. (1957) *Are You Listening*? McGraw-Hill, New York.
Novey, T.B. (1976) *TA for Management*, MAR Press, Sacramento CA.
Osborn, A. (1984) *Your Creative Power*, Dell, New York.
Quirke, B. (1996) Only communication will reforge the links of corporate loyalty, *Professional Manager*, January, pp. 12–14.
Rockey, E.H. (1977) *Communicating in Organizations*, Winthrop Publishers Inc., Cambridge MA.
Rogers, C.R. and Roethlisberger, F.J. (1991) Barriers and gateways to communication, *Harvard Business Review*, November–December pp. 105–111.
Scannell, E.E. (1970) *Communications for Leadership*, McGraw-Hill, New York.
Scott, B. (1986) *The Skills of Communicating*, Gower, Andover.
Scott, W. (1984) *Communication for Professional Engineers*, Thomas Telford, London.
Tubbs, S.L. and Moss, S. (1974) *Human Communication*, Random House, New York.
Wolf, M.P. and Aurner, R.R. (1974) *Effective Communication in Business*, Southwestern Publishing, Cincinnati OH.
Wortman, L.A. (1981) *Effective Management for Engineers and Scientists*, John Wiley, New York.
Young, M. and Post, J.E. (1994) How leading companies communicate with employees, *IEEE Engineering Management Review*, Spring, pp. 24–31.

20

Time management

Introduction

This chapter considers the importance of time, the major causes of ineffectual time management and the techniques available for improving this.

20.1 The importance of time

What is time? Everyone understands what time is; the problem comes when one tries to define it! Perhaps time should be considered to be a passage of events in one's life, until the final great event! Put that way one is sure to agree with Benjamin Franklin in his book *Advice to a Young Tradesman* (1745): 'Time is money' and 'Do not squander time, for this is the stuff life is made of'.

Engineering managers should consider time to be a valuable resource, and just as they would not squander resources such as money, capital equipment and skilled technical people, so they should be careful as to how they use the time available to them. The same criteria should be applied for time investment as for other tasks, such as return on investment (ROI). The tasks which show the greatest payback should be tackled first.

There are, however, several differences between time and other resources:

- Time is obtained free of charge, just like the air we breath. Each morning 24 hours' worth of time is given to us. 'The chief beauty about the constant supply of time is that you cannot waste it in advance' (Bennett, 1920). Each new second is a fresh second and can be used in any way, irrespective of how the previous second was used.
- One can never have more time than anyone else, no matter how powerful or important one may be. This means that the small company is on an equal level with its largest competitor in respect of this resource, and can use it to develop competitive advantage. In spite of the equal amount of time given to all it is clear that some people achieve infinitely more than others in the same period.
- The saying 'be efficient and save time' is incorrect; one cannot save time to the extent that parts of it can be saved and used at a later date. We can, however, use less time to achieve a given task. Even if no activity is undertaken in a given time period, that period is used up with doing nothing. Note, however, that even if time could be saved it should not be. No resource, even money, should be saved; it should be utilized to achieve corporate aims.
- There is no substitute for time in the way that there are substitutes for other resources, such as using capital instead of labour or steel instead of iron. It is therefore one of the most precious of resources.
- Every activity, no matter how small or large, requires use of the commodity called time. Therefore time is the most indispensable commodity (Drucker, 1970).
- Only you can make better use of your own time; others cannot do so, neither can they steal it from you. However, they can waste your time (if you let them) by preventing you from utilizing your time to achieve maximum payback.

A normal working day can be considered to have three elements:

- Core time, which is the time spent in the office during normal working hours. This may include a small element of overtime and is the time when the bulk of the organization's work is done.
- Extra time, which is excessive overtime, normally done at one's place of work or at home. This should be kept to a minimum, since few managers can operate at full efficiency if they regularly go home late at night with bulging briefcases.
- Twilight time, which is the subconscious time when one has flashes of inspiration. It may be at home, when one is half-asleep in bed, or often when one is travelling to or from work. Twilight time is important in that it represents the few instances when the manager is relaxed and not under day-to-day trivial pressures.

It is one of management's most fundamental duties to help staff manage their time more effectively. One cannot, however, force change in time management; it must be adopted by individuals themselves. Furthermore, one cannot manage others until one can manage oneself; hence this chapter concentrates on helping engineering managers to manage their own time. The sections which follow first consider the characteristics of management tasks and the time elements involved in these tasks, and then introduce techniques for managing the use of time.

20.2 Characteristics of management tasks

There are many similarities between management tasks done at different levels within an organization and in different industries (Mintzberg, 1973). The further up the organization one goes, however, one finds that:

- Very important tasks often arrive at unpredictable times, requiring instant attention, and trivial tasks often take up a disproportionately large amount of time and energy.
- Most tasks are of short duration and are not connected.
- Very little routine work is involved.
- Senior executives have little control over their own work priorities, which are usually set by unexpected interruptions.

Management tasks usually encompass the following:

- Seeking and giving information. Managers need to have their finger on the pulse of their departments, to know when things are going wrong and when corrective action is needed. They also need to be aware of activities in other departments and even outside the company, and the impact which this may have on their work. Most managers prefer oral communication since it represents more recent information and takes up less time than writing memos and reports. Oral communication can take the form of one-to-one discussions, larger meetings and presentations, and telephone calls. Problems exist with electronic mail, where the managers may find themselves on the distribution list of a large volume of irrelevant information, often generated by quite junior members within the organization and sent to a wide distribution.
- Planning and organizing, and determining the formal work of the department. Plans must be communicated to staff, actions agreed and results evaluated. Managers must ensure that they have time to look ahead and to formulate strategy which covers all aspects, including the process used.
- Analyzing problem areas and taking decisions.
- People-related activities, such as motivation, hiring, training and

development, appraisals and feedback, and promotions. People activities take time and cannot be rushed, even though managers may feel that they are wasting their time socializing. It is also important that managers find time to look after their own development.

- Acting as an ambassador for the department and for the company. This means interfacing with other departments, and one's own manager, in order to get their cooperation; interacting with outside organizations to learn and give information; dealing with customers; negotiating with suppliers and so on.

Any one of the above tasks can take up a lot of the manager's time and can be totally unproductive. Also, most tasks require a relatively long, continuous, uninterrupted time to make them viable, which is the manager's greatest problem. For example, if a report is to be written and it is estimated to take eight hours, it is difficult to write it on the basis of 15 minutes a day over 32 days.

20.3 Determining the time elements

20.3.1 The daily log

In order to be able to make the most effective use of time, it is first important to know how one spends one's time and to determine the time-wasting activities (Vliet, 1997). For this to be effective a written log should be kept of how time is being spent throughout the day. The method used is not critical; what is important is that times are written down. As Peter Drucker put it: 'if we rely on our memory we do not know how time has been spent' (Drucker, 1970).

The daily log, if kept for a few weeks, will soon identify the time-wasting elements encountered during the day. These can then be investigated to find the real cause of the problem. The clear question which needs to be answered in relation to every task done during the day is: 'what would happen if I did not do this activity?'. This includes the instances when the job is not done at all and when it is done by someone else (delegated).

In the section that follows the common time-wasting activities are introduced.

20.3.2 The time wasters

The key time wasters can be grouped into four areas:

- Goals and responsibilities not clear
- Poor self-discipline
- Lack of effective delegation
- Poor communications

Unclear goals and responsibilities

This is one of the key reasons for time being wasted; if the task which has to be done is not clearly defined, then time will be wasted in either doing nothing, or more likely in doing tasks that are not leading to the intended goal. (See the case study 'Defining the product'.)

This problem manifests itself in many ways; for example, no description existing for the job, or, equally bad, job descriptions which overlap significantly. This not only leads to duplication of work, but also makes it difficult for the job holders to know the true extent of their responsibilities, and so carry out the work.

Lack of clear responsibilities often leads to indecision, where the job holder is uncertain how to proceed for fear of failure or of treading on other people's toes. (See the case study 'The brief briefing'.) This is often as a result of poor management, where the candidate's manager has delegated responsibility but no authority to carry out the work.

Poor organizational structures, power struggles within the organization, and weak management, all lead to unclear responsibilities. This is often manifest as a muddled organization chart with a plethora of dotted reporting lines.

Case study

Defining the product

It was Peter Sell, Sales Director of the FloRoll Manufacturing Company, who first raised the issue of developing a new product for the Middle and Far East markets.

'It is a vast market', he said, 'stretching all the way from Saudi Arabia, through India and China, to Japan. But we are not going to win it on our existing products. We need to develop a more reliable product and preferably one that needs far less servicing. Price is important, mind you', he continued. 'There is not much money about and we may have to consider getting loan guarantees for some of the countries.'

After a considerable amount of discussion at Jake Topper's Executive Staff meetings, and several estimates on product development costs from Engineering and Manufacturing, it was decided to go ahead with the development of a new floroll.

'We are working to very tight time-scales', said Micro Dowling, Engineering Director. 'I will need to recruit a new team to work on the system specification right away. I need a product definition from Marketing to get them started.'

Graham Host, Product Line Manager, was given the task of developing the product marketing requirements, while Micro Dowling told Jill Tait, the Systems Design Manager, to recruit a team of three engineers for the product definition.

Recruitment was carried out relatively quickly, but the product marketing requirement specification was held up.

First there was disagreement between Sales and Marketing on the countries which should be covered, and whether the company had the resources to provide technical support to these geographically disparate regions if problems arose. Then Sales kept asking for more and more features, until Marketing decided that they would contact potential customers themselves in order to do a more effective market requirement analysis.

After waiting a month for the market requirement specification, Engineering decided to start work based on a memo from Marketing on what the specification may contain.

'I know we do not have a firm goal to work to, but it is better than not having anything', Jill Tait told her team of new recruits. 'The alternative is for us to carry on killing time until a proper marketing specification arrives, and we have wasted enough time as it is.'

The System Design team worked hard at their task and within two months they had produced the first pass at the design specification. Based on this the Hardware Design and Software Design teams produced better estimates of product development times and costs. This confirmed the original estimates.

'All we now need is a firm marketing specification before we start the design activities', Micro Dowling said at Jake Topper's meeting.

'I will send you a copy tomorrow', said Jack Fox, Director of Marketing. 'It took us longer than planned to get to interview the customers in China, but now we have got their requirements and the specification is in typing. It is worth the delay seeing that China is our largest potential market.'

When the System Design team went through the market requirement specification it was obvious to them that the requirements were very different from those that had been contained in the original marketing memo.

'This is ridiculous', said Micro, storming into Jack Fox's office. 'Why didn't you tell us earlier that there were such significant changes? We have wasted six person months of valuable engineering effort on this project.'

Jack Fox was apologetic but unrepentant. 'The bulk of the changes were made to accommodate China', he said, 'and that is too valuable a market for us not to meet.'

When the System Design team learned that their original design had to be abandoned, morale was low. 'That's the last time I authorize work to start until the goals have been clearly defined', reflected Jill Tait that evening, slumped in front of the television. Biggles, the labrador, whined in agreement and licked her hand in sympathy.

Case study

The brief briefing

The design tool set needed to develop the new series of florolls was getting out of date and David Laith, Manager of the Development Tools Group of the FloRoll

Manufacturing Company, knew that these had to be updated quickly, in order to meet the needs of a new contract which the company was expecting to receive. He decided to subdivide the work into three and to give a part to each of his three design sections.

David called each of his Section Leaders into his office in turn and explained the work he wanted done by the section. Because David was due to go out that evening the briefing sessions lasted less than an hour each, but David felt confident that he had given them enough information to carry out the designs.

As the work progressed each of the three groups found that they had incomplete information to do their part of the design and had to refer to each other. This wasted a considerable amount of time, and often the Section Leaders failed to take action because of uncertainty over their areas of responsibility. They were afraid of treading on each other's toes and problems fell between the cracks.

David Laith held separate, infrequent, brief reviews with his group leaders, to monitor progress. He failed to notice the divergence in their work, so that when the tool set was eventually brought together it did not work.

Lack of self-discipline

Lack of self-discipline often results in the absence of any plans for the work to be done (Horten, 1994). Sometimes the manager will rush into an interesting job in the mistaken belief that the sooner it is started the quicker the objective will be met. However, without an effective plan it is not possible to know the best method for achieving the end objectives. Lack of a plan results in a constant series of firefighting tasks, which makes the manager look busy but which rarely achieves anything useful.

Lack of an effective plan follows from managers failing to set priorities for their work. This means that interesting jobs, often of a low priority, will get done, while high-priority tasks will slip.

Lack of self-discipline also results in jobs being started and not finished; this can also happen if no deadlines are being set. Sometimes if deadlines are set they may be totally unrealistic and impossible to achieve.

It is important that the jobs that are to be done are carefully selected; only those that lead to the final goal should be done. The manager must learn to say 'no' to some tasks; he cannot hope to be involved in every activity that occurs within the organization. Control of interruptions is also important, and unscheduled meetings, visitors and phone calls need to be minimized.

Ineffective delegation

The importance of good delegation has already been discussed in Chapter 17. Ineffective delegation can occur when managers and their staff are poorly trained, and in departments that are under- or overstaffed.

Managers often believe in doing the important jobs themselves, feeling that they will do it better than others in their department. They worry that the subordinate will fail. This can be avoided by monitoring the job after it has been delegated. Very close monitoring, however, is also a cause of poor delegation. It is important to measure the results and not the many activities that are involved in attaining it.

The instructions provided when delegating a task must be clear and well understood by the recipient. However, the manager must ensure that the subordinate is not overloaded with work. The subordinate's load needs to be monitored and the manager must work with the subordinate to prioritize the tasks involved.

Sometimes tasks are not delegated because the manager enjoys doing hands-on work rather than managing. This is not necessarily bad, since a certain amount of detailed involvement keeps the manager in touch with the work of the team. However, this must not be at the expense of other management and team leader tasks, and the manager must never compete for work with the team. Also, some managers do not delegate out of fear that the subordinate may do a better job than themselves. Once again the managers need not fear, since they cannot expect to excel in every activity. It is the job of management to utilize the best skills available to them, to get the job done in the most effective way.

Poor communications

There are many aspects of poor communications, as described in Chapter 19, from poor use of the telephone to excessive travel instead of using facilities such as teleconferencing and videoconferencing.

It is important that managers understand the communications requirements placed on them. They should constantly ask the questions: what do my customers want from me? What format do they need the information in: spoken word (meetings, telephone, presentation) or written (memos, reports)?

The timing of information is also important, and delays often occur due to indecision or poor management, which can result in time being wasted in completing tasks.

Another cause of poor communications, and therefore of time wastage, is the difference in values between the parties involved in the communication. They may have different motivations and aims, which cause time delays in communications.

Meetings are a major source of time wastage (see Chapter 19). Telephone conversations can also waste a considerable amount of time. They usually interrupt an activity and since they are not planned they can take unnecessarily long while the conversation swings between social and work-related topics.

Control of telephone calls is usually best exercised by having a secretary take messages (or using voice mail) and then setting aside a certain amount of time for calling back all the people concerned, having first sketched out a brief (mental) agenda for the telephone call. Knowing the aim of the telephone call and the time which is to be allocated to it will ensure that the call is effective.

20.4 Time management techniques

Many of the techniques for avoiding the major time wasters were introduced in the last section. These can be defined as a set of hints or actions (Baker, 1997a, 1997b; Bliss, 1976; Fraser, 1992; Love, 1978; Mackenzie, 1975; Murdock, 1993) for improving time utilization, as follows:[1]

- Set goals and deadlines for every element of time. These may be short goals to cover the day, or longer goals for the week, month and year. The daily goals may just consist of a number of tasks to be done, in order of priority, compiled at the end of the previous day. These are then worked through, in order. Each task must be completed before the next one is started. At the end of the day the results are reviewed and the reason for any time wasters noted. A new list of priority tasks, which may include some from the previous day, is then made.
- Plan before starting any task, even if it is a telephone call. Aim for planning leverage (Figure 20.1), where extra time spent on planning will result in a much shorter implementation time, leading to a shorter overall time for the whole task. Take longer to complete in a shorter time.
- It is important to assess quickly the relevance of any new task. An example of this is skimming through the in-mail each day, putting aside those items which need to be read for later, and passing on or throwing away other mail. Ask the question: 'does this job need to be done?'. If the answer is 'yes' then the next question is: 'do I need to do this myself or can it be delegated?'. Resist the temptation to do every job and to be involved in every activity around the company.
- Group similar tasks and try to do these at the same time, such as making telephone calls or writing memos. Figure 20.2 illustrates the task staircase, which should be use for all tasks: plan the task; batch tasks of a similar type as much as possible (telephone calls, reading memos and so on); and then time the tasks to ensure that they are not taking unduly long. It is also better to have a routine during the day, for example: telephone calls always answered between 11:00 a.m. and 12:00 noon every day; meetings always scheduled for the latter half of the afternoon; and so on.

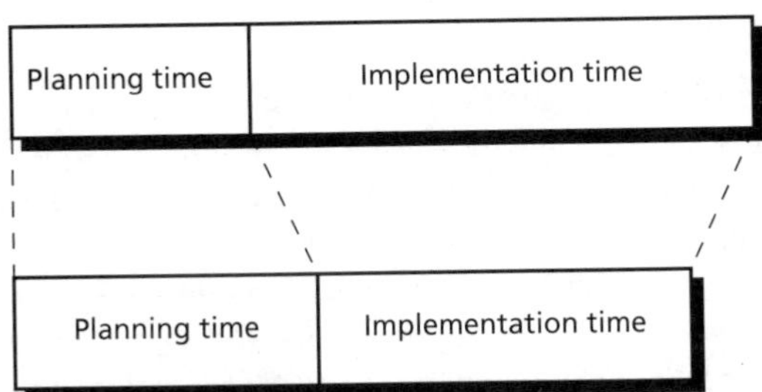

Figure 20.1 Planning leverage, in which a longer planning time results in a reduced overall task time.

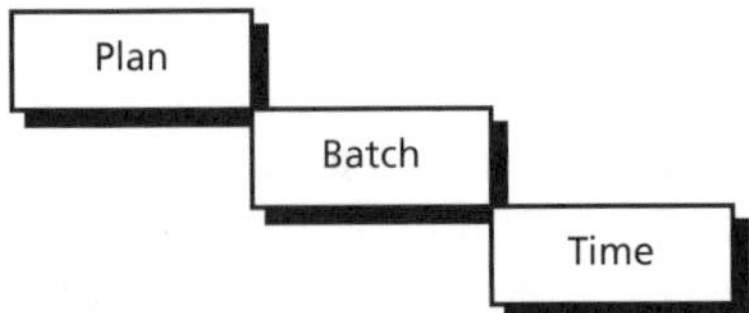

Figure 20.2 The task staircase.

- Guard against interruptions: telephone calls, unplanned visitors and so on. Be aware that the biggest culprit of all is yourself. It is so easy to interrupt yourself by breaking off in the middle of a boring piece of work to go for a coffee or a chat with a friend.
- Ensure that time is consolidated into large slices for the important tasks, such as planning and strategy making. Determine the time of day when you operate best and then reserve this time. There are many ways in which this can be done: for example, working from home one day a week or having meetings scheduled on four days a week only and keeping the fifth free.
- Recycle time: think of ways to use committed time (Adair, 1987). For example, use train journeys, time spent waiting at airports, and so on, for reading, writing memos and other tasks. Sometimes this can be the longest period of unbroken time that one is likely to get. Become a 'wait-watcher'.
- Pay attention to communication factors. Keep reports and memos short and to the point. Avoid long telephone discussions. Be selective as to which meetings to attend; send a representative if possible. Avoid overstaffing on tasks; this can lead to communication problems where more time is spent explaining than in getting the work done. (See the case study 'Subcontracting time'.) Teams should consist of people who are needed full time; other skills can be bought in.

Case study

Subcontracting time

The specification of the new floroll for the Middle and Far East markets was eventually agreed between the System Design and Marketing groups. It was now time for Engineering to commence work on the product. Time was of the essence, since the market window was relatively short.

A meeting of the design team was called to review the product design.

'In spite of the many false starts', said Jill Tait, System Design Manager, 'and all the trouble we had with Marketing earlier on, we now have a top-level design specification. It is in two parts, hardware and software. I suggest that we review these separately.'

Fred Doyle, Manager of the Network Software Group and Jane Furley, Manager of Hardware Design, looked over the top-level specification. It was clear that much work had to be done in a short time if the product was to meet its market requirements.

After the meeting Jane Furley called her team together to decide on the best way forward. The group was already heavily overloaded with modifications required on its existing floroll line, and it was soon evident that help was going to be needed.

'I think we should get in some contract staff for six months, to help with the early design work', said Jane.

The suggestion was soon adopted and a week later six new contractors joined the team. Problems were immediately evident. The new staff were unfamiliar with the basic design of the floroll, and engineers on the existing design teams had to stop what they were doing to train them. There was also a problem with the amount of design equipment available, owing to a delay in hiring them, so for much of the first few weeks the new recruits were hanging about, causing disruption to the team. Eventually the task of managing the new starters got so heavy that other projects started to slip. It was then that Micro Dowling, Director of Engineering, stepped in and suggested to Jane Furley that she follow Fred Doyle's example and subcontract the work.

Following the meeting with System Design, Fred Doyle also called his team together to discuss strategy. They were also very busy on existing developments, but decided to put three of their key software designers on the lower level design specification task, so that the work could then be passed out to a software subcontractor. It took three weeks to complete the lower level specifications and to subcontract the work. After that Fred Doyle held weekly review meetings with the software subcontractor to monitor progress.

The work went very well. The subcontractor provided its own tools and management resources, so that there was very little disruption to the design work being done within the FloRoll Manufacturing Company. When Jane held her meeting with Fred the project was eight weeks into design, and although hardware design had made very little progress, software design was well advanced. Even though it was very late, Jane decided to follow Fred's example and to subcontract the hardware design.

Summary

Time is an extremely valuable resource, even though it can be considered to be obtained free of charge. It is a limited resource; one can never have any more than anyone else. This means that small companies are on an equal level with their larger competitors, and if they can utilize this time more effectively they gain an important competitive advantage. Unlike other resources, time cannot be 'saved'; if it is not used it is wasted. Every activity uses this commodity called time and there is no substitute for it.

Many management tasks are of short duration and occur at unpredictable times. There is generally very little routine work involved, and managers usually have little control over their own work priorities, which are set by unexpected interruptions. In spite of this it is important that managers take control of their own time and avoid being driven completely by external events. To do this it is first important that they determine how their time is being spent, for example by keeping a daily log. This can be analyzed to identify the time wasters. Generally these can be grouped into four areas: work done when goals and responsibilities are not clear; poor self-discipline; lack of effective delegation; and poor communication.

There are several techniques for improving time utilization. For example, managers can set goals for themselves on a daily basis and analyze how they are performing against these, taking corrective action if too many time wasters are preventing the goals from being attained. All tasks should be carefully planned before they are started. A longer planning time will generally reduce the implementation time, resulting in a shorter overall time for the task, giving 'planning leverage'. All new tasks should be quickly assessed to determine the best method for performing them, such as not doing them at all or delegating them to others. If the tasks need to be done then smaller jobs of a similar type, such as telephone calls and memo writing, should be batched together and performed at a set time.

Managers need to be aware of interruptions, especially those that they create themselves, such as breaking off in the middle of a tedious piece of work to chat to a colleague. Time should also be recycled, so that committed time, such as waiting at airports, is used for reading or memo writing. It is important, however, that mangers seek to consolidate time into large slices for the important tasks, such as planning and strategy making. Communication also needs to be improved, such as keeping memos and telephone conversations short and attending meetings only if it is necessary to do so.

Endnotes

1. See also the following references: PM, 1977; Schwartz and Mackenzie, 1977.

Case study exercises

20.1 The number of projects within the FloRoll Manufacturing Group was now so large that Micro Dowling, Engineering Director, was finding it difficult keeping track of all the developments. Every section of every department within the company, including his own, produced a weekly report and he was on the distribution list for all of these. Most of them went unread, but Micro worried in case he should miss some important information.

He considered diverting some of these reports to his Group Managers and asking them to summarize the salient points in their own reports to him. However, he did not want to overload them and feared that they might miss important information. What are the alternatives open to Micro Dowling in order to maximize the use of his time?

20.2 Jack Fox, Marketing Director at the FloRoll Manufacturing Company, was asked by Jake Topper, Chief Executive, to produce a strategy plan for entering a new market for florolls. Jack knew that he needed long periods of uninterrupted time in order to do this plan, and feared that he would not get this at work due to interruptions. He considered several options, such as staying at home to do the work, or of delegating some of it to subordinates. What are the factors that would determine Jack Fox's best course of action?

20.3 Heather Bates, Customer Service Manager at the FloRoll Manufacturing Company, considered herself to be the most overworked person in the company, never having time to complete any job. At the first sign of a problem Heather would be there, so much so that she often felt like getting a small scooter to ride around the site on. Unfortunately, Heather never seemed to have time to solve any problems; as soon as she tackled one fire another would light up and she would rush off to put it out. What can Heather do to improve her utilization of time?

References

Adair, J. (1987) *How to Manage Your Time*, McGraw-Hill, Maidenhead.

Baker, H.K. (1977a) Invest time to save time, *Industrial Launderer*, July, pp. 34–7.

Baker, H.K. (1977b) The time budget: a personal planning tool, *Public Telecommunications Review*, May–June, pp. 27–30.

Bennett, A. (1920) *How to live on 24 hours a day*, Hodder & Stoughton, London.

Bliss, E. (1976) *Getting things done*, Bantam Books, New York.

Drucker, P.F. (1970) *The effective executive*, Pan Books, London.

Fraser, J. (1992) Take control of your time, *EDN*, May 7, pp. 264–8.

Horten, M. (1994) As time goes by, *Computing*, 6 October, pp. 40.

Love, S.F. (1978) *Mastery and Management of Time*, Prentice-Hall, New York.

Mackenzie, R.A. (1975) *New Time Management Methods for You and Your Staff*, Dartnell Corp., New York.

Mintzberg, H. (1973) *The Nature of Management Work*, Harper & Row, New York.

Murdock, A. (1993) *Personal Effectiveness*, Butterworth-Heinemann, Oxford.

PM (1977) How to put time in check, *Professional Manager*, March, pp. 24–5.

Schwartz, E.B. and Mackenzie, R.A. (1977) Time management for women, *Management Review*, September, pp. 19–25.

Vliet, A. van der (1997) Beat the time bandits, *Management Today*, May, pp. 90–2.

Index

Structural Mechanics

Ray Hulse
Deputy Dean
School of The Built Environment
Coventry University

Jack Cain
Formerly Senior Course Tutor
Civil Engineering
Coventry University

Second edition

palgrave
macmillan

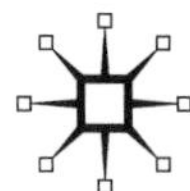

First edition 1990
Reprinted seven times
Second edition 2000
Published by
PALGRAVE MACMILLAN

Palgrave Macmillan in the UK is an imprint of Macmillan Publishers Limited, registered in England, company number 785998, of Houndmills, Basingstoke, Hampshire RG21 6XS.

Palgrave Macmillan in the US is a division of St. Martin's Press LLC, 175 Fifth Avenue, New York, NY 10010.

Palgrave Macmillan is the global academic imprint of the above companies and has companies and representatives throughout the world.

Palgrave® and Macmillan® are registered trademarks in the United States, the United Kingdom, Europe and other countries.

ISBN 978–0–333–80457–5 paperback

This book is printed on paper suitable for recycling and made from fully managed and sustained forest sources. Logging, pulping and manufacturing processes are expected to conform to the environmental regulations of the country of origin.

A catalogue record for this book is available from the British Library.

13
12

Printed and bound in Great Britain by
CPI Antony Rowe , Chippenham and Eastbourne